Emerging Applications of Plasma Science in Allied Technologies

Aamir Shahzad
Government College University, Faisalabad, Pakistan

Maogang He
Xi'an Jiaotong University, China

A volume in the Advances in Chemical and Materials Engineering (ACME) Book Series

IGI Global
PUBLISHER of TIMELY KNOWLEDGE

Published in the United States of America by
IGI Global
Engineering Science Reference (an imprint of IGI Global)
701 E. Chocolate Avenue
Hershey PA, USA 17033
Tel: 717-533-8845
Fax: 717-533-8661
E-mail: cust@igi-global.com
Web site: http://www.igi-global.com

Library of Congress Cataloging-in-Publication Data

Names: Shahzad, Aamir (College teacher), editor. | He, Maogang, 1970- editor.
Title: Emerging applications of plasma science in allied technologies / edited by Aamir Shahzad, Mao-Gang He.
Description: Hershey PA : Engineering Science Reference, [2024] | Includes bibliographical references. | Summary: "Nowadays, the advancements in industrial fabrication engineering and its desire on efficiency and feature in advanced manufacture technologies have become in complex and incorporated innovation processes"-- Provided by publisher.
Identifiers: LCCN 2023050854 (print) | LCCN 2023050855 (ebook) | ISBN 9798369309049 (hardcover) | ISBN 9798369309056 (ebook)
Subjects: LCSH: Plasma (Ionized gases)--Research.
Classification: LCC QC718.6 .E44 2024 (print) | LCC QC718.6 (ebook) | DDC 621.044--dc23/eng/20231127
LC record available at https://lccn.loc.gov/2023050854
LC ebook record available at https://lccn.loc.gov/2023050855

This book is published in the IGI Global book series Advances in Chemical and Materials Engineering (ACME) (ISSN: 2327-5448; eISSN: 2327-5456)

British Cataloguing in Publication Data
A Cataloguing in Publication record for this book is available from the British Library.

Advances in Chemical and Materials Engineering (ACME) Book Series

J. Paulo Davim
University of Aveiro, Portugal

ISSN:2327-5448
EISSN:2327-5456

Mission

The cross disciplinary approach of chemical and materials engineering is rapidly growing as it applies to the study of educational, scientific and industrial research activities by solving complex chemical problems using computational techniques and statistical methods.

The **Advances in Chemical and Materials Engineering (ACME) Book Series** provides research on the recent advances throughout computational and statistical methods of analysis and modeling. This series brings together collaboration between chemists, engineers, statisticians, and computer scientists and offers a wealth of knowledge and useful tools to academics, practitioners, and professionals through high quality publications.

Coverage

- Sustainable Materials
- Ductility and Crack-Resistance
- Toxicity of Materials
- Chemical engineering
- Numerical Techniques
- Composites
- Artificial Intelligence Methods
- Fatigue and Creep
- Computational methods
- Materials to Renewable Energies

IGI Global is currently accepting manuscripts for publication within this series. To submit a proposal for a volume in this series, please contact our Acquisition Editors at Acquisitions@igi-global.com or visit: http://www.igi-global.com/publish/.

The Advances in Chemical and Materials Engineering (ACME) Book Series (ISSN 2327-5448) is published by IGI Global, 701 E. Chocolate Avenue, Hershey, PA 17033-1240, USA, www.igi-global.com. This series is composed of titles available for purchase individually; each title is edited to be contextually exclusive from any other title within the series. For pricing and ordering information please visit http://www.igi-global.com/book-series/advances-chemical-materials-engineering/73687. Postmaster: Send all address changes to above address.

Titles in this Series

For a list of additional titles in this series, please visit: www.igi-global.com/book-series

Machining Polymer Matrix Composites Tools, Techniques, and Sustainability
Francisco Mata Cabrera (Universidad de Castilla-La Mancha, Spain) and Issam Hanafi (Université Abdelmalek Essaâdi, Moocco)
Engineering Science Reference • © 2024 • 320pp • H/C (ISBN: 9781668499276) • US $300.00

Building a Low-Carbon Society Through Applied Environmental Materials Science
Takaomi Kobayashi (Nagaoka University of Technology, apan)
Engineering Science Reference • © 2024 • 300pp • H/C (ISBN: 9798369300039) • US $255.00

FlexPDE and Finite Element Method Applications in Thermal Energy Storage and Cavities
Khashayar Hosseinzadeh (Sharif University of Technology, Iran) Mehdi Mahboobtosi (Babol Noshirvani University of Technology, Iran) and Davood Domairry Ganji (Babol Noshirvani University of Technology, Iran)
Engineering Science Reference • © 2024 • 300pp • H/C (ISBN: 9798369309322) • US $225.00

Advancements in Powder Metallurgy Processing, Applications, and Properties
Shashanka Rajendrachari (Bartin University, Turkey) and Rayappa Shrinivas Mahale (Jain College of Engineering and Research, India)
Engineering Science Reference • © 2024 • 300pp • H/C (ISBN: 9781668493854) • US $275.00

Innovative Multifunctional Nanomaterial for Photocatalysis, Sensing, and Imaging
Azad Kumar (Siddharth University, India)
Engineering Science Reference • © 2023 • 283pp • H/C (ISBN: 9781668487433) • US $270.00

Innovations in Materials Chemistry, Physics, and Engineering Research
Eugene de Silva (Virginia Research Institute, USA) and Pramudi Abeydeera (Polymeric Ltd., USA)
Engineering Science Reference • © 2023 • 446pp • H/C (ISBN: 9781668468302) • US $270.00

Handbook of Research on Advanced Functional Materials for Orthopedic Applications
R. Ranjith (SNS College of Technology, India) and J. Paulo Davim (University of Aveiro, Portugal)
Engineering Science Reference • © 2023 • 371pp • H/C (ISBN: 9781668474129) • US $325.00

Sustainable Utilization of Nanoparticles and Nanofluids in Engineering Applications
Sampath Boopathi (Mechanical Engineering, Muthayammal Engineering College, India) and J. Paulo Davim (University of Aveiro, Portugal)
Engineering Science Reference • © 2023 • 372pp • H/C (ISBN: 9781668491355) • US $270.00

IGI Global
PUBLISHER of TIMELY KNOWLEDGE

701 East Chocolate Avenue, Hershey, PA 17033, USA
Tel: 717-533-8845 x100 • Fax: 717-533-8661
E-Mail: cust@igi-global.com • www.igi-global.com

Table of Contents

Detailed Table of Contents

therapy, providing insights into how protein modifications affect cancer cell behavior and therapy responses. This knowledge advances NTP-based cancer treatments, offering the potential for personalized and targeted therapies in the future.

Section 2
Transport Properties of Plasmas

Spacecraft charging phenomenon at GEO altitudes is revisited by highlighting various spacecraft missions, spacecraft orbits, surface charge mitigating techniques, and particle distribution models. A comparative analysis is presented to study the onset of negative charging for spacecraft with single Maxwellian, Kappa, and q-nonextensive distributions. Since the plasma constitutes a mixture of two clouds having cold and hot electrons with their distinct thermal energies and densities, therefore it is important to investigate surface charging with double extensive and non-extensive distributions as well as analyzing the current-balance equations (CBEs) both analytically and numerically. The Whittaker function integral is employed to solve the power-law integrals. The q-distributed electrons show more pronounced energy tails to obey Tsallis statistics having nonstationary equilibrium states. Non-extensivity effect strongly modifies the CBE for spacecraft charging, supporting better in the limit $q<1$ in contrast to Maxwell and κ distributions. Numerically critical and anticritical thresholds are also identified to understand the onset of negative charging of various space-grade materials.

Equilibrium molecular dynamics (EMD) has been used to investigate diffusion coefficient (D) of 3D weakly coupled complex dusty plasma (WCCDP) by varying applied electric field (E). Velocity auto correlation function technique has been used to compute diffusion coefficient of dust particles. Impact of varying electric field has been tested for complete series of plasma parameters (Γ, κ) in a canonical ensemble. EMD simulation have been used to compute D for fixed number of particles ($N = 500$) at different values of plasma parameters ($0.1 \leq \Gamma \leq 0.9$ and ($1 \leq \kappa \leq 3$). Computed results showed that value of D decreases by increasing Γ and increases by increasing E and κ. Simulation results also exposed that result obtained through Green-Kubo has same drift as previous results. But this work has some advancement due to E.

A mesoscopic study of linear irreversible transport phenomena is proposed. This chapter takes a phenomenological and statistical approach to non-equilibrium phenomena. At thermodynamic equilibrium, the intensive physical quantities of a system are uniform in space and time. These quantities can be defined at any point and at any time, and are referred to as local thermodynamic equilibrium. Otherwise, the system is out of thermodynamic equilibrium. This is the case for all irreversible phenomena, which are generally induced by an external input of energy and/or matter to the system. In this chapter, we will be focusing on the phenomenon of transport, which is a key process in non-equilibrium physics. There are various transport phenomena. Each is characterized by macroscopic properties. A microscopic approach is taken to study the transport phenomena. However, we are particularly interested in the phenomenon of particle diffusion and of the thermal diffusion.

The interaction of beam driven solitons is studied in an unmagnetized plasma, comprising (r,q) distributed hot electrons, cold inertial electrons, warm electron beam and static ions. For this purpose, fluid equations are solved to obtain four distinct roots of linear dispersion relation which vary with beam speed. For small-amplitude limit, the standard Korteweg-de Vries (KdV) equation is derived that admits soliton solution and critical condition gives rise to modified KdV equation (mKdV). The overtaking interaction of KdV and mKdV solitons is studied by Hirota method, while the head-on collision (HOC) by the Poincare-Lighthill-Kuo (PLK) technique. The analytical expressions of phase shifts are examined for the plasma parameters, namely density, temperature, and velocity of beam electrons. In contrast to Hirota method, the PLK method provides the linear sum of the solitons at the point of interaction. However, despite this limitation, PLK technique is only analytical technique available to study HOC. The relevance of this model is interesting to understand tripolar electric field structures in Earth's Auroral region, which may be associated to the interaction of EASWs in space plasmas.

Electrorheological complex (dusty) plasmas (ER-CDPs) are type of plasmas with rheological behavior under external electric field (EEF). In this chapter, the nonequilibrium molecular dynamics simulations method is used to tune thermal conductivity for two-dimensional (2D) complex plasma liquids using EEF in different directions. Anisotropic thermal conductivity is investigated in three different cases, (i) Ex(x-axis), (ii) Ey(y-axis), and (iii) Exy(xy-axis) and with constant external perturbation force (Px = 0.02). The thermal conductivity under the influence of EEF is different in different directions. Obtained results are compared and discussed with previous known theoretical, simulation, and experimental data for 2D systems in constant EEF. The appropriate normalized thermal conductivity with Einstein frequency at constant EEF follows universal temperature scaling law. These comparisons and discussions show that algorithms of EEF of different cases with Yukawa potential have accuracy and consistency. These comparisons validated a new numerical model that can be used for variations of EEF along with different system sizes and plasma parameters.

This chapter provides a comprehensive exploration of the emerging applications of plasma in medicine and agriculture. Plasma, the fourth state of matter, offers unique properties that make it a versatile tool with transformative potential in these domains. The chapter delves into the advancements in plasma technology, discussing improved control over plasma parameters and the development of precise plasma sources. In the medical field, plasma has shown promise in wound healing, cancer treatment, sterilization, and drug delivery. Its non-invasive nature, antimicrobial properties, and ability to selectively target specific tissues make it an attractive option for tissue regeneration, infection control, and personalized therapy. The chapter emphasizes the potential benefits of plasma in enhancing patient outcomes and reducing healthcare costs. In agriculture, plasma-based technologies offer sustainable solutions for seed treatment, plant disease control, and crop improvement. Plasma treatments have proven effective in reducing microbial contamination, improving seed germination, stimulating plant growth, and enhancing crop resilience against diseases and environmental stresses. The chapter underscores the potential contributions of plasma-based technologies to increased agricultural productivity, reduced pesticide usage, and enhanced food security. The chapter further explores the emerging applications of cold atmospheric plasma (CAP), which operates at near-ambient temperatures. CAP has shown promise in wound healing, cancer therapy, plant disease control, and crop improvement. The underlying mechanisms of CAP's effects and its versatility in addressing diverse challenges in medicine and agriculture are discussed. Additionally, the chapter investigates the interactions between plasma and nanoparticles, highlighting their potential in targeted drug delivery, wound healing, and plant disease control. The ability to activate or modify nanoparticles using plasma opens new avenues for enhanced therapeutic and agricultural interventions.

Plasma-supported biochemical reactions have emerged as a promising and environmentally friendly approach for synthesizing microalgae-based biofuels. Microalgae, as single-celled organisms, possess the unique capability of efficiently converting sunlight into organic matter through photosynthesis, positioning them as potential green energy sources for biofuel and biomass production. However, traditional microalgae cultivation methods encounter challenges concerning environmental and economic sustainability, primarily due to their high water and fertilizer consumption. In this context, plasma technology has emerged as a powerful tool supporting biochemical reactions and optimizing microalgae cultivation processes, offering a greener and more sustainable alternative for biomass and biofuel production. This chapter explores the transformative potential of plasma-supported biochemical reactions in microalgal biofuel synthesis. It aims to provide insights into the intricate mechanisms underlying this innovative technology and its implications for the renewable energy landscape.

Bioprocess engineering and plasma technology have ushered in a new era of medicine production, transforming the pharmaceutical industry, and improving healthcare outcomes. This chapter explores the latest advancements in these fields and their significant impact on revolutionizing medicine production. The authors delve into key developments in bioprocess engineering, such as genome editing techniques, high-throughput screening, continuous bioprocessing, and synthetic biology. Additionally, the authors examine the diverse applications of plasma technology, including sterilization, plasma-enhanced chemical vapour deposition, disinfection of biological fluids, and plasma activation of surfaces. Together, these innovative approaches have accelerated drug development, enhanced drug delivery systems, and paved the way for personalized medicine.

Preface

This book summarizes the most significant applications of plasma science in allied technologies. The emphasis is on highlighting the most recent developments and important advances in the field of plasma science and technology. Other aspects covered include the fundamental physics, mechanism, design, allied technologies and novel methods. We hope to provide insight into some new directions in this field by providing a relatively widely description of current technologies and improved knowledge. The most promising plasma based devices and nano-plasma development allow us to change the approaches in industries, medicine/ clinical, agricultural and its controlled modes due to their thermophysical (transport and thermodynamics), wave behaviors, as well as external electric and magnetic, acoustic, light, temperature and other properties. This book on plasma based technologies with recent advances in energy materials, semiconductor systems, textile, plasma medicine, agricultural plasma and food processing, green plasma fusion energy, space science, anti bacterial and anti viral applications, plasma technology for conversion of CO2 as Eco-friendly environmental and green energy, and many other fields, has resulted in this book. Moreover, advances in life sciences are gradually more sourced by deployment of discrete technologies and information and plasma technology becomes a key advancement in modern medicine and health care.

This book will provide the introductory and current research contents on plasma science and allied developments trained by several world-wide universities and institutes. Nowadays, the advancements in industrial fabrication engineering and its desire on efficiency and feature in advanced manufacture technologies have become in complex and incorporated innovation processes. The contents indentify the key challenges including basic interactions of plasma based technology with materials, living tissues, fine particle plasmas that contain chaotic and stochastic processes, very high density plasma and micro-plasmas, high temperature and pressure plasmas. Advancement in life sciences is gradually more sourced by deployment of discrete technologies and information and fine particle plasma technology becomes a key advancement in modern medicine and health care. Very recently, plasma science in agriculture, medicine and plasma catalytic is rather new field that rose from the research in application of plasma in bioengineering. Plasma science is innovative and promising multidisciplinary novel field of research covering plasma physics, life sciences, clinical medicine to apply physical plasma for therapeutic applications, food safety and technology, and seeds generation and their production for sustainable food chains.

This book highlights the needs for alternative plasma technology in different fields that discuss theoretical, numerical and experimental techniques, and can available for all countries. One of the most significant problems facing civilization today is how to generate alternative clean technology sustainably for health care and in sufficient quantity to supply a growing global production. Conventional fuel burning machines pollute the atmosphere resulting in global warming and cause a health risk. Plasma energy does not produce any carbon dioxide and this book will provide the number challenges that have to be solved on before it may become the alternative source of technology and health care solutions in future. This book includes the theoretical and experimental observations of plasma technology for agricultural, medical, bioengineering, clinical and industrial applications. Moreover, the postgraduate researchers/ students from different universities/ institutions, industrial developers, biomedical engineering, agricultural seed production, plasma-based clinical practices in hospitals can benefits from this book.

The significant topics that are included in this book, the short overviews of chapters are as follow: First two chapters offer the plasma reactors as a sustainable solution for carbon dioxide (CO2) conversion. This chapter discus the various types of plasma reactors employed for CO2 conversion and provide a comprehensive comparison of their designs. The analysis of different reactor designs aims to assist in selecting the most suitable plasma reactor for specific CO2 conversion applications, emphasizing their potential impact on mitigating greenhouse gas emissions and contributing to a more environmentally friendly future. Moreover, it also explains the carbon dioxide (CO2) is one of the major greenhouse gases that contribute to global warming and environmental variations. Plasma technology, with its unique features and versatile capabilities, has shown immense potential for the conversion of CO2 into valuable products and fuels. The third Chapter incorporates the fascinating realm of plasma-assisted carbon dioxide (CO2) conversion, focusing on its applications, challenges, and environmental impacts. This chapter also explores the environmental benefits of this technology and evaluates its economic feasibility and potential market opportunities. In fourth chapter, the authors elaborate the non-thermal plasma (NTP) that holds promise in altering biological matter, particularly in cancer treatment. This book chapter reviews recent computer simulations and experimental findings exploring NTP's impact on proteins in cancer therapy, providing insights into how protein modifications affect cancer cell behavior and therapy responses. In the fifth chapter, the authors tackle the spacecraft charging phenomenon at GEO altitudes and it is revisited to highlight spacecraft missions, charging mitigation, and distribution models for energetic electrons. In the sixth chapter, researchers discuss the diffusion coefficient (D) of 3D weakly coupled complex dusty plasma using equilibrium molecular dynamics (EMD) with varying applied electric field (E). The seventh chapter explains a phenomenological and statistical approach to non-equilibrium phenomena. In this chapter, we will be focusing on the phenomenon of transport, which is a key process in non-equilibrium physics. eighth chapter disuses the electron acoustic waves (EAWs) in unmagnetized, four component plasma system, comprising (r,q) distributed hot electrons, cold inertial electrons, warm electron beam and static ions. The application of this model might be interesting to understand tripolar electric field structures, which are associated to the interaction of EASWs in space plasmas. The nineth chapter provides the electrorheological CDPs (ER-CDPs) are type of CDPs with rheological behavior under external electric field (EEF). In this chapter, the nonequilibrium molecular dynamics simulations method is used to tune thermal conductivity for 2D complex plasma liquids using EEF in different directions. anisotropic thermal conductivity is investigated in three different cases. The tenth chapter provides a comprehensive exploration of the emerging applications of plasma in medicine and agriculture. The chapter delves into the advancements in plasma technology, discussing improved control over plasma parameters and the development of precise plasma sources. In the medical field, plasma has shown promise in wound healing, cancer treatment, sterilization, and drug delivery. In the eleventh chapter, the authors study the plasma-supported biochemical reactions that have emerged as a promising and environmentally friendly approach for synthesizing microalgae-based biofuels. This chapter explores the transformative potential of plasma-supported biochemical reactions in microalgal biofuel synthesis. In last chapter gives the information regarding the bioprocess engineering and plasma technology have shared in a new era of medicine production, transforming the pharmaceutical industry and improving healthcare outcomes. This chapter delves into key developments in bioprocess engineering, such as genome editing techniques, high-throughput screening, continuous bioprocessing, and synthetic biology.

CONCLUSION

This book highlights the requirements for plasma science in allied fields that discuss theoretical, simulation and experimental methods, and may easily approachable for all communities. One of the most important issues facing civilization today is how to make green technologies sustainably for agriculture, industries, environment and health care and in enough quantity to supply an emergent global production. Plasma science is most promising field of science and technology that can provide carbon free approach for recent global warming issues and health risk applications.

Conventional fuel burning machines pollute the atmosphere resulting in global warming and cause a health risk. Plasma energy does not produce any carbon dioxide and this book will provide the number challenges that have to be solved on before it may become the alternative source of technology and health care solutions in future. This book includes the theoretical and experimental observations of plasma technology for agricultural, medical and industrial applications.

Aamir Shahzad
Modeling and Simulation Laboratory, Department of Physics, Government College University Faisalabad, Pakistan

Maogang HE
School of Energy and Power Engineering, Xi'an Jiaotong University, Xi'an, China

Section 1
Green Plasma Technology

Chapter 1
Plasma Reactors: A Sustainable Solution for Carbon Dioxide Conversion

Erfan Nouri
https://orcid.org/0000-0002-6558-7848
University of Guilan, Iran

Alireza Kardan
University of Guilan, Iran

Vahid Mottaghitalab
https://orcid.org/0000-0001-7211-3141
University of Guilan, Iran

ABSTRACT

The chapter explores the potential of plasma reactors as a sustainable solution for carbon dioxide (CO_2) conversion. The diverse types of plasmas used in CO_2 conversion are discussed. Mechanisms of plasma-based CO_2 conversion are examined, with a specific focus on direct CO_2 dissociation assisted by plasma, plasma-catalytic processes, and electrochemical CO_2 reduction using plasma. In addition, the chapter delves into the various types of plasma reactors employed for CO_2 conversion and provides a comprehensive comparison of their designs. The analysis of different reactor designs aims to assist in selecting the most suitable plasma reactor for specific CO_2 conversion applications. Furthermore, the chapter delves into the future perspectives of plasma reactors for CO_2 conversion. The chapter concludes by summarizing the essential findings and highlighting the importance of plasma reactors as a sustainable solution for CO_2 conversion, emphasizing their potential impact on mitigating greenhouse gas emissions and contributing to a more environmentally friendly future.

DOI: 10.4018/979-8-3693-0904-9.ch001

1. INTRODUCTION

Global climate change is a pressing issue in our modern world. Human activities, primarily the emission of greenhouse gases, particularly CO_2, are the leading cause of this problem. In order to mitigate the impacts of climate change, it is essential to adopt a comprehensive approach that focuses on reducing carbon emissions and decreasing the existing excessive amount of CO_2 in the atmosphere. Undoubtedly, there is growing interest in exploring the potential of carbon capture and utilization (CCU) as a means of converting CO_2 into valuable fuels and chemicals (Norhasyima & Mahlia, 2018). The cradle-to-cradle principle emphasizes the reduction of reliance on fossil fuels by utilizing waste gas as a renewable feedstock, thereby facilitating technological advancement (Vertongen et al., 2022). Various alternative technologies are under development for industry electrification, such as plasma technology (Snoeckx & Bogaerts, 2017).

The composition of plasma is a collection of differently-formed entities, including but not limited to electrons, ions, excited species, radicals, and molecules. When an electric field is utilized to produce a plasma, such as in the case of gas breakdown, the application of electric power prioritizes the heating of electrons in the plasma over other particles. As a result, the gas molecules are stimulated through electron impact reactions. The division of energy towards favorable chemical processes, such as the splitting of CO_2 (Fridman, 2008), varies according to the discharge condition and parameters involved. Due to its easily adjustable on/off feature, a plasma reactor holds considerable promise in converting intermittent renewable energy into chemical storage. Additionally, it offers significant advantages such as independence from rare resources and the ability to scale up according to the energy market's demands (Fridman, 2008).

Various forms of plasma reactors have been studied in the context of CO_2 conversion (Fridman, 2008). The structure of dielectric barrier discharges (DBDs) is straightforward, allowing them to function effectively in normal atmospheric conditions. These discharges have the potential to achieve significant conversion rates, reaching up to 30%. However, their energy efficiency is relatively restricted, typically between 5% and 10% (Aerts et al., 2015; Uytdenhouwen et al., 2018). Microwave (MW) plasmas achieve significantly greater energy efficiencies, primarily around 50%, occasionally reaching as high as 80-90%. However, these efficiencies are generally only observed when operating at lower pressures (Bongers et al., 2017). The drawbacks associated with gliding arc reactors can be effectively addressed. These reactors are capable of operating at atmospheric pressure, resulting in favorable energy efficiencies of approximately 30%. Nonetheless, the conversion rate of these reactors is confined to 8-9% due to the limitation in the fraction of gas that flows through the reactor (Ramakers et al., 2017; Trenchev et al., 2017). The utilization of atmospheric pressure glow discharges (APGD) is becoming increasingly popular for both fundamental research, such as the investigation of the transition from glow to arc (Boyle & Haworth, 1956; Bruggeman & Brandenburg, 2013), and practical applications like gas conversion (Wang et al., 1999; Wanten et al., 2022) and surface modification (Cernak et al., 1995; Shaw et al., 2016).

In summary, the book chapter explores the potential of plasma technology in addressing the challenge of CO_2 conversion. The chapter begins by discussing the types of plasmas used in CO_2 conversion, highlighting their properties and methods of generation. The mechanisms of plasma-based CO_2 conversion are then examined, with a focus on understanding the chemical reactions and physical processes involved. Subsequently, attention turns to plasma reactors for CO_2 conversion, where different reactor types and their operational parameters are explored. Furthermore, a comparison and analysis of these reactor designs is conducted, considering their respective advantages and disadvantages. Lastly, future perspectives and a concluding section are presented, outlining potential advancements and emerging technologies in plasma-based CO_2 conversion. Overall, this chapter contributes to the knowledge of sustainable energy and environmental science by providing insights into plasma reactors and their role in CO_2 conversion. The aim is to inspire further research, foster innovation, and promote the adoption of plasma technology for sustainable CO_2 conversion.

2. TYPES OF PLASMAS USED IN CO_2 CONVERSION

2.1. Thermal Plasmas

There are two possible methods to generate thermal plasmas: one is by subjecting the plasma to high temperatures, which usually fall within the range of 4000 K to 20 000 K and vary based on the ionization process, and the other is by exposing it to high gas pressure. The latter can be described in the following manner. In the beginning, electrons absorb energy from the electric field while moving between collisions, and they release a fraction of this energy when they collide with larger particles such as gas atoms or molecules. Following clashes of this kind, also referred to as 'Joule heating,' may result in the temperatures undergoing alterations to attain a state of equilibrium between the temperature of the electrons and that of the larger particles. In situations of elevated pressure, the average distance traveled by particles diminishes, resulting in an increased frequency of collisions. Consequently, this leads to a heightened effectiveness in the transfer of energy between the electrons and the larger particles. To be more precise, the temperature variation in gas discharges is directly proportional to the square of the ratio between the electric field (E) and the pressure (p), referred to as $(E/p)^2$ (Fridman, 2008).

Thermal plasmas possess several benefits when compared to conventional technologies as a result of their fascinating attributes, such as intense non-ionizing radiation, elevated temperature, and substantial energy density. The heat source is additionally oriented in a particular direction, featuring distinct boundaries and abrupt temperature variations that can be manipulated autonomously from the chemical aspects. Contrary to the maximum temperature of 2300 K achievable when burning fossil fuels, thermal plasmas, as previously stated, can surpass 20,000 K or even higher. A diverse array of applications already employ this form of plasma, including the management of hazardous waste materials, extractive metallurgy (like cutting and welding), the synthesis of fine powders, and the coating industry (Gomez et al., 2009).

Conversely, the unsuitability of thermal plasmas for the efficient conversion of CO_2 arises from their intrinsic characteristics. The temperature governs the ionization and chemical reactions in thermal plasmas. Consequently, the thermodynamic equilibrium efficiency is a restriction for achieving the utmost energy efficiency, with conversion rates at 3500 K amounting to 47% and 80%, respectively. In contrast to NTPs, lab-scale efficiencies of up to 90% have previously been documented (Fridman, 2008).

2.2. Non-Thermal Plasmas (NTP)

NTP is formed by applying a voltage between two parallel electrodes in a gas-filled reactor or by the electrodes themselves acting as walls of the reactor. The existence of this voltage variation generates an electrical field, leading to the occurrence of gas breakdown, wherein the gas somewhat separates into positively charged ions and electrons. In contrast, the majority of gas particles remain without charge.

The electric field drives the electrons toward the positively charged electrode (anode), causing them to gain acceleration. Ionization, dissociation, and excitation can occur when gas molecules collide. Ionization collisions in the system generate fresh electrons and ions. These ions, then propelled by the electric field, move towards the cathode or negative electrode. Upon reaching the cathode, they trigger a phenomenon known as secondary electron emission. The plasma achieves self-sustainability through additional ionization collisions, which occur when new electrons are generated through cathodic ionization or secondary electron emission. The wrecks of excitement create excited molecules, and these molecules can transition back to the ground state or a lower level. As a result of this transition, they release light, which is a distinguishing characteristic of plasmas. Ultimately, the breakup collisions produce radicals, which tend to effortlessly generate fresh compounds, serving as the foundation for utilizing NTPs in gas conversion processes (George et al., 2021; Moreau et al., 2008; Snoeckx & Bogaerts, 2017).

While this method of generating plasma, which involves applying a voltage difference between two electrodes, is merely one option for plasma creation, the concept of energizing gas through electrons to generate reactive species (such as excited species, ions, and radicals) applies to various other types of plasmas as mentioned in Section 4.1. Primarily, the efficient energy conversion of CO_2 through NTPs can be attributed to the gas activation caused by electrons. This aspect will be elaborated on later (George et al., 2021; Moreau et al., 2008; Snoeckx & Bogaerts, 2017).

NTPs exhibit a notable distinction in temperature between the electrons and the heavier particles, resulting in the occurrence of non-local thermodynamic equilibrium (non-LTE), as discussed earlier. Due to the vast range of species, the intricate interplay between their respective temperatures can become rather complicated. However, the electron temperature (Te) stands as the highest, trailed by the temperature of vibrationally excited molecules (Tv). In contrast, the lowest temperature is attributed to neutral species (T0, also known as the gas temperature, Tg), ions (Ti), and the rotational degrees of freedom of molecules (Tr) (Fridman, 2008). Typically, the electron temperature is approximately 1 electron volt (equivalent to around 10,000 K), while the temperature of the gas remains in the vicinity of ambient room temperature. The electron temperature becomes elevated because of the relatively low mass of the electrons, which enables them to undergo rapid acceleration under the influence of electromagnetic fields. In contrast, the heavier particles, including ions, experience less ease in acceleration. In addition, owing to the substantial disparity in mass, the electrons experience minimal energy loss when engaging in elastic interactions with heavier entities, enabling them to effortlessly retain the considerable energy they acquire from the electric field (George et al., 2021; Moreau et al., 2008; Snoeckx & Bogaerts, 2017).

As stated earlier, the electrons can be seen as the catalysts of the highly responsive chemical blend. Undoubtedly, one of the primary benefits of NTP technology resides in its ability to stimulate gases, such as CO_2, which are typically unresponsive, at ambient temperature through the utilization of exceedingly energetic electrons. Based on this, heating the entire reactor or the gas is unnecessary, as the discharge and related reactions can be easily triggered by applying an electromagnetic field. The second significant benefit of NTP technology for converting CO_2 is its high level of flexibility, often described as a turnkey process. It is due to its ability to be rapidly activated or deactivated, with the time required to stabilize conversion and product yield typically less than half an hour. Moreover, the capacity for adjusting and scaling its energy usage is effortless. Consequently, it possesses the innate potential to emerge as an ideal solution for harnessing surplus intermittent renewable energy (such as power obtained from wind turbines or solar panels) and converting it into chemical potential. Indeed, it is highly compatible with both peak reduction and additionally, grid stabilization endeavors due to its capacity to adjust accordingly to the inconsistent provision of renewable power. Thus, fundamentally, the ongoing shift towards sustainable energy sources not only grants plasma processes an eco-friendly power source but also presents an opportunity for plasma-driven CO_2 transformation technology to address the energy supply and demand gliding arc plasma (GAP) by allowing the storage of electrical energy in a preferred chemical state based on market requirements (George et al., 2021; Moreau et al., 2008; Snoeckx & Bogaerts, 2017).

One notable advantage in comparison to other developing methods for converting CO_2 is the affordability of the reactors, which do not require rare earth materials. Furthermore, these reactors have proven to be easily scalable from small-watt to large-megawatt applications, as exemplified by the successful creation of ozone generators. This aspect contributes to their overall appeal and potential for widespread use (Kogelschatz, 2003). The applications can differ significantly in size and type, ranging from small devices like vehicle exhaust treatment (Chen, Lee, Chen et al, 2009; Koeta et al., 2012) or indoor air treatment in households to medium-sized installations like modular containerized plant concepts (Patil et al., 2016) or in situ CO production (Foote et al., 2016), and finally to large-scale industrial plants, all thanks to their great scalability. Plasma technology possesses a remarkable ability to adapt to various locations, which greatly contributes to its scalability and application versatility. Hence, the feasibility of both on-demand production and remote production is achieved, thereby serving as a crucial aspect for the implementation of CCU methods (Snoeckx & Bogaerts, 2017).

In contrast, when comparing NTP technology to other innovative technologies, one limiting factor for its overall efficiency in converting solar energy to fuel is its reliance on indirect solar energy, specifically electricity, instead of technologies that directly harness solar power. Nevertheless, taking a more comprehensive view, the capacity of plasma technology to depend on various energy sources once again offers a significant benefit in terms of adaptability to different locations, especially considering that it can operate nonstop (24/7), even without sunlight (George et al., 2021; Moreau et al., 2008; Snoeckx & Bogaerts, 2017).

Despite its ability to generate a state of imbalance by activating stable molecules, NTP lacks selectivity in producing desired outcomes. In particular, the active particles resulting from the electrons' activity undergo chemical reactions based on principles of chemical kinetics. Consequently, they combine to form numerous distinct substances due to varying reaction circumstances. It is worth noting that these resulting products have the potential to be eliminated by subsequent electron collisions. There is not a significant problem when it comes to splitting only pure CO_2, as the primary products that can be formed are O_2 and CO (along with minimal amounts of O_3). By mixing CO_2 with different substances like H_2, H_2O, or CH_4, an extensive range of results can arise. These can encompass syngas, which consist of CO and H_2, and more complex hydrocarbons like C_4Hz, C_3Hy, and C_2Hx. Additionally, numerous oxygenated compounds such as carboxylic acids, acetaldehyde, dimethyl ether (DME), ethanol, formaldehyde, and methanol can also be produced. Because of this, plasma technology finds adding catalysis highly advantageous as it aims for the targeted, exclusive synthesis of particular compounds (George et al., 2021; Petitpas et al., 2007; Snoeckx & Bogaerts, 2017).

2.3. Warm Plasmas

Traditional thermal and nonthermal discharges fail to deliver a substantial level of non-equilibrium and a high electron density concurrently. However, many potential plasma chemical uses demand both characteristics: elevated power (which correlates to a high electron density) to enhance reactor efficiency and a significant state of nonequilibrium to specifically activate particular degrees of freedom like vibrationally excited states (Fridman et al., 2005). New research findings have indicated a potentially groundbreaking plasma known as warm discharge or warm plasma. This plasma, operating at the boundary between thermal and NTPs, exhibits properties that make it highly favorable for the conversion of CO_2. Non-equilibrium discharges are characterized by their capacity to not only furnish (re)active particles but also provide a regulated degree of translational temperature. Even though the temperature of this transforming gas is still considerably lower than that of the electrons, it is notably higher than the temperature found in a typical room. It has the potential to reach as high as 2000–3000 K. As a result, these hot plasmas can generate the benefit of an unbalanced state, all while exerting an impact on the chemical reactions as a result of their elevated gas temperature (Snoeckx & Bogaerts, 2017).

When it comes to using CO_2 conversion, the focus is not on higher temperatures but instead on the unique electron energy distribution in these discharges. This distribution results in a substantial amount of electron energy being directed towards the vibrational excitation of CO_2. Further details will be explored later on. One can acknowledge that the latter is recognized for its ability to offer the highest level of energy efficiency, thereby making it a crucial pathway for the dissociation of CO_2 (Fridman, 2008; Kozák & Bogaerts, 2014). Recently conducted modeling studies indicate that the undesired consequence of warm discharges is linked to the increased temperature of the gas (Berthelot & Bogaerts, 2017; Bogaerts et al., 2017; Kozák & Bogaerts, 2014). Therefore, the advancement of "colder" heat releases could prove advantageous in further enhancing the conversion of CO_2 and energy efficiency.

3. MECHANISMS OF PLASMA-BASED CO_2 CONVERSION

3.1. Direct CO_2 Dissociation Assisted by Plasma

The concentrated research endeavor that commenced roughly a decade ago regarding plasma-based CO_2 recycling is rooted in the understanding of CO_2 plasmas garnered during the 1960s and 1970s. In addition to the " warm" plasmas encountered during planetary entry, low-temperature CO_2 plasmas were initially explored in the context of developing CO_2 lasers by Witteman (2013) and Moore et al. (1967). The studies primarily concentrated on the vibrational dynamics of the initial energy states of CO_2 to minimize its dissociation. The investigation into the dissociation of CO_2 began in the 1970s, prompted by the inability to explain the experimental dissociation rates of CO_2 through the sole process of direct electron impact dissociation. It led to the works of Andreev et al. (1975) and Capezzuto et al. (1976). During that period, specific experiments demonstrated energy efficiencies of up to 90% in the dissociation process within supersonic MW plasmas, as observed in the works of Azizov et al. (1983). Despite the extensive research conducted by multiple research groups, the replication of these efficiencies has not been achieved. The highest recorded values, as observed in the works of Bongers et al. (2017), are approximately 45%. It corresponds to the theoretical maximum energy efficiency for the thermal dissociation of the CO_2 molecule.

In this section, we will provide a concise overview of the fundamental characteristics of CO_2 plasma as documented in the literature. Subsequently, the following section will delve into various methodologies currently employed to explore CO_2 plasmas.

3.1.1. Main CO_2 Dissociation Mechanisms

CO_2 plasma conversion can be achieved electronically or vibrationally stimulating the molecule, thereby enhancing its reactivity with other molecules. Another approach involves breaking down CO_2 into CO through dissociation.

$CO_2 \rightarrow CO + O$, ΔH=5.5 eV/mol

The process is highly endothermic, and the thermal degradation of pure CO_2 commences only at temperatures surpassing 2000 K, as supported by Gordon et al. (1994). In plasmas operating at low temperatures, there are two primary methods of dissociation: electron impact dissociation, as studied by Itikawa and Mason (2005), Hake and Phelps (1967), and Polak and Slovetsky (1976), and vibrational up-pumping, also known as ladder climbing, as discussed by Fridman (2008). The energy threshold for direct electron impact dissociation is higher than 7 eV, leading to the production of CO or O in an excited electronic state (the formation of ground state $CO(X^1\Sigma^+) + O(^3P)$ is not allowed due to spin restrictions). The construction of $CO(X^1\Sigma^+)$ and $O(^1D)$ occurs through the lowest energy process, while higher energy processes, such as the formation of $O(^1S)$ as discussed by LeClair and McConkey (1994), or $CO(a^3\Pi_r)$ as studied by Itikawa and Mason (2005), Cosby and Helm (1992), require energies higher than 10 eV. Note that electron impact dissociation can also occur from molecules in a state of vibrational excitation. It would decrease the energy threshold required for such processes, but it would also result in the generation of electronically excited states.

The mechanism known as vibrational up-pumping is an alternative pathway for dissociation. In this process, the energy carried by low-energy electrons (approximately ^{1}eV) is transferred to the asymmetric stretch vibration of CO_2, as shown in the work by Armenise and Kustova (2018). The transfer of vibrational energy between molecules can result in the accumulation of vibrational levels with significant excitation, culminating in the occurrence of a potential intersection between the electronically excited $CO_2(^3B_2)$ and the ground state of CO_2 ($CO_2(^1\Sigma^+)$) state at an approximate energy of 5.5 eV. Both ground state CO molecules and O atoms are generated simultaneously, as depicted in a schematic representation in Figure 1, as described by Fridman (2008). It is essential to acknowledge that this depiction is an idealized representation, as CO_2 is a molecule composed of three atoms. Consequently, its description is based on potential surfaces rather than potential curves, as explained by Huang et al. (2012). In this depiction, the equilibrium angles of the electronically excited states are graphed, which differ from the equilibrium angle of the ground state, as explained by Vargas et al. (2021).

Fridman (2008) has suggested that the most effective route for CO_2 dissociation is through the vibrational up-pumping of the asymmetric-stretch vibrational mode in CO_2. According to Rusanov et al. (1981), a significant portion of the energy released is transported by electrons with low energy levels, facilitating the transfer of this energy into the excitation of CO_2 and various other molecules' vibrations. In the graph provided as Figure 2, Grofulovic et al. (2016) illustrate the distribution of electron energy transfer into various channels at different reduced electric field levels. These channels include electronic excitation, vibrational excitation, ionization, and dissociation. However, in numerous plasma conditions, the primary method for CO_2 dissociation is electron collision-driven dissociation, particularly in discharges featuring an elevated reduced electric field like DBDs or nano-second discharges. Nevertheless, in a state of dissociation that is predominantly governed by electron impact, the progress of vibration could heavily impact the rates at which dissociation occurs through its effects on the electron distribution function (EEDF) and subsequent chemical reactions following the discharge, as noted by Capitelli et al. (2017).

Figure 1. Plot of CO2 potential energy as a function of CO-O distance
Sources: (Fridman, 2008; Klarenaar et al., 2018).

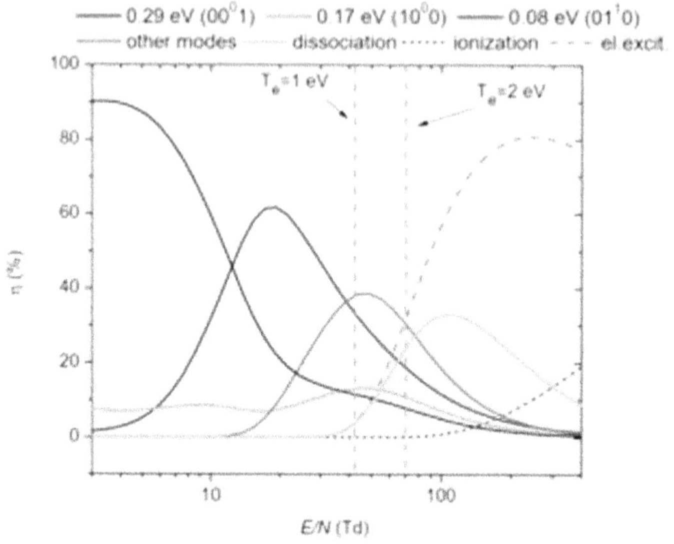

Figure 2. Fraction of electron energy transferred to electronic excitation, vibrational excitation, ionization, and dissociation as a function of the decreased electric field
Figure extracted from (Grofulović et al., 2016).

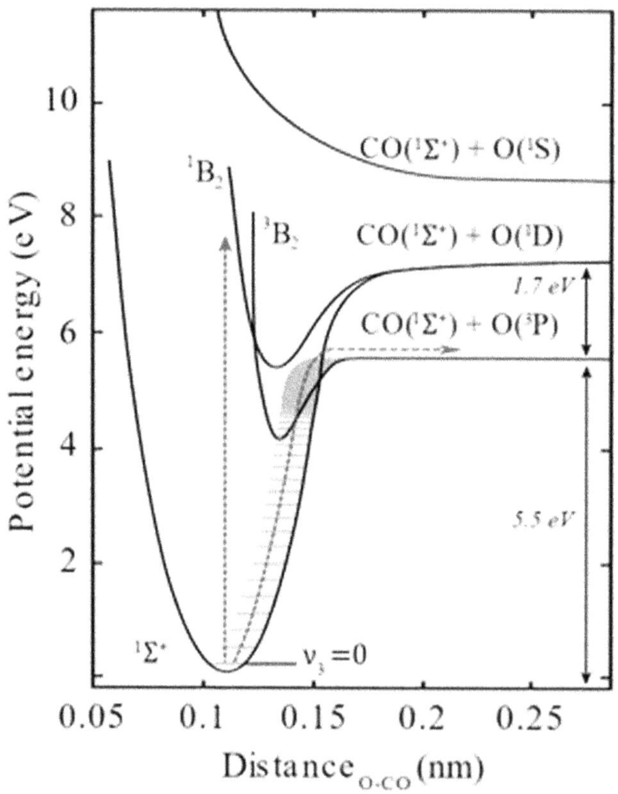

Numerous different organisms can directly or indirectly contribute to the separation of CO_2. To attain optimal energy efficiency during the conversion of CO_2, it is crucial for the O atoms resulting from the initial CO_2 dissociation process to combine with vibrationally stimulated CO_2 molecules, resulting in the production of O_2 and CO, as discussed by Fridman (2008). Chemical reactions can occur as a result of electronic excitations, causing dissociation such as $CO(a^3\Pi_r) + CO_2$ yielding 2CO + O, or these excited states can have a significant influence on the vibrational and electron kinetics. The forthcoming section will encompass a detailed description of these particular species (Morillo-Candas, 2019).

3.1.2. The Complexity of CO_2 Plasmas

The CO_2 plasma consists of a wide range of elements, such as neutrals in their initial condition or activated states (electronically or via vibrations), positively and negatively charged ions, radicals, electrons, and transient intermediates with short lifespans. Table 1 summarizes the significant species found in the gaseous state. Figure 3 displays a simple energy diagram depicting several key neutral species. The indicated levels represent the dissociation and ionization stages. This simplified illustration shows the numerous species in the CO_2 plasma kinetics, excluding ion levels and vibrational levels of electronic states.

Table 1. Incomplete inventory of the primary gas phase components found in CO_2 plasmas

Neutrals	**C, C_2, CO, CO_2, O, O_2, O_3, C_2O,...**
Vibrationally excited states	$CO(\nu)$, $CO_2\,(v_1 v_2^l v_3)$, $O_2(\nu)$, $X^e(\nu)^*$, $X^{+/-}(\nu)$,*
Electronically excited states	$O(^1S)$, $O(^1D)$, $CO(a^3\Pi_r)$, $O_2(a^1\Delta g)$, $O_2(b^1\Sigma_g^+)$...
Negative ions	$O^-, O_2^-, O_3^-, O_4^-, CO_3^-, CO_4^-$
Positive ions	$O^+, O_2^+, O_4^+, C^+, C_2^+, CO^+, CO_2^+, CO_4^+, C_2O_2^+, C_2O_3^+, C_2O_4^+, \ldots$

Source: (Morillo-Candas, 2019)

Figure 3 displays the electronic excited states of CO_2 in the leftmost column. The pure vibrational modes are represented by the three remaining sub-columns. The symmetric stretch, bending, and asymmetric stretch modes of CO_2 can be described by the quantum numbers ν_1, ν_2, and ν_3, making it a linear molecule with three vibrational modes. The CO_2 molecule's linear symmetry results in the doubly degenerate nature of the bending mode. As such, the description of the angular momentum contribution from ν_2 involves the incorporation of an additional quantum number l2. This quantum number can assume values ranging from ν_2, $\nu_2 - 2$, $\nu_2 - 4$, to 1 or 0. The approximated method for calculating the vibrational levels of CO_2 is through the anharmonic oscillator approximation, as presented by Treanor et al. (1968). The proximity between the energy of a single ν_1 quantum and that of two ν_2 quanta is exceptionally high. Hence, a robust Fermi resonance exists between states characterized by frequencies (ν_1, ν_2^{l2}, ν_3) and (($\nu_1 - 1$), $(\nu_2 + 2)^{l2}$, ν_3) (Morillo-Candas, 2019).

Figure 3. Energy level graph of principal electronic and vibrational states of neutral components in CO_2 plasmas
Source: (Morillo-Candas, 2019)

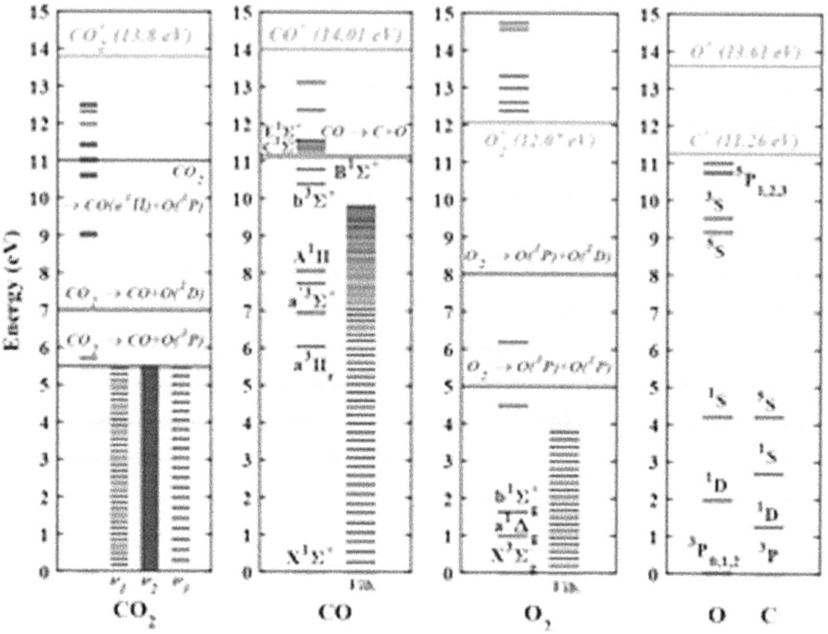

In the case of a glow discharge, where the average electron energy is approximately 2 eV, most electrons possess sufficient power solely to occupy vibrational states, as depicted in Figure 4. even though the energy range is low, there exist numerous vibrational levels of CO_2 that can be occupied due to its three vibrational modes. The occupation of various vibrational levels of CO_2 is not solely determined by its low energy range but also by multiple mechanisms, including direct electron impact (Morillo-Candas, 2019).

- Super-elastic collisions with low-energy electrons lead to the de-excitation of electrons caused by electron impacts.
- Reactions involving the exchange of vibrational energy, known as V-V' transfers, occur when vibrational quanta are exchanged between different modes of CO_2 vibration or with other molecules. One example is the quasi-resonant energy levels of CO that can interact with the asymmetric method of CO_2.
- Vibrational-translational/rotational relaxation, also known as V-T, occurs when particles or the walls of the reactor collide and cause a reduction in vibrational, translational, or rotational energy.
- The collision of molecules with electronically excited states can lead to vibrational excitations, as described by Harding et al. (1988), Huestis et al. (2002), and Chen et al. (2009)
- One potential way for chemical reactions is through the involvement of vibrationally excited molecules. In some instances, the vibrational energy may assist in surpassing the activation energy barrier for the response. A notable example is the reaction of $CO_2(\nu*)$ with O atoms, which can lead to the formation of CO, as proposed by Fridman (2008).

If the rates of energy transfers between vibrational and translational states are slower compared to the rates of energy exchanges within vibrational states only, it is possible for the energy to quickly redistribute within a specific vibrational mode. The rapid relaxation of the V-V, combined with the anharmonic energy levels, results in the vibrational distribution known as the Treanor distribution, as noted by Treanor et al. (1968). An effective V-T relaxation is facilitated in CO_2 through an asymmetric mode where the proximity of energy levels associated with bending and symmetric stretch allows for a readily attainable Treanor distribution.

In CO_2 plasmas, significant electronic states can be occupied, even at low energy levels of approximately 2 eV, which is representative of the majority of electrons in an NTP. The efficient population of multiple excited states of O_2 and O can be observed in the rightmost columns of Figure 4. Certain conditions, namely $O_2(a^1\Delta_g)$, $O(^1D)$, or $O_2(b^1\text{£}_g^+)$, can have a significant impact on chemical reactions and the transmission of energy to vibrations, as stated by Zhu et al. (1990). In a glow discharge, it is uncommon for a more significant proportion of electrons to possess sufficient energy (7 eV and higher) to directly cause the dissociation of CO_2. Near this energy threshold, a different electronic state known as $CO(a^3\Pi_r)$ might also become significant. This particular state exhibits a comparatively extended duration of existence, approximately lasting 2.5-7.5 milliseconds (Gilijamse et al., 2007; Mori et al., 2002), and can be occupied through electron collisions or transitions caused by radiative cascades. In a higher energy spectrum ($\geq$10 eV), it becomes challenging to predict the primary ions due to the similar ionization potentials of CO_2, CO, and O. The identification will heavily rely on the specific composition of the gas mixture and, consequently the dissociation rate. It is relatively more superficial to ionize O_2 than pure CO_2. However, the density of O_2 cannot exceed half of CO. In this higher energy range, it has been suggested that electronically excited states, though not as prevalent as lower levels, have a potential impact on electron kinetics through super-elastic collisions. For example, Capitelli et al. (2017) reported the involvement of the 10.5 eV electronic state of CO_2 in this phenomenon.

In order to explore particular aspects of CO_2 plasma kinetics, valuable data can be obtained from various domains of research encompassing past and current studies. The vast body of information regarding the advancement of CO_2 lasers is a crucial reference in the field. The functioning of these lasers relies on the imbalance of the number of particles in the first level of the asymmetric mode (ν_3=1). Consequently, these studies furnish highly significant data regarding the initial vibrational tiers of CO_2 as demonstrated by Huddleston and Weitz (1981), Dang et al. (1982) (1983), Zasavitskii et al. (1990), Toebaert et al. (1995), and Witteman (2013). Their focus is frequently on examining continuous plasmas at low pressure in gas mixtures containing CO_2-He-N_2 to reduce CO_2 dissociation. Research studies on combustion and shock tube experiments related to atmospheric entry yield valuable data, such as reaction rates, as demonstrated by Bykova et al. (1997) and Kovalev et al. (2004). Investigating the chemical processes triggered by the breakdown of molecules and the ionization caused by light in the higher regions of Venus, Mars, and the early Earth provides valuable insights into the behavior of excited electronic states. This research sheds light on the reactivity of these states and has been explored by various scientists, such as Perri et al. (2004), Donahue (1971), Krauss et al. (1971), and Fox and Dalgarno (1979). The analysis of planetary science frequently drives researchers to conduct experiments using electron beams and/or molecular beams to study chemical reactions or specific excitations (Mumma et al., 1975; Perri et al., 2004; Zhu et al., 1990). Conducting beam experiments is highly beneficial and yields information regarding reaction rates or the lifespan of entities. However, these experiments do not directly elucidate the function of excited states within a gas discharge. The documentation regarding the vibrational dynamics of CO_2 beyond the initial significant levels for CO_2 lasers is still inadequate. Hence, despite the abundance of valuable literature, the intricate nature of CO_2 plasmas and the challenges in accurately depicting the factors that impact CO_2 dissociation under plasma conditions necessitate a substantial amount of additional experimental and modeling efforts (Morillo-Candas, 2019).

3.2. Plasma-Catalytic Processes for CO_2 Conversion

NTP technology presents a compelling substitute for traditional methods in transforming unreactive carbon emissions. There have been various forms of plasmas utilized to reduce CO_2, such as corona discharges, GAP, glow discharges, MW discharges, radio frequency (RF) discharges, and DBDs (Chen, Britun, Godfroid, Delplancke-Ogletree et al, 2017; Chen, Britun, Godfroid, Georgieva et al, 2017; Chen et al., 2016; Duan et al., 2015; Mei et al., 2016; Mikoviny et al., 2003; Silva et al., 2014; Snoeckx et al., 2016; Spencer & Gallimore, 2011; Sun et al., 2017; Wen & Jiang, 2001). Presented in this section are the releases that are commonly utilized for the conversion of CO_2. DBDs have been recognized for over a hundred years. They were initially documented in 1857 by Siemens for their application in generating ozone and were originally referred to as 'silent' discharges (Siemens, 1857). The DBD is the most commonly utilized discharge method for converting CO_2 into other forms using plasma. This method is preferred due to its simple handling, low-cost equipment, and ability to operate under atmospheric pressure (Neyts et al., 2015). Even though the attainable conversion efficiencies in DBDs are typically quite modest (Mei et al., 2016; Ozkan, Dufour, Bogaerts et al, 2016; Paulussen et al., 2010; Snoeckx et al., 2016), the ability to operate at atmospheric pressure while maintaining non-equilibrium conditions remains a significant and compelling benefit of these discharges. When combined with plasma catalysis, these discharges have the possibility of improving the targeted production of the required substances (Chen et al., 2018).

A discharge known as a GAP can be created in an atmospheric pressure setting when two electrodes shaped like knives are placed parallel to each other, with a gaseous substance flowing in between. These emissions are appropriate for uses that necessitate substantial gas volumes (multiple liters per minute). The thermal and nonthermal modes of operation for the GAP can vary based on the power and flow rate used. Moreover, the arc can function within the phase of transition, wherein it undergoes development from the thermal condition to the nonthermal condition. The discharge energy efficiency for gas treatment is optimized by this transitional regime. Nunnally et al. documented a considerable increase in energy efficiency with the reverse vortex flow gliding arc discharge method, reporting a rate of 43% for CO2 decomposition. This value significantly surpasses the efficiency achieved using DBDs, which typically hover around 10% (Nunnally et al., 2011). Non-equilibrium vibrational excitation of CO_2 and a substantial temperature gradient between the gliding arc and the adjacent gas contributes significantly to the exceptional level of efficiency observed, resulting in rapid quenching.

MW plasmas, denoted as plasmas produced by introducing MW energy within the range of 100 MHz to 10 GHz, are often referred to as plasmas generated by injecting electromagnetic waves (Lebedev, 2010). Frequencies of 2.45 and 0.915 GHz are commonly employed to create MW discharges. A broad range of pressures, from low mTorr to atmospheric pressure, can be utilized to operate them. The characteristics of the MW discharges functioning under atmospheric conditions are similar to those of thermal plasma. Nevertheless, the MW discharges exhibit significant differences from thermodynamic equilibrium when operating under low pressure. The effectiveness of the CO_2 dissociation process in a MW discharge is greatly influenced by factors such as the power and operating pressure of the plasma. In a study, it was found that a MW plasma with supersonic gas flows achieved the most effective conversion of pure CO_2, with an energy efficiency of approximately 90% (Azizov et al., 1983). MW discharges can generate a robust non-equilibrium atmosphere that contains CO_2 molecules in highly vibrational states. These states are particularly advantageous for the energy-efficient decomposition of CO_2 (Chung & Chang, 2016). The remarkable effectiveness of MW plasmas is achieved because electrons absorb the power applied to them to a large extent, and there is also a relatively high stimulation of the CO2 asymmetric mode (Silva et al., 2014), which has a crucial function in the decomposition of CO_2 (Azizov et al., 1983). In situations with reduced pressure, MW plasmas are commonly distinguished by a gas temperature that remains under 1500 K and an electron temperature in the range of 1 to 2 electron volts. In these circumstances, calculations have shown that approximately 95% of the discharge energy is imparted from the plasma electrons to the CO_2 molecules, mainly focusing on their non-uniform vibrational pattern (Chung & Chang, 2016; Silva et al., 2014).

Figure 4. The percentage of electron energy diverted to various modes of excitation in relation to the reduced electric field (E/n)
Adapted from (Snoeckx & Bogaerts, 2017)

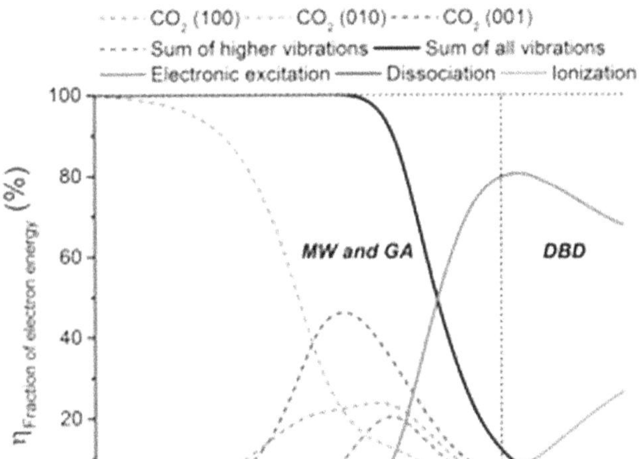

In their study, Bogaerts et al. have provided valuable information on how electron energy gets transferred to various channels, leading to excitation, dissociation, or ionization of CO_2 molecules (Aerts, 2014; Snoeckx & Bogaerts, 2017). In the discharge, Figure 4 displays the relationship between the reduced electric field (E/n) and the proportion of energy transferred from electrons to various channels involving excitation, ionization, and dissociation of CO_2. The calculations for this chart are derived from the cross-sectional measurements obtained from electron impact reactions discussed in references (Aerts, 2014; Bogaerts et al., 2015; Snoeckx & Bogaerts, 2017). In MW plasma, the electrified field is usually approximately 50 Td, best suited for stimulating the vibrations of CO_2 molecules. According to Fridman's research findings, it has been demonstrated that nearly all of the nonthermal discharge energy, up to 97%, can be effectively transmitted from the electrons in the plasma to induce vibrational excitement in CO_2 molecules. This phenomenon occurs when the electron temperature is approximately 1-2 eV or when the reduced electric field (E/n) reaches around 20-40 Td (Chung & Chang, 2016; Fridman & Rusanov, 1994). The curve known as the 'total vibrational sum' depicted in Figure 4 serves as an indication of this. In Figure 4, the significance of the purple curve should not be overlooked. It corresponds to the initial vibrational state of the asymmetrical vibration of CO_2, which is the primary pathway for dissociation (Aerts, 2014). The process of breaking down CO_2 in a DBD plasma has a relatively low level of energy efficiency, as observed in several studies (Aerts et al., 2015; Ozkan, Dufour, Silva et al, 2016). The electron temperature within a DBDs typically ranges from 2 to 3 electron volts (eV), a value that is relatively elevated when it comes to effectively excited CO_2 vibrational states. In the figure, the designated area known as the 'DBD region' often exhibits electric field values commonly around 200 Td or even more remarkable. The findings of previous research investigating plasma-induced CO_2 decomposition indicate that more significant pressures and reduced electric field values favor the vibrational excitation mechanism over the electronic excitation mechanism. It, in turn, helps to explain the enhanced energy efficiency observed in discharges such as GAP and MW (Bogaerts et al., 2015; Indarto et al., 2007; Van Rooij et al., 2015).

3.3. Electrochemical CO_2 Reduction Using Plasma

Using plasma in electrochemical CO_2 reduction (CO_2RR) has become an increasingly popular and promising approach. Plasma, which consists of highly ionized particles, offers the potential to enhance the conversion of CO_2 into valuable products. Compared to traditional electrochemical methods, plasma-based CO_2 reduction brings several advantages. One of the key benefits is its ability to generate highly reactive species like radicals and ions, which play a crucial role in activating CO_2 molecules. This activation step is essential due to the inherent thermodynamic stability of CO_2, making it challenging to convert into desired products. Another advantage is that plasma-induced reactions occur at lower temperatures, making the process more energy-efficient than conventional high-temperature reduction methods. By utilizing plasma, researchers can precisely control the reaction conditions and tailor the selectivity of product formation. Recent studies have centered on exploring the impact of plasma-generated species, such as excited states of molecules and reactive radicals, on improving the efficiency and selectivity of CO_2RR. Plasma treatment has been applied to a range of catalysts, including metal nanoparticles, metal oxides, and carbon-based materials, to enhance their performance for CO_2 reduction.

The electroreduction of CO_2 has the potential to be a sustainable route for the synthesis of industrial feedstocks and fuels, but efficient and selective catalysts are needed. During their investigation, Mistry et al. (Mistry et al., 2016) identified that subjecting silver foil to a plasma oxidation pre-treatment amplifies the quantity of catalytically active sites with low coordination. Consequently, this process results in a notable reduction in overpotential and an elevation in the activity of CO_2 electroreduction, specifically towards the production of CO. A pre-oxidized silver foil showed an impressive Faradaic efficiency of over 90% for CO production. Furthermore, Density-functional theory (DFT) calculations demonstrated that the catalyst's surface, rich in defects, significantly reduced the overpotential necessary for electroreducing CO_2. The results of this research have significant consequences for creating eco-friendly methods of storing energy and effectively managing carbon emissions.

Reducing CO_2 using electricity to create valuable chemicals and fuels shows significant potential for sustainable energy storage and mitigating the impact of greenhouse gas emissions. Gao et al. (2017) created copper nanocube catalysts with adjustable facets and ion content using low-pressure plasma pretreatment, resulting in highly effective and specific catalysts. The nanocube catalysts exhibited enhanced selectivity for ethanol, ethylene (C_2H_4), and n-propanol, as well as reduced overpotentials, ultimately leading to a maximum Faradaic efficiency of roughly 73% for C_2 and C_3 products. It was discovered that the existence of oxygen components in the surface and subsurface areas of the nanocube catalysts played a critical role in attaining superior performance and selectivity, underscoring the significance of plasma treatments in catalyst development for CO_2 electroreduction.

Mistry et al. (2017) performed a pioneering investigation into copper catalysts activated by plasma, which exhibited exceptional selectivity in converting CO_2 into C_2H_4. Through plasma treatments, they were able to enhance the electroreduction of CO_2 using oxidized copper catalysts. This innovation resulted in catalysts with reduced overpotentials and an unprecedented 60% selectivity towards C_2H_4. To gain a deeper understanding of the catalysts' enhanced effectiveness, the scientists incorporated electrochemical analyses alongside microscopic and spectroscopic methods. During the investigation, it was discovered that copper oxides exhibited unexpected stability against reduction, resulting in the retention of copper$^+$ species on the surface of the catalyst throughout the reaction. The existence of copper$^+$ proved to be vital in reducing the onset potential and increasing the selectivity of C_2H_4. Furthermore, the roughness of oxide-derived copper catalysts exhibited a partial influence on their catalytic performance.

Gao et al. (2017) have investigated how the design of electrolytes can enhance the efficiency of plasma-activated copper catalysts during the electroreduction of CO_2. The team of researchers discovered that the presence of halide ions, specifically I^-, can cause a remarkable process of nanostructuring on the surface of oxidized copper. This process ultimately leads to the creation of distinct and precise crystals made of Cu. Furthermore, it was observed that the inclusion of chloride ions (Cl^-), bromide ions (Br^-), and iodide ions (I^-) resulted in a reduction in overpotential and an enhancement in the rate of electroreduction of CO_2, all while maintaining the ability to selectively produce C_2-C_3 products. The enhancement is credited to the unique attachment of halides to the oxidized copper surface during the chemical reaction, aiding the creation and steadfastness of the carboxyl intermediate. These discoveries have significant implications for the advancement of eco-friendly methods for storing energy and completing the carbon cycle.

In their research, Pou et al. (2018) explored the potential of using a fluidized plasma reactor (FPR) in combination with a photovoltaic system to reduce CO_2 to CO at ambient pressure and low temperatures. The results of the study revealed that increasing the number of discharge points, voltage, and frequency at the anode led to greater conversions of CO_2 to CO, mainly due to a higher occurrence of NTP discharges within the FPR. Furthermore, enlarging the gap between the electrodes also resulted in enhanced conversions. The incorporation of small amounts of copper in the alumina fluidized bed had a minor positive impact on energy efficiency and CO_2 conversion. The maximum CO_2 conversion was 41%, while the highest energy efficiency reached 2.1%. Ultimately, the study inferred that using photovoltaic energy in conjunction with FPR technology could enhance the sustainability of the process and offset approximately 67% of the CO_2 emissions generated throughout the process.

The investigation conducted by Scholten et al. (2019) focused on utilizing low-pressure oxygen plasmas to alter the structure of dendritic Cu catalysts used in CO_2 electroreduction. According to their findings, the presence of oxides in the catalysts before the reaction had a minimal impact on their catalytic performance compared to the overall morphology of the catalyst. By employing operando spectroscopic techniques, the researchers were able to observe an improved selectivity towards C_2 and C_3 products. This study provides valuable insights into the relationship between catalyst structure and chemical state selectivity in CO_2RR, and it also suggests the potential for designing more efficient catalysts through plasma pretreatments.

In the research conducted by Han et al. (2021), they explored the application of plasma treatments in creating oxygen vacancies in amorphous MnO_x catalysts for the electrochemical reduction of CO_2 (CO_2RR). The results revealed that the catalysts with higher levels of oxygen vacancies demonstrated greater Faradaic efficiency and partial current density in the production of CO, even when operating at lower overpotentials. The authors attributed this improvement to the abundance of oxygen vacancies, which facilitate the adsorption and activation of CO_2 and ensure efficient charge transfer with the catalyst. This study emphasizes the potential of utilizing oxygen vacancy engineering to intelligently design CO_2RR catalysts with outstanding performance.

The study conducted by Du et al. (2021) focused on exploring the potential of In_2O_3 films modified with plasma assistance and oxygen vacancy engineering for improving the electrochemical reduction of CO_2. Their study found that magnetron sputtering with high-energy plasma treatment can generate oxygen vacancies in In_2O_3 films, which exhibit excellent performance in HCOOH production. An enhancement in the electronic density on the surface arising from the presence of oxygen vacancies has been observed to facilitate the generation of *COOH intermediates and lessen the formation barrier. As a result, a considerable amount of C_1 products is produced. This investigation emphasizes the possibility of fabricating catalysts with abundant defects for electrochemical CO_2RR, utilizing straightforward and manageable techniques to create oxygen vacancies.

Yang et al. (2022) investigated using plasma vulcanization activation to improve the electrocatalytic performance of silver electrodes for CO2 reduction. The investigation determined that the sulfide derived silver porous microrods (SD-AgPMRs) formed from sulfide with the help of the treatment displayed notable Faradaic efficiency and partial current density in the production of CO under low potentials. The enhanced performance was ascribed to the greater specific surface area provided by the porous microrod structure and the heightened adsorption energy of crucial intermediates, which were achieved by introducing sulfur atoms after activating the plasma vulcanization process. This research introduces an innovative approach to enhance the efficacy of metal electrocatalysts in CO_2 reduction and other catalytic uses.

Experimental studies have shown that plasma treatment leads to modified catalyst surfaces, such as the creation of oxygen vacancies or the formation of active sites, which can promote CO_2 activation and subsequent reduction. Plasma-derived species can interact with the catalyst surface, Plasma-based CO_2 reduction has the capability to modify the electronic structure of catalysts, aiding in the binding and conversion of CO_2 molecules. This process has shown potential in generating valuable products such as C_2H_4, methane (CH_4), formic acid (HCOOH), and CO. By adjusting plasma conditions, catalyst properties, and reaction parameters, it is possible to influence the selectivity towards specific products. Despite these advancements, there are challenges in understanding the complex interactions between plasma and catalysts, as well as optimizing the process for large-scale applications. Further research is necessary to explore the mechanism of plasma-induced CO_2 reduction and develop catalysts that exhibit high selectivity, stability, and scalability. Overall, CO_2RR using plasma offers a promising avenue for sustainable and efficient carbon utilization. Through the utilization of plasma-generated species and tailored catalysts, it is possible to achieve improved CO_2 conversion rates and produce valuable chemical feedstocks. Continued progress in this field can contribute to addressing global climate change and facilitating the transition towards a carbon-neutral society.

4. PLASMA REACTORS FOR CO_2 CONVERSION

4.1. Different Types of Plasma Reactors Used for CO_2 Conversion

Plasma technology has a variety of uses, and there are many different types of plasma reactors utilized for various applications. In addition to transforming CO_2 into CO, this technology is employed in hydrogen production through hydrocarbon reforming (Chen et al., 2008), as well as the medical field (Kaushik et al., 2019), thin-film deposition (Bräuer, 2003), surface cleaning and conditioning (Thanu et al., 2019), and material synthesis/modification. The discussion centers on the features and applications of different plasma reactors, including GAP, MW plasma, and dielectric barrier discharge reactors.

4.1.1. Microwave Plasma Reactor

MW discharges are generated using electromagnetic waves with frequencies exceeding 300 MHz. Recently, there has been considerable focus on utilizing MW plasma for various gas conversion applications. The MW plasma can either be atmospheric or low-pressure. Vesel et al. (2011) conducted a study on the impact of pressure on CO_2 corruption in a MW plasma system. They found that lower pressure resulted in higher levels of decomposition. MW plasma stands out from other plasma reactors due to its adaptability to different pressure conditions. The pressure range in which MW plasma can operate spans from 10–5 torr to 760 torr (2010), offering a wide range of possibilities. The MW plasma setup has the potential to be customized and adapted for various applications, including those requiring large-volume reactors. The system can function at both low and high power levels. The primary components of the MW plasma reactor system typically comprise a magnetron-type MW power source and a MW plasma applicator. The conversion of MWs into plasma is the primary purpose of an applicator. Various applicators, such as slow-wave plasma generators, surface wave plasma generators, and cavity MW plasma generators, exemplify this function.

There are a variety of elements that can impact the effectiveness of the conversion in a MW system. The pivotal role of CO_2 molecule vibrations is significant in the direct decomposition procedure. The efficiency of reactor conversion can be influenced by the power density and CO_2 concentration in the incoming stream (Heijkers et al., 2015). In any given MW system, there is a constant compromise between the amount of energy used and the level of conversion efficiency achieved. The catalyst surface may also experience a counteractive response due to the plasma. Hence, the choice of catalyst has significant importance in this regard (Spencer & Gallimore, 2012). The system can achieve a maximum of around 45% energy efficiency, significantly lower than the thermodynamic equilibrium level. Studies have reported improved conversion efficiencies at lower pressure levels (Van Rooij et al., 2015). MW plasma is significantly used in transforming CO_2 into products that have added value. However, its efficiency is hindered in industrial settings due to the requirement of lower pressures to achieve higher conversion rates. In their most recent publication, Berthelot and Bogaerts (2017) provided insights into various approaches that can be employed to enhance both the conversion and energy efficiencies of MW plasma. Even though chemical industries have successfully implemented MW plasma reactors, there are still numerous obstacles that must be confronted both operationally and technically. There are various tasks to be undertaken to enhance the operation of MW plasma technology at power levels exceeding 100 kW. These tasks involve integrating multiple MW generators to improve process stability and parameter control, selecting suitable materials to ensure stable plasma operation and increased power output, and conducting extensive research on kinetic modeling to achieve a better understanding of the processes involved and optimize their efficiency (de la Fuente et al., 2017).

4.1.2. Dielectric Barrier Discharge Plasma Reactor

The technology called "silent discharge," or dielectric barrier discharge is utilized in various industrial operations on a large scale. Various DBD reactors with a frequency range of 1 kHz to 10 MHz and an operating pressure of 10-500 kPa are accessible. The dielectric discharge mechanism involves an electrode positioned outside the plasma region, resulting in no direct contact between the two. The DBD reactor is also known as an ozonizer discharge, owing to its primary usage for ozone production (Vezyris et al., 2014). The DBD plasma uses alternating current (AC) instead of direct current (DC) and requires at least one dielectric material, typically ceramic or silica glass. These reactors can generate RF discharge at lower operating pressures, commonly used in the semiconductor industry. Initially, DBD reactors were utilized for various purposes, such as surface treatment, gas liquefaction, and removing gaseous pollutants and VOCs. The categorization of DBD reactors can be done based on the presence or absence of packing materials. Unpacked plasma reactors have been found to have a CO_2 conversion efficiency of 20-30%, as per the studies conducted by researchers (Kogelschatz, 2003). Despite the benefits of incorporating dielectric materials in packed-bed plasma reactors, such as increased average electron energy (Wang et al., 2012), the DBD reactor suffers from low conversion rates even with high voltage (2-5 kV) applied. However, ongoing research aims to investigate the broad of DBD reactors in decomposing CO_2 on a large scale.

4.1.3. Gliding Arc Plasma Reactor

The GAP plasma is categorized as a medium between thermal and NTP due to its unique properties. The GAP can generate extremely intense plasma, leading to elevated electron temperature and decreased gas temperature. The GAP has been utilized for advanced oxidation methods due to its cost-effectiveness and enhanced efficiency, allowing for increased rates of chemical reactions at both high-power and low-pressure input, all without any accompanying external hazards (El-Aragi, 2009). One instance of a typical reactor system for generating hydrogen is using pulsating GAP reactors combined with the capability of spraying water or various alcoholic solutions (Burlica & Locke, 2008). Different factors can modify the properties and transformation in a sliding arc setup. The efficiency of conversion in sliding arc plasma is affected by the existence of additional gases by their specific usage and gas composition (Indarto et al., 2007). The GAP system can experience enhanced conversion due to the vibrational excitations and temperature disparity between the surrounding gas and the arc (Nunnally et al., 2011). The GAP reactor is preferred for the CO_2 decomposition process due to its efficient energy usage compared to thermal cycles. This preference is based on optimizing kinetic studies and process parameters (Li et al., 2019). Unstable operational parameters, such as voltage, frequency, and current, are the primary constraints in GAP. Furthermore, the operational difficulties of GAP are exacerbated by factors such as the flow rate of gas, pressure levels, erosion of electrodes, as well as the composition, humidity, and products of the gas, all of which pose challenges to its performance in real-time conditions (Komarzyniec & Aftyka, 2020).

4.1.4. Radio Frequency (RF) Plasma

The RF plasma reactor operates by enabling chemical reactions to take place at low pressures without the need for an external magnetic field. Within the reactor, the electron density is approximately 1017 per cubic meter, and the average temperature of the electrons is between 2 and 4 eV. The energy efficiency of the reactor is around 3%, even though it can convert up to 90% of CO_2 (Spencer & Gallimore, 2011). There are two kinds of capacitive RF discharge: alpha and gamma. The alpha form exhibits lower conductivity in the electrode layer than the gamma form (Savinov et al., 1999). The electron impact in an RF plasma reactor can induce the conversion of CH_4 and CO_2, even at pressures as low as 5 torrs, resulting in their decomposition. The conversion rate in the RF plasma demonstrates an upward trend as the power input increases, thanks to the incorporation of diluent gases (Liao et al., 2001).

4.1.5. Transient Spark (TS) Discharge Plasma

here is limited research on applying plasma catalytic CO_2 conversion in plasma reactors, particularly in the case of transient spark (TS) discharge plasma. The self-pulsing discharge known as the TS is typically powered by DC and normally operates at frequencies between 1 and 10 kHz. The momentary flashes come from a streaker and transform into brief pulses of spark current. One drawback of spark discharge is its inability to maintain a steady arc as the input power decreases. Consequently, there will be a sudden cessation of the thermal arc. The utilization of TS discharge has been observed to reduce NOx emissions (Janda et al., 2016). Additional research is required to investigate the potential implementation of TS discharge studies in the plasma-aided catalytic reduction of CO_2. The temporary electric discharge produced by the spark is typically more expensive than other plasma systems, leading to its restricted use for the synthesis of high-value products and situations that require the elimination of pollutants.

4.1.6. Glow Discharge Plasma

Using glow discharge plasma is primarily observed in the DRM because of its elevated levels of electron density, plasma temperature, and electron temperature (Wang et al., 1999). The temperature of glow discharge plasma is higher than that of DBD plasma but lower than the thermal plasma reactors. The absence of vacuum in APGD plasma makes it a favorable option compared to standard glow discharge plasma reactors. The release can occur in either a stable homogeneous state or a state of filamentary glow, depending on the reactant mixture and electrode configurations. According to reports, working in a uniform state of illumination is said to have benefits in the context of applications involving gas treatment. The application of APGD in the DRM is still restricted, even though it resembles both dielectric barrier discharge plasma and corona discharge. One of thc primary difficulties encountered when implementing APGD is the issue of scaling up to effectively handle significant amounts of feedstock (Snoeckx & Bogaerts, 2017).

4.1.7. Corona Discharge Plasma

The corona discharge plasma reactor falls under a different classification of plasma reactors. In addition to the breakdown of CO_2 (Mikoviny et al., 2003), applications for surface treatment or modification (Khamsen et al., 2016), enhancement of agricultural produce quality (Lee et al., 2018), the utilization of plasma catalysis finds uses in the creation of nanocomposite coatings (Tyurikov et al., 2020), and various other spheres. The use of glow discharge plasma is primarily seen in the process of plasma reactors in the form of AC fans (Brock et al., 1999), CH_4 dry reforming (Wang et al., 1999), plasma generated in capillary tubes (Mori et al., 2006), plasma jet with a tunnel structure (Kobayashi & Hamanaka, 2006), and the process of decomposing CO_2.

4.2. Comparison of Reactor Designs and Their Advantages/Disadvantages

In a recent study, Snoeckx et al. (2017) provided an overview of the current status of plasma-based CO2 conversion techniques. Many investigations have focused on using DBDs due to their inherent NTP nature and convenient operation under atmospheric pressure conditions. Nevertheless, the energy efficiency of these systems generally tends to be constrained, typically staying below the 20% threshold. The conversion rates can reach a maximum of 40%, which is regulated by adjusting the specific energy input (or the residence time of the gas within the reactor), but this comes at the cost of energy efficiency. The constrained effectiveness is attributed to an elevated average electron energy, the primary driving force for the dissociation pathways.

MW discharges span a wide range of parameters regarding effectiveness and conversion, achieving the utmost energy efficiencies. The discharge category is recognized as the most appropriate for directing discharge power toward vibrational modes and possesses optimal capabilities for leveraging vibrational excitation. As previously mentioned, the most favorable outcomes (efficiency >60%) can be traced back to the initial research (Azizov et al., 1983; Maltsev et al., 1967; Rusanov et al., 1981; Semiokhin et al., 1964), and their replication has not been achieved in recent times. All recent studies fall within the spectrum of thermal equilibrium conversion (Spencer & Gallimore, 2012) and measurements of gas temperature indicate that temperatures surpassing 3000 K highlight the significance of thermal conversion (den Harder et al., 2017; Van Rooij et al., 2015). Since elevated temperature areas have been observed in earlier research as well, it appears that a compelling avenue for further enhancement lies in the integration of thermal conversion with nonequilibrium chemistry on the outskirts.

A different form of plasma discharge, the gliding arc discharge, has shown great potential for generating significant levels of vibrational excitation. Indeed, it is not unusual to observe energy efficiencies of up to 50% (Nunnally et al., 2011), and there have even been instances where record-breaking values of 65% have been documented (Kim et al., 2014). According to the model calculations, it was found that GA discharges also lead to higher temperatures, suggesting that thermal conversion is likely to have a notable impact (Wang et al., 2016). The potential for success lies in the combination of thermal modification and the super-ideal quenching induced by vibrational excitation, as has been acknowledged in the case of MW discharges.

5. FUTURE PERSPECTIVES AND CONCLUSION

Finding ways to convert CO_2, preferably into products with added value, is seen as a significant hurdle we face in the present era. Over the previous ten years, significant advancements have been achieved in various innovative technologies. In this assessment, we aim to present the reader with a comprehensive and unbiased evaluation of a relatively overlooked developing technology: the conversion of CO_2 using plasma chemical methods. To attain this objective, we offered a comprehensive overview of the fundamental principles underlying plasma and showcased its effectiveness as a technique for converting gas. Ultimately, we situated plasma within the larger framework of comparable technologies.

Without a doubt, it is clear that one outcome stands out from the discussed advancements in the field of CO_2 conversion in this review. The inquiry is not about whether any of these innovative technologies will obtain industrial competitiveness, but rather about the timing and identification of which one(s) will assume the predominant position. To shed a more intense spotlight on - the ultimate pressing inquiry - whether plasma technology might hold the key - or at least have a role to play - we provide a concise overview of the emerging possibilities and obstacles for plasma technology in the realm of CO_2 transformation. There are several notable advantages of plasmas when compared to other emerging technologies: (i) they can function at standard temperatures utilizing any form of renewable electricity, (ii) they offer significant adaptability in terms of the materials that need to be treated, (iii) they provide a remarkably versatile process that facilitates efficient energy storage, peak shaving, and grid stabilization, (iv) the reactors have minimal investment and operational expenses, (v) they can easily be scaled in terms of size and application, and (vi) most importantly, the technology is not reliant on scarce rare earth materials, thus distinguishing it from others currently available. The exceptional amalgamation of attributes provides plasma chemical conversion with exceptionally extensive adaptability, rendering it an exceedingly practical and invaluable technology for CCU.

One limitation of using plasma technology for solar-to-fuel conversion is its dependence on indirect solar energy in the form of electricity. This sets it apart from other technologies that can directly harness solar energy, resulting in a decrease in the efficiency of the conversion process. However, this feature also enables plasma technology to depend on alternative sources of sustainable energy. Furthermore, it significantly contributes to the adaptability of plasma technology in terms of both geographical location and operational procedures, as it is not reliant on the presence of sunlight. Nonetheless, the immediate concern pertains to the separation of products. After analyzing the extensive information presented in various sources, it is evident that the conversion of CO_2, whether with or without the presence of hydrogen, consistently results in the formation of a diverse range of substances post-reaction. This indicates the requirement for a separation step after processing, which can be costly at times.

After considering the comparison, it becomes evident that plasma technology can undeniably have a significant impact in the realm of CO_2 conversion. Therefore, it is well within our reach to contemplate its future application for commercial purposes, whether on a large or small level. However, as usual, only a few individuals appear to be appropriate for the particular undertaking, as we clarify in this account.

The discussion revolved around two primary approaches for converting CO_2: (i) splitting CO_2 directly and (ii) combining CO_2 with a co-reactant like H_2O, H_2, or CH_4 to serve as a hydrogen source. This combination leads to processes known as artificial photosynthesis, CO_2 hydrogenation, and DRM, respectively. Currently, the inefficient operation of CO_2 conversion via the latter two methods in a plasma-exclusive configuration is not only undesirable on an industrial level but also poses risks associated with the creation of potentially hazardous combinations that need to be managed. In the coming years, the resolution of these problems may be achieved by merging plasma with supplementary methods like catalysis or electrolysis. In contrast, various plasma reactors have already exceeded the targeted energy efficiency of 60% for pure CO_2 splitting and DRM. Consequently, they have achieved a solar-to-fuel conversion efficiency, specifically for producing syngas, that surpasses the minimum required efficiency of 20%.

There is excellent potential for pure CO_2 splitting in the short term when utilizing energy-efficient techniques such as GAP and MW discharges. In this particular scenario, it is possible to obtain a stream of pure CO and O_2 after undergoing a separation process. Alternatively, if in situ trapping of the O_2 is effectively executed, a pure stream of CO can be generated. Achieving this can be accomplished through various methods, such as incorporating other technologies or introducing scavengers.

Over time, it appears clear that the most promising approach for the efficient generation of valuable materials, including esters, alcohols, aldehydes, and acids, is the utilization of CO_2 conversion alongside a co-reactant rich in hydrogen. The potential of the direct oxidative pathway is significant, as it bypasses the need for the intermediary syngas step. Currently, there is a considerable requirement for further extensive research in this domain to enhance the specificity for these valuable large quantities of chemicals in comparison to the presently generated syngas. The attainment of this goal will likely require the utilization of catalysts. In the present day, GAP and APGD discharges are already offering an efficient energy-saving option for generating syngas in any desired proportion via DRM.

Plasma technology, without a doubt, is not a disregarded mythical creature – despite its exceptional qualities for CCU, it is important to acknowledge that, as discussed numerous times in this study, countless significant obstacles need to be addressed, just like any other developing technologies in this domain. The challenges that need to be addressed include the requirement for additional research in crucial areas. These areas include understanding the plasma chemical processes in warm discharges, such as MW and GAP, which stimulate vibrational excitation. It is essential to further exploit the non-equilibrium character of these plasmas to maximize energy efficiency, particularly by utilizing the most efficient dissociation pathway. Additionally, exploring the combination of plasma with catalysis can offer synergetic effects and selective production of value-added compounds. To further enhance conversions and yields, it is worth considering in situ O/O_2 trapping through scavengers or combining plasma with other technologies like catalysis or solid oxide electrolysis cells (SOEC). Furthermore, successful scale-up demonstrations for discharges other than DBDs are necessary, as well as the development and optimization of efficient power supplies for plasma generation and sustainment.

Furthermore, it is possible to identify several significant research challenges specific to the discharge. Regarding MW plasmas, the ultimate objective should be to attain equivalent energy efficiency at atmospheric pressure as what is presently achieved at reduced pressure for CO_2 splitting. Additionally, further research on DRM is crucial. Additional research in this area is required. Moreover, a few alternative plasma reactors exhibit satisfactory efficiency; however, there remains a scarcity of information and understanding, necessitating further investigation to unlock their complete capabilities. Naturally, included in each critique is an assessment that reflects the most advanced level of development. The field of CO_2 conversion using plasma technology is relatively recent, and it is rapidly progressing. Similar to any new technology, there is plenty of potential for improvement and new interdisciplinary advancements.

To attain efficient CO_2 conversion, the ideal plasma system would possess a high density of electrons (N_e) paired with a low electron temperature (T_e). These conditions can be achieved by maintaining a common reduced electric field (E/n). The combination of high Ne, low T_e, and a low E/n is crucial for stimulating and achieving a significant level of vibrational excitation. Simultaneously, this system should function at a decreased gas temperature (T_g) to reverse recombination reactions and minimize losses due to VT relaxation. To garner industrial appeal, it is necessary to accomplish this under standard atmospheric conditions.

The potential for future studies lies in exploring the direct generation of oxygenates through oxidative reactions, alongside investigating multi-reforming techniques that utilize a blend of gases (such as O_2, H_2O, CH_4, or CO_2) or multiple sequential reforming processes with plasma technology. These avenues hold great promise and merit further exploration in research efforts. In addition, it is highly desirable to have further interdisciplinary studies on integrating plasmas with other innovative technologies. Regarding catalysis, the current fusion is proving to be quite attractive, and we believe that the fusion with electrolysis, despite presenting even greater difficulties, could also result in synergistic outcomes. In the present era, it is highly likely that the future will depend on a genuine combination of various technologies and solutions, each catering to its specific tasks, rather than adhering to the outdated notion of one solution or technology that applies universally.

Based on the present evaluation, it is clear that the industrial execution of the direct separation of CO_2 can be accelerated significantly. However, using a co-reactant, such as hydrogen, in CO_2 conversion deserves continued investigation. This approach has the potential to produce an extensive range of valuable chemicals and fuels from a common starting material. Moreover, it allows for the ability to adjust the output to meet the demands of the market, making it a promising long-term option. For this reason, we think that plasmas might constitute, in some capacity, the illuminating source that radiates upon our prospects.

REFERENCES

Aerts, R. (2014). *Experimental and computational study of dielectric barrier discharges for environmental applications*. Universiteit Antwerpen.

Aerts, R., Somers, W., & Bogaerts, A. (2015). Carbon dioxide splitting in a dielectric barrier discharge plasma: A combined experimental and computational study. *ChemSusChem*, *8*(4), 702–716. doi:10.1002/cssc.201402818 PMID:25641832

Andreev, Y., & Voronkov, Y. (1975). I. Semiokhin, Dissociation Of Carbon-Dioxide In A Pulse Discharge. *Журнал физической химии*, *49*, 1178–1181.

Armenise, I., & Kustova, E. (2018). Mechanisms of coupled vibrational relaxation and dissociation in carbon dioxide. *The Journal of Physical Chemistry A*, *122*(23), 5107–5120. doi:10.1021/acs.jpca.8b03266 PMID:29781278

Azizov, R. I., Vakar, A. K., Zhivotov, V., Krotov, M. F., Zinovev, O., Potapkin, B., Rusanov, A., Rusanov, V. D., & Fridman, A. A. e. (1983). *Nonequilibrium plasmachemical process of CO2 decomposition in a supersonic microwave discharge*. Akademiia Nauk SSSR Doklady.

Berthelot, A., & Bogaerts, A. (2017). Modeling of CO2 splitting in a microwave plasma: How to improve the conversion and energy efficiency. *The Journal of Physical Chemistry. C, Nanomaterials and Interfaces*, *121*(15), 8236–8251. doi:10.1021/acs.jpcc.6b12840

Bogaerts, A., Berthelot, A., Heijkers, S., Kolev, S., Snoeckx, R., Sun, S., Trenchev, G., Van Laer, K., & Wang, W. (2017). CO2 conversion by plasma technology: Insights from modeling the plasma chemistry and plasma reactor design. *Plasma Sources Science & Technology*, *26*(6), 063001. doi:10.1088/1361-6595/aa6ada

Bogaerts, A., Kozák, T., Van Laer, K., & Snoeckx, R. (2015). Plasma-based conversion of CO 2: Current status and future challenges. *Faraday Discussions*, *183*, 217–232. doi:10.1039/C5FD00053J PMID:26391907

Bongers, W., Bouwmeester, H., Wolf, B., Peeters, F., Welzel, S., van den Bekerom, D., den Harder, N., Goede, A., Graswinckel, M., Groen, P. W., Kopecki, J., Leins, M., van Rooij, G., Schulz, A., Walker, M., & van de Sanden, R. (2017). Plasma-driven dissociation of CO2 for fuel synthesis. *Plasma Processes and Polymers*, *14*(6), 1600126. doi:10.1002/ppap.201600126

Boyle, W., & Haworth, F. (1956). Glow-to-arc transition. *Physical Review*, *101*(3), 935–938. doi:10.1103/PhysRev.101.935

G. Bräuer, G. (2003). *Magnetron-Sputtering*. ScienceDirect.

Brock, S. L., Shimojo, T., Marquez, M., Marun, C., Suib, S. L., Matsumoto, H., & Hayashi, Y. (1999). Factors Influencing the Decomposition of CO2in AC Fan-Type Plasma Reactors: Frequency, Waveform, and Concentration Effects. *Journal of Catalysis*, *184*(1), 123–133. doi:10.1006/jcat.1999.2440

Bruggeman, P., & Brandenburg, R. (2013). Atmospheric pressure discharge filaments and microplasmas: Physics, chemistry and diagnostics. *Journal of Physics. D, Applied Physics*, *46*(46), 464001. doi:10.1088/0022-3727/46/46/464001

Burlica, R., & Locke, B. R. (2008). Pulsed plasma gliding-arc discharges with water spray. *IEEE Transactions on Industry Applications*, *44*(2), 482–489. doi:10.1109/TIA.2008.916603

Bykova, N., Vasil'evskii, S., Gordeev, A., Kolesnikov, A. F., Fershin, I., & Yakushin, M. (1997). Determination of the effective probabilities of catalytic reactions on the surfaces of heat shield materials in dissociated carbon dioxide flows. *Fluid Dynamics*, *32*(6), 876–886. doi:10.1007/BF03374545

Capezzuto, P., Cramarossa, F., D'Agostino, R., & Molinari, E. (1976). Contribution of vibrational excitation to the rate of carbon dioxide dissociation in electrical discharges. *Journal of Physical Chemistry*, *80*(8), 882–888. doi:10.1021/j100549a024

Capitelli, M., Colonna, G., D'Ammando, G., & Pietanza, L. (2017). Self-consistent time dependent vibrational and free electron kinetics for CO2 dissociation and ionization in cold plasmas. *Plasma Sources Science & Technology*, *26*(5), 055009. doi:10.1088/1361-6595/aa6427

Cernak, M., Van Veldhuizen, E., Morva, I., & Rutgers, W. (1995). Effect of cathode surface properties on glow-to-arc transition in a short positive corona gap in ambient air. *Journal of Physics. D, Applied Physics*, *28*(6), 1126–1132. doi:10.1088/0022-3727/28/6/015

Chen, G., Britun, N., Godfroid, T., Delplancke-Ogletree, M., & Snyders, R. (2017). *Role of plasma catalysis in the microwave plasma-assisted conversion of CO2, Green Chemical Processing and Synthesis*. IntechOpen London, UK.

Chen, G., Britun, N., Godfroid, T., Georgieva, V., Snyders, R., & Delplancke-Ogletree, M.-P. (2017). An overview of CO2 conversion in a microwave discharge: The role of plasma-catalysis. *Journal of Physics. D, Applied Physics*, *50*(8), 084001. doi:10.1088/1361-6463/aa5616

Chen, G., Georgieva, V., Godfroid, T., Snyders, R., & Delplancke-Ogletree, M.-P. (2016). Plasma assisted catalytic decomposition of CO2. *Applied Catalysis B: Environmental*, *190*, 115–124. doi:10.1016/j.apcatb.2016.03.009

Chen, G., Wang, L., Godfroid, T., & Snyders, R. (2018). *Progress in plasma-assisted catalysis for carbon dioxide reduction, Plasma Chemistry and Gas Conversion*. IntechOpen London, UK.

Chen, H.-F., Chiang, H.-C., Matsui, H., Tsuchiya, S., & Lee, Y.-P. (2009). Distribution of vibrational states of CO2 in the reaction O (1D)+ CO2 from time-resolved fourier transform infrared emission spectra. *The Journal of Physical Chemistry A*, *113*(15), 3431–3437. doi:10.1021/jp807501c PMID:19354311

Chen, H. L., Lee, H. M., Chen, S. H., Chang, M. B., Yu, S. J., & Li, S. N. (2009). Removal of volatile organic compounds by single-stage and two-stage plasma catalysis systems: A review of the performance enhancement mechanisms, current status, and suitable applications. *Environmental Science & Technology*, *43*(7), 2216–2227. doi:10.1021/es802679b PMID:19452866

Chen, H. L., Lee, H. M., Chen, S. H., Chao, Y., & Chang, M. B. (2008). Review of plasma catalysis on hydrocarbon reforming for hydrogen production—Interaction, integration, and prospects. *Applied Catalysis B: Environmental*, *85*(1-2), 1–9. doi:10.1016/j.apcatb.2008.06.021

Chung, W.-C., & Chang, M.-B. (2016). Review of catalysis and plasma performance on dry reforming of CH4 and possible synergistic effects. *Renewable & Sustainable Energy Reviews*, *62*, 13–31. doi:10.1016/j.rser.2016.04.007

Cosby, P. & Helm, H. (1992). *Dissociation rates of diatomic molecules*. Wright Laboratory, Air Force Materiel Command, United States Air Force.

Dang, C., Reid, J., & Garside, B. (1982). Detailed vibrational population distributions in a CO2 laser discharge as measured with a tunable diode laser. *Applied Physics. B, Photophysics and Laser Chemistry*, *27*(3), 145–151. doi:10.1007/BF00694640

Dang, C., Reid, J., & Garside, B. (1983). Dynamics of the CO 2 upper laser level as measured with a tunable diode laser. *IEEE Journal of Quantum Electronics*, *19*(4), 755–764. doi:10.1109/JQE.1983.1071913

de la Fuente, J. F., Moreno, S. H., Stankiewicz, A. I., & Stefanidis, G. D. (2017). On the improvement of chemical conversion in a surface-wave microwave plasma reactor for CO2 reduction with hydrogen (The Reverse Water-Gas Shift reaction). *International Journal of Hydrogen Energy*, *42*(18), 12943–12955. doi:10.1016/j.ijhydene.2017.04.040

den Harder, N., van den Bekerom, D. C., Al, R. S., Graswinckel, M. F., Palomares, J. M., Peeters, F. J., Ponduri, S., Minea, T., Bongers, W. A., van de Sanden, M. C., & van Rooij, G. J. (2017). Homogeneous CO2 conversion by microwave plasma: Wave propagation and diagnostics. *Plasma Processes and Polymers*, *14*(6), 1600120. doi:10.1002/ppap.201600120

Donahue, T. (1971). Aeronomy of CO 2 atmospheres: A review. *Journal of the Atmospheric Sciences*, *28*(6), 895–900. doi:10.1175/1520-0469(1971)028<0895:AOCAAR>2.0.CO;2

Du, X., Qin, Y., Gao, B., Wang, K., Li, D., Li, Y., Ding, S., Song, Z., Su, Y., & Xiao, C. (2021). Plasma-assisted and oxygen vacancy-engineered In2O3 films for enhanced electrochemical reduction of CO2. *Applied Surface Science*, *563*, 150405. doi:10.1016/j.apsusc.2021.150405

Duan, X., Li, Y., Ge, W., & Wang, B. (2015). Degradation of CO2 through dielectric barrier discharge microplasma. *Greenhouse Gases: Science and Technology*, *5*(2), 131–140. doi:10.1002/ghg.1425

El-Aragi, G. (2009). *Gliding arc discharge (GAD) experiment*. Plasma Physics and Nuclear Fusion Dept., Nuclear Research Center.

Foote, A., Dedrick, J., O'Connell, D., North, M., & Gans, T. (2016). Conversion of CO 2 to CO using radio-frequency atmospheric pressure plasmas. *APS Annual Gaseous Electronics Meeting*. APS.

Fox, J. L., & Dalgarno, A. (1979). Ionization, luminosity, and heating of the upper atmosphere of Mars. *Journal of Geophysical Research*, *84*(A12), 7315–7333. doi:10.1029/JA084iA12p07315

Fridman, A. (2008). *Plasma chemistry*. Cambridge university press. doi:10.1017/CBO9780511546075

Fridman, A., Chirokov, A., & Gutsol, A. (2005). Non-thermal atmospheric pressure discharges. *Journal of Physics. D, Applied Physics*, *38*(2), R1–R24. doi:10.1088/0022-3727/38/2/R01

Fridman, A., & Rusanov, V. (1994). Theoretical basis of non-equilibrium near atmospheric pressure plasma chemistry. *Pure and Applied Chemistry*, *66*(6), 1267–1278. doi:10.1351/pac199466061267

Gao, D., Scholten, F., & Roldan Cuenya, B. (2017). Improved CO2 electroreduction performance on plasma-activated Cu catalysts via electrolyte design: Halide effect. *ACS Catalysis*, *7*(8), 5112–5120. doi:10.1021/acscatal.7b01416

Gao, D., Zegkinoglou, I., Divins, N. J., Scholten, F., Sinev, I., Grosse, P., & Roldan Cuenya, B. (2017). Plasma-activated copper nanocube catalysts for efficient carbon dioxide electroreduction to hydrocarbons and alcohols. *ACS Nano*, *11*(5), 4825–4831. doi:10.1021/acsnano.7b01257 PMID:28441005

George, A., Shen, B., Craven, M., Wang, Y., Kang, D., Wu, C., & Tu, X. (2021). A Review of Non-Thermal Plasma Technology: A novel solution for CO2 conversion and utilization. *Renewable & Sustainable Energy Reviews*, *135*, 109702. doi:10.1016/j.rser.2020.109702

Gilijamse, J. J., Hoekstra, S., Meek, S. A., Metsälä, M., van de Meerakker, S. Y., Meijer, G., & Groenenboom, G. C. (2007). The radiative lifetime of metastable CO (aΠ3, v=). *The Journal of Chemical Physics*, *127*(22), 127. doi:10.1063/1.2813888

Gomez, E., Rani, D. A., Cheeseman, C., Deegan, D., Wise, M., & Boccaccini, A. (2009). Thermal plasma technology for the treatment of wastes: A critical review. *Journal of Hazardous Materials*, *161*(2-3), 614–626. doi:10.1016/j.jhazmat.2008.04.017 PMID:18499345

Gordon, S., & McBride, B. J. (1994). Computer program for calculation of complex chemical equilibrium compositions and applications. Part 1. *Analysis*.

Grofulović, M., Alves, L. L., & Guerra, V. (2016). Electron-neutral scattering cross sections for CO2: A complete and consistent set and an assessment of dissociation. *Journal of Physics. D, Applied Physics*, *49*(39), 395207. doi:10.1088/0022-3727/49/39/395207

Hake, R. Jr, & Phelps, A. (1967). Momentum-Transfer and Inelastic-Collision Cross Sections for Electrons in O 2, CO, and C O 2. *Physical Review*, *158*(1), 70–84. doi:10.1103/PhysRev.158.70

Han, H., Jin, S., Park, S., Kim, Y., Jang, D., Seo, M. H., & Kim, W. B. (2021). Plasma-induced oxygen vacancies in amorphous MnOx boost catalytic performance for electrochemical CO2 reduction. *Nano Energy*, *79*, 105492. doi:10.1016/j.nanoen.2020.105492

Harding, D. R., Weston, R. E. Jr, & Flynn, G. W. (1988). Energy transfer to CO (v) in the O (1 D)+ CO (1Σ+ g) reaction. *The Journal of Chemical Physics*, *88*(6), 3590–3598. doi:10.1063/1.453908

Heijkers, S., Snoeckx, R., Kozák, T., Silva, T., Godfroid, T., Britun, N., Snyders, R., & Bogaerts, A. (2015). CO2 conversion in a microwave plasma reactor in the presence of N2: Elucidating the role of vibrational levels. *The Journal of Physical Chemistry. C, Nanomaterials and Interfaces*, *119*(23), 12815–12828. doi:10.1021/acs.jpcc.5b01466

Huang, X., Schwenke, D. W., Tashkun, S. A., & Lee, T. J. (2012). An isotopic-independent highly accurate potential energy surface for CO2 isotopologues and an initial 12C16O2 infrared line list. *The Journal of Chemical Physics*, 136. doi:10.1021/jp2104374 PMID:22462861

Huddleston, R., & Weitz, E. (1981). A laser-induced fluorescence study of energy transfer between the symmetric stretching and bending modes of CO2. *Chemical Physics Letters*, *83*(1), 174–179. doi:10.1016/0009-2614(81)80314-4

Huestis, D., Marschall, J., Billing, G., & Maclagan, R. (2002). Theoretical and Experimental Studies of O-CO 2 Collisions. *AGU Fall Meeting Abstracts*. AGU.

Indarto, A., Yang, D. R., Choi, J.-W., Lee, H., & Song, H. K. (2007). Gliding arc plasma processing of CO2 conversion. *Journal of Hazardous Materials*, *146*(1-2), 309–315. doi:10.1016/j.jhazmat.2006.12.023 PMID:17223261

Itikawa, Y., & Mason, N. (2005). Cross sections for electron collisions with water molecules. *Journal of Physical and Chemical Reference Data*, *34*(1), 1–22. doi:10.1063/1.1799251

Janda, M., Martišovitš, V., Hensel, K., & Machala, Z. (2016). Study of transient spark discharge focused at NOx generation for biomedical applications. *Journal of Physics: Conference Series*. IOP Publishing.

Kaushik, N. K., Kaushik, N., Linh, N. N., Ghimire, B., Pengkit, A., Sornsakdanuphap, J., Lee, S.-J., & Choi, E. H. (2019). Plasma and nanomaterials: Fabrication and biomedical applications. *Nanomaterials (Basel, Switzerland)*, *9*(1), 98. doi:10.3390/nano9010098 PMID:30646530

Khamsen, N., Akkarachanchainon, A., Teerakawanich, N., & Srisonphan, S. (2016). Organic and bio material surface modification via corona discharge induced atmospheric-cold plasma. *Procedia Computer Science*, *86*, 325–328. doi:10.1016/j.procs.2016.05.088

Kim, S. C., Lim, M. S., & Chun, Y. N. (2014). Reduction characteristics of carbon dioxide using a plasmatron. *Plasma Chemistry and Plasma Processing*, *34*(1), 125–143. doi:10.1007/s11090-013-9499-8

Klarenaar, B., Grofulović, M., Morillo-Candas, A.-S., Van Den Bekerom, D., Damen, M., Van De Sanden, M., Guaitella, O., & Engeln, R. (2018). A rotational Raman study under non-thermal conditions in a pulsed CO2 glow discharge. *Plasma Sources Science & Technology*, *27*(4), 045009. doi:10.1088/1361-6595/aabab6

Kobayashi, A., & Hamanaka, H. (2006). Decomposition characteristics of carbon dioxide by gas tunnel-type plasma jet. *Vacuum*, *80*(11-12), 1294–1298. doi:10.1016/j.vacuum.2006.01.064

Koeta, O., Blin-Simiand, N., Faider, W., Pasquiers, S., Bary, A., & Jorand, F. (2012). Decomposition of acetaldehyde in atmospheric pressure filamentary nitrogen plasma. *Plasma Chemistry and Plasma Processing*, *32*(5), 991–1023. doi:10.1007/s11090-012-9388-6

Kogelschatz, U. (2003). Dielectric-barrier discharges: Their history, discharge physics, and industrial applications. *Plasma Chemistry and Plasma Processing*, *23*(1), 1–46. doi:10.1023/A:1022470901385

Komarzyniec, G., & Aftyka, M. (2020). Operating problems of arc plasma reactors powered by AC/DC/AC converters. *Applied Sciences (Basel, Switzerland)*, *10*(9), 3295. doi:10.3390/app10093295

Kovalev, V., Afonina, N., & Gromov, V. (2004). Heat transfer modelling to catalytic protection systems of space vehicles entering into martian atmosphere. *Shock Waves: Proceedings of the 24th International Symposium on Shock Waves*. Springer.

Kozák, T., & Bogaerts, A. (2014a). Splitting of CO2 by vibrational excitation in non-equilibrium plasmas: A reaction kinetics model. *Plasma Sources Science & Technology*, *23*(4), 045004. doi:10.1088/0963-0252/23/4/045004

Kozák, T., & Bogaerts, A. (2014b). Evaluation of the energy efficiency of CO2 conversion in microwave discharges using a reaction kinetics model. *Plasma Sources Science & Technology*, *24*(1), 015024. doi:10.1088/0963-0252/24/1/015024

Krauss, M., Mielczarek, S., Neumann, D., & Kuyatt, C. (1971). Mechanism for production of the Fourth Positive Band System of CO by electron impact on CO2. *Journal of Geophysical Research*, *76*(16), 3733–3737. doi:10.1029/JA076i016p03733

Lebedev, Y. A. (2010). *Microwave discharges: generation and diagnostics, Journal of physics: conference series*. IOP Publishing.

LeClair, L. R., McConkey, J., & On, O. (1994). (1S) and CO (a 3 Pi) production from electron impact dissociation of CO2. *Journal of Physics. B, Atomic, Molecular, and Optical Physics*, *27*(17), 4039–4055. doi:10.1088/0953-4075/27/17/026

Lee, T., Puligundla, P., & Mok, C. (2018). Intermittent corona discharge plasma jet for improving tomato quality. *Journal of Food Engineering*, *223*, 168–174. doi:10.1016/j.jfoodeng.2017.11.004

Li, L., Zhang, H., Li, X., Kong, X., Xu, R., Tay, K., & Tu, X. (2019). Plasma-assisted CO2 conversion in a gliding arc discharge: Improving performance by optimizing the reactor design. *Journal of CO2 Utilization, 29*, 296-303.

Liao, W.-T., Lee, W.-J., Chen, C.-Y., & Shih, M. (2001). Decomposition of ethylene oxide in the RF plasma environment. *Environmental Technology*, *22*(2), 165–173. doi:10.1080/09593332208618293 PMID:11349375

Maltsev, A., Eremin, E., & Ivanter, L. (1967). Dissociation of carbon dioxide in glow discharge. *Russ. J. Phys. Chem*, *41*, 633.

Mei, D., He, Y. L., Liu, S., Yan, J., & Tu, X. (2016). Optimization of CO2 conversion in a cylindrical dielectric barrier discharge reactor using design of experiments. *Plasma Processes and Polymers*, *13*(5), 544–556. doi:10.1002/ppap.201500159

Mikoviny, T., Kocan, M., Matejcik, S., Mason, N., & Skalny, J. (2003). Experimental study of negative corona discharge in pure carbon dioxide and its mixtures with oxygen. *Journal of Physics. D, Applied Physics*, *37*(1), 64–73. doi:10.1088/0022-3727/37/1/011

Mistry, H., Choi, Y. W., Bagger, A., Scholten, F., Bonifacio, C. S., Sinev, I., Divins, N. J., Zegkinoglou, I., Jeon, H. S., Kisslinger, K., Stach, E. A., Yang, J. C., Rossmeisl, J., & Roldan Cuenya, B. (2017). Enhanced carbon dioxide electroreduction to carbon monoxide over defect-rich plasma-activated silver catalysts. *Angewandte Chemie*, *129*(38), 11552–11556. doi:10.1002/ange.201704613 PMID:28710839

Mistry, H., Varela, A. S., Bonifacio, C. S., Zegkinoglou, I., Sinev, I., Choi, Y.-W., Kisslinger, K., Stach, E. A., Yang, J. C., Strasser, P., & Cuenya, B. R. (2016). Highly selective plasma-activated copper catalysts for carbon dioxide reduction to ethylene. *Nature Communications*, *7*(1), 12123. doi:10.1038/ncomms12123 PMID:27356485

Moore, C. B., Wood, R. E., Hu, B. L., & Yardley, J. T. (1967). Vibrational energy transfer in CO2 lasers. *The Journal of Chemical Physics*, *46*(11), 4222–4231. doi:10.1063/1.1840527

Moreau, M., Orange, N., & Feuilloley, M. (2008). Non-thermal plasma technologies: New tools for bio-decontamination. *Biotechnology Advances*, *26*(6), 610–617. doi:10.1016/j.biotechadv.2008.08.001 PMID:18775485

Mori, S., Akatsuka, H., & Suzuki, M. (2002). Numerical analysis of carbon isotope separation by plasma chemical reactions in carbon monoxide glow discharge. *Journal of Nuclear Science and Technology*, *39*(6), 637–646. doi:10.1080/18811248.2002.9715244

Mori, S., Yamamoto, A., & Suzuki, M. (2006). Characterization of a capillary plasma reactor for carbon dioxide decomposition. *Plasma Sources Science & Technology*, *15*(4), 609–613. doi:10.1088/0963-0252/15/4/003

Morillo-Candas, A. S. (2019). *Investigation of fundamental mechanisms of CO2 plasmas.* Université Paris Saclay (COmUE).

Mumma, M., Stone, E., & Zipf, E. (1975). Nonthermal rotational distribution of CO ($A^1\Pi$) fragments produced by dissociative excitation of CO2 by electron impact. *Journal of Geophysical Research*, *80*(1), 161–167. doi:10.1029/JA080i001p00161

Neyts, E. C., Ostrikov, K., Sunkara, M. K., & Bogaerts, A. (2015). Plasma catalysis: Synergistic effects at the nanoscale. *Chemical Reviews*, *115*(24), 13408–13446. doi:10.1021/acs.chemrev.5b00362 PMID:26619209

Norhasyima, R. & Mahlia, T. (2018). Advances in CO_2 utilization technology: A patent landscape review. *Journal of CO2 Utilization, 26*, 323-335.

Nunnally, T., Gutsol, K., Rabinovich, A., Fridman, A., Gutsol, A., & Kemoun, A. (2011). Dissociation of CO2 in a low current gliding arc plasmatron. *Journal of Physics. D, Applied Physics*, *44*(27), 274009. doi:10.1088/0022-3727/44/27/274009

Ozkan, A., Dufour, T., Bogaerts, A., & Reniers, F. (2016). How do the barrier thickness and dielectric material influence the filamentary mode and CO2 conversion in a flowing DBD? *Plasma Sources Science & Technology*, *25*(4), 045016. doi:10.1088/0963-0252/25/4/045016

Ozkan, A., Dufour, T., Silva, T., Britun, N., Snyders, R., Bogaerts, A., & Reniers, F. (2016). The influence of power and frequency on the filamentary behavior of a flowing DBD—Application to the splitting of CO2. *Plasma Sources Science & Technology*, *25*(2), 025013. doi:10.1088/0963-0252/25/2/025013

Patil, B., Cherkasov, N., Lang, J., Ibhadon, A., Hessel, V., & Wang, Q. (2016). Low temperature plasma-catalytic NOx synthesis in a packed DBD reactor: Effect of support materials and supported active metal oxides. *Applied Catalysis B: Environmental*, *194*, 123–133. doi:10.1016/j.apcatb.2016.04.055

Paulussen, S., Verheyde, B., Tu, X., De Bie, C., Martens, T., Petrovic, D., Bogaerts, A., & Sels, B. (2010). Conversion of carbon dioxide to value-added chemicals in atmospheric pressure dielectric barrier discharges. *Plasma Sources Science & Technology*, *19*(3), 034015. doi:10.1088/0963-0252/19/3/034015

Perri, M. J., Van Wyngarden, A. L., Lin, J. J., Lee, Y. T., & Boering, K. A. (2004). Energy dependence of oxygen isotope exchange and quenching in the O (1D)+ CO2 reaction: A crossed molecular beam study. *The Journal of Physical Chemistry A*, *108*(39), 7995–8001. doi:10.1021/jp0485845

Petitpas, G., Rollier, J.-D., Darmon, A., Gonzalez-Aguilar, J., Metkemeijer, R., & Fulcheri, L. (2007). A comparative study of non-thermal plasma assisted reforming technologies. *International Journal of Hydrogen Energy*, *32*(14), 2848–2867. doi:10.1016/j.ijhydene.2007.03.026

Polak, L., & Slovetsky, D. (1976). Electron impact induced electronic excitation and molecular dissociation. *International Journal for Radiation Physics and Chemistry*, *8*(1-2), 257–282. doi:10.1016/0020-7055(76)90070-X

Pou, J., Colominas, C., & Gonzalez-Olmos, R. (2018). CO2 reduction using non-thermal plasma generated with photovoltaic energy in a fluidized reactor. *Journal of CO2 Utilization, 27,* 528-535.

Ramakers, M., Trenchev, G., Heijkers, S., Wang, W., & Bogaerts, A. (2017). Gliding arc plasmatron: Providing an alternative method for carbon dioxide conversion. *ChemSusChem*, *10*(12), 2642–2652. doi:10.1002/cssc.201700589 PMID:28481058

Rusanov, V. D., Fridman, A., & Sholin, G. V. e. (1981). The physics of a chemically active plasma with nonequilibrium vibrational excitation of molecules. *Soviet Physics - Uspekhi*, *24*(6), 447–474. doi:10.1070/PU1981v024n06ABEH004884

Savinov, S. Y., Lee, H., Song, H. K., & Na, B.-K. (1999). Decomposition of methane and carbon dioxide in a radio-frequency discharge. *Industrial & Engineering Chemistry Research*, *38*(7), 2540–2547. doi:10.1021/ie980492c

Scholten, F., Sinev, I., Bernal, M., & Roldan Cuenya, B. (2019). Plasma-modified dendritic Cu catalyst for CO2 electroreduction. *ACS Catalysis*, *9*(6), 5496–5502. doi:10.1021/acscatal.9b00483 PMID:31205798

Semiokhin, I., Andreev, Y. P., & Panchenkov, G. (1964). Dissociation of carbon dioxide circulating in a silent electric Discharge, Russ. *Journal of Physical Chemistry*, *38*, 1126.

Shaw, D., West, A., Bredin, J., & Wagenaars, E. (2016). Mechanisms behind surface modification of polypropylene film using an atmospheric-pressure plasma jet. *Plasma Sources Science & Technology*, *25*(6), 065018. doi:10.1088/0963-0252/25/6/065018

Siemens, W. (1857). Ueber die elektrostatische Induction und die Verzögerung des Stroms in Flaschendrähten. *Annalen der Physik*, *178*(9), 66–122. doi:10.1002/andp.18571780905

Silva, T., Britun, N., Godfroid, T., & Snyders, R. (2014). Optical characterization of a microwave pulsed discharge used for dissociation of CO2. *Plasma Sources Science & Technology*, *23*(2), 025009. doi:10.1088/0963-0252/23/2/025009

Snoeckx, R., & Bogaerts, A. (2017). Plasma technology–a novel solution for CO 2 conversion? *Chemical Society Reviews*, *46*(19), 5805–5863. doi:10.1039/C6CS00066E PMID:28825736

Snoeckx, R., Heijkers, S., Van Wesenbeeck, K., Lenaerts, S., & Bogaerts, A. (2016). CO 2 conversion in a dielectric barrier discharge plasma: N 2 in the mix as a helping hand or problematic impurity? *Energy & Environmental Science*, *9*(3), 999–1011. doi:10.1039/C5EE03304G

Spencer, L., & Gallimore, A. (2012). CO2 dissociation in an atmospheric pressure plasma/catalyst system: A study of efficiency. *Plasma Sources Science & Technology*, *22*(1), 015019. doi:10.1088/0963-0252/22/1/015019

Spencer, L. F., & Gallimore, A. D. (2011). Efficiency of CO 2 dissociation in a radio-frequency discharge. *Plasma Chemistry and Plasma Processing*, *31*(1), 79–89. doi:10.1007/s11090-010-9273-0

Sun, S., Wang, H., Mei, D., Tu, X., & Bogaerts, A. (2017). CO2 conversion in a gliding arc plasma: Performance improvement based on chemical reaction modeling. *Journal of CO2 Utilization, 17,* 220-234.

Thanu, D., Srinadhu, E., Zhao, M., Dole, N., & Keswani, M. (2019). Fundamentals and applications of plasma cleaning. *Developments in surface contamination and cleaning: applications of cleaning techniques*, 289-353.

Toebaert, D., Muys, P., & Desoppere, E. (1995). Spatially resolved measurement of the vibrational temperatures of the plasma in a DC-excited fast-axial-flow CO/sub 2/laser. *IEEE Journal of Quantum Electronics*, *31*(10), 1774–1778. doi:10.1109/3.466051

Treanor, C. E., Rich, J., & Rehm, R. (1968). Vibrational relaxation of anharmonic oscillators with exchange-dominated collisions. *The Journal of Chemical Physics*, *48*(4), 1798–1807. doi:10.1063/1.1668914

Trenchev, G., Kolev, S., Wang, W., Ramakers, M., & Bogaerts, A. (2017). CO2 conversion in a gliding arc plasmatron: Multidimensional modeling for improved efficiency. *The Journal of Physical Chemistry. C, Nanomaterials and Interfaces*, *121*(44), 24470–24479. doi:10.1021/acs.jpcc.7b08511

Tyurikov, K., Alexandrov, S., & Iankevich, G. (2020). Corona discharge plasma application for the deposition of nanocomposite coatings. *Materials Today: Proceedings*, *30*, 404–407. doi:10.1016/j.matpr.2019.12.385

Uytdenhouwen, Y., Van Alphen, S., Michielsen, I., Meynen, V., Cool, P., & Bogaerts, A. (2018). A packed-bed DBD micro plasma reactor for CO2 dissociation: Does size matter? *Chemical Engineering Journal*, *348*, 557–568. doi:10.1016/j.cej.2018.04.210

Van Rooij, G., Van Den Bekerom, D., Den Harder, N., Minea, T., Berden, G., Bongers, W., Engeln, R., Graswinckel, M., Zoethout, E., & Van De Sanden, M. (2015). Taming microwave plasma to beat thermodynamics in CO 2 dissociation. *Faraday Discussions, 183*, 233–248. doi:10.1039/C5FD00045A PMID:26388308

Vargas, J., Lopez, B., & Lino da Silva, M. (2021). Heavy particle impact vibrational excitation and dissociation processes in CO2. *The Journal of Physical Chemistry A, 125*(2), 493–512. doi:10.1021/acs.jpca.0c05677 PMID:33406355

Vertongen, R, Trenchev, G., Van Loenhout, R., & Bogaerts, A. (2022). Enhancing CO2 conversion with plasma reactors in series and O2 removal. *Journal of CO2 utilization, 66*, 102252.

Vesel, A., Mozetic, M., Drenik, A., & Balat-Pichelin, M. (2011). Dissociation of CO2 molecules in microwave plasma. *Chemical Physics, 382*(1-3), 127–131. doi:10.1016/j.chemphys.2011.03.015

Vezyris, C., Kavvadias, I., Papoutsis-Kiachagias, E., & Giannakoglou, K. (2014). *Unsteady continuous adjoint method using POD for jet-based flow control.* 11th World congress on computational mechanics, Barcelona, Spain.

Wang, J.-Y., Xia, G.-G., Huang, A., Suib, S. L., Hayashi, Y., & Matsumoto, H. (1999). CO2 decomposition using glow discharge plasmas. *Journal of Catalysis, 185*(1), 152–159. doi:10.1006/jcat.1999.2499

Wang, S., Zhang, Y., Liu, X., & Wang, X. (2012). Enhancement of CO 2 conversion rate and conversion efficiency by homogeneous discharges. *Plasma Chemistry and Plasma Processing, 32*(5), 979–989. doi:10.1007/s11090-012-9386-8

Wang, W., Berthelot, A., Kolev, S., Tu, X., & Bogaerts, A. (2016). CO2 conversion in a gliding arc plasma: 1D cylindrical discharge model. *Plasma Sources Science & Technology, 25*(6), 065012. doi:10.1088/0963-0252/25/6/065012

Wanten, B., Maerivoet, S., Vantomme, C., Slaets, J., Trenchev, G., & Bogaerts, A. (2022). Dry reforming of methane in an atmospheric pressure glow discharge: Confining the plasma to expand the performance. *Journal of CO2 utilization, 56*, 101869.

Wen, Y., & Jiang, X. (2001). Decomposition of CO2 using pulsed corona discharges combined with catalyst. *Plasma Chemistry and Plasma Processing, 21*(4), 665–678. doi:10.1023/A:1012011420757

Witteman, W. J. (2013). *The CO2 laser*. Springer.

Yang, J., Du, H., Yu, Q., Zhang, W., Zhang, Y., Ge, J., Li, H., Liu, J., Li, H., & Xu, H. (2022). Porous silver microrods by plasma vulcanization activation for enhanced electrocatalytic carbon dioxide reduction. *Journal of Colloid and Interface Science, 606*, 793–799. doi:10.1016/j.jcis.2021.08.061 PMID:34419818

Zasavitskii, I., Islamov, R. S., Kerimkulov, M., Konev, Y. B., Ochkin, V., Savinov, S. Y., Sobolev, N., Spiridonov, M., & Shotov, A. (1990). Kinetics of formation of the CO2 molecule distributions over the vibrational-rotational levels in the active medium of a waveguide CO2 laser. *Journal of Soviet Laser Research, 11*(4), 361–375. doi:10.1007/BF01120633

Zhu, L., Kreutz, T. G., Hewitt, S. A., & Flynn, G. W. (1990). Diode laser probing of vibrational, rotational, and translational excitation of CO2 following collisions with O (1D). I. Inelastic scattering. *The Journal of Chemical Physics*, *93*(5), 3277–3288. doi:10.1063/1.458861

ABBREVIATIONS

APGD: Atmospheric Pressure Glow Discharge
CCU: Carbon Capture And Utilization
CO_2RR: Electrochemical CO_2 Reduction
DBDs: Dielectric Barrier Discharges
DRM: Dry Reforming Of Methane
DME: Dimethyl Ether
FPR: Fluidized Plasma Reactor
GAP: Gliding Arc Plasma
MW: Microwave
NTP: Non-Thermal Plasma
RF: Radio Frequency
TS: Transient Spark

Chapter 2
Plasma Technology for Carbon Dioxide Conversion

Mudassar Shahzad
Government Postgraduate College Khanewal, Bahauddin Zakariya University, Multan, Pakistan

Noor-ul-Huda Altaf
University of Agriculture, Faisalabad, Pakistan

Muhammad Ayyaz
University of Agriculture, Faisalabad, Pakistan

Sehrish Maqsood
University of Agriculture, Faisalabad, Pakistan

Tayyba Shoukat
University of Electronic Science and Technology of China, Chengdu, China

Mumtaz Ali
Karakoram International University, Gilgit, Pakistan

Muhammad Yasin Naz
University of Agriculture, Faisalabad, Pakistan

Shazia Shukrullah
University of Agriculture, Faisalabad, Pakistan

ABSTRACT

Carbon dioxide (CO_2) is one of the major greenhouse gases that contributes to global warming and environmental variations. The increasing concentration of CO_2 in the atmosphere is posing severe threats to human health and the environment. With the increasing concerns about climate change and CO_2 levels, the need for advanced and effective technologies to mitigate CO_2 emissions is more critical than ever. Plasma technology, with its unique features and versatile capabilities, has shown immense potential for the conversion of CO_2 into valuable products and fuels. This proposed chapter aims to explore the recent advances in plasma-based CO_2 conversion processes. The chapter covers the fundamental principles, plasma generation systems, reaction mechanisms, and implications of plasma technology in CO_2 conversion, presenting a comprehensive understanding of this transformative field.

DOI: 10.4018/979-8-3693-0904-9.ch002

1. INTRODUCTION

As the world confronts the increasingly urgent challenge of climate change, the search for innovative technologies to combat rising CO_2 levels and reduce greenhouse gas emissions has gained paramount importance. One such groundbreaking approach is the use of plasma technology for CO_2 conversion a process that holds the promise of addressing both environmental and energy sustainability concerns (Legg, 2021). Developing a sustainable and low carbon-based energy economy is crucial for reducing dependence on non-renewable fossil fuels. In addition to exploring renewable energy sources like solar, hydro, and wind, recycling and utilizing CO_2 to synthesize high-value products offer an alternative solution to combat climate change. Plasma-based CO_2 decomposition, particularly using microwave discharge technology, has garnered significant interest due to its exceptional capability to efficiently produce non-equilibrium plasma with high ionization power. This technology shows promise for converting CO_2 efficiently and sustainably into valuable resources. (Ong, Nomanbhay, Kusumo, & Show, 2022). This chapter serves as an introductory exploration of the fusion of plasma technology and CO_2 conversion, aiming to shed light on the potential of this transformative field.

1.1 Plasma Technology

Plasma, often referred to as "fourth state of matter," is a remarkable physical state distinct from solids, liquids, and gases. Unlike these more conventional states, plasma is characterized by its ionized nature, where electrons and positively charged ions coexist in abundance. This ionization results in a highly conductive medium capable of generating immense energy and diverse chemical reactions. The utilization of plasma technology spans a wide spectrum of applications, ranging from industrial processes such as metal cutting and welding to cutting-edge fields like aerospace propulsion and medical treatments. Thermal and non-thermal plasma are the two main subcategories of plasma. Thermal plasmas, characterized by their high temperatures, are commonly used in industrial applications, while non-thermal or cold plasmas operate at lower temperatures, making them ideal for various scientific and environmental applications, including CO_2 conversion (López et al., 2019; Sardella, Palumbo, Camporeale, & Favia, 2016).

1.2 Rising and Mitigating CO_2 Levels: A Global Challenge

The unprecedented increase in atmospheric CO_2 levels due to human activities, primarily deforestation, industrial processes, and burning of fossil fuels, has culminated in a global environmental crisis-climate change. The scientific consensus, as articulated by bodies like Intergovernmental Panel on Climate Change (IPCC), underscores the role of elevated CO_2 levels in driving rising sea levels, global warming, and extreme weather events. This global issue necessitates immediate and substantial action to mitigate its adverse effects on ecosystems, economies, and human well-being (Lackner, 2003).

While traditional approaches to addressing CO_2 emissions, such as carbon capture and storage (CCS), have made significant strides, they still face challenges in terms of cost, scalability, and sustainability. To meet ambitious climate targets and transition toward a low-carbon future, there is a growing need for innovative technologies capable of not just capturing CO_2, but transforming it into valuable resources. This is where plasma technology emerges as a game-changing solution (Ashford & Tu, 2017).

1.3 Objectives and Scope of the Chapter

The primary objectives of this chapter intend to offer comprehensive overview of plasma technology role in mitigating CO_2 emissions through conversion processes. Specifically, the chapter will:

- Introduce the fundamentals of plasma technology, including its types, generation techniques, and unique properties.
- Examine the global issue of rising CO_2 levels and its implications for climate change, highlighting the urgency of sustainable solutions.
- Discuss the need for innovative technologies to combat CO_2 emissions and their limitations in existing frameworks.
- Explore the potential of plasma technology in addressing these challenges, with a focus on its applications in carbon dioxide conversion.
- Present recent advancements and case studies showcasing the feasibility and advantages of CO_2 conversion based on plasma.
- Highlight future research directions and plasma technology's contribution to the development of a sustainable, low-carbon future.

By delving into these facets, this chapter aims to provide a foundation for understanding the transformative potential of plasma technology in addressing a critical issue of CO_2 emissions and climate change.

2. CLASSIFICATION OF PLASMA TECHNOLOGY

Plasma is a fascinating and unique physical state that distinguishes itself from the more familiar gas, solid, and liquid phases in which a substantial fraction of the particles, typically electrons and ions, are no longer bound together in neutral atoms or molecules. Instead, they exist as charged entities, making plasma an electrically conductive medium (Grill, 1994). The ionization of particles in plasma occurs due to the supply of energy, such as heat, electrical discharge, or radiation, which provides the necessary energy to strip electrons from atoms or molecules. Unlike solids, liquids, and gases, plasma does not have a fixed shape or volume, and it can exhibit unique properties such as self-organization, the generation of electric and magnetic fields, and the emission of light, often in the form of vibrant colors. These properties make plasma a matter of immense interest and significance in various scientific and industrial applications (Fridman, 2008). Plasma can be broadly categorized into two primary types: non-thermal plasma and thermal plasma, each with distinct characteristics and applications as summarized in Table 11.1.

2.1 Thermal Plasma

Extremely high temperatures are a characteristic of thermal plasmas, often exceeding thousands of degrees Celsius. These plasmas are typically created by heating a gas or material to such high temperatures that the electrons are stripped away from atoms or molecules, resulting in a fully ionized state. Some common methods for generating thermal plasma include arc discharge, induction heating and flame plasma.

Table 1. Plasma classification

Plasma type	State	Examples
Thermal Plasma		
High Temperature Plasma (Thermally Equilibrium)	$T_i \approx T_e \approx T_r \approx T_g \approx T_v = 10^6$ - 10^8 K, $n_e \geq 10^{20}$ m^{-3}	Thermonuclear Fusion, Solar Core
Low Temperature Plasma (Quasi Equilibrium)	$T_i \approx T_e \approx T_r \approx T_g \approx T_v \leq 2 \times 10^4$ K, $n_e \geq 10^{20}$ m^{-3}	RF inductively coupled discharge, Arc plasma, Lightning
Non-Thermal Plasma (Non-Equilibrium)	$T_v \approx T_e > T_g \approx T_i \approx T_r = 300$ - 1000 K $n_e \geq 10^{10}$ m^{-3}	Plasma needle, Dielectric barrier discharge (DBD), Micro hollow cathode discharge (MHCD), Corona discharge, Glow discharge,
T_i = ion temperature, T_e = electron temperature, T_r = rotational temperature, T_g gas temperature, T_v = vibrational, and n_e = number density		

In an arc discharge, an electric current passes through a gas, creating a high-temperature plasma arc. This type of plasma is used in applications like welding, cutting, and metal refining. While induction heating uses electromagnetic fields to heat a material, creating a plasma. It is commonly employed in material processing, such as melting and metal hardening. Flame Plasma are commonly produced by combustion flames, especially in the presence of highly energetic reactions. Its applications include flame cutting and spraying (Graves, 2012).

2.2 Non-Thermal Plasma/Cold Plasma

Non-thermal plasmas, also known as cold plasmas, operate at significantly lower temperatures compared to their thermal counterparts, often near or at room temperature. These plasmas retain a substantial portion of their electrons in a non-ionized state, resulting in unique properties that are advantageous for various applications. Common methods for generating non-thermal plasma include Dielectric barrier discharge (DBD) plasma, glow discharge and corona discharge.

The DBD plasmas are applied to produce a high-frequency voltage across a dielectric barrier, leading to ionization of the gas. DBD is used in applications like surface treatments, sterilization, and pollution control (von Siemens, 1857). Dielectric discharges occur by creating an electrical discharge between a pair of electrodes separated by a dielectric barrier. The gap is typically quite small, usually just a few millimeters. To sustain these discharges, a voltage is applied (1 to 100 kV), along with frequencies spanning from 50 Hz to 1 MHz. Such type of non-thermal plasma is harnessed for several purposes, such as operating CO_2 lasers and generating excimer radiation within the UV and V-UV spectral regions. It is also applicable in processes like various thin-film deposition techniques, methanol production from methane/oxygen mixtures, and the treatment of exhaust gases (Kogelschatz & Eliasson, 1996; Salge, 1995).

Glow discharges are low-pressure plasmas that emit a visible glow. They are employed in analytical instruments like mass spectrometers and are used for plasma etching in microelectronics. Glow discharge plasma is primarily utilized in methane dry reforming due to its novel features, such as high electron density, electron temperature, and plasma temperature (L. Wang, Yi, Guo, & Tu, 2018). When compared to DBD plasma, glow discharge exhibits higher temperature, but lower than that of thermal plasma reactors. The atmospheric pressure glow discharge (APGD) plasma has promising applications due to its operation at atmospheric pressure, eliminating the need for a vacuum as required in conventional glow discharge plasma reactors. Depending on the reactants mixture and electrode configurations, the discharge can manifest either as a stable homogeneous state or as a filamentary glow state. Operating in a homogeneous glow state is found to be advantageous in gas treatment applications. APGD closely resembles dielectric barrier discharge plasma and corona discharge, yet its applications in the dry reforming of methane remains somewhat limited. One major challenge associated with APGD is its scalability issue when dealing with large feedstock volumes (Snoeckx & Bogaerts, 2017)

Corona discharge consists of large streamers originating from the electrode's tip and extending into the surrounding atmosphere, ultimately fading away. When the voltage approaches threshold limit, it creates an intense electric field in the vicinity of the electrode, resulting in ionization of gas molecules by accelerated electrons, leading to an electron avalanche. This avalanche triggers the ion generation, subsequently amplifying the electric field and extending the length of the streamers. Subsequently, an ionization zone is formed around the electrode, which give rise to electric discharges. Corona discharge is often used for ozone generation and air purification. It operates at atmospheric pressure and is characterized by a faint, visible corona. These distinct types of plasma have diverse applications across fields such as materials science, environmental science, aerospace engineering, and healthcare (Bhatnagar, 2019).

3 PLASMA GENERATION TECHNIQUES

Numerous setups have been developed to produce different types of plasma. Some popular and efficient setups are elaborated below.

3.1. Arc Discharge

Arc discharge plasma generation involves the creation of an electrical arc between two electrodes, typically using high-current electricity. This method yields high-temperature plasmas, making it suitable for applications such as materials processing, welding, and cutting. However, it demands substantial electrical power input, leading to high operating costs. Additionally, electrode wear and maintenance are common challenges in arc discharge systems (F. Chen et al., 2021).

3.2. Radiofrequency (RF) Discharge (MHz)

RF discharges use radiofrequency electromagnetic fields to ionize gases. These discharges can operate at lower temperatures than arc discharges, making them well-suited for controlled, low-temperature plasma applications. The advantage of reduced electrode wear compared to arc discharges is balanced by the need for specialized equipment for RF power generation, and complex setup and tuning requirements (F. Chen et al., 2021; Fridman, 2008).

Radio frequency oscillating electric fields is an advantageous approach in plasma science, as it is primarily influenced by displacement currents, allowing for the use of insulating electrodes. Two distinct types of this method are available:

3.3. Capacitive Coupled Plasma (CCP)

The plasma confinement setup involves two electrodes, which together work as a capacitor within the system's circuit. Typically, one of these electrodes is grounded, while the other one is energized by an RF generator at voltages usually in the range of several hundred volts. Typically, RF-generator connected to the electrode by using a matching unit having an LC-network. In the case of inductively coupled plasmas (ICP), the electric fields are introduced into the plasma through induction phenomenon, attained by applying a variable magnetic field. In this setup, a coil is linked to the RF generator, and the variable magnetic field produces an oscillating electric field within the plasma, which ultimately drives an RF plasma current. In this setup, the plasma effectively acts as a secondary coil in a transformer system in which primary external coil can take the form of few windings coiled around a quartz tube used for plasma confinement or a planar antenna positioned behind a quartz window (Raizer, Shneider, & Yatsenko, 2017).

3.4. Microwave Discharge

Microwave discharges leverage microwave radiation to induce ionization in gases. They exhibit high efficiency in ionizing gases and offer precise control over plasma properties. Microwave plasmas are suitable for low-pressure environments and chemical processes. However, specialized equipment for microwave generation is essential, and operating conditions may be sensitive to gas composition and pressure (F. Chen et al., 2021; Fridman, 2008). Numerous microwave (MW) plasma discharge setups have been developed and utilized for a range of applications, including cavity-induced discharges, electron cyclotron resonance, surface wave discharge, and freely expanding atmospheric plasma torches. -Among these, the surface wave discharge setup is most the frequently employed for CO_2 conversion. Typically, surface wave discharge is generated by passing gas through a quartz tube intersected by a waveguide (G. Chen et al., 2017). **Figure 1** illustrates of a waveguide-supplied, nozzleless, coaxial line-based MW plasma system employed in different chemical reactions.

Several types of microwave reactors have been developed, which are commonly categorized into three distinct groups: those occurring within closed structures, open structures, and resonance structures with a magnetic field.-Closed structured plasma chambers are enclosed by metallic walls, while the resonance structures involve high-quality resonant cavities with intense electric fields which boost up the discharge process, even at higher pressures. Open structures include slow wave structures, microwave torches, and surfatrons. Electron cyclotron resonance (ECR) plasmas is a common example of microwave plasma configuration influenced by magnetic fields (Ferreira & Moisan, 2013).

3.5. Direct Current (DC) Discharge

DC discharge plasma generation involves applying continuous electrical current between two electrodes immersed in a gas. This method is known for its simplicity and versatility, making it suitable for various applications. However, DC discharges are limited by specific gas pressures and compositions, and they may require electrode maintenance due to erosion. They also typically operate at lower electron temperatures compared to RF and microwave discharges (F. Chen et al., 2021).

Figure 1. Schematic and image of a MW discharge plasma
Reproduced from (Snoeckx & Bogaerts, 2017) with permission from the Royal Society of Chemistry.

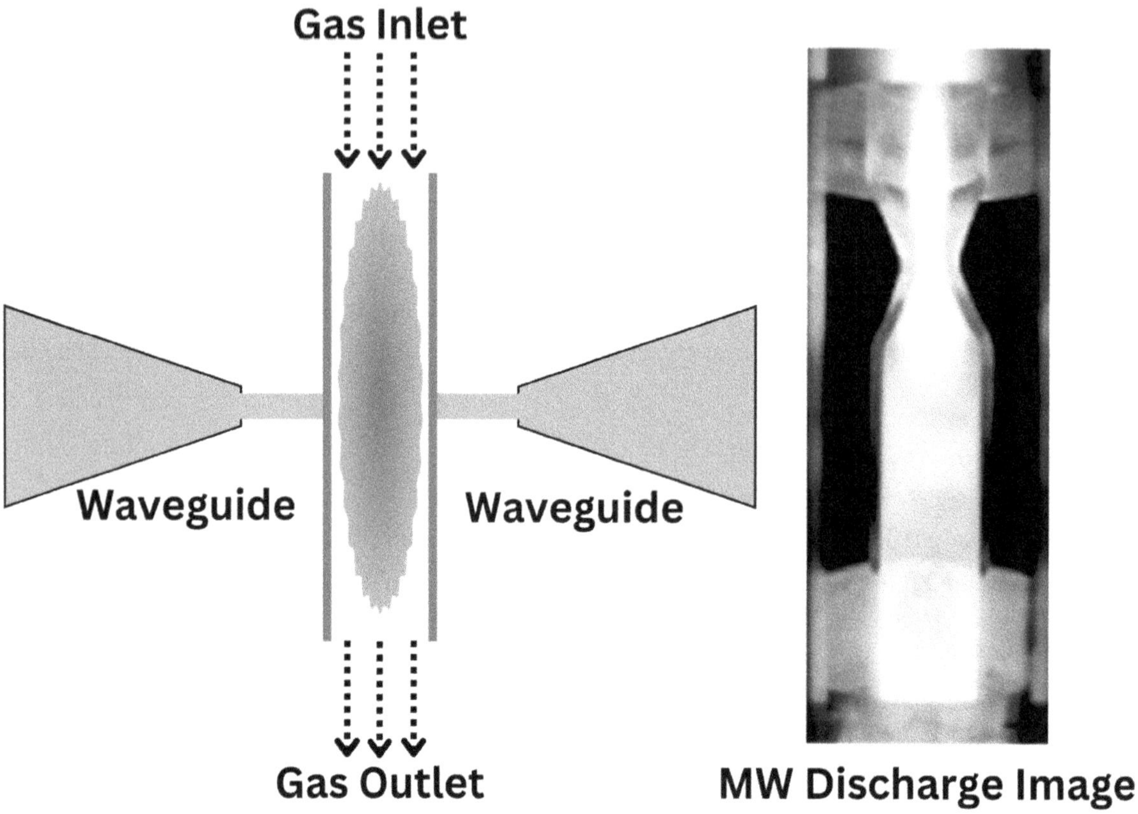

3.6. Alternating Current (AC) Discharge (kHz)

AC discharges entail applying alternating current to the gas, causing electrons to move back and forth, leading to ionization and plasma formation. Similar to DC discharges, AC discharges offer simplicity in setup and versatility. They can be used in a range of plasma applications, but they may exhibit lower plasma density compared to RF or microwave discharges. However, they mitigate some of the electrode wear issues associated with arc discharges (F. Chen et al., 2021).

3.7. Wave Heated Plasmas (GHz and Optical)

The production of GHz plasma is made possible by applying electric field of high frequency so that electrons starts to follow the applied electric field. Since electromagnetic waves are unable to completely penetrate into plasma due to smaller value of skin depth in conductive medium, thus an additional magnetic field is also applied to made resonant heating of electrons possible. DBD, MW, and GA reactors are the most commonly employed plasmas setups for CO_2 conversion as illustrated in **Figure 2**.

Figure 2. Schematic of the most commonly employed plasmas setups for CO_2 conversion, (a) cylindrically arranged GA discharge reactor, called gliding arc plasmatron (GAP); (b) MW plasma reactor; and (c) DBD reactor

Reproduced from (Snoeckx & Bogaerts, 2017) with permission from the Royal Society of Chemistry.

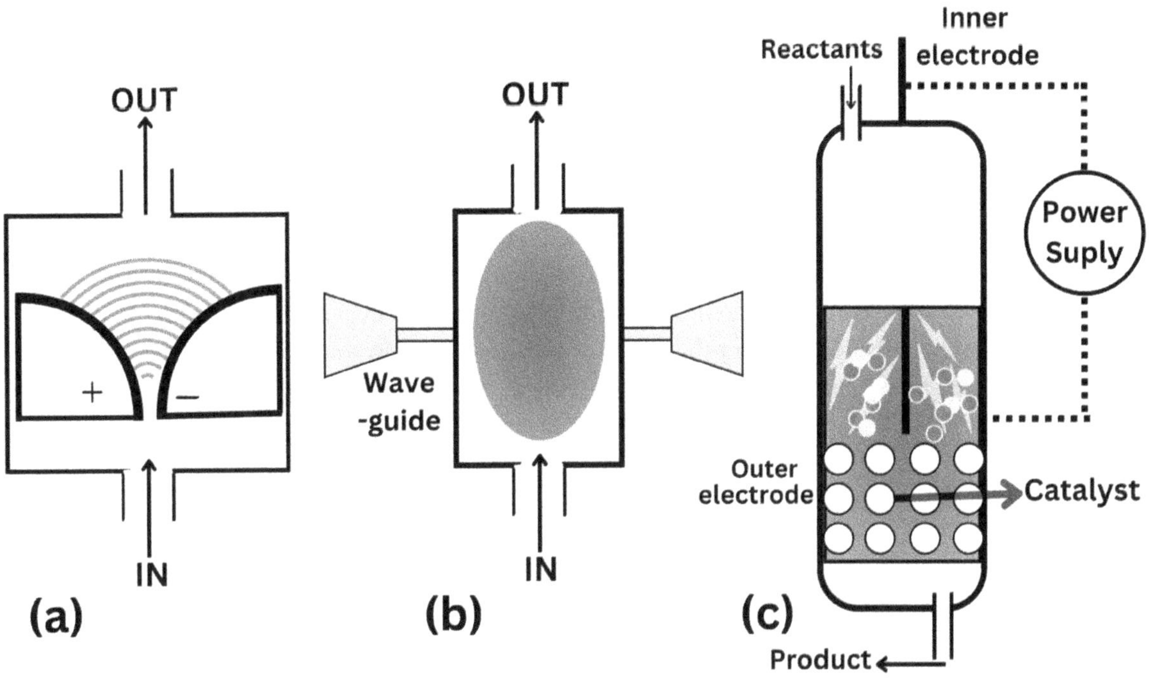

Each of these plasma generation techniques has its place in diverse applications, from materials processing to chemical synthesis and environmental control. The choice of method depends on a specific requirements of application, considering factors such as temperature, energy efficiency, and cost-effectiveness.

4. PLASMA CHARACTERISTICS ESSENTIAL FOR CO_2 CONVERSION

Due to CO_2's high stability and inertness, breaking the C-O bond requires a significant amount of energy. CO_2 dissociation is a process that is endothermic. Thermal decomposition is an energetic inefficient process since the CO_2 conversion thermodynamic equilibrium limit is between 40 and 50% [6]. Thermodynamically undesirable chemical reactions can, however, occur in plasma at low temperatures (L. Li et al., 2019), It might offer a CO_2 conversion pathway that is more energy-efficient. As a result, much effort has been put into using plasma technology to convert CO_2 into value-added chemicals and fuels. Plasmas exhibit several characteristics that are particularly relevant to their application in carbon dioxide (CO_2) conversion processes. Understanding these plasma properties is essential for harnessing their potential in transforming CO_2 into valuable products. Here are the key characteristics of plasmas relevant to CO_2 conversion:

4.1. High Energy and Ionization

Plasmas are characterized by their high energy levels, often resulting in temperatures ranging from tens and thousands of degrees Celsius for thermal plasmas. This high energy is essential for breaking the strong chemical bonds in CO_2 molecules, allowing for their transformation into other compounds. The ionization of particles in plasma enables the dissociation of CO_2 into its constituent atoms (carbon and oxygen), facilitating subsequent chemical reactions (Yin et al., 2021). The production and storage of renewable energy is the most challenging task of this century due to increasing demand of renewable electricity, e.g. wind and solar, which generate peak powers. The plasma reactors are ideally suited for production and storage of renewable energy due very small consumption of electricity as they are switched on/off very quickly. Moreover, they are proficient in conversion of greenhouse gases like CH_4 and CO_2 into renewable fuels and chemicals without increase of global warming issues. Indeed, plasma technology is a technique which can latch three birds with a single stone, i.e. 1) Reduction of greenhouse gases from atmosphere by conversion process, 2) production of renewable fuels and valuable chemical products from greenhouse gases rather than from fossil fuels which cause increase in CO_2 emission, 3) It uses renewable electricity to avoid CO_2 emission (X. Lu et al., 2022).

4.2. Reactive Species Generation

Plasmas generate a rich variety of reactive species, including electrons, electrons, free radicals, and excited-state molecules and atoms. These species are essential for turning CO_2 into the necessary products. Reactive oxygen and nitrogen species, in particular, can participate in chemical reactions that lead to formation of value-added compounds (Chung & Chang, 2016).

4.3. Non-Thermal Nature

Non-thermal or cold plasmas, which operate at or near room temperature, are particularly relevant to CO_2 conversion applications. Their non-thermal nature allows for selective and controlled reactions without excessive heat, making them suitable for processing temperature-sensitive compounds and reducing the risk of undesirable byproducts. NTP has been used to study a number of approaches for CO_2 conversion, including dry reforming of methane (R1), CO_2 hydrogenation (R2 and R4), reduction of CO_2 with water (R5-R6) and CO_2 breaks down into CO and O_2 (R7), as seen in **Figure 3**.

R1: $CO_2+CH_4 \rightarrow 2CO+2H_2 \quad \Delta H=247KJ/mol$
R2: $CO_2+4H_2 \rightarrow CH_4+2H_2O \quad \Delta H= -165KJ/mol$
R3: $CO_2+3H_2 \rightarrow CH_3OH+H_2O \quad \Delta H= -40kJ/mol$
R4: $CO_2+H_2 \rightarrow CO+H_2O \quad \Delta H=41.2kJ/mol$
R5: $CO_2+H_2O \rightarrow CO+H_2+\frac{1}{2}O_2 \quad \Delta H=525kJ/mol$
R6: $CO_2+2H_2O \rightarrow CH_3OH+\frac{3}{2}O_2 \quad \Delta H=676kJ/mol$
R7: $2CO_2 \rightarrow 2CO+O_2 \quad \Delta H=283kJ/mol$

Figure 3. CO_2 conversion routes
Reproduced from (George et al., 2021).

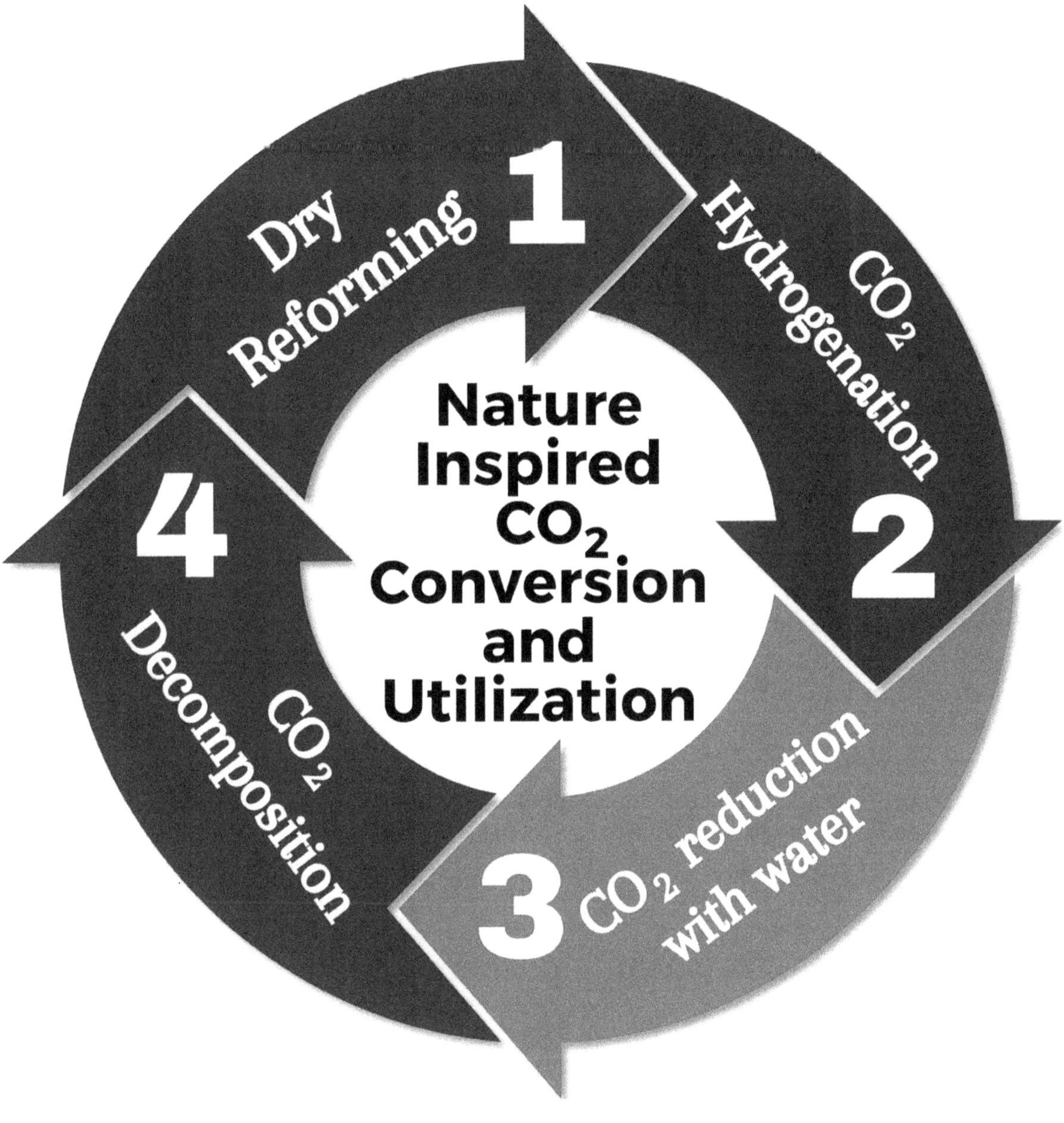

4.4. High Electron Density

Plasmas typically have a high electron density, which enhances electron impact reactions. In CO_2 conversion, high electron densities can facilitate electron attachment to CO_2 molecules, leading to the formation of negative ions that can undergo various chemical reactions, including reduction processes.

4.5. Electron Energy Distribution

Plasma electron energy distribution is typically non-Maxwellian, meaning that electrons have a wide range of energies. This distribution enables a variety of collision processes with CO_2 molecules, resulting in different reaction pathways. Some reactions may require high-energy electrons, while others are more efficient at lower energies, providing versatility in CO_2 conversion processes (George et al., 2021).

4.6. Electric Fields and Surface Effects

Plasmas can create electric fields and surface effects that influence the interaction of CO_2 with plasma-generated species. Electric fields can drive charged species towards electrodes or substrates, promoting surface reactions. Surface effects, such as dielectric barrier discharges, can enhance the energy transfer to CO_2 and improve reaction efficiency (George et al., 2021).

4.7. Scalability and Control

Plasmas can be generated at various scales, from laboratory benchtop setups to industrial-scale reactors. Their operation can be precisely controlled in terms of power input, gas composition, and residence time, allowing for tailoresd CO_2 conversion processes. This scalability and control make plasmas adaptable to diverse applications, from laboratory experimentation to large-scale industrial operations (Arpagaus, Oberbossel, & Rudolf von Rohr, 2018).

4.8. Environmental Compatibility

Non-thermal plasmas, in particular, are known for their environmental compatibility, as they can operate at atmospheric pressure without the need for high-temperature combustion. This characteristic aligns well with sustainable CO_2 conversion strategies, as it reduces energy consumption and greenhouse gas emissions associated with the process (Marcantonio, De Falco, & Bocci, 2022).

5. SOURCES OF CARBON DIOXIDE EMISSIONS

The primary cause of CO_2 emissions, which play a significant role in global climate change, is human activity. Identifying the main reasons behind CO_2 emissions is crucial for understanding the extent of the problem and devising strategies to mitigate it (Masson-Delmotte et al., 2022). Here are the main sources of CO_2 emissions:

5.1. Fossil Fuel Combustion

The burning of gasoline and diesel fuels for transportation in cars, trucks, ships, airplanes, and trains is a significant source of CO_2 emissions. This sector accounts for a substantial portion of global emissions due to widespread use of fossil fuels for mobility (Masson-Delmotte et al., 2022).

Electricity generation in power plants from coal, natural gas, and oil is a main source of CO_2 emissions. Coal fired power plants, in particular, are known for their high emissions, although cleaner technologies are becoming more widespread.

Various industries, including manufacturing, petrochemicals, and cement production, release CO_2 during their operations. This can occur through combustion, chemical reactions, and other industrial processes.

5.2. Deforestation and Land Use Changes

Deforestation, the removal of forests and trees for agriculture, urban development, and other purposes releases CO_2 stored in trees and vegetation into the atmosphere. Deforestation is a significant source of emissions, particularly in tropical regions.

Land use changes caused by conversion land for agriculture or urbanization can release stored carbon and alter natural carbon cycles, contributing to CO_2 emissions.

5.3. Residential and Commercial Energy Use

The use of fossil fuels for heating and cooling in residential and commercial buildings contributes to CO_2 emissions, as does the electricity consumption for lighting, appliances, and electronics (Gilfillan & Marland, 2021).

5.4. Waste Management

The decomposition of organic waste in landfills generates methane (CH_4), a potent greenhouse gas. While not CO_2 emissions per se, CH_4 emissions are often converted to CO_2 equivalents for reporting.

6. CHALLENGES OF CAPTURING AND STORING CO_2

While recognizing the sources of CO_2 emissions is essential, addressing this global challenge also requires effective strategies for capturing and storing CO_2. Here are the challenges associated with these processes:

6.1. Carbon Capture

Technologies for capturing carbon can be energy-intensive, potentially reducing the overall efficiency of power plants or industrial processes. The implementation of carbon capture systems can be expensive, making it a financial challenge for many industries.

6.2. Carbon Storage (Geological Sequestration)

Site selection in which identifying suitable geological formations for CO_2 storage can be challenging. The availability of appropriate sites varies regionally. Ensuring the long-term integrity of storage reservoirs to prevent CO_2 leakage is essential for avoiding environmental risks. Ongoing monitoring and verification of stored CO_2 are necessary to assess containment and address potential issues (Metz, Davidson, De Coninck, Loos, & Meyer, 2005).

6.3. Public Acceptance

Public acceptance and regulatory frameworks for carbon capture and storage (CCS) can be barriers to implementation (Le Quéré et al., 2018).

6.4. Alternative Solutions

Exploring alternative solutions for CO_2 reduction, such as renewable energy adoption and energy efficiency measures, can divert attention and resources from CCS (Haszeldine, 2009).

Addressing these challenges requires a combination of technological innovation, supportive policies, and international cooperation.

7. CARBON DIOXIDE CONVERSION PATHWAYS

Converting CO_2 into valuable products represents a promising approach to mitigate its environmental impact while simultaneously providing economic benefits. There are several pathways for CO_2 conversion, each offering unique advantages and applications (Artz et al., 2018). Here, we present various pathways and explain their potential environmental and economic benefits.

7.1. Carbon Capture and Utilization (CCU)

Mineralization process in which CO_2 can react with metal oxides to form stable carbonates. This pathway permanently sequesters CO_2 and can be used for carbon-negative building materials, such as concrete. CO_2 can also be converted into chemicals and fuels using renewable energy sources. Examples include methanol and synthetic natural gas production (Ho, Iizuka, & Shibata, 2019).

7.2. Biological CO_2 Conversion

Biological photosynthesis conversion of solar energy into valuable chemical energy is another way of biofuel production. It is also a proficient way for CO_2 conversion into fuels and other value-added chemicals. However, controversy is being seen related to first generation biofuels usage mainly due to their competition with agronomy and forming for arable use of land for food production which is impacting food markets globally (Schenk et al., 2008). Some common biological CO_2 conversions include: Algae and certain microorganisms that can capture CO_2 through photosynthesis and convert it into biofuels, bioplastics, and other valuable products, Some biocatalysts including enzymes and engineered microorganisms that can catalyze the conversion of CO_2 into chemicals and biofuels, providing environmentally friendly alternatives to traditional petrochemical processes (Salehizadeh, Yan, & Farnood, 2020).

7.3. Electrochemical CO_2 Reduction

Electrolysis is a process in which electrochemical cells use renewable electricity to reduce CO_2 into valuable chemicals such as ethylene, carbon monoxide, and formic acid (Bagger, Ju, Varela, Strasser, & Rossmeisl, 2017). Electrolysis is the closest competitors to plasma technology as both technologies rely on renewable electricity to initiate the conversion process. Electrochemical reduction of CO_2 is a novel technology which utilizes electricity to set-up a potential difference between two electrodes to initiate CO_2 reduction reactions under mild conditions (Albo, Alvarez-Guerra, Castaño, & Irabien, 2015; X. Zhang, Lee, Mingos, & Hayward, 2003). This conversion mechanism can is accomplished through a variety of steps depending upon the experimental conditions. The electrochemical conversion of CO_2 can be ensued via two-, four-, six- and eight-electron reduction routes under different electrode or cell configuration in aqueous, non-aqueous or gaseous phases (Albo et al., 2015; Budzianowski, 2012).

7.4. Hydrogenation and Methanation

Hydrogenation of CO_2 can be employed to produce methanol or higher hydrocarbons, which can serve as chemical feedstocks or clean-burning fuels (Choe, Kang, Kim, Park, & Park, 2005). Similarly, by methanation process CO_2 can be converted into methane, a valuable energy carrier, using renewable hydrogen.

8. POTENTIAL ENVIRONMENTAL BENEFITS OF CO_2 CONVERSION

- Reduced Greenhouse Gas Emissions: CO_2 conversion pathways can offset emissions by converting CO_2 into valuable products, thus reducing net emissions and helping combat climate change.
- Resource Efficiency: CO_2 conversion can reduce the reliance on fossil fuels and virgin resources by producing chemicals, fuels, and materials from waste CO_2 streams.
- Carbon Recycling: CO_2 conversion facilitates the recycling of carbon, promoting a circular economy and reducing the need for virgin carbon sources.
- Air Quality Improvement: Reducing CO_2 emissions can also lead to a reduction in co-emitted pollutants, enhancing air quality and public health (Ayodele, Mustapa, Ayodele, & Mohammad, 2020).

9. POTENTIAL ECONOMIC BENEFITS OF CO_2 CONVERSION

- Market Opportunities: The production of valuable products from CO_2 can create new markets and industries, driving economic growth and job creation.
- Energy Transition: CO_2 conversion technologies can complement the transition to renewable energy sources, providing a means to store excess renewable energy as chemical fuels.
- Resource Diversification: Diversifying the sources of raw materials for industries reduces dependence on fossil fuels, enhancing resource security and resilience.

- Sustainability Credentials: Companies and products utilizing CO_2 conversion pathways may gain a competitive advantage by demonstrating their commitment to sustainability and reducing their carbon footprint (Rafiee, Khalilpour, Milani, & Panahi, 2018). These pathways offer promising solutions to address the challenges of rising CO_2 levels and contribute to a sustainable and circular economy.

10. PLASMA-BASED CO_2 CONVERSION TECHNOLOGIES

The rising levels of CO_2 in the environment have led to a global effort to mitigate global warming by developing innovative technologies for CO_2 conversion and utilization. Among these technologies, plasma reactors have gained recognition as efficient tools for transforming CO_2 into valuable products while reducing greenhouse gas emissions. These reactors harness the power of plasma; a high-energy state of matter to initiate chemical reactions that can convert CO_2 into a variety of useful compounds (Dębek, Azzolina-Jury, Travert, & Maugé, 2019).

Plasma reactors are designed to create and sustain a plasma state, typically using electrical discharges or radiofrequency waves. Within the plasma, highly energetic electrons, ions, and radicals are generated, which can interact with CO_2 molecules in several ways (Dębek et al., 2019). A flowchart showcasing the components and design of a typical plasma reactor for CO_2 conversion in **Figure 4** and some involved processes are explained below:

Figure 4. Components and design of a typical plasma reactor for CO_2 conversion
Reproduced from (Anoop et al., 2021).

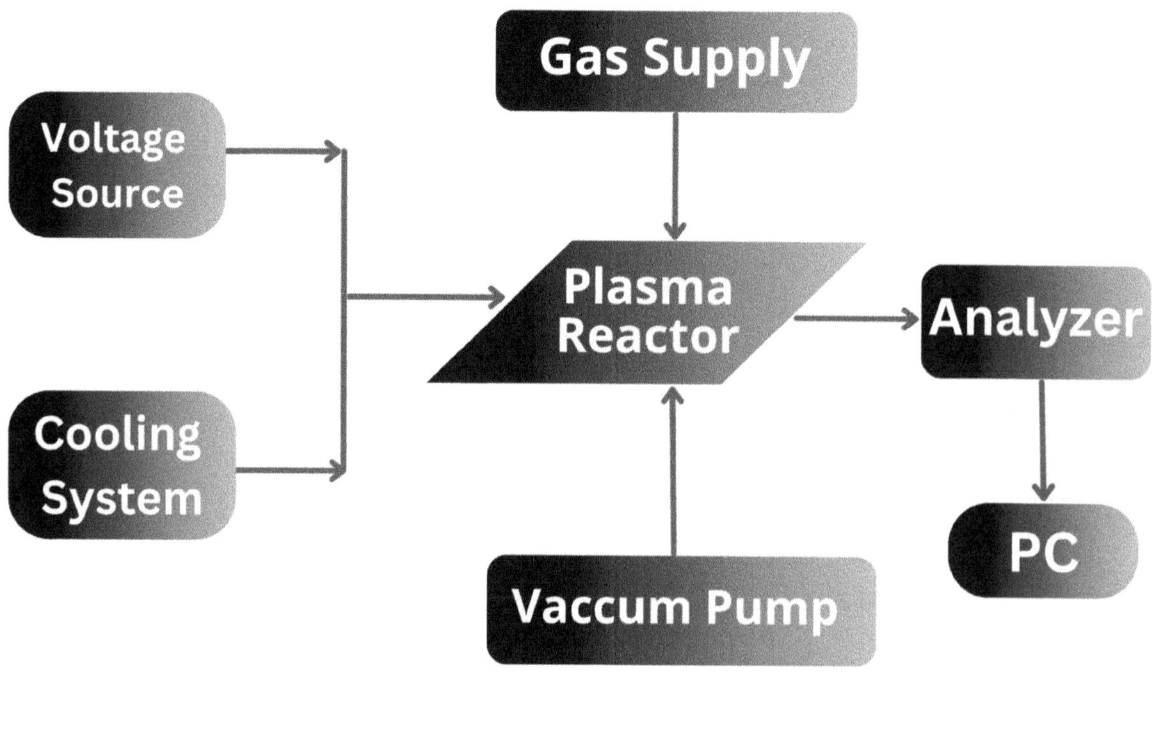

- *Electron Impact Dissociation*: High-energy electrons collide with CO_2 molecules, breaking them apart into carbon (C) and oxygen (O) atoms or other fragments, initiating chemical reactions.
- *Radical Formation*: Plasma generates highly reactive radicals, such as hydroxyl radicals (•OH) and atomic oxygen (O), which can react with CO_2 to form various products.
- *Energy Transfer*: The high energy in plasma can transfer to CO_2 molecules, increasing their internal energy and facilitating chemical reactions that might be otherwise energetically unfavorable.

These mechanisms open up a world of possibilities for CO_2 conversion, allowing plasma reactors to serve as versatile platforms for producing fuels, chemicals, and materials from CO_2 (George et al., 2021).

Plasma reactors represent a promising frontier in the quest to decreased CO_2 emissions and convert greenhouse gases into valuable resources, contributing to both environmental sustainability and the circular economy (A Bogaerts et al., 2017). Ongoing research and development efforts are expected to further enhance the efficiency and feasibility of these reactors for CO_2 conversion in coming years. **Figure 5** depicts some major pathways for CO_2 conversion using plasma reactors and their possible applications in different fields including:

- *Syngas Production*: Plasma can convert CO_2 and methane (CH_4) into syngas, a valuable mixture of carbon monoxide (CO) and hydrogen (H_2). Syngas is used in various industries, including production of chemicals, electricity, and fuels.
- *Methanol and Methane Synthesis*: Plasma can facilitate the conversion of CO_2 and hydrogen (H_2) into methanol (CH_3OH) or methane (CH_4). These compounds are essential as clean energy carriers and chemical feedstocks.
- *Hydrocarbon Synthesis*: Plasma technology can be employed to convert CO_2 into hydrocarbons, including methane (CH_4) and higher hydrocarbons. These hydrocarbons can serve as clean-burning fuels or chemical feedstocks (Tu & Whitehead, 2014).
- *Nanomaterial Synthesis*: Plasma reactors can synthesize carbon nanotubes, nanoparticles, and other nanomaterials from CO_2, which have applications in advanced materials and catalysis.
- *CO_2 Valorization in Industry*: Plasma-based CO_2 conversion technologies have the potential to be integrated into industrial processes, reducing carbon emissions and offering sustainable alternatives for various industries, including petrochemicals and metallurgy (Vadikkeettil et al., 2022).
- *Waste Gas Treatment*: Plasma reactors are employed to treat industrial waste gases containing CO_2 and volatile organic compounds (VOCs), reducing air pollution and greenhouse gas emissions.
- *Environmental Remediation*: Plasma reactors can be employed for environmental applications, such as CO_2 capture and conversion. By utilizing renewable energy sources, plasma-based CO_2 conversion can help reduce greenhouse gas emissions and mitigate climate change (Kim, 2004).

Plasma reactors offer innovative solutions for CO_2 conversion by harnessing the unique properties of plasma to initiate chemical reactions. There are many types of plasma reactors used in CO_2 conversion, each with distinct principles and mechanisms. Below, we discuss different types of plasma reactors, their principles, and relevant chemical reactions, supported by references.

Figure 3. Main pathways of CO_2 utilization
Reproduced from (Baena-Moreno et al., 2019).

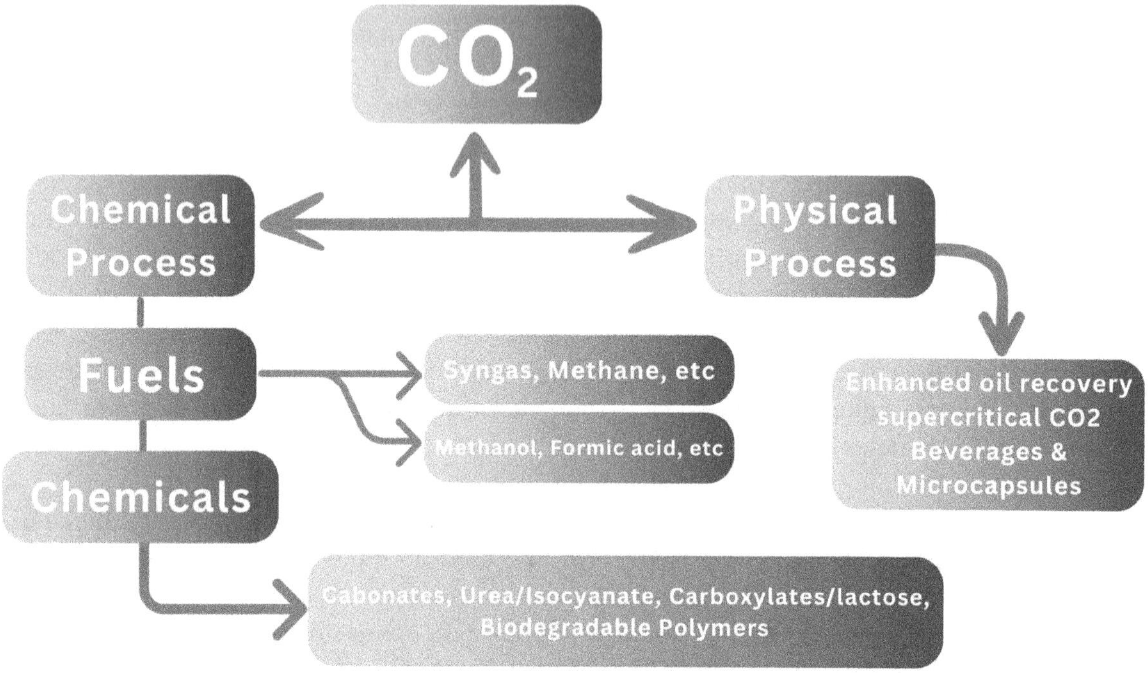

10.1. Non-Thermal Plasma (NTP) Reactors

NTP reactors create a non-equilibrium plasma at relatively low temperatures. High-voltage electrical discharges or radiofrequency waves are used to generate the plasma. In NTP, high-energy electrons and ions interact with CO_2 molecules, leading to electron impact dissociation and radical formation (George et al., 2021; Y. Zeng & Tu, 2015). Reactions involved are elaborated as:

Impact Dissociation of Electrons:

$CO_2 + e^- \rightarrow CO + O$

$CO_2 + e^- \rightarrow C + 2O$

Radical Formation:

$H_2O \rightarrow \bullet OH + \bullet H$

$O_2 \rightarrow 2\,\bullet O$

10.2. Dielectric Barrier Discharge (DBD) Reactors

DBD plasma reactors consist of two electrodes separated by a dielectric material. Plasma is created within the gap when electrodes are subjected to high voltage. DBD reactors produce a plasma rich in electrons, ions, and radicals, which interact with CO_2 molecules (Neyts, Ostrikov, Sunkara, & Bogaerts, 2015; Shi et al., 2021). Reactions involved are elaborated as:

Impact Dissociation of Electrons:

$CO_2+e^- \rightarrow CO+O$

$CO_2+e^- \rightarrow C+2O$

Radical Formation:

$H_2O \rightarrow \bullet OH+\bullet H$

$O_2 \rightarrow 2 \bullet O$

10.3. Gliding Arc Discharge Reactors

Gliding arc discharge reactors feature a continuous electrical discharge between two diverging electrodes. The gliding arc creates plasma, and CO_2 and other gases are exposed to this plasma, leading to dissociation, ionization, and radical formation (H. Zhang et al., 2018). Reactions involved are elaborated as:

Impact Dissociation of Electrons:

$CO_2+e^- \rightarrow CO+O$

$CO_2+e^- \rightarrow C+2O$

Radical Formation:

$H_2O \rightarrow \bullet OH+\bullet H$

$O_2 \rightarrow 2 \bullet O$

10.4. Plasma Electrolysis Reactors

Plasma electrolysis combines traditional electrolysis with plasma generation. Electrochemical cells are used to facilitate reactions. Plasma electrolysis can enhance the electrolysis of CO_2 into oxygen (O_2) and carbon monoxide (CO) or be used for CO_2 capture from flue gases (Snoeckx & Bogaerts, 2017). Reactions involved are elaborated as:

Electrolysis of CO_2:

$2CO_2+4H_2O \rightarrow 2CO+8H^++8e^-$ (at anode)

$8H^++8e^- \rightarrow 4H_2+4OH^-$ (at cathode)

$2CO_2+4H_2O \rightarrow 2CO+4H_2+4OH^-$ (overall)

CO_2 Capture:

$CO_2(g)+H_2O(l) \rightarrow H_2CO_3(aq)$

$$H_2CO_3(aq)+2OH^-(aq)\rightarrow 2HCO_{3^-}(aq)+H_2O(l)$$

These different kinds of plasma reactors offer unique advantages for CO_2 conversion, and researchers are continually working on optimizing these technologies to make them more efficient and economically viable for a wide range of applications.

11. CATALYSIS IN PLASMA-ASSISTED CO_2 CONVERSION: ENHANCING EFFICIENCY AND SELECTIVITY

Catalysis is really important in plasma-assisted CO_2 conversion by enhancing the efficiency and selectivity of chemical reactions as summarized in Figure 6. Plasma reactors create a highly reactive environment with a broad range of energetic species, but presence of catalysts further facilitates the transfiguration of CO_2 into valuable products. In this discussion, we will explore the role of catalysts in plasma assisted CO_2 conversion and provide examples of catalytic reactions within plasma systems.

- Lower Activation Energy: The activation energy needed for particular chemical reactions is reduced by catalysts, allowing them to occur more readily within the plasma environment. This acceleration of reaction rates is especially valuable in plasma systems where high-energy species are abundant.
- Selective Product Formation: Catalysts can steer the reaction pathway toward the desired product by promoting specific intermediate steps while inhibiting unwanted side reactions. This selectivity is essential for the controlled conversion of CO_2 into target compounds.
- Enhanced Efficiency: Catalysts can improve the overall effectiveness of plasma-assisted CO_2 conversion processes, reducing energy consumption and resource usage while increasing the yield of valuable products.

Figure 6. Plasma-catalysts synergism
Reproduced from (Khoja, Tahir, & Amin, 2019)

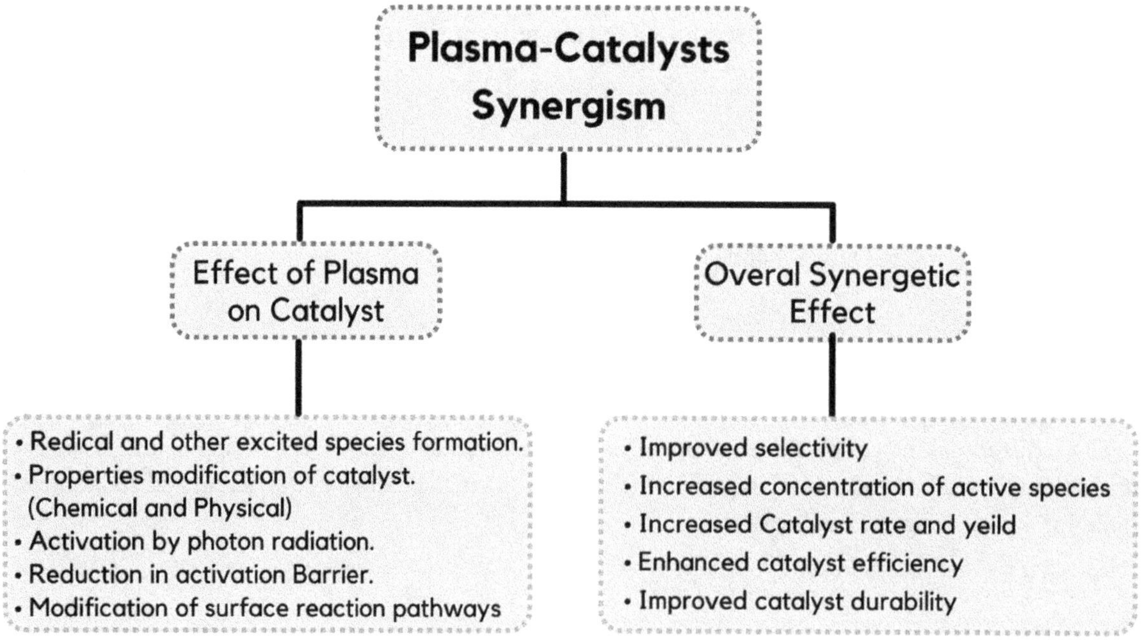

Numerous scientific studies have demonstrated that integration plasma technology with a catalyst can significantly improve efficiency of both energy production and conversion rates. The non-thermal plasma systems supply necessary energy required for highly endothermic reaction, while the catalyst boost up the process selectivity rate. As illustrated in **Figure 7**, extensive research is being made on the 'Plasma based catalytic reduction of CO_2' via ferroelectric material ($BaTiO_3$) and a variety of catalyst oxides of Ti, Ni, Cu, Co etc (Anoop et al., 2021). There are two primary routes for incorporation of catalyst into the plasma systems. The first one route, known as in-plasma catalysis (IPC), involves introduction of the catalyst directly within the plasma discharge region which is also termed as single-stage plasma catalytic process. While the second route, also termed as post-plasma catalysis (PPC), the catalyst is introduced at some distance from the plasma discharges (Whitehead, 2016). The specific method by which the catalyst is introduced into the plasma systems also plays a critical role in understanding the physics involved in plasma-catalyst interactions within the system (Michielsen et al., 2017).

Chen and his co-workers contemplated the absence of carbon deposition during the process of CO_2 reduction. They reported a significant improvement in conversion efficiency, ranging from 23% to 42%, with increased energy production rate from 9.6% to 17.2% by using an Ar plasma pre-treated NiO/TiO_2 catalyst. Such improved efficacy was attributed to the formation of oxygen vacancies within the system (G. Chen, Georgieva, Godfroid, Snyders, & Delplancke-Ogletree, 2016). In another study, Ray et al. explored the effects of various catalyst with glass beads, Al_2O_3, TiO_2, and CeO_2 employed as packing materials in a DBD reactor for CO_2 conversion. The most proficient conversion rates were found in ascending order of DBD reactor without packing material $<$glass beads $<TiO_2 <Al_2O_3 <CeO_2$. The highest reported conversion rate reached 10.6%. The maximum efficiency achieved by CeO2 based system was 1.91 mmol/kJ (Ray & Subrahmanyam, 2016).

Lu et al. in his study also explored the synergistic effects of the g-C_3N_4 catalyst (stable and non-toxic metal compound), when combined with a plasma system for the reduction of CO_2 into CO. A cylindrical dielectric barrier discharge setup employed in this study yield a remarkable 47% increase in conversion efficacy (both with and without a cooling system) and a significant 157% boost in energy efficacy. The energy efficiency was found increased to 0.26 mmol/kJ. The introduction of the g-C_3N_4 catalyst into the discharge gap transformed the homogeneous gas-phase reaction into a heterogeneous one. Such transformation enhanced the electron energy and electric field, resulting in improved conversion rates. With the implementation of an opted cooling systems, a further 10% increase in conversion was also achieved, leading to a six-fold increase in energy efficacy. (N. Lu et al., 2018).

Here is an overview of catalytic reactions involved in plasma based CO_2 Conversion into other valuable products.

Sabatier Reaction (Methane Synthesis):

$$CO_2+4H_2 \rightarrow CH_4+2H_2O$$

Nickel (Ni) or other transition metals are commonly used catalysts in plasma generators for the methanation of CO_2. These catalysts facilitate activation and adsorption of CO_2 and H_2 molecules, promoting the formation of CH_4 (Gao et al., 2021).

Reverse Water Gas Shift Reaction (CO Synthesis):

$$2CO_2+2H_2 \rightarrow 2CO+2H_2O$$

Figure 7. Literature published on plasma-catalytic conversion of CO_2 by using different catalysts Reproduced from (Anoop et al., 2021).

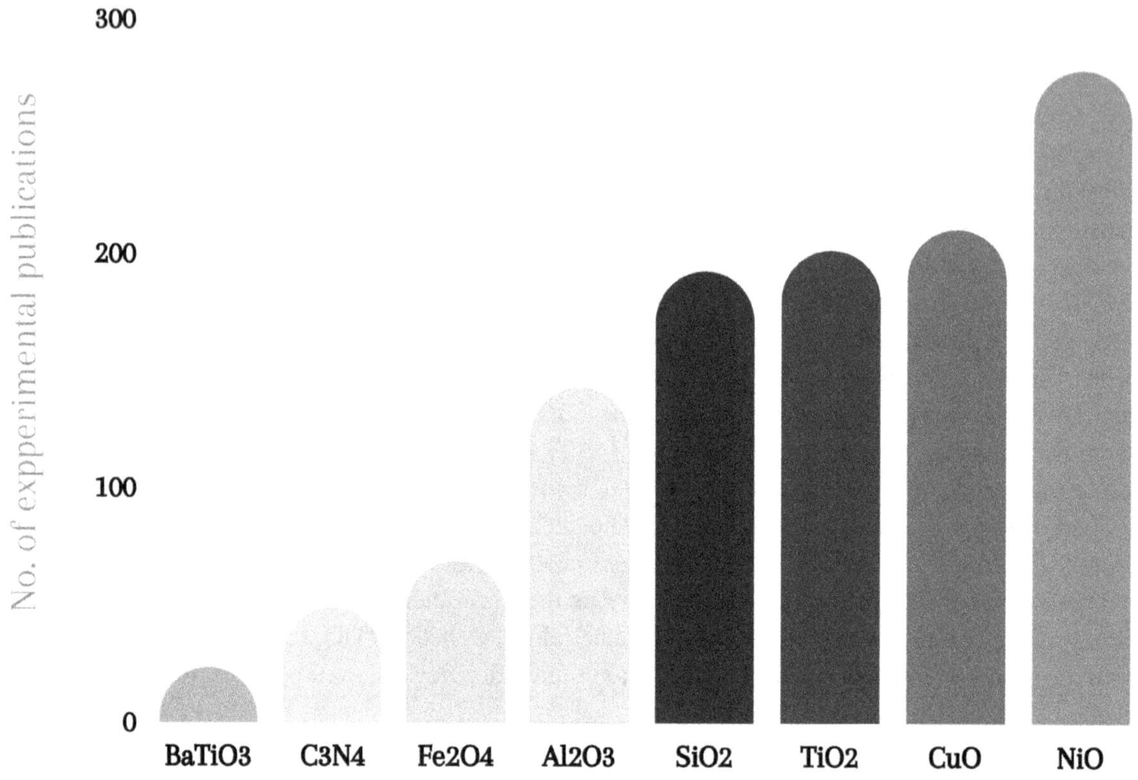

Various catalysts, such as ceria-based materials or metal catalysts like ruthenium (Ru), can be employed in plasma-assisted reverse water-gas shift reactions to enhance the conversion of CO_2 into carbon monoxide (CO) (Ahasan, Hossain, Ding, & Wang, 2023).

Fischer-Tropsch Synthesis (CO_2 Hydrogenation):

$$CO_2+4H_2 \rightarrow CH_4+2H_2O$$

Catalysts like iron (Fe) or cobalt (Co) nanoparticles are employed in plasma-assisted Fischer-Tropsch synthesis, which converts CO_2 and H_2 into hydrocarbons, including CH_4 (S. Wang et al., 2023).

Hydrogen Peroxide Production Reaction:

$$CO_2+H_2O+O_2 \rightarrow H_2O_2+CO_2$$

Platinum (Pt) or palladium (Pd) catalysts can be used in plasma-assisted reactions to produce hydrogen peroxide (H_2O_2) from CO_2, water (H_2O), and oxygen (O_2) (Daligaux, Richard, & Manero, 2021).

These examples illustrate how catalysts can significantly enhance the efficiency and selectivity of plasma-assisted CO_2 conversion reactions. By carefully selecting and optimizing catalyst materials and operating conditions, researchers aim to develop sustainable processes for turning CO_2 into something useful chemicals and fuels while reducing greenhouse gas emissions.

13. PLASMA BASED CO_2 CONVERSION FOR HYDROCARBON SYNTHESIS

The transformation of CO_2 into hydrocarbons using plasma technology has gained considerable attention as a sustainable approach for addressing climate change and producing valuable chemicals and fuels. This innovative process offers a promising solution to reduce CO_2 emissions while simultaneously generating valuable hydrocarbon products. In this article, we will delve into the detailed process of converting CO_2 into hydrocarbons through plasma technology and explore the diverse potential applications, comparing them with previous research findings.

The process begins with the generation of a plasma, typically through the application of electrical discharges or radiofrequency waves in a controlled environment. This plasma state consists of highly energetic ions, electrons, radicals, and excited species.

A generated plasma provides the necessary energy to activate CO_2 molecules. The high-energy radicals, ions, and electrons, in plasma collide with CO_2 molecules, breaking the CO_2 bonds and initiating chemical reactions.

The dissociated carbon (C) and oxygen (O) atoms from CO_2, along with hydrogen (H_2) from an external source (e.g., methane or water vapor), combine to form various hydrocarbon compounds. These hydrocarbon synthesis reactions typically involve the recombination of carbon and hydrogen atoms. The reaction is as follows:

$$CO_2+4H_2 \rightarrow CH_4+2H_2O$$

After hydrocarbon synthesis, the resulting mixture may contain various hydrocarbon products, as well as byproducts and unreacted gases. Separation and purification techniques are employed to isolate the desired hydrocarbons.

14. POTENTIAL APPLICATIONS IN FUEL PRODUCTION

14.1. Methane Production

Methane (CH_4) is a natural gas significant component and can serve as a clean-burning fuel. Plasma technology enables the conversion of H_2 and CO_2 into methane, offering a renewable and sustainable source of natural gas. Previous studies have demonstrated efficient production of methane from CO_2 using non-thermal plasma reactors with catalysts such as nickel (Ni) (Legg, 2021). This process holds promise for natural gas production and energy storage.

14.2. Synthetic Liquid Fuels

Plasma-based CO_2 conversion can be used to synthesize synthetic liquid fuels, such as methanol (CH_3OH) and dimethyl ether (DME). These fuels are valuable for various applications, including transportation and as chemical feedstocks. Researchers have investigated the conversion of H_2 and CO_2 into methanol using dielectric barrier discharge (DBD) plasma reactors with catalysts like copper (Cu) and zirconium (Zr) (Annemie Bogaerts & Snoeckx, 2019). Methanol production can provide an alternative to traditional petrochemical processes.

14.3. Hydrogen Production

Hydrogen (H_2) is a clean energy carrier and is essential for numerous industrial processes. Plasma-based CO_2 conversion can produce hydrogen from CO_2 and water (H_2O), offering a renewable source of this valuable gas. Studies have explored the production of hydrogen and carbon monoxide (CO) from CO_2 and water using gliding arc discharge plasma with catalysts like ceria-zirconia (CeO_2-ZrO_2) (López et al., 2019). This approach can contribute to a hydrogen-based energy economy.

14.4. Syngas Generation

Syngas, a mixture of carbon monoxide (CO) and hydrogen (H_2), is a versatile feedstock for producing a wide range of chemicals and fuels. Plasma technology can efficiently transform CO_2 and methane (CH_4) into syngas. Researchers have investigated the production of syngas from CH_4 and CO_2 using gliding arc plasma technology, often with nickel (Ni)-based catalysts (Martin-del-Campo, Uceda, Coulombe, & Kopyscinski, 2021). Syngas can be used for the synthesis of chemicals and as an energy source.

14.5. Comparison with Previous Research

Previous research in the field of plasma-based CO_2 conversion to hydrocarbons has yielded valuable insights and demonstrated the feasibility of these processes. Researchers have explored various plasma types, reactor configurations, and catalysts to enhance the efficiency and selectivity of hydrocarbon synthesis. Some key findings from previous studies include:

14.6. Optimization of Plasma Parameters

Researchers have focused on fine-tuning temperature, pressure, and discharge power to achieve better control over reaction kinetics and product distributions (Britun et al., 2018).

14.7. Catalyst Development

Catalysts play a crucial role in enhancing hydrocarbon synthesis reactions. Previous work has explored a range of catalyst materials, including transition metals like nickel (Ni), copper (Cu), and ceria-zirconia (CeO_2-ZrO_2), to improve conversion rates and product selectivity (Chadha et al., 2022).

14.8. Reactor Design

Different varieties of plasma reactors, such as gliding arc discharge and dielectric barrier discharge (DBD), have been employed and optimized for hydrocarbon production from CO_2 (Gao, Liang, Wu, Wu, & Kawi, 2022).

14.9. Scaling Up and Industrial Applications

While much of the early research focused on laboratory-scale experiments, recent studies have begun to explore the scalability of plasma-based CO_2 conversion technologies for potential industrial applications (N. Li, Mo, & Unluer, 2022).

Overall, previous research has laid the foundation for the development of plasma-based CO_2 conversion processes for hydrocarbon synthesis. These efforts have provided valuable insights into reactor design, catalyst selection, and process optimization, bringing us closer to realizing the potential of plasma technology for sustainable fuel production and greenhouse gas mitigation.

In conclusion, plasma technology offers a promising pathway for converting CO_2 into hydrocarbons with a range of potential applications in fuel production. Research in this field continues to advance, aiming to make these processes more efficient, economically viable, and environmentally sustainable. Plasma-based CO_2 conversion holds great promise for addressing climate change and meeting the growing demand for clean energy and chemicals.

15. FUTURE PROSPECTS AND RESEARCH DIRECTIONS IN PLASMA-BASED CO_2 CONVERSION

Plasma-based CO_2 conversion holds immense potential as an innovative and sustainable technology for mitigating climate change and producing valuable chemicals and fuels. Ongoing research and development in this field are focused on optimizing existing processes, exploring novel reactor designs, and enhancing catalyst performance to make plasma-based CO_2 conversion more efficient, cost-effective, and scalable. Here, we discuss the future prospects and research directions in plasma-based CO_2 conversion, highlighting the crucial role of innovation in addressing climate change.

15.1. Enhanced Efficiency and Product Selectivity

Future research efforts will aim to make better the overall efficiency and selectivity of plasma based CO_2 conversion processes. This includes fine-tuning plasma parameters (e.g., temperature, pressure, discharge power) to augment yield of desired products while reducing undesirable byproducts. Catalyst development will continue to play pivotal part in achieving higher transformation rates and product selectivity (Shi et al., 2021).

15.2. Scaling Up for Industrial Applications

While many studies have demonstrated the feasibility of plasma-based CO_2 conversion at the laboratory scale, the next frontier is scaling up these processes for industrial applications. Researchers will focus on reactor design and engineering to facilitate large-scale CO_2 conversion. This involves addressing challenges related to energy efficiency, cost-effectiveness, and process stability (George et al., 2021).

15.3. Integration With Renewable Energy Sources

To make plasma based CO_2 conversion more sustainable, researchers are exploring ways to integrate it with renewable energy fount. Wind and solar power can provide the necessary electricity for plasma generation, reducing the procedure's carbon footprint. This approach aligns with a broader goal of transitioning to a green energy economy (K. Zeng & Zhang, 2010).

15.4. Valorization of Byproducts

Efforts will also focus on the valorization of byproducts generated during plasma-based CO_2 conversion. For example, carbon nanoparticles or nanotubes produced as byproducts can find applications in various industries, including materials science and electronics (Xia et al., 2020).

15.5. Carbon Capture and Utilization (CCU)

Plasma-based CO_2 conversion can play a vital role in carbon capture and utilization (CCU) strategies. Research will continue to explore ways to catch CO_2 from industrial processes and transform it into valuable products, thereby reducing emissions and promoting a circular carbon economy (Ashford & Tu, 2017).

15.6. Interdisciplinary Collaboration and Policy Support

Addressing climate change through plasma-based CO_2 conversion requires interdisciplinary collaboration between scientists, engineers, policymakers, and industry stakeholders. Governments and international organizations can support research and development efforts through incentives, funding, and regulatory frameworks that encourage sustainable practices and the adoption of innovative technologies (Metz et al., 2005).

In conclusion, plasma-based CO_2 conversion represents a promising frontier in the fight against climate change and the quest of chemical and energy production that is sustainable. Ongoing research and innovation will be instrumental in realizing the full potential of this technology. By optimizing efficiency, scalability, and incorporating with renewable sources of energy, plasma based CO_2 conversion can contribute significantly to reducing greenhouse gas emissions and transitioning toward a more sustainable and circular economy.

16. CONCLUSION

In this chapter, we embarked on a journey into the innovative realm of plasma-based CO_2 conversion, a technology that holds immense promise in mitigating greenhouse gas emissions and addressing the urgent global challenge of climate change. We explored the intricacies of this cutting-edge approach, delving into its processes, applications, and the crucial role it plays in advancing sustainable energy and chemical production.

We began by dissecting the process of converting CO_2 into valuable hydrocarbons using plasma technology. Through the generation of highly reactive plasma, CO_2 molecules are dissociated and recombined with hydrogen (H_2) to produce hydrocarbon compounds such as methane (CH_4), methanol (CH_3OH), and synthetic liquid fuels.

The possible applications of plasma based CO_2 conversion are diverse and impactful. We explored production of methane, synthetic liquid fuels, hydrogen, and syngas, highlighting their relevance in clean energy production, transportation, and chemical manufacturing. We drew upon previous research findings to underscore the progress made in this field. Catalyst development, reactor design, and optimization of plasma parameters have been central to enhancing the efficiency and selectivity of hydrocarbon synthesis from CO_2.

The future of plasma-based CO_2 conversion is bright and multifaceted. Research directions include optimizing efficiency and selectivity, scaling up for industrial applications, integrating renewable energy sources, valorizing byproducts, and contributing to carbon capture and utilization (CCU) strategies.

Addressing climate change through plasma technology requires collaboration across disciplines and the support of policymakers and industry stakeholders. It aligns with the broader goal of transitioning to a sustainable and circular carbon economy.

Overall, plasma technology serves as a beacon of hope, illuminating a path towards a sustainable, low-carbon future. By embracing innovation and scientific inquiry, we have the power to mitigate CO_2 emissions, combat climate change, and pave the way for a brighter, more sustainable world. The journey ahead is challenging, but it is also filled with promise and the potential to shape a more sustainable and prosperous future for generations to come.

REFERENCES

Ahasan, M. R., Hossain, M. M., Ding, X., & Wang, R. (2023). Non-equilibrium plasma-assisted dry reforming of methane over shape-controlled CeO 2 supported ruthenium catalysts. *Journal of Materials Chemistry. A, Materials for Energy and Sustainability*, *11*(20), 10993–11009. doi:10.1039/D3TA01196H

Albo, J., Alvarez-Guerra, M., Castaño, P., & Irabien, A. (2015). Towards the electrochemical conversion of carbon dioxide into methanol. *Green Chemistry*, *17*(4), 2304–2324. doi:10.1039/C4GC02453B

Anoop, N., Sundaramurthy, S., Jha, J. M., Chandrabalan, S., Singh, N., Verma, J., Parvatalu, D., & Katti, S. (2021). Plasma catalysis: A feasible solution for carbon dioxide valorization? *Clean Technologies and Environmental Policy*, *23*(10), 2789–2811. doi:10.1007/s10098-021-02203-y

Arpagaus, C., Oberbossel, G., & Rudolf von Rohr, P. (2018). Plasma treatment of polymer powders–from laboratory research to industrial application. *Plasma Processes and Polymers*, *15*(12), 1800133. doi:10.1002/ppap.201800133

Artz, J., Müller, T. E., Thenert, K., Kleinekorte, J., Meys, R., Sternberg, A., Bardow, A., & Leitner, W. (2018). Sustainable conversion of carbon dioxide: An integrated review of catalysis and life cycle assessment. *Chemical Reviews*, *118*(2), 434–504. doi:10.1021/acs.chemrev.7b00435 PMID:29220170

Ashford, B., & Tu, X. (2017). Non-thermal plasma technology for the conversion of CO2. *Current Opinion in Green and Sustainable Chemistry*, *3*, 45–49. doi:10.1016/j.cogsc.2016.12.001

Ayodele, F. O., Mustapa, S. I., Ayodele, B. V., & Mohammad, N. (2020). An overview of economic analysis and environmental impacts of natural gas conversion technologies. *Sustainability (Basel)*, *12*(23), 10148. doi:10.3390/su122310148

Baena-Moreno, F. M., Rodríguez-Galán, M., Vega, F., Alonso-Fariñas, B., Vilches Arenas, L. F., & Navarrete, B. (2019). Carbon capture and utilization technologies: A literature review and recent advances. *Energy Sources. Part A, Recovery, Utilization, and Environmental Effects*, *41*(12), 1403–1433. doi:10.1080/15567036.2018.1548518

Bagger, A., Ju, W., Varela, A. S., Strasser, P., & Rossmeisl, J. (2017). Electrochemical CO2 reduction: A classification problem. *ChemPhysChem*, *18*(22), 3266–3273. doi:10.1002/cphc.201700736 PMID:28872756

Bhatnagar, P. (2019). Cold plasma technology: A new hope to food industry. *Octa Jf Biosci*, *7*, 108–112.

Bogaerts, A., Berthelot, A., Heijkers, S., Kolev, S., Snoeckx, R., Sun, S., Trenchev, G., Van Laer, K., & Wang, W. (2017). CO2 conversion by plasma technology: Insights from modeling the plasma chemistry and plasma reactor design. *Plasma Sources Science & Technology*, *26*(6), 063001. doi:10.1088/1361-6595/aa6ada

Bogaerts, A., & Snoeckx, R. (2019). Plasma-Based CO 2 Conversion. *An Economy Based on Carbon Dioxide and Water: Potential of Large Scale Carbon Dioxide Utilization*, 287-325.

Britun, N., Silva, T., Chen, G., Godfroid, T., van der Mullen, J., & Snyders, R. (2018). Plasma-assisted CO2 conversion: Optimizing performance via microwave power modulation. *Journal of Physics. D, Applied Physics*, *51*(14), 144002. doi:10.1088/1361-6463/aab1ad

Budzianowski, W. M. (2012). Negative carbon intensity of renewable energy technologies involving biomass or carbon dioxide as inputs. *Renewable & Sustainable Energy Reviews*, *16*(9), 6507–6521. doi:10.1016/j.rser.2012.08.016

Chadha, U., Selvaraj, S. K., Ashokan, H., Hariharan, S. P., Mathew Paul, V., Venkatarangan, V., & Paramasivam, V. (2022). Complex nanomaterials in catalysis for chemically significant applications: From synthesis and hydrocarbon processing to renewable energy applications. *Advances in Materials Science and Engineering*, *2022*, 1–72. doi:10.1155/2022/1552334

Chen, F., Zhou, Y., Zhu, Y., Zhu, R., Guan, P., Fan, J., & Saribatir, E. (2021). Recent progress in artificial synaptic devices: Materials, processing and applications. *Journal of Materials Chemistry. C, Materials for Optical and Electronic Devices*, *9*(27), 8372–8394. doi:10.1039/D1TC01211H

Chen, G., Georgieva, V., Godfroid, T., Snyders, R., & Delplancke-Ogletree, M.-P. (2016). Plasma assisted catalytic decomposition of CO2. *Applied Catalysis B: Environmental*, *190*, 115–124. doi:10.1016/j.apcatb.2016.03.009

Chen, G., Godfroid, T., Britun, N., Georgieva, V., Delplancke-Ogletree, M.-P., & Snyders, R. (2017). Plasma-catalytic conversion of CO2 and CO2/H2O in a surface-wave sustained microwave discharge. *Applied Catalysis B: Environmental*, *214*, 114–125. doi:10.1016/j.apcatb.2017.05.032

Choe, S. J., Kang, H. J., Kim, S.-J., Park, S.-B., & Park, D. H. (2005). Adsorbed carbon formation and carbon hydrogenation for CO2 methanation on the Ni (111) surface: ASED-MO study. *Bulletin of the Korean Chemical Society*, *26*(11), 1682–1688. doi:10.5012/bkcs.2005.26.11.1682

Chung, W.-C., & Chang, M.-B. (2016). Review of catalysis and plasma performance on dry reforming of CH4 and possible synergistic effects. *Renewable & Sustainable Energy Reviews*, *62*, 13–31. doi:10.1016/j.rser.2016.04.007

Daligaux, V., Richard, R., & Manero, M. (2021). Deactivation and Regeneration of Zeolite Catalysts Used in Pyrolysis of Plastic Wastes—A Process and Analytical Review. *Catalysts*, *2021*(11), 770. doi:10.3390/catal11070770

Dębek, R., Azzolina-Jury, F., Travert, A., & Maugé, F. (2019). A review on plasma-catalytic methanation of carbon dioxide–Looking for an efficient catalyst. *Renewable & Sustainable Energy Reviews*, *116*, 109427. doi:10.1016/j.rser.2019.109427

Ferreira, C. M., & Moisan, M. (2013). *Microwave discharges: fundamentals and applications* (Vol. 302). Springer Science & Business Media.

Fridman, A. (2008). *Plasma chemistry*. Cambridge university press. doi:10.1017/CBO9780511546075

Gao, X., Liang, J., Wu, L., Wu, L., & Kawi, S. (2022). Dielectric barrier discharge plasma-assisted catalytic CO2 hydrogenation: Synergy of catalyst and plasma. *Catalysts*, *12*(1), 66. doi:10.3390/catal12010066

Gao, X., Lin, Z., Li, T., Huang, L., Zhang, J., Askari, S., Dewangan, N., Jangam, A., & Kawi, S. (2021). Recent developments in dielectric barrier discharge plasma-assisted catalytic dry reforming of methane over Ni-based catalysts. *Catalysts*, *11*(4), 455. doi:10.3390/catal11040455

George, A., Shen, B., Craven, M., Wang, Y., Kang, D., Wu, C., & Tu, X. (2021). A Review of Non-Thermal Plasma Technology: A novel solution for CO2 conversion and utilization. *Renewable & Sustainable Energy Reviews*, *135*, 109702. doi:10.1016/j.rser.2020.109702

Gilfillan, D., & Marland, G. (2021). CDIAC-FF: global and national CO< sub> 2</sub> emissions from fossil fuel combustion and cement manufacture: 1751–2017. *Earth System Science Data*, *13*(4), 1667–1680. doi:10.5194/essd-13-1667-2021

Graves, D. B. (2012). The emerging role of reactive oxygen and nitrogen species in redox biology and some implications for plasma applications to medicine and biology. *Journal of Physics. D, Applied Physics*, *45*(26), 263001. doi:10.1088/0022-3727/45/26/263001

Grill, A. (1994). *Cold plasma in materials fabrication* (Vol. 151). IEEE Press. doi:10.1109/9780470544273

Haszeldine, R. S. (2009). Carbon capture and storage: How green can black be? *Science*, *325*(5948), 1647–1652. doi:10.1126/science.1172246 PMID:19779187

Ho, H.-J., Iizuka, A., & Shibata, E. (2019). Carbon capture and utilization technology without carbon dioxide purification and pressurization: A review on its necessity and available technologies. *Industrial & Engineering Chemistry Research*, *58*(21), 8941–8954. doi:10.1021/acs.iecr.9b01213

Khoja, A. H., Tahir, M., & Amin, N. A. S. (2019). Recent developments in non-thermal catalytic DBD plasma reactor for dry reforming of methane. *Energy Conversion and Management*, *183*, 529–560. doi:10.1016/j.enconman.2018.12.112

Kim, H. H. (2004). Nonthermal plasma processing for air-pollution control: A historical review, current issues, and future prospects. *Plasma Processes and Polymers*, *1*(2), 91–110. doi:10.1002/ppap.200400028

Kogelschatz, U., & Eliasson, B. (1996). The applications of silent discharges. *Physikalische Blätter*, *52*(4), 360–362. doi:10.1002/phbl.19960520409

Lackner, K. S. (2003). A guide to CO2 sequestration. *Science*, *300*(5626), 1677–1678. doi:10.1126/science.1079033 PMID:12805529

Le Quéré, C., Andrew, R. M., Friedlingstein, P., Sitch, S., Hauck, J., Pongratz, J., & Canadell, J. G. (2018). Global carbon budget 2018. *Earth System Science Data Discussions*, *2018*, 1–3.

Legg, S. (2021). IPCC, 2021: Climate change 2021-the physical science basis. *Interaction*, *49*(4), 44–45.

Li, L., Zhang, H., Li, X., Kong, X., Xu, R., Tay, K., & Tu, X. (2019). Plasma-assisted CO2 conversion in a gliding arc discharge: Improving performance by optimizing the reactor design. *Journal of CO2 Utilization*, *29*, 296-303.

Li, N., Mo, L., & Unluer, C. (2022). Emerging CO2 utilization technologies for construction materials: A review. *Journal of CO2 Utilization*, *65*, 102237.

López, M., Calvo, T., Prieto, M., Múgica-Vidal, R., Muro-Fraguas, I., Alba-Elías, F., & Alvarez-Ordóñez, A. (2019). A review on non-thermal atmospheric plasma for food preservation: Mode of action, determinants of effectiveness, and applications. *Frontiers in Microbiology*, *10*, 622. doi:10.3389/fmicb.2019.00622 PMID:31001215

Lu, N., Sun, D., Zhang, C., Jiang, N., Shang, K., Bao, X., Li, J., & Wu, Y. (2018). CO2 conversion in non-thermal plasma and plasma/g-C3N4 catalyst hybrid processes. *Journal of Physics. D, Applied Physics*, *51*(9), 094001. doi:10.1088/1361-6463/aaa919

Lu, X., Bruggeman, P. J., Reuter, S., Naidis, G., Bogaerts, A., Laroussi, M., & Liu, D. (2022). Grand challenges in low temperature plasmas. *Frontiers in Physics*, 1036.

Marcantonio, V., De Falco, M., & Bocci, E. (2022). Non-Thermal Plasma Technology for CO2 Conversion—An Overview of the Most Relevant Experimental Results and Kinetic Models. *Energies*, *15*(20), 7790. doi:10.3390/en15207790

Martin-del-Campo, J., Uceda, M., Coulombe, S., & Kopyscinski, J. (2021). Plasma-catalytic dry reforming of methane over Ni-supported catalysts in a rotating gliding arc–Spouted bed reactor. *Journal of CO2 Utilization*, *46*, 101474.

Masson-Delmotte, V., Zhai, P., Pörtner, H.-O., Roberts, D., Skea, J., & Shukla, P. R. (2022). *Global Warming of 1.5 C: IPCC special report on impacts of global warming of 1.5 C above pre-industrial levels in context of strengthening response to climate change, sustainable development, and efforts to eradicate poverty*. Cambridge University Press.

Metz, B., Davidson, O., De Coninck, H., Loos, M., & Meyer, L. (2005). *IPCC special report on carbon dioxide capture and storage*. Cambridge University Press.

Michielsen, I., Uytdenhouwen, Y., Pype, J., Michielsen, B., Mertens, J., Reniers, F., Meynen, V., & Bogaerts, A. (2017). CO2 dissociation in a packed bed DBD reactor: First steps towards a better understanding of plasma catalysis. *Chemical Engineering Journal*, *326*, 477–488. doi:10.1016/j.cej.2017.05.177

Neyts, E. C., Ostrikov, K., Sunkara, M. K., & Bogaerts, A. (2015). Plasma catalysis: Synergistic effects at the nanoscale. *Chemical Reviews*, *115*(24), 13408–13446. doi:10.1021/acs.chemrev.5b00362 PMID:26619209

Ong, M., Nomanbhay, S., Kusumo, F., & Show, P. (2022). Application of microwave plasma technology to convert carbon dioxide (CO2) into high value products: A review. *Journal of Cleaner Production*, *336*, 130447. doi:10.1016/j.jclepro.2022.130447

Rafiee, A., Khalilpour, K. R., Milani, D., & Panahi, M. (2018). Trends in CO2 conversion and utilization: A review from process systems perspective. *Journal of Environmental Chemical Engineering*, *6*(5), 5771–5794. doi:10.1016/j.jece.2018.08.065

Raizer, Y. P., Shneider, M. N., & Yatsenko, N. A. (2017). *Radio-frequency capacitive discharges*. CRC press. doi:10.1201/9780203741337

Ray, D., & Subrahmanyam, C. (2016). CO 2 decomposition in a packed DBD plasma reactor: Influence of packing materials. *RSC Advances*, *6*(45), 39492–39499. doi:10.1039/C5RA27085E

Salehizadeh, H., Yan, N., & Farnood, R. (2020). Recent advances in microbial CO2 fixation and conversion to value-added products. *Chemical Engineering Journal*, *390*, 124584. doi:10.1016/j.cej.2020.124584

Salge, J. (1995). Plasma-assisted deposition at atmospheric pressure. *Le Journal de Physique IV, 5*(C5), C5-583-C585-592.

Sardella, E., Palumbo, F., Camporeale, G., & Favia, P. (2016). Non-equilibrium plasma processing for the preparation of antibacterial surfaces. *Materials (Basel)*, *9*(7), 515. doi:10.3390/ma9070515 PMID:28773637

Schenk, P. M., Thomas-Hall, S. R., Stephens, E., Marx, U. C., Mussgnug, J. H., Posten, C., Kruse, O., & Hankamer, B. (2008). Second generation biofuels: High-efficiency microalgae for biodiesel production. *BioEnergy Research*, *1*(1), 20–43. doi:10.1007/s12155-008-9008-8

Shi, C., Wang, S., Ge, X., Deng, S., Chen, B., & Shen, J. (2021). A review of different catalytic systems for dry reforming of methane: Conventional catalysis-alone and plasma-catalytic system. *Journal of CO2 Utilization, 46*, 101462.

Snoeckx, R., & Bogaerts, A. (2017). Plasma technology–a novel solution for CO 2 conversion? *Chemical Society Reviews*, *46*(19), 5805–5863. doi:10.1039/C6CS00066E PMID:28825736

Tu, X., & Whitehead, J. C. (2014). Plasma dry reforming of methane in an atmospheric pressure AC gliding arc discharge: Co-generation of syngas and carbon nanomaterials. *International Journal of Hydrogen Energy*, *39*(18), 9658–9669. doi:10.1016/j.ijhydene.2014.04.073

Vadikkeettil, Y., Subramaniam, Y., Murugan, R., Ananthapadmanabhan, P., Mostaghimi, J., Pershin, L., Batiot-Dupeyrat, C., & Kobayashi, Y. (2022). Plasma assisted decomposition and reforming of greenhouse gases: A review of current status and emerging trends. *Renewable & Sustainable Energy Reviews, 161*, 112343. doi:10.1016/j.rser.2022.112343

von Siemens, W. (1857). Ueber die elektrostatische Induction und die Verzögerung des Stroms in Flaschendrähten. *Annalen der Physik, 102*(9), 66–122. doi:10.1002/andp.18571780905

Wang, L., Yi, Y., Guo, H., & Tu, X. (2018). Atmospheric pressure and room temperature synthesis of methanol through plasma-catalytic hydrogenation of CO2. *ACS Catalysis, 8*(1), 90–100. doi:10.1021/acscatal.7b02733

Wang, S., Rohani, V., Ye, T., Dupont, P., Pagnon, S., Sennour, M., & Fulcheri, L. (2023). Effect of K-promoter use in iron-based plasma-catalytic conversion of CO2 and CH4 into higher value products. *Applied Catalysis A, General, 663*, 119315. doi:10.1016/j.apcata.2023.119315

Whitehead, J. C. (2016). Plasma–catalysis: The known knowns, the known unknowns and the unknown unknowns. *Journal of Physics. D, Applied Physics, 49*(24), 243001. doi:10.1088/0022-3727/49/24/243001

Xia, Y., Lu, N., Li, J., Jiang, N., Shang, K., & Wu, Y. (2020). Combined steam and CO2 reforming of CH4 for syngas production in a gliding arc discharge plasma. *Journal of CO2 Utilization, 37*, 248-259.

Yin, Y., Yang, T., Li, Z., Devid, E., Auerbach, D., & Kleyn, A. W. (2021). CO 2 conversion by plasma: How to get efficient CO 2 conversion and high energy efficiency. *Physical Chemistry Chemical Physics, 23*(13), 7974–7987. doi:10.1039/D0CP05275B PMID:33464254

Zeng, K., & Zhang, D. (2010). Recent progress in alkaline water electrolysis for hydrogen production and applications. *Progress in Energy and Combustion Science, 36*(3), 307–326. doi:10.1016/j.pecs.2009.11.002

Zeng, Y., & Tu, X. (2015). Plasma-catalytic CO 2 hydrogenation at low temperatures. *IEEE Transactions on Plasma Science, 44*(4), 405–411. doi:10.1109/TPS.2015.2504549

Zhang, H., Li, L., Li, X., Wang, W., Yan, J., & Tu, X. (2018). Warm plasma activation of CO2 in a rotating gliding arc discharge reactor. *Journal of CO2 Utilization, 27*, 472-479.

Zhang, X., Lee, C. S.-M., Mingos, D. M. P., & Hayward, D. O. (2003). Carbon dioxide reforming of methane with Pt catalysts using microwave dielectric heating. *Catalysis Letters, 88*(3/4), 129–139. doi:10.1023/A:1024049403422

Chapter 3
Plasma-Assisted Carbon Dioxide Conversion:
Applications, Challenges, and Environmental Impacts

Erfan Nouri
https://orcid.org/0000-0002-6558-7848
University of Guilan, Iran

Alireza Kardan
University of Guilan, Iran

Vahid Mottaghitalab
https://orcid.org/0000-0001-7211-3141
University of Guilan, Iran

ABSTRACT

This book chapter explores the fascinating realm of plasma-assisted carbon dioxide (CO_2) conversion, focusing on its applications, challenges, and environmental impacts. The authors begin with an introduction, followed by an examination of key parameters that influence the efficiency of plasma-based CO_2 conversion. Next, the authors delve into various applications and products derived from this process, including plasma-catalytic CO_2 conversion and the synthesis of fuels and value-added chemicals. They then address the challenges and limitations surrounding plasma-based CO_2 conversion, such as cost considerations, catalyst selection, and scaling-up for industrial applications. Finally, the authors explore the environmental benefits of this technology and evaluate its economic feasibility and potential market opportunities. This chapter serves as a comprehensive overview of the field and aims to shed light on the potential of plasma-assisted CO_2 conversion in contributing to a sustainable future.

DOI: 10.4018/979-8-3693-0904-9.ch003

1. INTRODUCTION

In order to mitigate environmental risks, there is a need for innovative CO_2 management in the process of transitioning towards a sustainable society (Valluri et al., 2022). The main objective of carbon capture and utilization (CCU) is to capture CO_2 from sources that emit it and simultaneously utilize it as a raw material for cleaner processes (Hepburn et al., 2019). Numerous pathways for CCU are currently being developed. Extensive research has been conducted on diverse techniques for converting CO_2 into liquid fuels, such as photochemical, electrochemical, and thermal conversion routes (Al-Mamoori et al., 2017; Mikulčić et al., 2019; Sabri et al., 2021; Saleh, 2022; Wanten et al., 2023).

Plasma technology has become of great interest in recent decades as a means of converting CO_2 (Snoeckx & Bogaerts, 2017). Plasma is a type of gas that has been ionized. It has the ability to transform stable molecules into valuable chemicals, such as CO_2 conversion (Chen et al., 2021; Snoeckx & Bogaerts, 2017), and also convert molecular nitrogen for the production of fertilizers (Jardali et al., 2021; Rouwenhorst et al., 2021). The capability of this technology to handle various gases as input and provide immediate control over the process allows for its effective use in electrified production alongside the variable supply of renewable electricity (Snoeckx & Bogaerts, 2017). In addition, there is no need for the utilization of limited resources (Snoeckx & Bogaerts, 2017), and the technique has already been proven effective in a range of industrial contexts, such as generating ozone and operating arc plasma furnaces in the steelmaking industry (Delikonstantis et al., 2022; Wanten et al., 2023).

The review paper conducted by Snoeckx and Bogaerts (2017) extensively discussed the performance of plasma technology in comparison to other CCU technologies. It is evident from their research that there is a range of plasma reactors utilized for gas conversion, which vary in terms of design, applicable currents, flow rates, pressures, voltages, and other factors. A comprehensive analysis of these reactors and their effectiveness in converting CO_2 is outlined in their research paper and other recent literature references (e.g. Refs.(George et al., 2021; Liu et al., 2020; Ong et al., 2022; Vadikkeettil et al., 2022)). This evaluation holds great significance in the advancement of this technology. In order to conduct this evaluation, it is crucial to possess a defined, accurate, and reliable procedure for assessing the various measures of performance.

This book chapter provides a comprehensive exploration of plasma-based CO_2 conversion. It covers the important parameters affecting efficiency, various applications and products, challenges and limitations, and environmental benefits. The chapter aims to contribute to a comprehensive understanding of this technology and its implications for addressing environmental challenges.

2. KEY PARAMETERS AFFECTING THE EFFICIENCY OF PLASMA-BASED CO_2 CONVERSION

Numerous attempts have been made to enhance the efficiency of the plasma catalytic system by modifying various operational factors to achieve maximum conversion and energy effectiveness. In their research conducted in 2016, Mei et al. (2016) examined how reducing the flow rate of the feed gas can elongate the time the gas remains in the reactor, thereby leading to a higher conversion rate of the reactant. According to reports, the energy efficiency is inversely related to the specific energy density. In the subsequent section, we will delve into an array of additional elements that impact energy efficiency.

These factors include the rate at which gas flows, the influence of diluting gases, the size of particles, the dimensions of the gap, and various others.

2.1. Diluting Gas

Research has been carried out to investigate the impact of the addition of gases like argon and nitrogen on the conversion of CO_2. In their study, de la Fuente et al. (2017) also examined how the conversion rate is influenced by the operating pressure and the flow rate of the coolant. New findings suggest that using N_2 as a cooling agent hampers the conversion process by reducing the reaction zone due to the negative impact of increased nitrogen flow rate. Hence, the design of the plasma reactor is instrumental in enhancing the system's overall efficiency.

In their recent research, Nagassou et al. (2020) explored how introducing diluted gases, specifically water, nitrogen, and CH_4, impacts the efficiency of conversion and energy in a solar-gliding arc plasma (GAP) reactor. Incorporating nitrogen as a diluting gas and CO_2 in a proportion of 7:1 resulted in a conversion rate of 45% and an energy efficiency of 70%. Dry reforming of methane (DRM) has higher conversion efficiency but lower energy efficiency compared to other systems. A combination of CO_2 and nitrogen has been found to be optimal in terms of both energy and conversion efficiencies, thanks to the stimulation of vibrations in CO_2 molecules. Plasma systems commonly use argon gas as a diluent (Zhang, 2018).

Experiments were carried out by Xu et al. (2017) using a plasma reactor packed with barium titanate. The gas inlet was mixed with either 80% N_2 or Ar, leading to an increase in conversion rates to 35% and 36%, respectively, compared to the initial value of 19%. Combining diluent gases decreases energy efficiency, especially with N_2 gas dilution. However, adjusting the amounts of Ar and N_2 can effectively regulate the formation of ozone and nitrogen oxides. Dilution with argon decreases ozone formation but increases nitrogen oxides as by-products. The production of ozone is influenced by the amount of energy introduced (up to 36 kJ/L) and the presence of diluent gases. Subsequently, it relies solely on the amount of energy inputted. In a study conducted by Snoeckx et al. (2016), it was found that increasing levels of nitrogen dilution can potentially decrease both the overall energy efficiency and the conversion of CO_2. There will be an elevation in the production of nitrogen oxides, which has the potential to exceed 100 ppm, leading to environmental risks associated with pollution. The rise in conversion rates can be attributed to the unstable condition of nitrogen molecules, which is a counterbalance to the lack of CO_2 levels within the reactor. The findings validate the earlier deduction regarding the decline in energy efficacy.

In their research, Ray et al. (2017) investigated the impact of carrier gas on a plasma reactor that employed a dielectric barrier discharge. The study involved exploring the effects of three different carrier gases: argon, He, and nitrogen. Notably, the highest conversion rate, reaching 19.5%, was achieved when a mixture of argon and CO_2 was used in a 2:1 ratio. Unlike this, Wang et al. (2017) found that dilution with N_2 resulted in improved conversion rates compared to dilution with Argon. Increased carrier gas concentration enhances the efficacy of this transformation. The investigations were performed at a consistent flow rate of 0.2 L/min and a power of 50 W. The rise in conversion was a consequence of the decrease in the overall CO_2 flow rate caused by the addition of dilution gas, leading to a significant enhancement in collisions. Furthermore, N_2, as a diatomic substance, can provoke more efficient clashes for CO_2 division in comparison to Ar, which belongs to the category of monoatomic elements. Continued dilution reduced selectivity for CO, as it caused additional breakdown of CO into different substances.

Argon, widely employed as a carrier gas in plasma systems across various applications, has demonstrated promising outcomes in prior research (Mertens et al., 2019; Wang, Wandell, Tachibana et al, 2018). The impact of diluent gases in a micro-hollow cathode discharge plasma reactor was explored by Taylan and Berberoglu (2014) in their 2015 study. The tests were carried out under consistent conditions, maintaining a voltage level of 4.5 kV and a flow rate of 0.1 L/min. The concentration of CO_2 in the mixture ranged from 9.95% to 89.6% when Argon was used as the carrier gas. It has been observed that Argon has a lower breakdown voltage compared to CO_2. It leads to quick ionization of Argon, resulting in more electrons available for collision with the incoming CO_2 molecules. As a result, the energy efficiency is improved.

The impact of auxiliary gases like air, nitrogen, and oxygen in a GAP was examined by Indarto et al. (2007) under a consistent flow rate of 2 L/min. Adding nitrogen to the gas stream increases CO_2 conversion by 2.5 times compared to pure CO_2. Nitrogen's vibrational excitation enhances collisions and conversion but also forms nitrogen oxides. Introducing oxygen is not beneficial due to CO_2's higher ionization tendency and the potential for CO and oxygen to impede the process. Similar results are observed when mixing with air.

The incorporation of argon and He gases can improve both conversion procedures. Even though both gases have a comparable impact on conversion rates, more favorable outcomes are achieved with a more significant proportion of Argon in comparison to He as a diluting gas (Ramakers et al., 2015). In a plasma reactor using a dielectric barrier discharge, the conversion of CO_2 was 41% with 95% argon dilution, whereas only 25% conversion was achieved with He dilution. The conversion patterns are akin to approximately 70% of the diluted gas. The energy efficiency is significantly decreased for both gases at higher dilution due to the limited effective conversion fraction.

2.2. Input Power

The CO_2 conversion rates are significantly influenced by the applied potential, as well as the selectivity towards CO. In their study, Lu et al. (2018) examined how power discharge impacts the selectivity towards CO and the conversion of CO_2. The report highlights that the transformation of CO_2 and selectivity towards CO initially increases up to a discharge power of 20 W, but then decreases due to elevated gas temperature causing recombination. However, using a cooling water system and increasing power input can lead to a 10% rise in CO_2 conversion and an 8% increase in CO yield. Without the cooling system, energy efficiency declines as power increases, but incorporating cooling can enhance conversion and energy efficiency.

The plasma resulting from the micro-hollow cathode discharge was examined by Taylan and Berberoglu (2014) in their 2015 study, with an applied voltage spanning from 2.5 to 4.5 kilovolts. The measurement was represented by the concentration of CO_2 in the outgoing flow. As the voltage was adjusted from 2.5 to 4.5 kilovolts, and with a gas flow rate of 100 mL/min, the CO_2 concentration in the outgoing stream decreased from 99.4% to 98.3%. One can deduce that when the power increases, the energy efficiency decreases. Modifying the voltage to adjust the power does not improve the strength of the electric field.

The impact of power on the conversion of CO_2 in a pulsed dielectric barrier discharge plasma was studied by Wang et al. (2017). The experiment maintained a frequency of 700 Hz and a steady flow rate of 0.15 L/min. The researchers performed tests using pure CO_2 and gas blends of CO_2/N_2 and CO_2/Ar with an equal amount of moles. The findings indicate that as the flow rate increases, the conversion of CO_2 also increases. However, after a certain point, the rate of conversion experiences a slight decline.

The conversion rate was higher for the gas mixtures than pure CO_2 gas. Within the range of 50 W power, there was an initial rise observed in the preferential treatment of CO, which was subsequently followed by a subsequent decrease.

The text highlights that as power input increases, the conversion rate also increases up to a certain point due to high-energy electrons. However, higher gas temperatures can reduce CO_2 conversion rates due to recombination effects. To overcome this, a cooling system can be incorporated into the reactor design. However, increasing power input decreases energy efficiency, making the process impractical (Brock et al., 1999).

2.3. Gas Flow Rate

The efficiency of the process is significantly impacted by the rate at which pure CO_2 gas flows. The investigation conducted by Lu et al. (2018) explored the correlation between the efficiency of CO_2 conversion and the gas flow rate while keeping the power constant at 40 W. The findings reveal a 63% decline in CO_2 conversion as the flow rate rose from 0.025 to 0.1 L/min. The reduced collision between highly energetic plasma electrons and CO_2 gas in the reaction system causes a decrease in the residence time, leading to a lower conversion rate. When the flow rate increases from 0.025 L/min to 0.1 L/min, the energy efficiency shifts from 0.06 mmol/kJ to 0.09 mmol/kJ. Considering both energy and conversion efficiency, it was recommended that a gas flow rate of 0.1 L/min at a power supply of 40 W would be cost-effective.

The investigation conducted by Taylan and Berberoglu (2014) examined the elements that impact plasma catalysis within a reactor that utilizes micro-hollow cathode plasma discharge technology. As the gas flow rate was adjusted within the range of 0.01 to 0.1 L/min alongside a voltage supply of 4.5 kV, there was an observed correlation between the flow rate and the concentration of CO_2 in the outlet stream. It can be attributed to decreased gas residence time within the reactor as the flow rate increased. Additionally, it has been documented that as the gas flow rate increases, there is a corresponding decrease in specific energy input. Table 1 provides information on the efficiency variations in relation to different flow rates.

In a study conducted by Wang et al. (2017), the conversion of CO_2 to CO was investigated using a plasma reactor equipped with a dielectric barrier discharge. The experimentation examined different flow rates while maintaining a constant power input of 50 W. It was noted that as the flow rate of CO_2 gas increased, there was a decrease in the conversion of CO_2 to CO. At a flow rate of 0.05 L/min, the highest yield reported was 23.31%. This finding reinforces the understanding that a higher flow rate, which leads to reduced residence time, is responsible for decreased conversion.

The impact of operational variables on the conversion and energy efficiencies in a GAP system was investigated by Zhang et al. (2018). The research was carried out in two sections: one involved a flow rate higher than 6 L/min with a specific energy input below 3.5 kJ/L, and the other focused on a flow rate below 6 L/min where the clear energy input exceeded 3.5 kJ/L. According to the findings, the conversion rate steadily rises with flow rates up to 6 L/min, reaching a peak of 4.5%. Additionally, the research reported an optimum energy efficiency of approximately 17.5%. The variation in results can be attributed to numerous factors, including the kinetics of vibrational excitation, retention time, gas temperature, and specific energy input.

2.4. Gap Size and Particle Size

Several researchers have investigated the impact of particle size and the discharge space on conversion in packed-bed plasma reactors. In a packed-bed reactor with a dielectric barrier, Michielsen et al. (2017) analyzed how the size of beads used as packing materials and the catalysts employed influence the outcomes. Different sizes of beads made of Al_2O_3, $BaTiO_3$, SiO_2, and ZrO_2, ranging from 2 to 2.24 mm, 1.6 to 1.8 mm, and 1.25 to 1.4 mm, were used in the experiments. Regardless of the size of the beads, $BaTiO_3$ (a ferroelectric material) demonstrated higher levels of conversion and energy efficiency compared to the conversion achieved in an unoccupied tower. The enhanced conversion and energy efficiency were exemplified by the increased size of the Al_2O_3 beads, which demonstrated the link between the compensation for reduced residence time and the packing material. The assertion that a reduced feed flow rate is advantageous for achieving increased conversions is further supported by this evidence. The energy efficiencies and maximum conversion are documented as 3.5% and 20%, respectively (Van Laer & Bogaerts, 2017).

In their research, Butterworth et al. (2016) examined how the packing size influenced the performance of a packed-bed reactor. They employed Al_2O_3 and $BaTiO_3$ beads and found that reducing the particle size enhanced conversion rates. At the ideal voltage level, there is a nearly 70% increase in the conversion rate. The most significant conversion level was observed when using Al_2O_3 particles with a size range of 180 to 300 micrometers. In addition to particle size, the conversion rate is influenced by factors such as the electric field, catalytic properties, and various other elements. According to research findings, it has been indicated that a decrease in particle size has a positive impact on the breakdown voltage. Additionally, the occurrence of partial discharge is observed during the formation of plasma within the reactor.

The impact of the size of the gap on the conversion and energy efficiency in a microplasma reactor utilizing a packed-bed dielectric barrier discharge was analyzed by Uytdenhouwen et al. (2018) in their 2018 study. According to the research findings, it was concluded that decreasing the gap size can lead to an improvement in conversion efficiency while having no impact on energy efficiency. According to the MATLAB script, the smallest gap size of 268 μm achieved a maximum conversion rate of approximately 34% in the case of an empty tower reactor. By introducing glass beads, the conversion rate experienced a significant boost of 50–55%, which then reached a stable state as the gap size decreased further. This equilibrium point for conversion was achieved due to the power-driven mechanism. Hence, the conversion rate tends to decrease as the gap size increases, except for ZrO_2 and SiO_2, which exhibit higher conversion rates at larger gap sizes, specifically at 1230 μm. The cause of this unusual behavior is currently under investigation.

3. APPLICATIONS AND PRODUCTS

Plasma possesses the capability to facilitate chemical reactions that would typically be thermodynamically impractical due to its non-equilibrium characteristics, low energy demand, and ability to trigger physical and chemical reactions at a comparatively lower temperature (Li, 2019). Extensive research has been undertaken regarding implementing plasma technology in the context of converting CO_2. The splitting of CO_2 was found to have an efficiency of 43% and a conversion rate of 18% by Nunnally et al. (2011). The achieved level of effectiveness can be attributed to the stimulation of CO_2 that is not in

Table 1. The impact of different factors on the CO_2 reduction process in plasma reactors

Parameters	Parameter Range	Reactor Type	Energy Efficiency (mmol/kJ)	Conversion (%)	Ref
Diluting gas	Ar N_2	Packed-Bed	Decreases	36 35	(Xu et al., 2017)
	Ar He	Dielectric Barrier Discharge	3.57×10^{-2} - 7.14×10^{-2} 3.57×10^{-2} - 7.14×10^{-2}	41 25	(Ramakers et al., 2015)
	Ar N_2		- -	19.37 21.32	(Wang et al., 2017)
	He		0.945	19.5	(Ray et al., 2017)
	N_2		Decreases	decreases for N_2 > 50% Increases for N_2 > 50%	(Snoeckx et al., 2016)
	N_2	Gliding Arc Plasma	-	three-fold increase	(Indarto et al., 2007)
	N_2	Solar-Gliding Arc Plasma Reactor	2.5	44	(Nagassou, 2020)
Power Input	10–50 W	Dielectric Barrier Discharge	1.07	13-14	(Lu et al., 2018)
	2.5–4.5 kV	Micro- hollow Cathode Discharge	0.12	CO_2 outlet concentration 98.3–99.4%	(Taylan & Berberoglu, 2014)
	30–55 W	Plasma Reactor	Decreases	two-fold increase	(Wang et al., 2017)
Flow rate	50–300 mL/min	Dielectric Barrier Discharge	-	13.11-23.31	(Wang et al., 2017)
	0.025- 0.100 L/min		0.06	63	(Lu et al., 2018)
	10–800 mL/min	Micro- hollow Cathode Discharge	-	13.7	(Taylan & Berberoglu, 2014)
	6–7 L/min	Gliding Arc Plasma	0.57 – 0.61	4-4.4	(Zhang, 2018)
	10 L/min		1.07	8.6	(Ramakers et al., 2015)
	2 L/min		-	35	(Indarto et al., 2007)
Gap Size	268 μm	Dielectric Barrier Discharge Micro	-	33	(Uytdenhouwen et al., 2018)
Particle Size	180–300 μm	Packed-Bed	-	70% enhancement	(Butterworth et al., 2016)
	2–2.24 mm	Dielectric Barrier Discharge	0.125	20	(Michielsen et al., 2017)

equilibrium and the presence of a high-temperature difference between gliding arc discharge (GAD) and the surrounding gas, resulting in rapid extinguishing. In a study conducted by Tu et al. (2014), they achieved a 60% level of energy efficiency with a reported conversion range of 8-16% for DRM utilizing a GAD. The findings revealed that the energy efficiency of DRM with GAD is significantly higher than dielectric barrier discharge (DBD) processing or corona discharge (CD), giving rise to a difference in magnitudes. The reason behind this is primarily attributable to the increased density of electrons (around 1023 per cubic meter) produced within the GAD. In their research, Lu et al. (2018) investigated the conversion of CO_2 under specific conditions of atmospheric pressure and low temperature using a DBD reactor. The system achieved a CH_4 conversion rate of 41% and a CO_2 conversion rate of 55% when operated at a temperature of 600 degrees Celsius without a catalyst. The dissociation of CO_2 in a radio frequency (RF) plasma and microwave (MW) plasma was conducted under low gas pressure conditions, with both systems displaying characteristics of non-thermal plasma (NTP). The highest achieved level of energy efficiency was 59.3% (Huang et al., 2017). The conversion rate of CO_2 to CO was minimal, as observed in experiments utilizing a wire-to-cylinder reactor at a negative polarity without a catalyst. These experiments were conducted at a flow rate of 100 ml/min and voltage range of 4.5-6.5 kV, both in pure CO_2 and a mixture of CO_2 and O_2 (Steinberg & Dang, 1977). When the percentage of O_2 in the gas mixture was elevated, there was an observed rise in the production of CO. A study explored the application of direct current wire-plate CD for CO_2 decomposition, it was found that the highest level of deterioration of CO_2 was 10.91% when the flow rate was set at 30 ml/min (Xu et al., 2004). The affirmation has been made that utilizing direct current plasma, specifically CD plasma, can result in greater efficiency in converting CO_2 into CO and O_2. The crucial aspect of converting electrical energy into chemical energy in CO_2 molecules is combining low discharge power and elevated flow rates (Xu et al., 2004). The highest CO_2 conversion achieved without a catalyst using a coplanar DBD was 1% at an energy density of 0.190 kJ/L. Furthermore, by employing diluted CO_2 in plasma operation, the energy efficiency can be enhanced, as evidenced by attaining the highest CO yield at a distinct input energy of 0.210 kJ/L, which amounted to 0.7% (Schiorlin et al., 2016). According to research, the current record for energy efficiency in MW stands at 90% when operating within the pressure range of 100 Torr to 200 Torr (Fridman, 2008). The information presented in Table 2 outlines the findings from multiple studies conducted on non-catalytic plasma CO_2.

3.1. Plasma-Catalytic CO_2 Conversion

Utilizing a catalyst in conjunction with NTP shows excellent potential in addressing the conversion of CO_2 (Azzolina-Jury, 2017). The fusion of NTP with catalysts has the potential to enhance reaction energy efficiency by over 60%, produce UV emission, boost the creation of long-lived intermediate species (100 μ s) and short-lived radicals (Gholami et al., 2018), and facilitate adjustments to the catalyst surface (Talebizadeh et al., 2014). The utilization of various catalysts within the plasma region is commonly performed in NTP for DRM to achieve elevated conversion rates and better selectivity towards H_2 while concurrently decreasing carbon deposition quantities (Abbas & Daud, 2010; Jiang et al., 2002). Experiments were conducted to study DRM using a Pt-Ni/ZSM-5 catalyst prepared through the impregnation method. The obtained results were then compared with the theoretical conversions achieved without a catalyst (Pawelec et al., 2007). The significance of the quantity of the nickel precursor in bimetallic catalysts has been found to significantly impact the surface characteristics and catalytic properties. The enhanced efficiency and durability attained through utilizing bimetallic catalysts were connected to

Table 2. Converting plasma CO_2 without the need for a catalyst

Plasma Type	Energy Efficiency (%)	Conversion (%)	Specific Input Energy (eV/molecule)	Ref
Microwave	90	10	0.3	(Asisov, 1981)
	82	3	0.1	(Mitsingas et al., 2016)
	50	9	0.9	(Mitsingas et al., 2016)
	20	20	2.9	(Chen, Britun, Godfroid et al, 2017)
	20	20	3	(de la Fuente et al., 2016)
	24	83	10.3	(Bongers et al., 2017)
	51	11	0.6	(Bongers et al., 2017)
	35	47	3.9	(Bongers et al., 2017)
	20	10	1.4	(Spencer & Gallimore, 2012)
	6	80	39	(Silva et al., 2014)
Dielectric Barrier Discharge	3.8	18	13	(Duan et al., 2015)
	1	30	87	(Paulussen et al., 2010)
	8	14.3	5.2	(Mei, He, Liu et al, 2016)
Gliding Arc Discharge	19	15	2.3	(Indarto et al., 2007)
	46	4.3	0.3	(Nunnally et al., 2011)
Radio Frequency	3	20	19	(Spencer & Gallimore, 2011)
Corona Discharge	2	11	16	(Mikoviny et al., 2003)

the amplified dispersion of nickel. In an investigation on CO_2 conversion at reduced temperature and atmospheric pressure employing a DBD reactor, the conversion rate achieved by aiming a plasma/g-C_3N_4 catalyst was double the conversion rate accomplished without the presence of the g-C_3N_4 catalyst, using only plasma. Integrating the g-C_3N_4 catalyst into the plasma setup enhanced the energy efficacy by 157% (Lu et al., 2018). The conversion of CO_2 adsorbed on the catalyst's surface to CO_{ad} and O_{ad} occurred as a result of the plasma-injected high-energy electrons being introduced into the conduction band of g-C_2N_4. It was discovered that applying a robust electrical charge increased the distance between the holes and the electrons generated on the catalyst's surface. The plasma discharge was made even more by modifying the discharge mode using the g-C_3N_4 catalyst. Michielsen et al. (Michielsen et al., 2017) presented a study on how different catalyst materials and packing methods affect the conversion of CO_2 in a DBD plasma reactor. The collaboration between the plasma and catalyst resulted in a reduction in the operational temperature of the catalyst, enhancing the efficiency, selectivity, stability, yield, and energy utilization of the CO_2 conversion process. $BaTiO_3$ proved to be the most effective catalyst for CO2 conversion and energy efficiency, reaching a maximum of 25% and 4.5%, respectively. This study demonstrated that modifying the active sites of the catalyst and modifying the characteristics of the support (such as packing and material) offer significant possibilities for enhancing the efficiency of a CO_2 conversion-packed bed plasma reactor. Mei et al. observed that the utilization of Ni/Al_2O_3 in a cylindrical DBD reactor for DRM led to a significant enhancement in the production of the desired output ($CO+H_2$), surpassing the results obtained from using only the catalyst or NTP (Mei et al., 2017). The remarkable capability and harmonious interaction between the NTP and catalyst were believed to be the cause behind the reduction in catalyst operating temperature and enhancement in both activity and stability of the catalyst. The enhancement of product yield was achieved through significant improvements in reactant conversion, product selectivity, and energy efficiency. The information presented in Table 3 offers a concise overview of the different catalysts employed for plasma-based CO_2 conversion processes.

There are two primary forms of catalysis utilized in plasma catalytic processes: catalysis based on noble metals (such as Rh and Ru) and catalysis based on transition metals (including Cu, Ni, and Fe). Utilizing noble metals as catalysts is highly effective in the production of formates and formic acids. However, their utilization in industry is hindered by the prohibitively expensive nature of these precious metals (Wang et al., 2019). On the other hand, catalysts utilizing transition metals are significantly more affordable but typically encounter challenges associated with limited selectivity and yields. Therefore, it is crucial that innovations are implemented to enhance the catalytic systems in order for the plasma catalytic procedure to attain economic viability (Snoeckx & Bogaerts, 2017; Wang et al., 2011).

Table 3. Summary of different catalysts employed in plasma-assisted CO_2 conversion

Catalyst	Power (W)	Catalyst Loading (g)	Conversion (%)	Ref
Zeolite	74	1	50	(Nguyen & Kim, 2015)
g-C_3N_4	40	0.2	12	(Lu et al., 2018)
$BaTiO_3$	100	0.3	25	(Michielsen et al., 2017)
Ni/Al_2O_3	50	0.25	13	(Tu et al., 2011)
Pt12Ni	34.6	0.05	73	(Pawelec et al., 2007)
$LaNiO_3$@SiO_2	150	0.2	78	(Zheng et al., 2014)
K-Ni/MgOZrO_2	-	0.2	88	(Nagaraja et al., 2011)
$BaZr_{0.75}T_{0.25}O_3$	-	4.2	43	(Chung et al., 2014)
KuCN/AO	60	-	20.6	(Lu et al., 2021)
Pd/ZnO	20	0.5	36.7	(Sun et al., 2022)
Ni/CeZrO_x	14.5	-	73.5	(Mikhail et al., 2021)

3.1.1. CO_2 Dissociation to CO and O_2

Considerable attention has been focused on dissociating CO_2 into CO and O_2. It is essential to highlight that achieving this reaction at low temperatures using readily available catalysis remains a challenge despite the significance of CO as a vital precursor for fuel synthesis and the manufacturing of various chemicals (Ashford & Tu, 2017; Zhu et al., 2017). An intriguing approach to exploit CO_2 and prevent the formation of byproducts from hydrocarbons is the direct decomposition of CO_2 (Yap et al., 2015). The reaction is considerably enhanced by NTP, despite its highly endothermic nature. A variety of researchers have performed investigations and simulations utilizing different types of plasma. The most frequently employed reactor is the coaxial DBD reactor, as indicated by previous studies (Ozkan, 2016; Pinhão et al., 2011). Furthermore, the MW and GAD methods have also been extensively researched (Chen, Godfroid, Britun et al, 2017; Crim, 2008). The potential impacts of NTP on the system can be linked to the subsequent responses: activating the electrons in molecules, causing gas heating through the joule effect due to vibration, ionization, and directly breaking down CO_2 molecules through electron interaction.

The state denoted as v* is the condition where vibrations are heightened. It has been observed that nearly 97% of the complete energy present in NTP can be transferred to the vibrationally excited state of CO_2 in cases where the NTP discharge possesses an electron temperature varying between 1 and 2 electron volts (eV), or has a reduced electric field (E/N) ranging from 20 to 40 Td (Bahri & Haghighat, 2014; Mei, Zhu, Wu et al, 2016). In the case of temporary and spatially transient DBD plasmas, the value of E/N is primarily influenced by the microdischarge. According to research findings, when utilizing DBD technology, the typical range for energy efficiency in converting CO_2 is typically between 4% and 8% (Snoeckx et al., 2015).

When high-energy electrons collide with CO_2, they undergo dissociation to create O_2, CO, and C through a range of reactions that depend on the electron energy levels (Michielsen et al., 2017). The specific input energy is a critical factor in the dissociation of CO_2 since it impacts both the efficiency of energy utilization and the conversion rate of CO_2 (Aerts et al., 2015). When the input energy was decreased, there was an enhancement in energy efficiency by 9% while witnessing a corresponding decrease in conversion by 8%. Packed bed DBDs reactors have been employed to dissociate CO_2, utilizing various catalysts including Ag, Pd, and Ni are coated with a zeolite 3A or Al_2O_3 support (Michielsen et al., 2017). Zeolite 3A showed an improved preference for hydrogen, leading to the production of liquid hydrocarbons (C_xH_y) (Gallon et al., 2012) as a result of its selective shape properties. The interaction between the catalysts and plasma is contingent on the changes in plasma properties caused by the specific catalyst material and packaging. The potential impact, whether positive or negative, relies on various factors, such as alterations in the types of plasma species generated, the influence of sorption effects, the presence of electric fields, the kind of discharge utilized, and the length and power of the release (Tu et al., 2011). The observation was made that the primary factor influencing the efficiency of the process energy is the flow rate in the utilization of a cylindrical Dielectric Barrier Discharge reactor for the conversion of CO_2 (Aerts et al., 2015). Simultaneously, when the power and length of discharge were augmented, there was a corresponding rise in the conversion rate of CO_2. In addition, the impact of the frequency used was inconsequential regarding energy efficiency and conversion rates. However, the plasma's appearance is more characterized by filaments when operating at a high frequency of 75 kHz, as opposed to a low frequency of 6 kHz. The investigation also covered the examination of the influence of carrier gases such as He, nitrogen (N_2), and argon (Ar) on the dissociation of CO_2 through NTP. The utilization of Ar and He led to an observed rise in the conversion of CO_2. The efficiency of this conversion process was diminished as a result of the decreased presence of CO_2 in the mixture, thus resulting in inefficiency in mitigating the decline in CO_2 fraction (Leiweke & Ganguly, 2007; Moss et al., 2017; Tu et al., 2011). According to observations, the inclusion of carrier gas led to higher expenses. However, it has been proven that DBD can achieve success even without the presence of carrier gas. It has been discovered that the gas flow rate is the most significant factor since it impacts the production of CO and the transformation of CO_2. Enhancement of energy efficiency and CO_2 conversion can be achieved through various methods, such as optimizing the dielectric material and reactor geometry. Incorporating a catalyst bed in the discharge zone also contributes to these improvements (Paulussen et al., 2010). The process of CO_2 decomposition in DBD plasma is mainly attributed to electron impact dissociation, while in GAD and MW discharges, a significant role is played by a more energy-efficient pathway involving vibrational excitation (Wang et al., 2016).

In the study conducted by (Sun, 2017), using a dielectric material, specifically BaTiO3, in the DBD process enhanced the energy efficiency and decomposition of CO_2. At a CO2 feed rate of 50 standard cubic centimeters per minute and power input of 20 watts, the achieved rates of CO_2 conversion and energy efficiency were 13.5% and 7%, respectively. Recent studies have provided evidence that the decomposition of CO_2 can be efficiently achieved through MW discharge of Ar/CO_2 or He/CO_2. Among these carrier gas options, Ar has proven to be the most efficient for the low-temperature transformation of CO_2 into O_2 (Huang et al., 2017). One popular method for breaking down CO_2 is using a surface guide. This technique commonly employs two MW frequencies: 2.45 GHz and 915 GHz (Tsuji et al., 2001). The potential cause of this phenomenon might be the prevalence of MW, which often corresponds to the utmost absorption of the supplied energy by the electrons. Moreover, it can also be attributed to the slightly intensified stimulation of the CO_2's asymmetrical mode, which plays a vital role in the decomposition of CO_2.

In their study, Wang et al. (1999) conducted CO_2 decomposition by employing a glow discharge and obtained a 30% conversion of CO_2. Researchers discovered that increasing the voltage of the inputs had a favorable impact on the reaction rate. Nevertheless, the best energy efficiency was achieved by utilizing gas that included CO_2 while maintaining high flow rates, a high concentration of CO_2, and reducing the input voltage. It came at the cost of diminishing the conversion of CO_2. Enhancing the reaction rate and reactor efficiency can be achieved by selecting high flow rates, as this allows for effective heat removal from the reactor. A study was conducted to evaluate the effectiveness of dissociating CO_2 by applying RF discharge with a plasma power of 1 kW and a frequency of 13.56 MHz. As a result, a conversion rate of 90% was obtained (Spencer & Gallimore, 2011), which can be attributed to the utilization of a plasma system operating at low temperatures.

In their study, Huang et al. (2017) investigated the dissociation of CO_2 using two distinct setups. The first involved an inductively coupled RF plasma operating at a power of 2 kW and a frequency of 27.12 MHz, while the second utilized a MW plasma also running at a capacity of 2 kW. The two systems exhibited characteristics of NTP, with the maximum achieved energy efficiency being 59.3%, surpassing the thermodynamic equilibrium's peak value of around 45%. The presence of electron-triggered vibrational stimulation is an evident indication of its notable contribution to the dissociation of CO_2 in NTP. During the process of CO_2 dissociation using GAD at normal pressure, the introduction of gases resulted in enhancements to the conversion reaction. It can be attributed to the presence of excited N_2 levels, which contribute positively to the overall outcome. On the other hand, the company of air and oxygen potentially had an adverse impact, potentially attributable to the reversal of the reaction between carbon and CO, resulting in the production of CO_2 (Indarto et al., 2006; Indarto et al., 2007).

3.1.2. Dry Reforming of Methane

The use of plasma to catalyze the DRM has shown great potential for transforming natural gas into valuable products at low temperatures and pressures. Nickel has emerged as a highly investigated catalyst in the DRM procedure. The utilization of NTP in conjunction with a catalyst has demonstrated considerable advancements in recent times. For DRM, the catalysts' selectivity is a significant issue. Even though it generates a mixture of CO and hydrogen using CO_2 reduction, resembling the process observed in the direct separation of CO_2, the enthalpy of the DRM reaction exhibits a slightly reduced value. Researchers have experimented with various supports for catalysts to enhance the catalytic efficiency of DRM. Al_2O_3, TiO_2, ZrO_2, SiO_2, MgO, etc., are the primary catalysts employed. Abiev et al. (Abiev et al., 2020) documented the utilization of various combinations of catalysts and supports in the dry reforming process, providing insights into their effectiveness.

In their research conducted in 2020, Nagassou et al. (2020) examined how the dilution of gases, specifically nitrogen, CH_4, and water, impacts the conversion and energy efficiencies of a solar-GAP reactor. The DRM process is influenced by operating parameters, with nickel-based catalysts being the preferred choice due to their cost-effectiveness. The use of ferroelectric substances and enhancers like cerium oxide and magnesium oxide improves effectiveness and reduces coke buildup. Catalyst preparation method is important, with the co-impregnation technique resulting in superior performance compared to sequential impregnation. Surfactants during co-precipitation also enhance support material adherence. The utilization of plasma treatment to enhance the interaction between metals and support has been explored by some researchers; however, it has implications for the overall cost-effectiveness of the procedure (Usman et al., 2015). Numerous scientists have conducted investigations on diverse permutations of perovskites, trimetallic catalysts, bimetallic catalysts, and metal support to enhance performance, as evidenced by recent studies. The cost-effective nature of these perovskite materials makes them a promising alternative to noble metals for future applications. Consideration must be given to the design of catalysts when carrying out the DRM process, as it is crucial to minimize the formation of undesired byproducts (Xu, 2017). Aramouni et al. (2018) have conducted extensive research on Cobalt, which is among the metals currently under active scrutiny for DRM. However, it is imperative to conduct further investigations to determine the optimal combination of support material and promoters. Plasma catalytic DRM encounters a significant hurdle known as coke deposition.

3.1.3. Hydrogenation

Table 4 lists different pathways for converting CO_2 into CH_4, with hydrogenating CO_2 to produce CH_4 being a thermodynamically favorable method. Extensive research on nickel catalysts has shown impressive selectivity of over 90% at 700 Kelvin, but sustainable sources of hydrogen are required for financial feasibility. Currently, commercially available hydrogen is mainly derived from the expensive steam methane reforming (SMR) process, making the combined SMR and hydrogenation reaction economically unviable. Another option is using copper-based catalysts to convert CO_2 into methanol, with the primary focus being the selectivity of this process (Snoeckx & Bogaerts, 2017).

Table 4. Different reaction types and the enthalpies associated with them

Reaction	Stoichiometry	Enthalpy (kJ/mol)
Hydrogenation	$CO_2 + 4H_2 \rightarrow CH_4 + 2H_2O$ $CO_2 + H_2 \rightarrow CO + H_2O$ $CO_2 + 2H_2 \rightarrow C + 2H_2O$ $CO + H_2 \rightarrow C + H_2O$	-165 41.2 -90 -131
Water-gas shift	$CO + H_2O \rightarrow CO_2 + H_2$	-41.2
Reverse water gas shift	$CO_2 + H_2 \rightarrow CO + H_2O$	41.2
CO_2 splitting	$CO_2 \rightarrow CO + 0.5O_2$	280
Dry reforming of methane	$CH_4 + CO_2 \rightarrow 2CO + 2H_2$	247

Source: (Anoop et al., 2021)

The research conducted by Benrabbah et al. (2017) focused on investigating the conversion of CO_2 into CH_4 through plasma catalytic hydrogenation, utilizing a catalyst that is based on nickel. According to the report, catalysts that underwent pre-plasma treatment demonstrated a remarkable selectivity exceeding 80% toward CH_4. A substantial quantity of CO is generated as a byproduct when there is a utilization of a powerful energy source with an output of approximately 25 watts. Despite the potential enhancement of the catalyst-support (zirconia) interaction through pre-treatment, there is a downside regarding the economic feasibility of the process. The less potential feasibility of using hydrogen in the process is due to its cyclic loop nature.

Plasma reactors show potential for converting CO_2 and hydrogen into methanol more effectively than forming methane. However, high pressure and temperature are required. Copper and platinum catalysts are effective in this process. Methanol production depends on conversion and selectivity, resulting in a combination of CO, methanol, methane, and reactants that haven't fully transformed. Increasing hydrogen and CO_2 amounts enhances methanol selectivity, while conversion remains steady (Wang, Yi, Guo et al, 2018). Separating various products becomes challenging due to the combination of multiple substances. Despite the potential for enhancement in laboratory and modeling studies, the current findings suggest a promising outlook for the route's industrial application. In recent times, several investigations have analyzed how copper surfaces facilitate the production of methanol to uncover potential uses of this process in industry (Michiels et al., 2020).

3.2. Synthesizing Fuels and Value-Added Chemicals From CO_2

In theory, the conversion of CO_2 using hydrogen or water has the potential to generate a wide range of substances, including various hydrocarbons, CO, and oxygenates such as formic acid, formate, and methanol. These different chemicals can be produced through diverse reaction routes (Wang et al., 2015). Based on the most recent reports (Ashford & Tu, 2017; Wang et al., 2015; Zeng & Tu, 2017), the main outputs resulting from the process of CO_2 hydrogenation include formic acid, methanol, CH_4, and CO.

The methods used to achieve CO_2 hydrogenation are similar to those used for CO_2 direct dissociation, such as electrochemical, photochemical, traditional thermal catalytic, and plasma-assisted techniques. Plasma techniques have the advantage of being able to carry out the reaction at ambient pressure and temperature, eliminating the need for extreme reaction conditions.

The field of CO_2 hydrogenation using H_2O for the production of H_2O_2 and oxalic acid has seen a significant lack of research since the initial investigation by Ihara et al. (1994) with a MW plasma reactor. Commencing in 2013 (Mahammadunnisa et al., 2013), there was an emergence of research topics regarding plasma-assisted hydrogenation of CO_2. Researchers primarily address the process of CO_2 methanation, aiming to enhance CO_2 conversion and CH_4 selectivity by adjusting various factors like reactor configuration, gas flow rate, CO/H_2 ratio, and employing diverse plasma sources. Various techniques have been reported for the methanation of CO_2, including MW (Chen, 2015), Cu-Mn/γ-Al_2O_3 (Zeng & Tu, 2015), dielectric barrier discharge with different catalysts (Mora et al., 2016; Zeng & Tu, 2017), Ni/$MgAl_2O_4$ (Fan et al., 2015), Ni/zeolite pellets (Jwa et al., 2013), Ni-CexZr1-xO_2 (Benrabbah et al., 2017; Nizio, Albarazi, Cavadias et al, 2016; Nizio, Benrabbah, Krzak et al, 2016), low pressure magnetic field (Arita & Iizuka, 2015), and pulsed corona discharges (Hoeben et al., 2015). The findings in these reports suggested that a greater ratio of H_2/CO_2 is advantageous for CO_2 conversion, but it hurts the selectivity of CH_4. More specifically, in the DBD reactor, employing a dielectric barrier with a higher permittivity improved the CO_2 conversion (Mora et al., 2016). The current body of literature does not provide a comprehensive understanding of the synergistic relationship between plasma and catalyst in the plasma-catalytic process, as there is limited systematic research in this area.

Two additional publications (Ref. (De Bie et al., 2016; Su et al., 2017)) explored the explanation of the prospective process in CO_2 hydrogenation with hydrogen. De Bie et al. (2016) presented a study employing a fluid model to comprehend the influence of residence time and gas mixing ratio on the primary reaction steps leading to various product formations (CH_2, CO, CH_2O, CH_3OH), thereby aiding in optimizing the operational parameters for specific product targets. According to their suggestions, the gas conversions are mainly influenced by radical reactions rather than electron impact excitation, ionization, and dissociation reactions. They emphasized that the increased density of CH_2 and CH_3 radicals is crucial in enhancing CO_2 conversion. This study stresses the effectiveness of plasma-assisted CO_2 hydrogenation without a catalyst for generating CO, CH_4, and H_2O. These products exhibited considerable selectivity and were deemed the most suitable options compared to other potential outcomes. It is also the reason why the prior investigations mainly concentrated on the conversion of CO_2 into CH_4, aiming for the production of CH_4.

An examination conducted by Su et al. (2017) focused on the utilization of plasma to facilitate the conversion of CO_2 to dimethyl ether (DME) through the introduction of hydrogen. The catalyst used in the process was Cu-Fe-Ce/HZSM-5, and the study employed an intrinsic kinetics model in a two-stage plasma-catalysis system. According to the findings of the inherent kinetic investigation, the reverse water-gas shift reaction, the activation energy of methanol synthesis, and methanol dehydration to DME could be decreased by incorporating a plasma source into the reaction system. The achievement of DME selectivity and CO_2 conversion reached 69.5% and 24.3%, respectively. Intrinsic kinetics examination shows that adding appropriate catalysts to plasma-catalytic CO_2 hydrogenation can increase selectivity for producing DME, various hydrocarbon and oxygenates. However, without a catalyst, plasma-assisted CO_2 hydrogenation is better for producing CH_4. Further research is needed to improve this area.

Wang et al. (2021) delves into the difficulties associated with transforming CO_2 into fuels and value-added chemicals using catalysis. They introduce a catalytic process, facilitated by plasma, that accomplishes the hydrogenation of CO_2 into C_2^+ hydrocarbons in a single step. This process takes place at low temperatures and atmospheric pressure within a packed-bed reactor employing a DBD. Through the combination of plasma and an alumina-supported Co catalyst, along with modifications to the catalyst-bed arrangement within the plasma discharge zone, the authors achieved a substantial selectivity of C_2^+ hydrocarbons, reaching 46% with approximately 74% CO_2 conversion. The article also explores the potential sources of C_2^+ formation and highlights the significance of the catalyst-bed configuration.

Ashford et al. (2022) address the pressing need for carbon-neutral processes in their article "Plasma-catalytic CO_2 hydrogenation to ethane in a dielectric barrier discharge reactor." They highlight the potential of plasma catalysis for CO_2 conversion and utilization, particularly in the production of green chemicals and fuels using renewable hydrogen. Unlike previous studies focused on CH_4 production, this work centers on the NTP-catalytic conversion of CO_2 into C_2H_6 using a DBDs reactor. The authors evaluate the influence of different alumina-supported metal catalysts (Ru, Cu, Ni, and Fe) on the plasma-catalytic CO_2 hydrogenation to ethane, and find that the Ru catalyst demonstrates the highest selectivity at nearly 40%. Additionally, the Ru catalyst improves the energy efficiency of the process and achieves the highest H_2 conversion at 29%. These findings showcase the promise of plasma-assisted production of ethane as a carbon-neutral approach for CO_2 utilization.

In their article, Farahani et al. (2022) explore the potential of NTP for methanol synthesis through CO_2 hydrogenation. They extensively review recent studies on the mechanism of this reaction under NTP, reactor configuration modifications, and catalyst design. The authors analyze the strengths and weaknesses of each study and offer insights into future directions for sustainable heterogeneous catalysis. This comprehensive review serves as a valuable reference for researchers interested in carbon-neutral processes for addressing greenhouse gas emissions.

4. CHALLENGES AND LIMITATIONS

4.1. Cost Considerations and Energy Efficiency

The efficiency of producing specific products through plasma-catalytic activation of CO_2, especially in converting CO_2 into liquid fuels and conducting dry reforming, is significantly lower than thermal catalysis. This is primarily due to the limited understanding of how to design catalysts that are both highly effective and stable in promoting the plasma-catalytic conversion of CO_2. An approach frequently employed in plasma catalysis is the utilization of the most suitable catalysts in the corresponding reactions carried out under thermal conditions (Whitehead, 2016). Regrettably, the catalysts that exhibit good performance in thermal catalytic CO_2 activation may not be effective in plasma-catalytic reactions due to the contrasting operating conditions. Plasma catalysis involves different factors, such as high pressure or ambient pressure, as well as high temperature or low temperature (Mehta et al., 2019; Zeng et al., 2018). The inclusion of plasma in catalytic processes can modify the catalyst's surface and create new substances, resulting in changes to adsorption and reaction pathways. Plasma catalytic CO_2 hydrogenation produces CO and excited CO_2 molecules that are easily absorbed by the catalyst, leading to the formation of new reaction pathways not found in traditional methods. In some cases, catalysts that are inefficient in thermal catalysis show high effectiveness in plasma-assisted CO_2 activation (Zeng et al., 2018). Hence, the task lies in choosing and developing the most effective catalysts for a designated CO_2 activation procedure within set circumstances to yield desired outputs.

One of the significant obstacles in the plasma-catalytic conversion of CO_2 is the need to improve the energy efficiency of the procedure while simultaneously upholding a high level of CO_2 conversion and selectivity in the product. Using MW and GAP systems, it is possible to attain an energy efficiency of approximately 50% to 60% when converting CO_2 into CO (Snoeckx & Bogaerts, 2017). Yet, the level of CO_2 transformation remains insufficient (~20%), and the economical separation of both CO and O_2 could potentially present a significant obstacle in the direct splitting of CO_2. There is considerable potential to reduce the energy cost of methanol production by improving the plasma-catalytic process, specifically by increasing the methanol selectivity to 70-80% and the CO_2 conversion to 50-60%. This would be an improvement over the current state-of-the-art method of plasma-catalytic CO_2 hydrogenation to methanol, which has an energy cost of around 100 kWh kg^{-1} (Wang, Yi, Guo et al, 2018). The plasma catalytic hydrogenation method would become more appealing when compared to thermal catalytic CO_2 hydrogenation due to the considerably lowered expenses, resulting in the avoidance of high-pressure apparatus. It should be noted that the power-to-X process incurs various expenses beyond just the energy or electricity consumption associated with plasma-catalytic CO_2 hydrogenation.

The lack of knowledge about plasma-catalytic activation of CO_2 and the surface reactions involved is a major challenge. Understanding how catalyst surfaces change when exposed to plasma is essential. Impurities in CO_2 streams, including captured CO_2, have been studied for their impact on CO_2 decomposition, but little research has focused on how various impurities affect catalyst durability and product selectivity in different CO_2 conversion pathways using plasma-catalysis. Scaling up from a small-scale laboratory setup to an industrial operation also presents unique difficulties. Most current studies on plasma-catalytic CO_2 activation use low flow rates of CO_2 with DBDs setups (Mei, Zhu, Wu et al, 2016; Van Laer & Bogaerts, 2015; Wang, Yi, Guo et al, 2018).

4.2. Catalyst Selection

Plasma catalytic activation of CO_2 is a promising technique for fuel and chemical production, but it requires cost-effective and efficient catalysts. Compared to thermal catalysis, plasma catalysis offers a more attractive alternative by eliminating the need for high temperatures and pressures. To optimize the process, research focuses on enhancing plasma-induced surface reactions by integrating high-throughput catalyst screening methods. By manipulating the coupling of plasma and catalyst, researchers aim to identify the best interactions to activate CO_2 efficiently. Additionally, advanced and customized reactor designs are being developed to improve the overall efficiency. For example, using a water-cooled DBD reactor with a water electrode proved to be more efficient for producing methanol from CO_2 than a conventional DBD reactor without water cooling (Wang, Yi, Guo et al, 2018). Additionally, a metallic foam electrode in a DBD reactor has been proven to significantly increase energy efficiency and CO_2 conversion during the plasma splitting of CO_2 (Shengjie, 2019). Enhancing the energy efficiency of CO_2 plasma catalytic activation could be accomplished by globally optimizing the process parameters, power supply, catalyst design, and reactor design.

Gaining a new understanding of the reaction mechanism and pathways in plasma-catalytic CO_2 activation is crucial for improving catalyst design and enhancing the performance of CO_2 conversion. This can be achieved through the use of in situ surface probing techniques like in situ infrared spectroscopy, along with the combination of in situ and ex-situ catalyst characterization and plasma catalysis modeling. These methods allow for investigating processes such as plasma-assisted adsorption and desorption on the catalyst surface, providing valuable insights at atomic to macroscopic scales (Bogaerts et al., 2020).

Modular plasma-catalytic systems have potential for large-scale CO_2 conversion. It is crucial to examine various aspects, including CO_2 sources, capture methods, plant sizes, hydrogen sources, product applications, conversion routes, and transportation/storage infrastructure. Social and environmental impacts need consideration for the development of plasma-catalytic CO_2 conversion at a larger scale to achieve negative or zero carbon emissions.

4.3. Scaling Up for Industrial Applications: Challenges and Solutions

Implementing energy efficiencies in practical applications has been challenging, but studying scaling criteria such as conversion rates and overall energy efficiency is crucial. Limited research has shown progress in using plasma to convert CO_2. An innovative concept being tested involves combining electrolysis of water and oxidative DRM within a GAP, which operates at an impressive overall energy efficiency level of around 80% (Li et al., 2018). This has the potential to act as an innovative method for generating syngas for the synthesis of methanol. To evaluate the technological feasibility of converting

CO_2 using plasma, it is necessary to incorporate pilots that combine a plasma reactor with separation processes. Separating products is frequently essential due to restrictions imposed by reductions in efficiency, thermodynamics, and concerns about safety, such as the potential formation of an explosive combination during the CO_2 splitting reaction. Currently, there is ongoing research into plasma membrane reactors as a possible alternative for improving separation and mass transfer (Hessel et al., 2013; Mori et al., 2016). An alternative approach for eliminating products involves utilizing oxygen-absorbing substances, including metals with elemental properties. To mitigate the risks of explosions, inert gases like He and argon can be introduced. However, this approach entails more costly processes for recovery and separation, as well as increased energy input (Bogaerts & Snoeckx, 2019).

In certain instances, industrial processes driven by plasma-based conversion have reached maturity. A noteworthy illustration of this is the production of ozone (Kogelschatz, 2003). The initial exploration of ozone generation through electrical sparks and DBD reactors was conducted by Martinus van Marum and Werner von Siemens between 1785 and 1857, respectively (Hackmann, 1972; Siemens, 1857). In the present day, it is feasible to achieve production rates of several hundred kilograms per hour in a single setup, with conversion rates ranging from 5% to 15% in a single pass (Kogelschatz, 2003). By implementing a filtration process for separation, it is possible to achieve a nearly complete conversion of oxygen, reaching close to 100% (Hong et al., 2018). In a conventional ozone generator used in industries, the reactor dimensions usually range from 1 to 3 meters long and have diameters between 20 and 50 millimeters (Kogelschatz, 2003). When it comes to the inner electrode, it is generally characterized by a width of 0.5 to 1 millimeter (Kogelschatz, 2003). Therefore, commercially utilized DBD reactors bear a resemblance to the ones employed in laboratory settings, and their expansion primarily relies on parallelization. It is essential to mention that ozone generation occurs without a catalyst (Hong et al., 2018), whereas CO_2 conversions driven by plasma may necessitate the presence of a catalyst within the discharge gap.

5. ENVIRONMENTAL AND ECONOMIC IMPLICATIONS

5.1. Environmental Benefits of Plasma-Based CO_2 Conversion

Igniting plasma, specifically NTPs, is a challenging task in plasma technologies. NTPs offer various advantages such as high energy density, high electron temperature, and low background temperature. These advantages, along with the absence of vacuum equipment, make NTPs a potential solution for environmentally friendly and industrial energy applications. Chemical applications in the early 1900s saw the initial use of the gliding arc discharge to produce fertilizer with a nitrogen base (Naville et al., 1905). In industrial and laboratory chemical processes, Czernichowski showcased several practical benefits of the gliding arc in 1994 (Czernichowski, 1994). Our laboratory has been working on the development of gliding arc reactors patented in China since 2003, to utilize them for a more sustainable and environmentally friendly energy and engineering application (Du, 2007; Du et al., 2007; Yan, 2004).

Gliding arc discharges occur between two electrodes in a fast-moving gas stream, creating a unique non-thermal form of plasma. When the voltage reaches a certain level, a breakdown occurs in the gas flow, creating an arc that is propelled downstream until it extinguishes. After the arc fades, a new breakdown occurs, repeating the cycle. These discharges transition between arc and release phases (Czernichowski, 1994). The gliding arc discharges possess several notable benefits (Czernichowski, 1994):

(i) Electrodes of gliding arc discharges are self-maintained, which helps to avoid chemical erosion and corrosion.
(ii) The electrodes in gliding arc discharges are not susceptible to chemical corrosion or erosion due to their self-maintenance. Additionally, there is no need to cool the electrodes as the electric energy is efficiently transferred to the gas being processed.
(iii) It is feasible to effortlessly set up multi-electrodc systcms within extensive gas pipelines.
(iv) It is possible to handle various gases and vapors, even if they are contaminated or contain moisture, mists, or powders. The gas can be at any temperature or pressure.

Gliding arc discharge, initially designed in a flat shape, has evolved into a cylindrical form for different uses. This has led to the creation of various gliding arc reactors, such as rotating gliding arc and gas-liquid gliding arc discharge, with novel configurations based on their two-dimensional or three-dimensional geometry (Du et al., 2008; Lu et al., 2012; Ren et al., 2013). It is possible to create multiple electrodes (consisting of three phases, n phases, or a combination of parallel, serial, or mixed mounting) for industrial gliding arc reactors using AC or DC power (Czernichowski, 1994).

GAP reactors can handle various gases, including but not limited to CO_2, nitrogen, oxygen, air, gas-liquid mixtures, water vapor, argon, hydrocarbons, hydrogen sulfide, CO, and their combinations, without causing any significant decrease in pressure. Numerous reactors have been created for various purposes, such as environmental control and energy engineering, and have been successfully implemented in laboratory and industrial settings (Czernichowski, 1994; Du, 2007; Du, 2007; Du et al., 2012; Du et al., 2008; Du et al., 2007; Yan, 2004; Yan, 2005).

- Removing contamination from soil or industrial sands.
- Generating UV light;
- Controlling the release of volatile organic compounds (such as tetrachloroethylene, xylene, toluene, and heptane), organic nitrates, phenols, NH_3, formaldehyde, diluted H2S or mercaptans, etc. from industrial sources;
- Igniting propellants;
- Raising the temperature of oxygen, steam, and other flames or gases beyond their normal levels;
- Dissociating CO_2 molecules;
- Decomposition and burning concentrated Freons;
- Transforming heavy residues from petroleum into usable products;
- Converting natural gas into syngas, which consists of H_2 and CO;
- Incinerating concentrated mixtures of H_2S or $H_2S + CO_2$, either completely or partially;
- Managing emissions of soot, polyaromatic hydrocarbons, sulfur oxides (SO_x), and nitrogen oxides (NO_x).

5.2. Evaluating Economic Feasibility and Potential Market Opportunities

Limited research has been conducted on the energy usage and economic implications of NTP technologies, especially on a large scale. However, some studies suggest that as plasma technology grows, it requires a substantial amount of energy input. Nonetheless, this energy can be sourced from renewable sources like hydro, solar, and wind power, which would benefit both the environment and the economy. Even though the existing literature has a scarcity of economic analysis research, Rooji et al. (2017) conducted a practical process design of MW plasma to convert CO_2 into pure CO. Researchers conducted this study to determine the economic feasibility of expanding a facility that produces pure CO. They assembled 264 reactors with a capacity of 0.5 MW each to meet the necessary power demand. The equipment costs for the facility amounted to \$30.2 million, but advancements in technology are expected to reduce costs in the future. With this reduction, the total expense would be approximately \$9 million, nearly 50% lower than a conventional CO plant. The study also considered the electricity cost, which makes up nearly 50% of the overall operational expenses.

In their study, Pou et al. (2018) examined the effectiveness of a plasma reactor that utilizes fluidized particles in converting CO_2 into CO. They considered two different situations: one where the reactor is powered by electricity sourced from a combination of renewable and fossil fuel energy on the grid and another where it relies solely on photovoltaic-generated electricity. The objective of the research was to exhibit the financial advantage of utilizing renewable energy sources to supply plasma. The flow rate of CO_2 was set at 10 mL per minute, while the conversion of CO_2 amounted to 41%. In scenario one, where a combination of electricity was employed to power the plasma reactor, the mitigation of CO_2 emissions amounted to 3%. This implies that only a mere 3% of the CO_2 emissions can be offset through this process, while the energy consumption was recorded at 0.11 kWh for every gram of CO_2 transformed into CO. The CO_2-eq/kWh emissions linked to electricity usage amounted to 32 grams. However, in the alternate scenario where photovoltaic power was utilized to fuel the plasma reactor, there was a notable 67% reduction in the offsetting of CO_2 emissions. The sustainability of the process is magnified when renewable energy sources are combined with NTP technology, leading to a substantial improvement compared to the initial situation. The amount of CO_2 equivalent emissions resulting from electricity use was 22 grams per hour.

The research conducted by Butterworth et al. (2016) proved to be highly significant in examining energy consumption. The power usage of a DBD plasma reactor in a small-scale laboratory setting was observed while altering the materials used in the packing bed. In their experiment, it was found that the plasma power remained unchanged while using a pure CO_2 feed gas flow rate of 100 mL/min. However, the only variable affected by the packing bed was the conversion of CO_2. In one experiment, the utilization of $BaTiO_3$ as a spherical substance, and in another experiment, the use of Al_2O_3 resulted in a wide range of diameters, varying from 180 to 2000 μm. Interestingly, the plasma power remained unaffected by the size and the type of material used in the packed bed. However, it was observed that the efficiency of CO_2 conversion was heavily reliant on these particular parameters.

- When spherical $BaTiO_3$ particles with diameters ranging from 1400 to 2000 μm are utilized, a plasma power of approximately 13 MW leads to a maximum CO2 conversion rate of 2%.
- With the utilization of spherical Al_2O_3 (with diameters ranging from 1400 to 2000 μm), the highest achieved CO_2 conversion is 1.4%, which aligns with a plasma power of approximately 13.
- By employing perfectly spherical BaTiO3 particles with sizes ranging from 850 to 1400 μm, a plasma power of approximately 13 MW was found to yield a CO_2 conversion rate of up to 1.2%.
- Utilizing Al_2O_3 spheres (with diameters ranging from 850 to 1400 μm), a plasma power of approximately 13 MW achieves a CO_2 conversion rate of 0.5% at maximum.

In their study, Biset-Peirò et al. (2020) examined the relationship between plasma power and temperature within a DBD plasma reactor. The research compared two setups, pseudo-adiabatic and adiabatic, in a plasma reactor. The temperature and CO_2 conversion increased as plasma power rose. The pseudo-adiabatic setup achieved a 75% CO_2 conversion at 20W, while the adiabatic setup reached the same conversion with lower power. The experiments showed that using a non-adiabatic configuration in the plasma reactor improves energy efficiency by 20% and reduces costs compared to quasi-adiabatic setups. Adiabatic conditions are less energy-intensive due to the interaction between plasma and thermal activation.

CONCLUSION

Plasma-assisted carbon dioxide conversion exhibits great promise in addressing the challenges of CO_2 emissions and environmental degradation. Throughout this chapter, we have explored the various applications, challenges, and environmental impacts associated with this technology. Key parameters affecting efficiency have been identified, while the potential for plasma-catalytic CO_2 conversion, dry reforming of methane, and hydrogenation have been highlighted. We have also discussed the challenges of cost, catalyst selection, and scalability, offering potential solutions for overcoming these limitations. Moreover, we have emphasized the environmental benefits of plasma-based CO_2 conversion, illuminating its role in mitigating greenhouse gas emissions. Lastly, we have evaluated the economic feasibility and market opportunities, indicating a promising avenue for future development. As efforts continue to combat climate change and transition towards sustainable energy systems, plasma-assisted CO_2 conversion presents itself as a valuable tool for creating a greener and more sustainable planet.

REFERENCES

Abbas, H. F., & Daud, W. W. (2010). Hydrogen production by methane decomposition: A review. *International Journal of Hydrogen Energy*, *35*(3), 1160–1190. doi:10.1016/j.ijhydene.2009.11.036

Abiev, R. S., Sladkovskiy, D. A., Semikin, K. V., Murzin, D. Y., & Rebrov, E. V. (2020). Non-thermal plasma for process and energy intensification in dry reforming of methane. *Catalysts*, *10*(11), 1358. doi:10.3390/catal10111358

Aerts, R., Somers, W., & Bogaerts, A. (2015). Carbon dioxide splitting in a dielectric barrier discharge plasma: A combined experimental and computational study. *ChemSusChem*, *8*(4), 702–716. doi:10.1002/cssc.201402818 PMID:25641832

Al-Mamoori, A., Krishnamurthy, A., Rownaghi, A. A., & Rezaei, F. (2017). Carbon capture and utilization update. *Energy Technology (Weinheim)*, *5*(6), 834–849. doi:10.1002/ente.201600747

Anoop, N., Sundaramurthy, S., Jha, J. M., Chandrabalan, S., Singh, N., Verma, J., Parvatalu, D., & Katti, S. (2021). Plasma catalysis: A feasible solution for carbon dioxide valorization? *Clean Technologies and Environmental Policy*, *23*(10), 2789–2811. doi:10.1007/s10098-021-02203-y

Aramouni, N. A. K., Touma, J. G., Tarboush, B. A., Zeaiter, J., & Ahmad, M. N. (2018). Catalyst design for dry reforming of methane: Analysis review. *Renewable & Sustainable Energy Reviews*, *82*, 2570–2585. doi:10.1016/j.rser.2017.09.076

Arita, K., & Iizuka, S. (2015). Production of CH 4 in a low-pressure CO 2/H 2 discharge with magnetic field. *Journal of Materials Science and Chemical Engineering*, *3*(12), 69. doi:10.4236/msce.2015.312011

Ashford, B. (2022). Plasma-catalytic CO2 hydrogenation to ethane in a dielectric barrier discharge reactor. *Journal of CO2 Utilization, 57*, 101882.

Ashford, B., & Tu, X. (2017). Non-thermal plasma technology for the conversion of CO2. *Current Opinion in Green and Sustainable Chemistry*, *3*, 45–49. doi:10.1016/j.cogsc.2016.12.001

Asisov, R. (1981). *Carbon dioxide dissociation in non-equilibrium plasma. 5th International Symposium on Plasma Chemistry*, Edinburgh.

Azzolina-Jury, F. (2017). Chemical engineering aspects of plasma-assisted CO2 hydrogenation over nickel zeolites under partial vacuum. *Journal of CO2 Utilization, 22*, 97-109.

Bahri, M., & Haghighat, F. (2014). Plasma-B ased Indoor Air Cleaning Technologies: The State of the Art-R eview. *Clean (Weinheim)*, *42*(12), 1667–1680. doi:10.1002/clen.201300296

Benrabbah, R., Cavaniol, C., Liu, H., Ognier, S., Cavadias, S., Gálvez, M. E., & Da Costa, P. (2017). Plasma DBD activated ceria-zirconia-promoted Ni-catalysts for plasma catalytic CO2 hydrogenation at low temperature. *Catalysis Communications*, *89*, 73–76. doi:10.1016/j.catcom.2016.10.028

Biset-Peiró, M., Mey, R., Guilera, J., & Andreu, T. (2020). Adiabatic plasma-catalytic reactor configuration: Energy efficiency enhancement by plasma and thermal synergies on CO2 methanation. *Chemical Engineering Journal*, *393*, 124786. doi:10.1016/j.cej.2020.124786

Bogaerts, A. & Snoeckx, R. (2019). *Plasma-Based CO 2 Conversion.* An Economy Based on Carbon Dioxide and Water: Potential of Large Scale Carbon Dioxide Utilization, 287-325.

Bogaerts, A., Tu, X., Whitehead, J. C., Centi, G., Lefferts, L., Guaitella, O., Azzolina-Jury, F., Kim, H.-H., Murphy, A. B., Schneider, W. F., Nozaki, T., Hicks, J. C., Rousseau, A., Thevenet, F., Khacef, A., & Carreon, M. (2020). The 2020 plasma catalysis roadmap. *Journal of Physics. D, Applied Physics*, *53*(44), 443001. doi:10.1088/1361-6463/ab9048

Bongers, W., Bouwmeester, H., Wolf, B., Peeters, F., Welzel, S., van den Bekerom, D., den Harder, N., Goede, A., Graswinckel, M., Groen, P. W., Kopecki, J., Leins, M., van Rooij, G., Schulz, A., Walker, M., & van de Sanden, R. (2017). Plasma-driven dissociation of CO2 for fuel synthesis. *Plasma Processes and Polymers*, *14*(6), 1600126. doi:10.1002/ppap.201600126

Brock, S. L., Shimojo, T., Marquez, M., Marun, C., Suib, S. L., Matsumoto, H., & Hayashi, Y. (1999). Factors Influencing the Decomposition of CO2in AC Fan-Type Plasma Reactors: Frequency, Waveform, and Concentration Effects. *Journal of Catalysis*, *184*(1), 123–133. doi:10.1006/jcat.1999.2440

Butterworth, T., Elder, R., & Allen, R. (2016). Effects of particle size on CO2 reduction and discharge characteristics in a packed bed plasma reactor. *Chemical Engineering Journal*, *293*, 55–67. doi:10.1016/j.cej.2016.02.047

Chen, G. (2015). Simultaneous dissociation of CO2 and H2O to syngas in a surface-wave microwave discharge. *International Journal Of Hydrogen Energy, 40*(9), 3789-3796.

Chen, G., Britun, N., Godfroid, T., Georgieva, V., Snyders, R., & Delplancke-Ogletree, M.-P. (2017). An overview of CO2 conversion in a microwave discharge: The role of plasma-catalysis. *Journal of Physics. D, Applied Physics*, *50*(8), 084001. doi:10.1088/1361-6463/aa5616

Chen, G., Godfroid, T., Britun, N., Georgieva, V., Delplancke-Ogletree, M.-P., & Snyders, R. (2017). Plasma-catalytic conversion of CO2 and CO2/H2O in a surface-wave sustained microwave discharge. *Applied Catalysis B: Environmental*, *214*, 114–125. doi:10.1016/j.apcatb.2017.05.032

Chen, G., Snyders, R., & Britun, N. (2021). CO2 conversion using catalyst-free and catalyst-assisted plasma-processes: Recent progress and understanding. *Journal of CO2 Utilization, 49.*

Chung, W.-C., Pan, K.-L., Lee, H.-M., & Chang, M.-B. (2014). Dry reforming of methane with dielectric barrier discharge and ferroelectric packed-bed reactors. *Energy & Fuels*, *28*(12), 7621–7631. doi:10.1021/ef5020555

Crim, F. F. (2008). Chemical dynamics of vibrationally excited molecules: Controlling reactions in gases and on surfaces. *Proceedings of the National Academy of Sciences of the United States of America*, *105*(35), 12654–12661. doi:10.1073/pnas.0803010105 PMID:18765816

Czernichowski, A. (1994). Gliding arc: Applications to engineering and environment control. *Pure and Applied Chemistry*, *66*(6), 1301–1310. doi:10.1351/pac199466061301

De Bie, C., van Dijk, J., & Bogaerts, A. (2016). CO2 hydrogenation in a dielectric barrier discharge plasma revealed. *The Journal of Physical Chemistry. C, Nanomaterials and Interfaces*, *120*(44), 25210–25224. doi:10.1021/acs.jpcc.6b07639

de la Fuente, J. F., Moreno, S. H., Stankiewicz, A. I., & Stefanidis, G. D. (2016). Reduction of CO2 with hydrogen in a non-equilibrium microwave plasma reactor. *International Journal of Hydrogen Energy*, *41*(46), 21067–21077. doi:10.1016/j.ijhydene.2016.08.032

de la Fuente, J. F., Moreno, S. H., Stankiewicz, A. I., & Stefanidis, G. D. (2017). On the improvement of chemical conversion in a surface-wave microwave plasma reactor for CO2 reduction with hydrogen (The Reverse Water-Gas Shift reaction). *International Journal of Hydrogen Energy*, *42*(18), 12943–12955. doi:10.1016/j.ijhydene.2017.04.040

Delikonstantis, E., Cameli, F., Scapinello, M., Rosa, V., Van Geem, K. M., & Stefanidis, G. D. (2022). Low-carbon footprint chemical manufacturing using plasma technology. *Current Opinion in Chemical Engineering*, *38*, 100857. doi:10.1016/j.coche.2022.100857

Du, C. (2007a). *A rotary gliding arc discharge plasma device for volatile organic compounds gas treatment, China.* Patent CN201131940Y.

Du, C. (2007b). *A three phase gliding arc non equilibrium plasma device for water treatment, China.* Patent CN201099636Y.

Du, C., Liu, H., Xiao, M. D., Gao, D., Huang, D. W., Li, Z. Y., Chen, T. F., Mo, J. M., Wang, K., & Zhang, C. R. (2012). Adsorption of iron and lead ions from an aqueous solution by plasma-modified activated carbon. *Industrial & Engineering Chemistry Research*, *51*(48), 15618–15625. doi:10.1021/ie302011n

Du, C., Shi, T. H., Sun, Y. W., & Zhuang, X. F. (2008). Decolorization of Acid Orange 7 solution by gas–liquid gliding arc discharge plasma. *Journal of Hazardous Materials*, *154*(1-3), 1192–1197. doi:10.1016/j.jhazmat.2007.11.032 PMID:18162301

Du, C. M., Yan, J. H., & Cheron, B. (2007). Decomposition of toluene in a gliding arc discharge plasma reactor. *Plasma Sources Science & Technology*, *16*(4), 791–797. doi:10.1088/0963-0252/16/4/014

Duan, X., Li, Y., Ge, W., & Wang, B. (2015). Degradation of CO2 through dielectric barrier discharge microplasma. *Greenhouse Gases: Science and Technology*, *5*(2), 131–140. doi:10.1002/ghg.1425

Fan, Z., Sun, K., Rui, N., Zhao, B., & Liu, C. (2015). Improved activity of Ni/MgAl2O4 for CO2 methanation by the plasma decomposition. *Journal of Energy Chemistry*, *24*(5), 655–659. doi:10.1016/j.jechem.2015.09.004

Farahani, M. D., Zeng, Y., & Zheng, Y. (2022). The application of nonthermal plasma in methanol synthesis via CO2 hydrogenation. *Energy Science & Engineering*, *10*(5), 1572–1583. doi:10.1002/ese3.1107

Fridman, A. (2008). *Plasma chemistry*. Cambridge university press. doi:10.1017/CBO9780511546075

Gallon, H. J., Tu, X., & Whitehead, J. C. (2012). Effects of reactor packing materials on H2 production by CO2 reforming of CH4 in a dielectric barrier discharge. *Plasma Processes and Polymers*, *9*(1), 90–97. doi:10.1002/ppap.201100130

George, A., Shen, B., Craven, M., Wang, Y., Kang, D., Wu, C., & Tu, X. (2021). A Review of Non-Thermal Plasma Technology: A novel solution for CO2 conversion and utilization. *Renewable & Sustainable Energy Reviews*, *135*, 109702. doi:10.1016/j.rser.2020.109702

Gholami, R., Stere, C. E., Goguet, A., & Hardacre, C. (2018). Non-thermal-plasma-activated de-NOx catalysis. *Philosophical Transactions. Series A, Mathematical, Physical, and Engineering Sciences*, *376*(2110), 20170054. doi:10.1098/rsta.2017.0054 PMID:29175870

Hackmann, W. D. (1972). The researches of Dr. Martinus Van Marum (1750–1837) on the influence of electricity on animals and plants. *Medical History*, *16*(1), 11–26. doi:10.1017/S0025727300017221 PMID:4558435

Hepburn, C., Adlen, E., Beddington, J., Carter, E. A., Fuss, S., Mac Dowell, N., Minx, J. C., Smith, P., & Williams, C. K. (2019). The technological and economic prospects for CO2 utilization and removal. *Nature*, *575*(7781), 87–97. doi:10.1038/s41586-019-1681-6 PMID:31695213

Hessel, V., Anastasopoulou, A., Wang, Q., Kolb, G., & Lang, J. (2013). Energy, catalyst and reactor considerations for (near)-industrial plasma processing and learning for nitrogen-fixation reactions. *Catalysis Today*, *211*, 9–28. doi:10.1016/j.cattod.2013.04.005

Hoeben, W., van Heesch, E. J. M., Beckers, F. J. C. M., Boekhoven, W., & Pemen, A. J. M. (2015). Plasma-driven water assisted CO 2 methanation. *IEEE Transactions on Plasma Science*, *43*(6), 1954–1958. doi:10.1109/TPS.2015.2429316

Hong, J., Prawer, S., & Murphy, A. B. (2018). Plasma catalysis as an alternative route for ammonia production: Status, mechanisms, and prospects for progress. *ACS Sustainable Chemistry & Engineering*, *6*(1), 15–31. doi:10.1021/acssuschemeng.7b02381

Huang, Q., Zhang, D., Wang, D., Liu, K., & Kleyn, A. W. (2017). Carbon dioxide dissociation in non-thermal radiofrequency and microwave plasma. *Journal of Physics. D, Applied Physics*, *50*(29), 294001. doi:10.1088/1361-6463/aa754e

Ihara, T., Kiboku, M., & Iriyama, Y. (1994). Plasma reduction of CO2 with H2O for the formation of organic compounds. *Bulletin of the Chemical Society of Japan*, *67*(1), 312–314. doi:10.1246/bcsj.67.312

Indarto, A., Choi, J.-W., Lee, H., & Song, H. K. (2006). Conversion of CO2 by gliding arc plasma. *Environmental Engineering Science*, *23*(6), 1033–1043. doi:10.1089/ees.2006.23.1033 PMID:17223261

Indarto, A., Yang, D. R., Choi, J.-W., Lee, H., & Song, H. K. (2007). Gliding arc plasma processing of CO2 conversion. *Journal of Hazardous Materials*, *146*(1-2), 309–315. doi:10.1016/j.jhazmat.2006.12.023 PMID:17223261

Jardali, F., Van Alphen, S., Creel, J., Ahmadi Eshtehardi, H., Axelsson, M., Ingels, R., Snyders, R., & Bogaerts, A. (2021). NO x production in a rotating gliding arc plasma: Potential avenue for sustainable nitrogen fixation. *Green Chemistry*, *23*(4), 1748–1757. doi:10.1039/D0GC03521A

Jiang, T., Li, Y., Liu, C., Xu, G., Eliasson, B., & Xue, B. (2002). Plasma methane conversion using dielectric-barrier discharges with zeolite A. *Catalysis Today*, *72*(3-4), 229–235. doi:10.1016/S0920-5861(01)00497-7

Jwa, E., Lee, S. B., Lee, H. W., & Mok, Y. S. (2013). Plasma-assisted catalytic methanation of CO and CO2 over Ni–zeolite catalysts. *Fuel Processing Technology*, *108*, 89–93. doi:10.1016/j.fuproc.2012.03.008

Kogelschatz, U. (2003). Dielectric-barrier discharges: Their history, discharge physics, and industrial applications. *Plasma Chemistry and Plasma Processing*, *23*(1), 1–46. doi:10.1023/A:1022470901385

Leiweke, R. J., & Ganguly, B. N. (2007). Effects of pulsed-excitation applied voltage rise time on argon metastable production efficiency in a high pressure dielectric barrier discharge. *Applied Physics Letters*, *90*(24), 241501. doi:10.1063/1.2748324

Li, K., Liu, J.-L., Li, X.-S., Lian, H.-Y., Zhu, X., Bogaerts, A., & Zhu, A.-M. (2018). Novel power-to-syngas concept for plasma catalytic reforming coupled with water electrolysis. *Chemical Engineering Journal*, *353*, 297–304. doi:10.1016/j.cej.2018.07.111

Li, L. (2019). Plasma-assisted CO2 conversion in a gliding arc discharge: Improving performance by optimizing the reactor design. *Journal of CO2 Utilization, 29*, 296-303.

Liu, J.-L., Wang, X., Li, X.-S., Likozar, B., & Zhu, A.-M. (2020). CO2 conversion, utilisation and valorisation in gliding arc plasma reactors. *Journal of Physics. D, Applied Physics*, *53*(25), 253001. doi:10.1088/1361-6463/ab7c04

Lu, N., Liu, N., Zhang, C., Su, Y., Shang, K., Jiang, N., Li, J., & Wu, Y. (2021). CO2 conversion promoted by potassium intercalated g-C3N4 catalyst in DBD plasma system. *Chemical Engineering Journal*, *417*, 129283. doi:10.1016/j.cej.2021.129283

Lu, N., Sun, D., Zhang, C., Jiang, N., Shang, K., Bao, X., Li, J., & Wu, Y. (2018). CO2 conversion in non-thermal plasma and plasma/g-C3N4 catalyst hybrid processes. *Journal of Physics. D, Applied Physics*, *51*(9), 094001. doi:10.1088/1361-6463/aaa919

Lu, S. Y., Sun, X. M., Li, X. D., Yan, J. H., & Du, C. M. (2012). Decomposition of toluene in a rotating glidarc discharge reactor. *IEEE Transactions on Plasma Science*, *40*(9), 2151–2156. doi:10.1109/TPS.2012.2206119

Mahammadunnisa, S., Reddy, E. L., Ray, D., Subrahmanyam, C., & Whitehead, J. C. (2013). CO2 reduction to syngas and carbon nanofibres by plasma-assisted in situ decomposition of water. *International Journal of Greenhouse Gas Control*, *16*, 361–363. doi:10.1016/j.ijggc.2013.04.008

Mehta, P., Barboun, P., Go, D. B., Hicks, J. C., & Schneider, W. F. (2019). Catalysis enabled by plasma activation of strong chemical bonds: A review. *ACS Energy Letters*, *4*(5), 1115–1133. doi:10.1021/acsenergylett.9b00263

Mei, D., He, Y.-L., Liu, S., Yan, J., & Tu, X. (2016). Optimization of CO2 conversion in a cylindrical dielectric barrier discharge reactor using design of experiments. *Plasma Processes and Polymers*, *13*(5), 544–556. doi:10.1002/ppap.201500159

Mei, D., Liu, S., & Tu, X. (2017). CO2 reforming with methane for syngas production using a dielectric barrier discharge plasma coupled with Ni/γ-Al2O3 catalysts: Process optimization through response surface methodology. *Journal of CO2 Utilization*, *21*, 314-326.

Mei, D., Zhu, X., Wu, C., Ashford, B., Williams, P. T., & Tu, X. (2016). Plasma-photocatalytic conversion of CO2 at low temperatures: Understanding the synergistic effect of plasma-catalysis. *Applied Catalysis B: Environmental*, *182*, 525–532. doi:10.1016/j.apcatb.2015.09.052

Mertens, J., Watson, S., Nisol, B., Wertheimer, M. R., & Reniers, F. (2019). Energetics of reactions in a dielectric barrier discharge with argon carrier gas: VII anhydrides. *Plasma Processes and Polymers*, *16*(5), 1800186. doi:10.1002/ppap.201800186

Michiels, R., Engelmann, Y., & Bogaerts, A. (2020). Plasma catalysis for CO2 hydrogenation: Unlocking new pathways toward CH3OH. *The Journal of Physical Chemistry. C, Nanomaterials and Interfaces*, *124*(47), 25859–25872. doi:10.1021/acs.jpcc.0c07632

Michielsen, I., Uytdenhouwen, Y., Pype, J., Michielsen, B., Mertens, J., Reniers, F., Meynen, V., & Bogaerts, A. (2017). CO2 dissociation in a packed bed DBD reactor: First steps towards a better understanding of plasma catalysis. *Chemical Engineering Journal*, *326*, 477–488. doi:10.1016/j.cej.2017.05.177

Mikhail, M., Costa, P. D., Amouroux, J., Cavadias, S., Tatoulian, M., Ognier, S., & Gálvez, M. E. (2021). Effect of Na and K impurities on the performance of Ni/CeZrOx catalysts in DBD plasma-catalytic CO2 methanation. *Fuel*, *306*, 121639. doi:10.1016/j.fuel.2021.121639

Mikoviny, T., Kocan, M., Matejcik, S., Mason, N. J., & Skalny, J. D. (2003). Experimental study of negative corona discharge in pure carbon dioxide and its mixtures with oxygen. *Journal of Physics. D, Applied Physics*, *37*(1), 64–73. doi:10.1088/0022-3727/37/1/011

Mikulčić, H., Ridjan Skov, I., Dominković, D. F., Wan Alwi, S. R., Manan, Z. A., Tan, R., Duić, N., Hidayah Mohamad, S. N., & Wang, X. (2019). Flexible Carbon Capture and Utilization technologies in future energy systems and the utilization pathways of captured CO2. *Renewable & Sustainable Energy Reviews*, *114*, 109338. doi:10.1016/j.rser.2019.109338

Mitsingas, C. M., Rajasegar, R., Hammack, S., Do, H., & Lee, T. (2016). High energy efficiency plasma conversion of CO 2 at atmospheric pressure using a direct-coupled microwave plasma system. *IEEE Transactions on Plasma Science*, *44*(4), 651–656. doi:10.1109/TPS.2016.2531641

Mora, E., Sarmiento, A., & Vera, E. (2016). Alumina and quartz as dielectrics in a dielectric barrier discharges DBD system for CO2 hydrogenation. Journal of Physics: Conference Series. IOP Publishing.

Mori, S., Matsuura, N., Tun, L. L., & Suzuki, M. (2016). Direct synthesis of carbon nanotubes from only CO 2 by a hybrid reactor of dielectric barrier discharge and solid oxide electrolyser cell. *Plasma Chemistry and Plasma Processing*, *36*(1), 231–239. doi:10.1007/s11090-015-9681-2

Moss, M., Yanallah, K., Allen, R. W. K., & Pontiga, F. (2017). An investigation of CO2 splitting using nanosecond pulsed corona discharge: Effect of argon addition on CO2 conversion and energy efficiency. *Plasma Sources Science & Technology*, *26*(3), 035009. doi:10.1088/1361-6595/aa5b1d

Nagaraja, B. M., Bulushev, D. A., Beloshapkin, S., & Ross, J. R. H. (2011). The effect of potassium on the activity and stability of Ni–MgO–ZrO2 catalysts for the dry reforming of methane to give synthesis gas. *Catalysis Today*, *178*(1), 132–136. doi:10.1016/j.cattod.2011.08.040

. Nagassou, D. (2020). Decomposition of CO2 in a solar-gliding arc plasma reactor: Effects of water, nitrogen, methane, and process optimization. *Journal of CO2 Utilization, 38*, 39-48.

Naville, A., Guye, C., & Guye, P. (1905). *Gas Reactions at the Temperature of the Electric Arc*. French, Patent FR.

Nguyen, H. H., & Kim, K.-S. (2015). Combination of plasmas and catalytic reactions for CO2 reforming of CH4 by dielectric barrier discharge process. *Catalysis Today*, *256*, 88–95. doi:10.1016/j.cattod.2015.04.034

Nizio, M., Albarazi, A., Cavadias, S., Amouroux, J., Galvez, M. E., & Da Costa, P. (2016). Hybrid plasma-catalytic methanation of CO2 at low temperature over ceria zirconia supported Ni catalysts. *International Journal of Hydrogen Energy*, *41*(27), 11584–11592. doi:10.1016/j.ijhydene.2016.02.020

Nizio, M., Benrabbah, R., Krzak, M., Debek, R., Motak, M., Cavadias, S., Gálvez, M. E., & Da Costa, P. (2016). Low temperature hybrid plasma-catalytic methanation over Ni-Ce-Zr hydrotalcite-derived catalysts. *Catalysis Communications*, *83*, 14–17. doi:10.1016/j.catcom.2016.04.023

Nunnally, T., Gutsol, K., Rabinovich, A., Fridman, A., Gutsol, A., & Kemoun, A. (2011). Dissociation of CO2 in a low current gliding arc plasmatron. *Journal of Physics. D, Applied Physics*, *44*(27), 274009. doi:10.1088/0022-3727/44/27/274009

Ong, M., Nomanbhay, S., Kusumo, F., & Show, P. L. (2022). Application of microwave plasma technology to convert carbon dioxide (CO2) into high value products: A review. *Journal of Cleaner Production*, *336*, 130447. doi:10.1016/j.jclepro.2022.130447

Ozkan, A. (2016). Splitting in a dielectric barrier discharge plasma: understanding of physical and chemical aspects. University of Antwerp.

Paulussen, S., Verheyde, B., Tu, X., De Bie, C., Martens, T., Petrovic, D., Bogaerts, A., & Sels, B. (2010). Conversion of carbon dioxide to value-added chemicals in atmospheric pressure dielectric barrier discharges. *Plasma Sources Science & Technology, 19*(3), 034015. doi:10.1088/0963-0252/19/3/034015

Pawelec, B., Damyanova, S., Arishtirova, K., Fierro, J. L. G., & Petrov, L. (2007). Structural and surface features of PtNi catalysts for reforming of methane with CO2. *Applied Catalysis A, General, 323*, 188–201. doi:10.1016/j.apcata.2007.02.017

Pinhão, N., Janeco, A., & Branco, J. (2011). Influence of helium on the conversion of methane and carbon dioxide in a dielectric barrier discharge. *Plasma Chemistry and Plasma Processing, 31*(3), 427–439. doi:10.1007/s11090-011-9294-3

Pou, J., Colominas, C., & Gonzalez-Olmos, R. (2018). CO2 reduction using non-thermal plasma generated with photovoltaic energy in a fluidized reactor. *Journal of CO2 Utilization, 27*, 528-535.

Ramakers, M., Michielsen, I., Aerts, R., Meynen, V., & Bogaerts, A. (2015). Effect of argon or helium on the CO2 conversion in a dielectric barrier discharge. *Plasma Processes and Polymers, 12*(8), 755–763. doi:10.1002/ppap.201400213

Ray, D., Saha, R., & Ch, S. (2017). DBD plasma assisted CO2 decomposition: Influence of diluent gases. *Catalysts, 7*(9), 244. doi:10.3390/catal7090244

Ren, Y., Li, X., Yu, L., Cheng, K., Yan, J., & Du, C. (2013). Degradation of PCDD/Fs in fly ash by vortex-shaped gliding arc plasma. *Plasma Chemistry and Plasma Processing, 33*(1), 293–305. doi:10.1007/s11090-012-9421-9

Rouwenhorst, K. H., Jardali, F., Bogaerts, A., & Lefferts, L. (2021). From the Birkeland–Eyde process towards energy-efficient plasma-based NO X synthesis: A techno-economic analysis. *Energy & Environmental Science, 14*(5), 2520–2534. doi:10.1039/D0EE03763J PMID:34046082

Sabri, M. A., Al Jitan, S., Bahamon, D., Vega, L. F., & Palmisano, G. (2021). Current and future perspectives on catalytic-based integrated carbon capture and utilization. *The Science of the Total Environment, 790*, 148081. doi:10.1016/j.scitotenv.2021.148081 PMID:34091328

Saleh, T. A. (2022). Nanomaterials and hybrid nanocomposites for CO 2 capture and utilization: Environmental and energy sustainability. *RSC Advances, 12*(37), 23869–23888. doi:10.1039/D2RA03242B PMID:36093256

Schiorlin, M., Klink, R., & Brandenburg, R. (2016). Carbon dioxide conversion by means of coplanar dielectric barrier discharges. *The European Physical Journal Applied Physics, 75*(2), 24704. doi:10.1051/epjap/2016160073

Shengjie, Z. (2019). Enhanced CO2 decomposition via metallic foamed electrode packed in self-cooling DBD plasma device. *Plasma Science & Technology, 21*(8), 085504. doi:10.1088/2058-6272/ab15e5

Siemens, W. (1857). Ueber die elektrostatische Induction und die Verzögerung des Stroms in Flaschendrähten. *Annalen der Physik, 178*(9), 66–122. doi:10.1002/andp.18571780905

Silva, T., Britun, N., Godfroid, T., & Snyders, R. (2014). Optical characterization of a microwave pulsed discharge used for dissociation of CO2. *Plasma Sources Science & Technology*, *23*(2), 025009. doi:10.1088/0963-0252/23/2/025009

Snoeckx, R., & Bogaerts, A. (2017). Plasma technology–a novel solution for CO 2 conversion? *Chemical Society Reviews*, *46*(19), 5805–5863. doi:10.1039/C6CS00066E PMID:28825736

Snoeckx, R., Heijkers, S., Van Wesenbeeck, K., Lenaerts, S., & Bogaerts, A. (2016). CO 2 conversion in a dielectric barrier discharge plasma: N 2 in the mix as a helping hand or problematic impurity? *Energy & Environmental Science*, *9*(3), 999–1011. doi:10.1039/C5EE03304G

Snoeckx, R., Zeng, Y. X., Tu, X., & Bogaerts, A. (2015). Plasma-based dry reforming: Improving the conversion and energy efficiency in a dielectric barrier discharge. *RSC Advances*, *5*(38), 29799–29808. doi:10.1039/C5RA01100K

Spencer, L., & Gallimore, A. (2012). CO2 dissociation in an atmospheric pressure plasma/catalyst system: A study of efficiency. *Plasma Sources Science & Technology*, *22*(1), 015019. doi:10.1088/0963-0252/22/1/015019

Spencer, L. F., & Gallimore, A. D. (2011). Efficiency of CO 2 dissociation in a radio-frequency discharge. *Plasma Chemistry and Plasma Processing*, *31*(1), 79–89. doi:10.1007/s11090-010-9273-0

Steinberg, M., & Dang, V.-D. (1977). Production of synthetic methanol from air and water using controlled thermonuclear reactor power—I. Technology and energy requirement. *Energy Conversion*, *17*(2-3), 97–112. doi:10.1016/0013-7480(77)90080-8

Su, T., Zhou, X., Qin, Z., & Ji, H. (2017). Intrinsic Kinetics of Dimethyl Ether Synthesis from Plasma Activation of CO2 Hydrogenation over Cu–Fe–Ce/HZSM-5. *ChemPhysChem*, *18*(3), 299–309. doi:10.1002/cphc.201601283 PMID:27898184

Sun, S. (2017). CO2 conversion in a gliding arc plasma: Performance improvement based on chemical reaction modeling. *Journal of CO2 Utilization, 17*, 220-234.

Sun, Y., Wu, J., Wang, Y., Li, J., Wang, N., Harding, J., Mo, S., Chen, L., Chen, P., Fu, M., Ye, D., Huang, J., & Tu, X. (2022). Plasma-catalytic CO2 hydrogenation over a Pd/ZnO catalyst: In situ probing of gas-phase and surface reactions. *JACS Au*, *2*(8), 1800–1810. doi:10.1021/jacsau.2c00028 PMID:36032530

Talebizadeh, P., Babaie, M., Brown, R., Rahimzadeh, H., Ristovski, Z., & Arai, M. (2014). The role of non-thermal plasma technique in NOx treatment: A review. *Renewable & Sustainable Energy Reviews*, *40*, 886–901. doi:10.1016/j.rser.2014.07.194

Taylan, O., & Berberoglu, H. (2014). Dissociation of carbon dioxide using a microhollow cathode discharge plasma reactor: Effects of applied voltage, flow rate and concentration. *Plasma Sources Science & Technology*, *24*(1), 015006. doi:10.1088/0963-0252/24/1/015006

Tsuji, M., Tanoue, T., Nakano, K., & Nishimura, Y. (2001). Decomposition of CO2 into CO and O in a microwave-excited discharge flow of CO2/He or CO2/Ar mixtures. *Chemistry Letters*, *30*(1), 22–23. doi:10.1246/cl.2001.22

Tu, X., Gallon, H. J., Twigg, M. V., Gorry, P. A., & Whitehead, J. C. (2011). Dry reforming of methane over a Ni/Al2O3 catalyst in a coaxial dielectric barrier discharge reactor. *Journal of Physics. D, Applied Physics*, *44*(27), 274007. doi:10.1088/0022-3727/44/27/274007

Tu, X., & Whitehead, J. C. (2014). Plasma dry reforming of methane in an atmospheric pressure AC gliding arc discharge: Co-generation of syngas and carbon nanomaterials. *International Journal of Hydrogen Energy*, *39*(18), 9658–9669. doi:10.1016/j.ijhydene.2014.04.073

Usman, M., Daud, W. W., & Abbas, H. F. (2015). Dry reforming of methane: Influence of process parameters—A review. *Renewable & Sustainable Energy Reviews*, *45*, 710–744. doi:10.1016/j.rser.2015.02.026

Uytdenhouwen, Y., Van Alphen, S., Michielsen, I., Meynen, V., Cool, P., & Bogaerts, A. (2018). A packed-bed DBD micro plasma reactor for CO2 dissociation: Does size matter? *Chemical Engineering Journal*, *348*, 557–568. doi:10.1016/j.cej.2018.04.210

Vadikkeettil, Y., Subramaniam, Y., Murugan, R., Ananthapadmanabhan, P. V., Mostaghimi, J., Pershin, L., Batiot-Dupeyrat, C., & Kobayashi, Y. (2022). Plasma assisted decomposition and reforming of greenhouse gases: A review of current status and emerging trends. *Renewable & Sustainable Energy Reviews*, *161*, 112343. doi:10.1016/j.rser.2022.112343

Valluri, S., Claremboux, V., & Kawatra, S. (2022). Opportunities and challenges in CO2 utilization. *Journal Of Environmental Sciences, 113*, 322-344.

Van Laer, K., & Bogaerts, A. (2015). Improving the conversion and energy efficiency of carbon dioxide splitting in a zirconia-packed dielectric barrier discharge reactor. *Energy Technology (Weinheim)*, *3*(10), 1038–1044. doi:10.1002/ente.201500127

Van Laer, K., & Bogaerts, A. (2017). How bead size and dielectric constant affect the plasma behaviour in a packed bed plasma reactor: A modelling study. *Plasma Sources Science & Technology*, *26*(8), 085007. doi:10.1088/1361-6595/aa7c59

Van Rooij, G., Akse, H. N., Bongers, W. A., & van de Sanden, M. C. M. (2017). Plasma for electrification of chemical industry: A case study on CO2 reduction. *Plasma Physics and Controlled Fusion*, *60*(1), 014019. doi:10.1088/1361-6587/aa8f7d

Wang, H., Wandell, R. J., Tachibana, K., Voráč, J., & Locke, B. R. (2018). The influence of liquid conductivity on electrical breakdown and hydrogen peroxide production in a nanosecond pulsed plasma discharge generated in a water-film plasma reactor. *Journal of Physics. D, Applied Physics*, *52*(7), 075201. doi:10.1088/1361-6463/aaf132

Wang, J., AlQahtani, M. S., Wang, X., Knecht, S. D., Bilén, S. G., Song, C., & Chu, W. (2021). One-step plasma-enabled catalytic carbon dioxide hydrogenation to higher hydrocarbons: Significance of catalyst-bed configuration. *Green Chemistry*, *23*(4), 1642–1647. doi:10.1039/D0GC03779F

Wang, J.-Y., Xia, G.-G., Huang, A., Suib, S. L., Hayashi, Y., & Matsumoto, H. (1999). CO2 decomposition using glow discharge plasmas. *Journal of Catalysis*, *185*(1), 152–159. doi:10.1006/jcat.1999.2499

Wang, L., Yi, Y., Guo, H., & Tu, X. (2018). Atmospheric pressure and room temperature synthesis of methanol through plasma-catalytic hydrogenation of CO2. *ACS Catalysis*, *8*(1), 90–100. doi:10.1021/acscatal.7b02733

Wang, W., Berthelot, A., Kolev, S., Tu, X., & Bogaerts, A. (2016). CO2 conversion in a gliding arc plasma: 1D cylindrical discharge model. *Plasma Sources Science & Technology*, *25*(6), 065012. doi:10.1088/0963-0252/25/6/065012

Wang, W., Mei, D., Tu, X., & Bogaerts, A. (2017). Gliding arc plasma for CO2 conversion: Better insights by a combined experimental and modelling approach. *Chemical Engineering Journal*, *330*, 11–25. doi:10.1016/j.cej.2017.07.133

Wang, W., Wang, S., Ma, X., & Gong, J. (2011). Recent advances in catalytic hydrogenation of carbon dioxide. *Chemical Society Reviews*, *40*(7), 3703–3727. doi:10.1039/c1cs15008a PMID:21505692

Wang, W.-H., Himeda, Y., Muckerman, J. T., Manbeck, G. F., & Fujita, E. (2015). CO2 hydrogenation to formate and methanol as an alternative to photo-and electrochemical CO2 reduction. *Chemical Reviews*, *115*(23), 12936–12973. doi:10.1021/acs.chemrev.5b00197 PMID:26335851

Wang, Y., Craven, M., Yu, X., Ding, J., Bryant, P., Huang, J., & Tu, X. (2019). Plasma-enhanced catalytic synthesis of ammonia over a Ni/Al2O3 catalyst at near-room temperature: Insights into the importance of the catalyst surface on the reaction mechanism. *ACS Catalysis*, *9*(12), 10780–10793. doi:10.1021/acscatal.9b02538 PMID:32064144

Wanten, B., Vertongen, R., De Meyer, R., & Bogaerts, A. (2023). Plasma-based CO2 conversion: How to correctly analyze the performance? *Journal of Energy Chemistry*, *86*, 180–196. doi:10.1016/j.jechem.2023.07.005

Whitehead, J. C. (2016). Plasma–catalysis: The known knowns, the known unknowns and the unknown unknowns. *Journal of Physics. D, Applied Physics*, *49*(24), 243001. doi:10.1088/0022-3727/49/24/243001

Xu, S. (2017). *Plasma-Assisted Conversion of CO 2*. The University of Manchester.

Xu, S., Whitehead, J. C., & Martin, P. A. (2017). CO2 conversion in a non-thermal, barium titanate packed bed plasma reactor: The effect of dilution by Ar and N2. *Chemical Engineering Journal*, *327*, 764–773. doi:10.1016/j.cej.2017.06.090

Xu, W., Li, M.-W., Xu, G.-H., & Tian, Y.-L. (2004). Decomposition of CO2 using DC corona discharge at atmospheric pressure. *Japanese Journal of Applied Physics*, *43*(12, 12R), 8310–8311. doi:10.1143/JJAP.43.8310

Yan, J. (2004). *Gliding arc discharge plasma device for organic waste water treatment, China.* Patent CN1557731A.

Yan, J. (2005). *A volatile organic compounds gas treatment device, China.* Patent CN2817959Y.

Yap, D., Tatibouët, J., & Batiot-Dupeyrat, C. (2015). Carbon dioxide dissociation to carbon monoxide by non-thermal plasma. *Journal of CO2 Utilization, 12,* 54-61.

Zeng, Y., & Tu, X. (2015). Plasma-catalytic CO 2 hydrogenation at low temperatures. *IEEE Transactions on Plasma Science*, *44*(4), 405–411. doi:10.1109/TPS.2015.2504549

Zeng, Y., & Tu, X. (2017). Plasma-catalytic hydrogenation of CO2 for the cogeneration of CO and CH4 in a dielectric barrier discharge reactor: Effect of argon addition. *Journal of Physics. D, Applied Physics*, *50*(18), 184004. doi:10.1088/1361-6463/aa64bb

Zeng, Y., Wang, L., Wu, C. F., Wang, J. Q., Shen, B. X., & Tu, X. (2018). Low temperature reforming of biogas over K-, Mg-and Ce-promoted Ni/Al2O3 catalysts for the production of hydrogen rich syngas: Understanding the plasma-catalytic synergy. *Applied Catalysis B: Environmental*, *224*, 469–478. doi:10.1016/j.apcatb.2017.10.017

Zhang, H. (2018). Warm plasma activation of CO2 in a rotating gliding arc discharge reactor. *Journal of CO2 Utilization*, *27*, 472-479.

Zheng, X., Tan, S., Dong, L., Li, S., & Chen, H. (2014). LaNiO3@ SiO2 core–shell nano-particles for the dry reforming of CH4 in the dielectric barrier discharge plasma. *International Journal of Hydrogen Energy*, *39*(22), 11360–11367. doi:10.1016/j.ijhydene.2014.05.083

Zhu, X., Liu, J.-H., Li, X.-S., Liu, J.-L., Qu, X., & Zhu, A.-M. (2017). Enhanced effect of plasma on catalytic reduction of CO2 to CO with hydrogen over Au/CeO2 at low temperature. *Journal of Energy Chemistry*, *26*(3), 488–493. doi:10.1016/j.jechem.2016.11.023

Chapter 4

Computer Simulation Studies of Non-Thermal Plasma in Cancer Treatment:

Understanding Protein Modifications for Enhanced Therapeutic Efficacy

Davronjon Abduvokhidov

R&D Center, New Uzbekistan University, Uzbekistan & Department of Information Technologies, Tashkent International University of Education, Uzbekistan & Institute of Material Sciences, Academy of Sciences, Uzbekistan

Zhitong Chen

Institute of Biomedical and Health Engineering, Shenzhen Institute of Advanced Technology, Chinese Academy of Sciences, China & Center for Advanced Therapy, National Innovation Center for Advanced Medical Devices, China

Jamoliddin Razzokov

https://orcid.org/0000-0002-3098-0797

Institute of Fundamental and Applied Research, National Research University TIIAME, Uzbekistan & School of Engineering, Central Asian University, Uzbekistan & Department of Biomedical Engineering, Tashkent State Technical University, Uzbekistan

ABSTRACT

Non-thermal plasma (NTP), the fourth state of matter, holds promise in altering biological matter, particularly in cancer treatment. NTP influences cell signaling by modifying key components, such as membranes, proteins, and DNA. It selectively targets cancer cells through the generation of reactive oxygen and nitrogen species (RONS), which modify amino acids within proteins. Understanding these molecular mechanisms is vital for optimizing NTP's potential in oncology. This book chapter reviews recent computer simulations and experimental findings exploring NTP's impact on proteins in cancer therapy, providing insights into how protein modifications affect cancer cell behavior and therapy responses. This knowledge advances NTP-based cancer treatments, offering the potential for personalized and targeted therapies in the future.

DOI: 10.4018/979-8-3693-0904-9.ch004

1. INTRODUCTION

Non-thermal plasma (NTP) as the 4th state of matter has emerged as a promising tool with significant impact on biological matter, particularly in cancer treatment. Exposure to NTP has been shown to influence the signaling pathways of cells by modifying cellular components, including membranes, proteins, and DNA. In particular, NTP has demonstrated potential in selectively targeting and killing cancer cells, making it an intriguing area of research in oncology.

The major impact of NTP on biological matter lies in the generation of reactive oxygen and nitrogen species (RONS). These RONS are crucial in causing damage and alterations to amino acids within proteins. Proteins, as versatile biomolecules, carry out a wide variety of functions in the cell, including catalytic, transporting, and signaling roles. Therefore, understanding the molecular-level mechanisms of protein modifications, such as oxidation and nitration induced by NTP, is crucial to obtain the full potential of this technology in cancer treatment.

Here, we aim to summarize the latest findings from computer simulation studies investigating the effects of NTP on proteins in cancer treatment. By exploring these simulations, we seek to gain deeper insights into the intricate interactions between NTP-derived RONS and protein structures. These insights will enhance our understanding of how specific protein modifications may contribute to altering cancer cell behavior and sensitivity to therapy.

By presenting an overview of computer simulation studies conducted to date in plasma cancer treatment, this chapter provides a comprehensive perspective on the role of NTP-induced protein modifications in cancer cell viability and therapeutic response. Such knowledge is essential for advancing NTP-based cancer treatment strategies and may pave the way for personalized and targeted therapies in the future.

2. POST-TRANSLATIONAL MODIFICATIONS AND THEIR IMPACT ON CATALYTIC PROTEINS

Proteins, particularly catalytic ones, play vital roles in various biochemical reactions and cellular pathways (Alberts, 2017). Their functionality can be affected by various internal and external factors, with RONS being one of the most significant external influencers (Dröge, 2002). This section sheds light on the effects of RONS on several catalytic proteins (Halliwell, 2006), particularly focusing on cytoglobin, catalase, peroxidase, and SOD, detailing their structure, function, and interactions under various conditions. Cytoglobin (CYGB) is a relatively newly discovered globin proposed to play a role in cellular protection against oxidative stress (Burmester et al., 2002; De Backer et al., 2018). One of the emerging tools in cancer therapy is NTP, which primarily generates RONS (Fridman et al., 2008). The interaction of these RONS with cellular proteins, especially redox-regulatory ones like CYGB, can determine the cell's fate (Smith et al., 2010). NTP treatment on CYGB has shown that while the protein undergoes chemical modifications, its secondary structure remains largely unaffected (De Backer et al., 2018). Spectroscopic analysis revealed the oxidations to mainly occur in sulfur-containing and aromatic amino acids (Stadtman & Levine, 2003). With prolonged NTP exposure, nitration of the heme group in CYGB was also observed. Furthermore, the two surface-exposed cysteine residues in CYGB were oxidized, leading to the formation of both intermolecular and potential intramolecular disulfide bridges.

Figure 1. The binding hotspot of $CYGB_{SH-SH}$ is illustrated in panels (a-b), while panels (c-d) depict $CYGB_{S-S}$, with these regions shown as red meshed areas. To enhance clarity, the helices within the structures are represented in subtle, pale colors. The heme group and specific histidine residues (81, 113, 117) are given in green and light purple, with licorice-style representations. It's evident that the access to the heme group is notably increased in $CYGB_{S-S}$ compared to $CYGB_{SH-SH}$.
Reprinted/adapted with permission from Ref.(De Backer et al., 2018). 2018, The Authors.

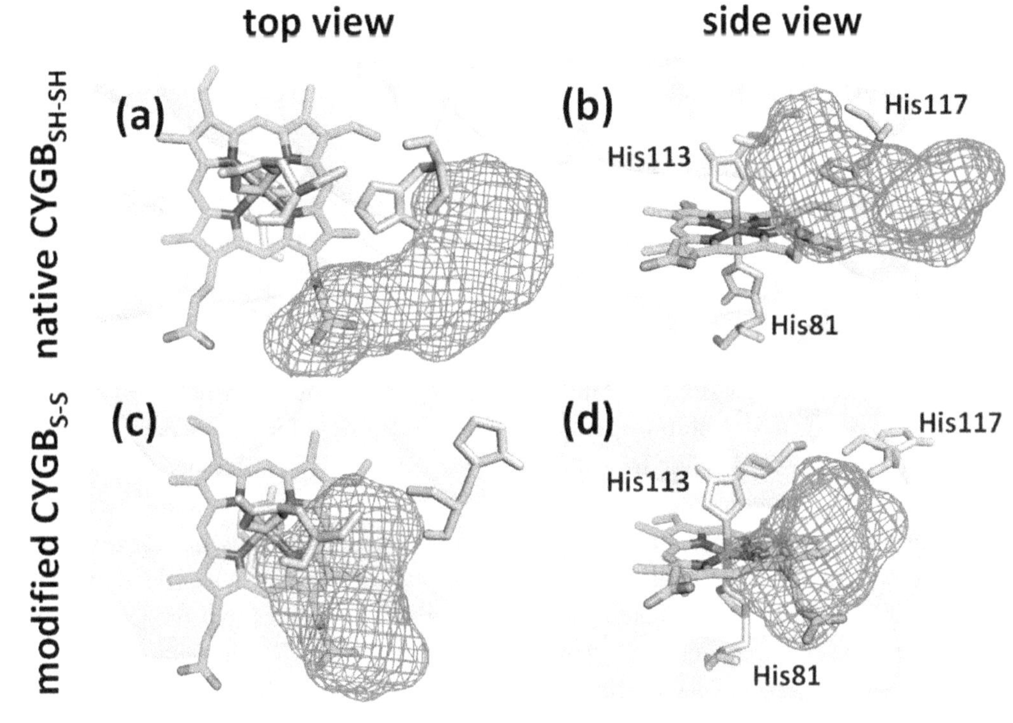

Molecular dynamics (MD) simulations and molecular docking techniques revealed that formation of disulfide bridge among cysteine residues (Cys_{38} and Cys_{83}) induce conformational change. This led to increase the accessibility of the binding pocket to CYGB's heme group, subsequently influencing its ability to bind ligands. These conformational alterations within CYGB resulted in distinct interactions with exogenous ligands, including drug-like compounds with potential implications for intracellular oxidative stress (see Figure 1). Moreover, in the recent study the mutation Lys_{80} to Ala (L80A) had the effect of increasing the accessibility of the CYGB binding pocket to the heme group, even in the absence of disulfide bond formation (Razzokov et al., 2021). However, the formation of the disulfide bond further enhanced the access of the binding pocket to the heme group. In contrast, a prior study they conducted (De Backer et al., 2018) revealed that in the absence of this mutation and disulfide bond formation, the CYGB binding pocket remained inaccessible to the heme group (Razzokov et al., 2021). In essence, the combined impact of the CYGB mutation and the creation of a disulfide bond between Cys_{38} and Cys_{83} through NTP-generated RONS amplified the enzymatic activity of CYGB, making it a more effective cytoprotective protein. This enhancement could hold potential benefits for cancer therapy.

Catalase is another essential enzyme that play a crucial role in protecting cells from oxidative damage caused by reactive oxygen species (ROS). Its ability to decompose hydrogen peroxide into water and oxygen makes it a key component in the cellular defence against oxidative stress (Chance et al., 1979).

The anti-cancer effect of NTP is postulated to be due to its ability to inactivate catalase, either directly or indirectly, leading to the reactivation of certain apoptotic cell signalling pathways (Bengtson & Bogaerts, 2020). However, a thorough analysis suggests that a catalase-dependent reactivation of these apoptotic pathways is unlikely to be the primary cause for the observed anti-cancer effect of NTP (Bengtson & Bogaerts, 2020; Yan et al., 2015). Thus, while the precise mechanisms remain under investigation, the potential of NTP in cancer treatment is evident.

Peroxidase enzymes are indispensable in cellular biochemistry, primarily for their role in catalysing the decomposition of peroxides, thus safeguarding cells against oxidative damage (Cui et al., 2022). These enzymes are pivotal in maintaining cellular redox homeostasis, a balance that is often disrupted in pathological conditions, including cancer (Marengo, 2016). In the context of plasma treatment, peroxidases gain significance due to their potential in modulating the effects of ROS, which are abundantly produced by NTP (Surowsky et al., 2013). The interaction between peroxidase activity and plasma-induced ROS highlights an intriguing area of research, particularly in understanding the therapeutic mechanisms and cellular responses in plasma-treated cancer cells (Zou et al., 2017).

In the realm of cancer treatment, NTP has shown synergistic effect in enhancing the sensitivity of certain drugs (Welz et al., 2015). NTP has been found to increase the sensitivity of three-dimensional glioblastoma spheroids to temozolomide, a standard drug for glioblastoma treatment. This increased sensitivity is believed to be due to the inhibition of the glutathione (GSH)/glutathione peroxidase 4 (GPX4) antioxidant machinery by NTP, leading to oxidative stress-mediated DNA damage (Shaw et al., 2021; Vandamme et al., 2012). The combined treatment of NTP with temozolomide demonstrated a significant reduction in spheroid growth, pointing towards a promising combination therapy for glioblastoma treatment (Keidar et al., 2011; Shaw et al., 2021). Pancreatic cancer remains one of the most challenging cancers to treat, due to the chemoresistance. However, studies have shown that oxidants derived from physical plasma can sensitize pancreatic cancer cells to ferroptotic cell death (Girard et al., 2016). By targeting the NRF2-HMOX1-GPX4 specific kinase signaling network, these oxidants show potential in providing a novel therapeutic strategy for pancreatic cancer (Kumar et al., 2021; Yan et al., 2017).

Oxidative modifications, especially those caused by the RONS, have profound effects on protein structure and function (Stadtman & Levine, 2003; Stadtman, 2006). Recent advancements in proteomic investigations have paved the way to study post-translational modifications (oxPTMs) due to ROS and RNS (Galli, 2012). These techniques enable a systematic approach to understand the changes in protein structure and function. Modern mass spectrometers have become indispensable tools in this endeavor, offering specific strategies to identify ROS/RNS-derived PTMs and their localization within the protein structure (Takai et al., 2014).

OxPTMs can have significant functional consequences. For instance, certain modifications, especially on cysteine residues, can lead to alterations in protein signaling, transcriptional regulation of genes, and other cellular processes (Galli, 2012). Understanding these modifications is vital as they can either promote or inhibit specific cellular pathways, impacting overall cell health and function (Radi, 2018). Endogenous antioxidant systems, including MnSOD and thioredoxin reductases, are instrumental in mitigating the impacts of oxidative stress. (Galli, 2012). When these systems are overwhelmed, oxidative modifications can prevail, leading to cellular dysfunction. Understanding the balance between oxidative stress and antioxidant systems is crucial to maintaining cellular health and preventing diseases.

In conclusion, catalytic proteins, while playing essential roles in various cellular processes, can undergo modifications upon interaction with RONS, affecting their function and conformation (Graves,

2012). Unraveling the modifications and their functional implications offers a deeper understanding of cellular dysregulation and provides avenues for potential therapeutic interventions (Galli, 2012).

3. SIGNALING PROTEINS

Signaling proteins play a vital role in various cellular processes, including cell growth, cell differentiation, and even the cell's response to external stimuli. Their structural and functional integrity is essential for the proper functioning of cells. This section discusses the impact of plasma oxidation on the structural features of two important signaling proteins: the human epidermal growth factor (hEGF) and the immune checkpoint CD47 and CD44.

The epidermal growth factor receptor (EGFR) is one of the important signalling protein in both cancer therapy (Giannoni et al., 2009) and wound healing (Kalay & Cevher, 2012), engaging multiple ligands for activation (Jorissen, 2003; Wee & Wang, 2017). This protein is essential for cell proliferation, survival, and differentiation processes that are beneficial in wound repair. In cancer cells, however, this mechanism hinders treatment and aids in the disease's advancement (Carpenter, 1981; Herbst, 2004; Iqbal, 2014; Razzokov et al., 2022; Yusupov et al., 2018). EGFR overexpression, a common occurrence in many cancer types, affects the cell cycle by initiating these mechanisms (Berlanga-Acosta et al., 2009; Iqbal, 2014). The hEGF is a primary activating ligand of EGFR (Henson & Gibson, 2006), indirectly promotes cancer growth by stimulating cytoprotection, chemotaxis, mutagenesis, and mitogenesis, although these actions are advantageous in wound healing (Stoscheck & King, 1986). Consequently, modifications to hEGF induced by RONS generated through NTP could act as a double edged sword for either wound healing or cancer treatment (Yusupov et al., 2018). The exposure time and specific RONS in NTP treatment are tuning parameters that determine the extent and type of hEGF modifications (Laroussi et al., 2021). Excessive nitro-oxidation of hEGF residues by RONS might significantly impair its normal function, an outcome beneficial in cancer therapy (Razzokov et al., 2022; Yusupov et al., 2018). On the other hand, minor structural alterations might preserve hEGF's regular functions, advantageous for wound healing (Yusupov et al., 2018). The interplay between hEGF and RONS affects the stability and flexibility of its structure, thus influencing on its binding modes to EGFR. Through simulations and experiments (Yusupov et al., 2018), including MD, PCA, and CD spectroscopy (see Figure 2), it is found that increased oxidation (represented by OX1, OX2, OX3 states) leads to greater structural flexibility (see Figure 2a and b), loss of β-sheet conformations (see Figure 2c and d), and decreased binding affinity with the hEGF receptor, as seen in the binding free energies and structural alignments. These changes are suggested to influence cellular signaling pathways, potentially explaining the differing effects of NTP in wound healing and cancer treatment. It was determined that an increment of oxidative stress on hEGF significantly reduces the binding affinity between EGFR making hEGF more unstable. The higher oxidation levels lead to hEGF denaturation, predominantly through the disintegration of disulfide bonds. The nitrosylation of hEGF was examined and determined the effect on the stability and conformation of structures (Razzokov et al., 2022). Despite the breaking of all three disulfide bonds at high nitrosylation levels, hEGF's structure remains relatively stable. Nitrosylated amino acids become more hydrophobic, resulting in a more compact structure compared to the native form, as they are less interactive with water. Thus, the low level of hEGF nitrosylation does not substantially perturb its structure or function, which can be advantageous for wound healing. However, the higher doses could potentially inhibit its ability to bind its receptor by, thereby inhibiting cancer cell proliferation (Razzokov et al., 2022).

Figure 2. A) MD Simulations. RMSD of the α-carbons in the native (represented in black) and oxidized hEGF proteins (denoted in red, blue, and cyan, respectively) B) The PCA outcomes derived from both the native and oxidized hEGF configurations were collated, drawing from the terminal 30 ns of the simulations C) Secondary structure evaluation of plasma-treated hEGF via CD spectroscopy D) The alignment of oxidation structures (red) with the Native hEGF (cyan) uses the final MD simulation snapshots to showcase the conformational changes in the oxidized hEGFs compared to the native hEGF.
Reprinted/adapted with permission from Ref.(Yusupov et al., 2018). 2018, The Authors..

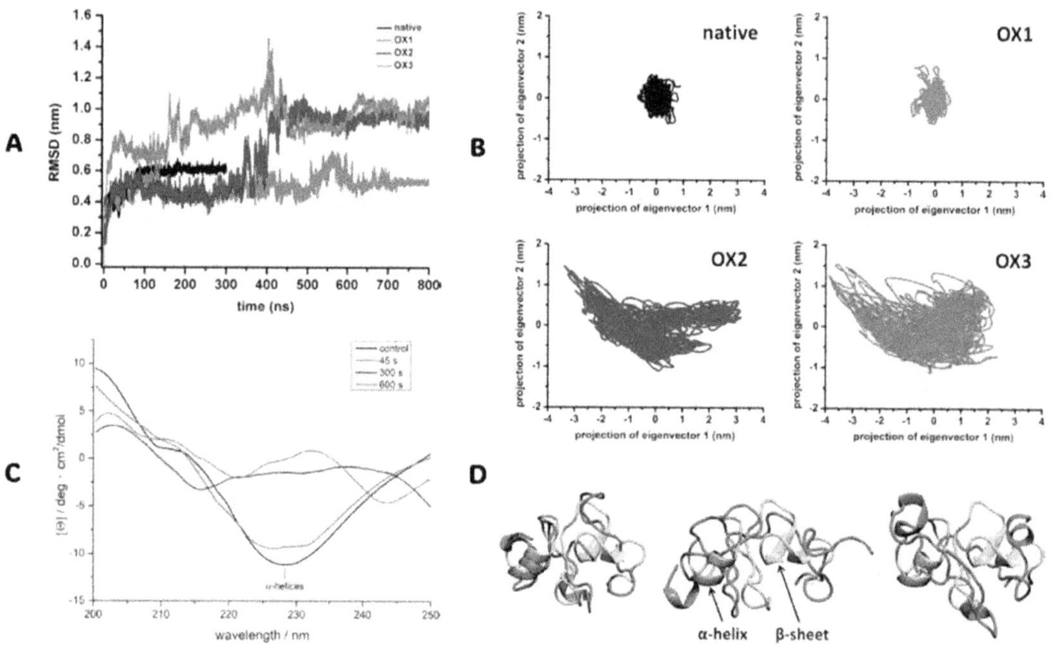

The cluster of differentiation 44 (CD44) is another signalling protein that notably overexpressed in a myriad of cancer cell types (Liu et al., 2019). As a main part of the tumour microenvironment the heightened expression correlates with the progression of cancer, metastasis, and an unfavorable prognosis. CD44 is integral to cell adhesion, migration, and signaling, with its overexpression playing a significant role in the invasion and metastasis of cancer cells. It aids in both cell-to-cell and cell-to-matrix interactions, activating cancer cell proliferation, differentiation, invasion, and migration (Thapa & Wilson, 2016). The principal ligand of CD44, hyaluronan (HA), found within the tumor's extracellular matrix, triggers signaling pathways that escalate cell metastasis (Misra et al., 2015). Yusupov et al., performed computational study showing that inhibiting the interaction between HA and CD44 via NTP treatment that impedes signaling pathways pivotal to tumor advancement (Yusupov et al., 2021). Investigating the oxidative damage to HA and CD44 caused by NTP-generated RONS, researchers used both reactive and non-reactive MD simulations. Notably, this oxidation markedly reduced their binding affinity, as demonstrated by the free energy profiles in Figure 3 and the significant binding energy decrease in the OX3 system (−22.1 kJ/mol, down by 11.3 kJ/mol from the native state, as shown in Table 1). This weakening of the HA-CD44 interaction, primarily due to structural destabilization from disrupted disulfide bonds, underscores a crucial mechanism in oxidative stress-based cancer therapies (Yusupov et al., 2021).

Figure 3. Free energy profiles (FEPs) depict the dissociation of HA from CD44 in four states: both non-oxidized (black curve), only HA oxidized (red curve), only CD44 oxidized (green curve), and both oxidized (blue curve)
Reprinted/adapted with permission from Ref.(Yusupov et al., 2021). 2021, The Authors.

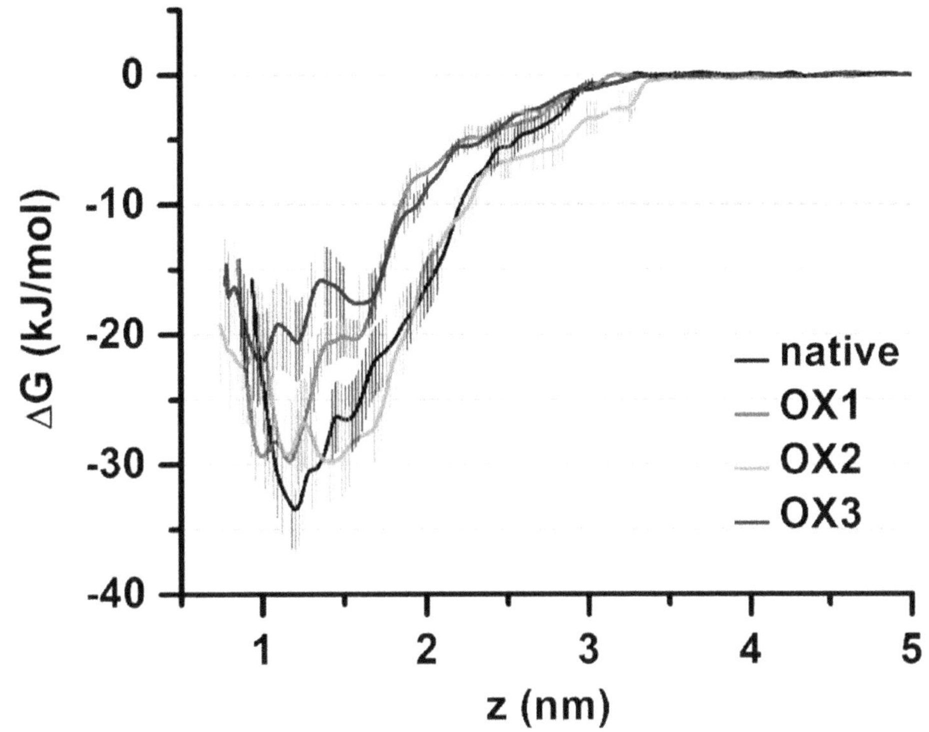

Table 1. Calculated binding free energies (ΔG) were determined for the native, OX1, OX2, and OX3 systems, with ΔΔG representing the energy loss relative to the native state

System	ΔG (kJ/mol)	ΔΔG (kJ/mol)
native (CD44–HA_8)	−33.4 ± 3.3	–
OX1 (CD44–$HA_{8\text{-}OX}$)	−29.8 ± 2.6	3.6
OX2 ($CD44_{OX}$–HA_8)	−29.8 ± 3.2	3.6
OX3 ($CD44_{OX}$–$HA_{8\text{-}OX}$)	−22.1 ± 3.6	11.3

Reprinted/adapted with permission from Ref.(Yusupov et al., 2021). 2021, The Authors.

CD47, an essential innate immune checkpoint, acts as a "don't eat me" signal, protecting cells from being engulfed and digested by macrophages. Overexpression of CD47 is a common feature in various cancer cells, making it a potential target for cancer treatments (Lin et al., 2021). Lin and collaborators studied via both experimental and computational methodologies, to elucidate the influence of NTP on CD47 (Lin et al., 2021). Their simulations revealed that the oxidation of CD47 reduced its affinity for binding to SIRPα on the surface of innate immune cells (see Figure 4). This decrease in affinity unmasks cancer cells to the immune system (Lin et al., 2021). The free energy landscapes, as delineated by umbrella sampling simulations, indicated that the free energy associated with the dissociation of CD47 from SIRPα decreased by approximately 45 kJ/mol subsequent to oxidation (Figure 4b and c). This reduction was attributed to structural alterations in the oxidized CD47, specifically the disruption of the Lys_{39}-Asp_{100} and Lys_{41}-Asp_{100} salt bridges between CD47 and SIRPα (Figure 4a). The authors assumed that the less favourable on binding of the oxidized CD47 to SIRPα causes the immunogenicity of tumors and bolsters the anti-cancer immune response, thereby underlining its pivotal role in NTP-facilitated cancer immunotherapy (Lin et al., 2021).

In conclusion, the modulation of signaling proteins by oxidation, especially through NTP or NTP treatments, holds significant potential for biomedical applications. Whether it's harnessing the effects for wound healing or tapping into the inhibitory effects for cancer treatments, understanding the intricate details of these interactions is paramount.

Figure 4. An in silico assessment was conducted to ascertain the impact of CD47 oxidation on its binding affinity with SIRPα. (A) In the native complex, salt bridges are established between Lys39-Asp100 and Lys41-Asp100, yet these connections are disrupted following the oxidation of CD47, as highlighted in the inset illustration. (B) Both CD47 and its oxidized variant, CD47OX, were subjected to a mechanical pulling process against SIRPα. (C) The FEPs of both the native (illustrated in black) and oxidized (depicted in red) CD47 structures, upon their dissociation from SIRPα, reveal the energy required for dissociation. The errors related to the sampling process are represented in a lighter shade.
Reprinted/adapted with permission from Ref.(Lin et al., 2021). 2021, The Authors

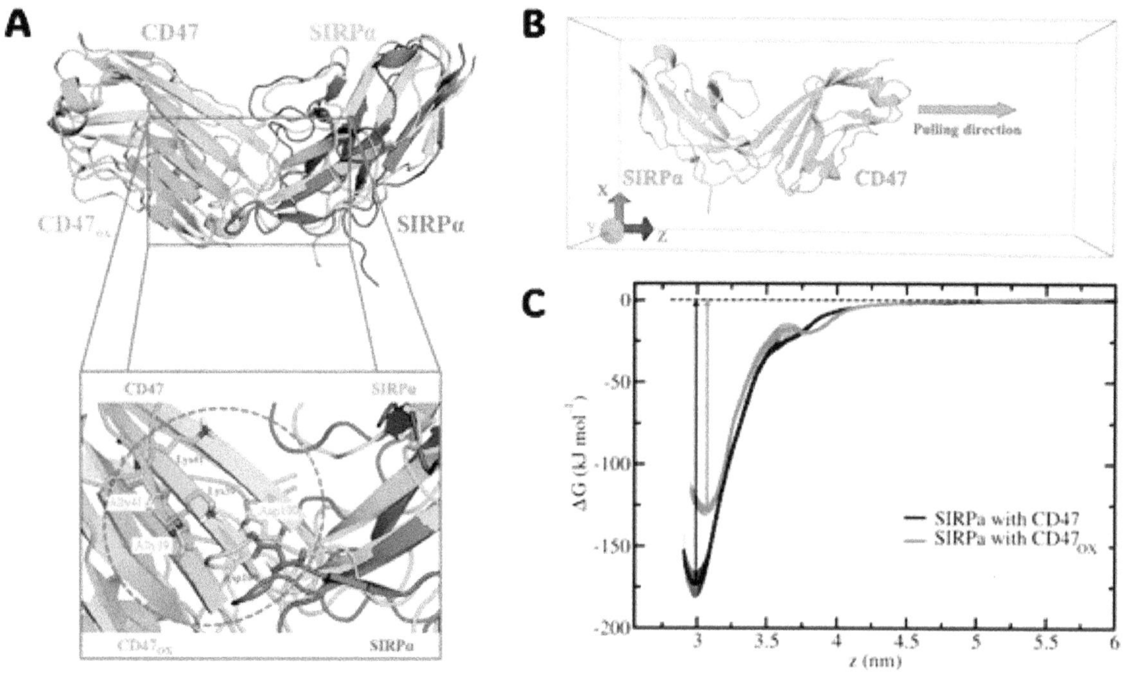

4. THE ROLE TRANSMEMBRANE PROTEIN IN NTP CANCER TREATMENT

Aquaporins (AQPs) are a class of integral membrane proteins that primarily facilitate the transport of water molecules across the cell membrane (Figure 5a, b, c and d). However, their function isn't limited to water transport alone; they also mediate the movement of other solutes, such as the RONS (Yusupov, 2019). Notably, RONS are produced by various sources, including NPT, and have significant implications in inducing oxidative stress within cells (Bienert et al., 2007). Such stress is particularly relevant in the context of cancer treatment, as it can inhibit the growth and proliferation of cancer cells (Trachootham et al., 2009). Despite the clear importance of AQPs in facilitating RONS transport, the molecular-level mechanisms underlying this process remain somewhat elusive (Papadopoulos & Saadoun, 2015). Yusupov et al. studied on elucidating the mechanisms of RONS permeation through AQPs with a specific focus on AQP1, employing MD simulations (Yusupov, 2019). Their study focused on both hydrophilic RONS (H_2O_2 and OH) and hydrophobic RONS (NO_2 and NO). The simulations revealed that all these RONS could permeate across the AQP1 pores. Interestingly, the permeation free energy barrier for OH and NO was found to be lower than that for H_2O_2 and NO_2. This suggests that OH and NO might have easier access to the pore's interior, allowing them to interact more freely with AQP1's amino acid residues.

Oxidation plays a pivotal role in the functioning of AQPs. Yusupov et al. further probed the effects of RONS-induced oxidation on both phospholipids and AQP1 itself, specifically focusing on the sulfenylation of Cys_{191}, a residue in AQP1 (Yusupov, 2019). Their findings indicated that both lipid and protein oxidation could elevate the free energy barrier for H_2O_2 and NO_2 permeation. However, for OH and NO, the effects of oxidation weren't as pronounced. The study provides insights into the intricate relationship between RONS and their transport mechanisms. For instance, while hydrophilic RONS like H_2O_2 might preferentially enter cells through AQPs, hydrophobic RONS like NO and NO_2 likely permeate primarily through the phospholipid bilayer. The study's findings have profound implications in the context of cancer therapy. AQPs are often more expressed in cancer cells. The facilitated transport of RONS, especially through AQPs, might explain the observed increase in RONS levels in cancer cells. This could shed light on the selective anticancer mechanisms of NTP, potentially offering new therapeutic avenues (Yusupov, 2019). In summary, AQPs are more complex and varied than previously understood. Beyond water transport, their involvement in mediating the transport of reactive species like RONS positions them at the forefront of research in oxidative stress and cancer therapy. Understanding the molecular mechanisms of RONS transport through AQPs can offer valuable insights, potentially leading to more targeted and effective therapeutic strategies against cancer.

Another protein prolifically expressed across diverse cancerous cell varieties, and essential for cellular sustenance, is the xC^- antiporter (Liu et al., 2021). This antiporter mediates the complex transport of cystine (the sophisticated oxidized dimer form of Cys) from extracellular environment into the intracellular space whilst simultaneously releasing glutamate into the external milieu (Bannai & Kitamura, 1981; Lo et al., 2008). In cellular environments, cystine is rapidly converted to Cys, significantly enhancing the biosynthesis of glutathione, a critical antioxidant combating oxidative stress within cells (Lo et al., 2008). Certain cancer cells, particularly those associated with lymphoma and leukemia, lack the ability to synthesize Cys autonomously. Consequently, these cells rely on the uptake of cystine via the xC^- transporter, a fact underscored by the overexpression of this transporter on their (Gout et al., 1997). The function of the xC^- transporter can be inhibited by NTP generated RONS. This inhibition leads to a reduction in intracellular Cys levels, especially in cancer cells, disrupting glutathione homeostasis. This disruption inhibits cell proliferation and triggers apoptosis (Lo et al., 2008). Detailed studies on the xC^- transporter have shown that a mutation replacing Cys_{327} (located near the protein channel and the extracellular space) with alanine disrupts normal cystine transport, highlighting the importance of Cys_{327} in this process (Jiménez-Vidal et al., 2004). Cys_{327} is prone to oxidation into cysteic acid in the presence of RONS (Attri et al., 2018). MD simulations conducted by Ghasemitarei et al. (2019) have demonstrated that the oxidation of Cys_{327} alters the structure of the transporter channel, impeding cystine uptake. Umbrella sampling simulations used to analyze cystine transport across the xC^- transporter indicated a significant energy barrier of 33.9 kJ/mol following the oxidation of Cys_{327}, reducing cystine influx (see Figure 6). Examination of the protein channel revealed a notable narrowing near the oxidized Cys_{327}, which presents a significant obstacle to cystine uptake. Thus, the inhibition of the xC^- transporter by NTP-derived RONS results in a marked decrease in intracellular cystine levels, leading to a reduction in glutathione stores. This decrease in glutathione levels causes an increase in intracellular oxidative stress, inhibiting cell proliferation and inducing apoptosis, particularly in cancer cells (Ghasemitarei et al., 2019).

Figure 5. (A) Overhead and (B) side views of the NAT model: the native AQP1 tetramer within a POPC phospholipid bilayer (gray) and encased in water layers (cyan in (B)). The top and bottom water layers symbolize extracellular and cytoplasmic zones. In (A), these layers are omitted for clarity. Panel (c) underscores specific AQP residues, particularly reactive to hydroxyl radicals, with Cys191 marked in red dashed circles (Xu & Chance, 2007). Chemical structures of cysteine and its oxidized form, cysteine sulfenic acid (Cys-SOH), alongside POPC and oxidized POPC-ALD, are detailed in (C) and (D) respectively. These form the OXL model, with Cys-SOH comprising the OXP model. Cys191, selected for oxidation in OXP, is noted for its easy oxidizability and location near the ar/R selectivity area (see E and F) Average FEP for hydrophilic (OH) and hydrophobic (NO2) RONS traversing AQP1 pores are depicted for native (NAT, black curve) and oxidized (OXL and OXP, red and green curves) models. Extracellular and cytoplasmic water phases are illustrated in light blue, with uncertainties shown in paler shades. The Asn-Pro-Ala (NPA) and aromatic/arginine (ar/R) selectivity zones are marked by brown dashed lines, centered on the NPA area. This notation is consistent across related figures.
Reprinted/adapted with permission from Ref. (Yusupov, 2019). 2021, The Authors

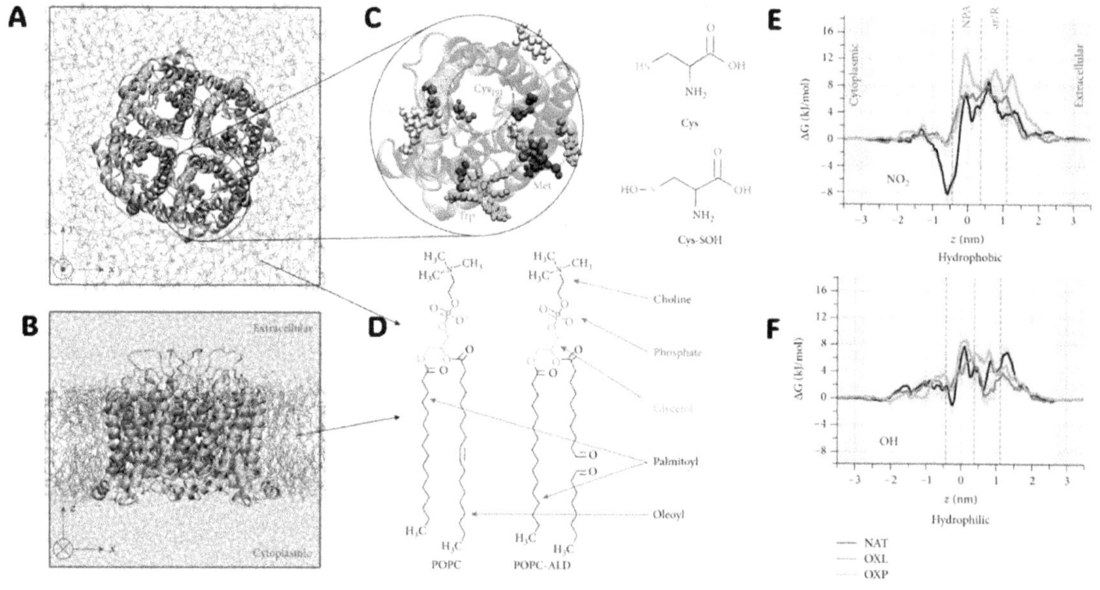

Figure 6. FEPs the translocation of cystine across the xC⁻ antiporter exhibit marked contrasts between its non-oxidized (represented by the black curve) and oxidized states (illustrated by the red curve)
Reprinted/adapted with permission from Refs. (Ghasemitarei et al., 2019). 2019, Elsevier.

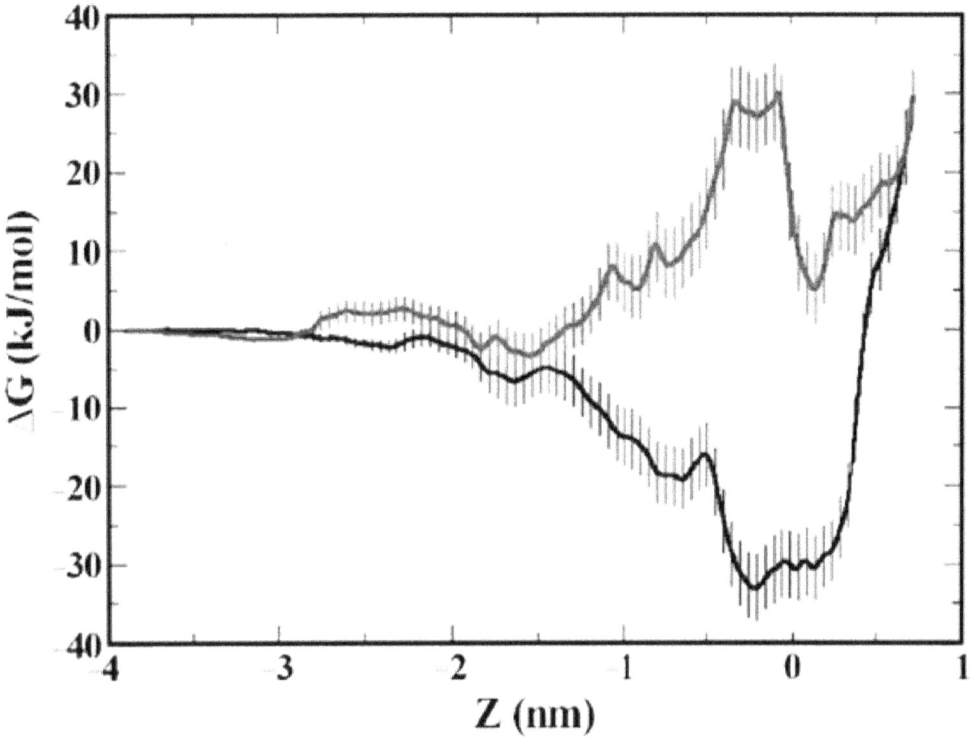

5. CONCLUSION

In conclusion, our investigation partially demonstrates the potential of NTP for the selective elimination of cancer cells. However, the exact mechanism of NTP's selectivity remains elusive due to the complex interactions between RONS and cellular components. In our recent computer simulation studies, we made strides toward explaining this selectivity. Notably, catalytic, signaling, transporter and immune checkpoint proteins emerge as key determinants in shaping cell fate through amino acid modifications via NTP treatment. While our work has made significant progress, further investigations are still needed to elucidate the precise molecular-level mechanisms responsible for cancer cell death following NTP treatment.

ACKNOWLEDGMENT

J.R. gratefully acknowledges the financial support received from the Agency for Innovative Development of the Republic of Uzbekistan, under grant number AL-59-21122141. Z.Ch. expresses appreciation for the funding received from the National Key Research and Development Program of China, with grant number 2022YFE0126000.

REFERENCES

Alberts, B. (2017). *Molecular biology of the cell.* Garland science.

Attri, P., Han, J., Choi, S., Choi, E. H., Bogaerts, A., & Lee, W. (2018). CAP modifies the structure of a model protein from thermophilic bacteria: Mechanisms of CAP-mediated inactivation. *Scientific Reports, 8*(1), 10218. doi:10.1038/s41598-018-28600-w PMID:29977069

Bannai, S., & Kitamura, E. (1981). Role of proton dissociation in the transport of cystine and glutamate in human diploid fibroblasts in culture. *The Journal of Biological Chemistry, 256*(11), 5770–5772. doi:10.1016/S0021-9258(19)69273-2 PMID:6113240

Bengtson, C., & Bogaerts, A. (2020). On the anti-cancer effect of cold atmospheric plasma and the possible role of catalase-dependent apoptotic pathways. *Cells, 9*(10), 2330. doi:10.3390/cells9102330 PMID:33096638

Berlanga-Acosta, J., Gavilondo-Cowley, J., López-Saura, P., González-López, T., Castro-Santana, M. D., López-Mola, E., Guillén-Nieto, G., & Herrera-Martinez, L. (2009). Epidermal growth factor in clinical practice–a review of its biological actions, clinical indications and safety implications. *International Wound Journal, 6*(5), 331–346. doi:10.1111/j.1742-481X.2009.00622.x PMID:19912390

Bienert, G. P., Møller, A. L. B., Kristiansen, K. A., Schulz, A., Møller, I. M., Schjoerring, J. K., & Jahn, T. P. (2007). Specific aquaporins facilitate the diffusion of hydrogen peroxide across membranes. *The Journal of Biological Chemistry, 282*(2), 1183–1192. doi:10.1074/jbc.M603761200 PMID:17105724

Burmester, T., Ebner, B., Weich, B., & Hankeln, T. (2002). Cytoglobin: A novel globin type ubiquitously expressed invertebrate tissues. *Molecular Biology and Evolution, 19*(4), 416–421. doi:10.1093/oxfordjournals.molbev.a004096 PMID:11919282

Carpenter, G. (1981). Epidermal growth factor. In *Tissue growth factors* (pp. 89–132). Springer. doi:10.1007/978-3-642-67986-5_3

Chance, B., Sies, H., & Boveris, A. (1979). Hydroperoxide metabolism in mammalian organs. *Physiological Reviews, 59*(3), 527–605. doi:10.1152/physrev.1979.59.3.527 PMID:37532

Cui, D., Yin, Y., Sun, H., Wang, X., Zhuang, J., Wang, L., Ma, R., & Jiao, Z. (2022). Regulation of cellular redox homeostasis in Arabidopsis thaliana seedling by atmospheric pressure cold plasma-generated reactive oxygen/nitrogen species. *Ecotoxicology and Environmental Safety, 240*, 113703. doi:10.1016/j.ecoenv.2022.113703 PMID:35659700

De Backer, J., Razzokov, J., Hammerschmid, D., Mensch, C., Hafideddine, Z., Kumar, N., van Raemdonck, G., Yusupov, M., Van Doorslaer, S., Johannessen, C., Sobott, F., Bogaerts, A., & Dewilde, S. (2018). The effect of reactive oxygen and nitrogen species on the structure of cytoglobin: A potential tumor suppressor. *Redox Biology, 19*, 1–10. doi:10.1016/j.redox.2018.07.019 PMID:30081385

Dröge, W. (2002). Free radicals in the physiological control of cell function. *Physiological Reviews, 82*(1), 47–95. doi:10.1152/physrev.00018.2001 PMID:11773609

Fridman, G., Friedman, G., Gutsol, A., Shekhter, A. B., Vasilets, V. N., & Fridman, A. (2008). Applied plasma medicine. *Plasma Processes and Polymers*, *5*(6), 503–533. doi:10.1002/ppap.200700154

Galli, F. (2012). Amino acid and protein modification by oxygen and nitrogen species. *Amino Acids*, *42*(1), 1–4. doi:10.1007/s00726-010-0670-8 PMID:20602126

Ghasemitarei, M., Yusupov, M., Razzokov, J., Shokri, B., & Bogaerts, A. (2019). Effect of oxidative stress on cystine transportation by xC^{-} antiporter. *Archives of Biochemistry and Biophysics*, *674*, 108114. doi:10.1016/j.abb.2019.108114 PMID:31557465

Giannoni, E., Fiaschi, T., Ramponi, G., & Chiarugi, P. (2009). Redox regulation of anoikis resistance of metastatic prostate cancer cells: Key role for Src and EGFR-mediated pro-survival signals. *Oncogene*, *28*(20), 2074–2086. doi:10.1038/onc.2009.77 PMID:19377510

Girard, P.-M., Arbabian, A., Fleury, M., Bauville, G., Puech, V., Dutreix, M., & Sousa, J. S. (2016). Synergistic effect of H2O2 and NO2 in cell death induced by cold atmospheric He plasma. *Scientific Reports*, *6*(1), 29098. doi:10.1038/srep29098 PMID:27364563

Gout, P., Kang, Y. J., Buckley, D. J., Bruchovsky, N., & Buckley, A. R. (1997). Increased cystine uptake capability associated with malignant progression of Nb2 lymphoma cells. *Leukemia*, *11*(8), 1329–1337. doi:10.1038/sj.leu.2400739 PMID:9264389

Graves, D. B. (2012). The emerging role of reactive oxygen and nitrogen species in redox biology and some implications for plasma applications to medicine and biology. *Journal of Physics. D, Applied Physics*, *45*(26), 263001. doi:10.1088/0022-3727/45/26/263001

Halliwell, B. (2006). Reactive species and antioxidants. Redox biology is a fundamental theme of aerobic life. *Plant Physiology*, *141*(2), 312–322. doi:10.1104/pp.106.077073 PMID:16760481

Henson, E. S., & Gibson, S. B. (2006). Surviving cell death through epidermal growth factor (EGF) signal transduction pathways: Implications for cancer therapy. *Cellular Signalling*, *18*(12), 2089–2097. doi:10.1016/j.cellsig.2006.05.015 PMID:16815674

Herbst, R.S. (2004). Review of epidermal growth factor receptor biology. *International Journal of Radiation Oncology* Biology* Physics*, *59*(2), S21-S26.

Iqbal, N. (2014). *Human epidermal growth factor receptor 2 (HER2) in cancers: overexpression and therapeutic implications.* Molecular biology international.

Jiménez-Vidal, M., Gasol, E., Zorzano, A., Nunes, V., Palacín, M., & Chillarón, J. (2004). Thiol modification of cysteine 327 in the eighth transmembrane domain of the light subunit xCT of the heteromeric cystine/glutamate antiporter suggests close proximity to the substrate binding site/permeation pathway. *The Journal of Biological Chemistry*, *279*(12), 11214–11221. doi:10.1074/jbc.M309866200 PMID:14722095

Jorissen, R.N. (2003). Epidermal growth factor receptor: mechanisms of activation and signalling. *The EGF receptor family*, 33-55.

Kalay, Z., & Cevher, S. C. (2012). Oxidant and antioxidant events during epidermal growth factor therapy to cutaneous wound healing in rats. *International Wound Journal*, *9*(4), 362–371. doi:10.1111/j.1742-481X.2011.00895.x PMID:22129466

Keidar, M., Walk, R., Shashurin, A., Srinivasan, P., Sandler, A., Dasgupta, S., Ravi, R., Guerrero-Preston, R., & Trink, B. (2011). Cold plasma selectivity and the possibility of a paradigm shift in cancer therapy. *British Journal of Cancer, 105*(9), 1295–1301. doi:10.1038/bjc.2011.386 PMID:21979421

Kumar, N., Perez-Novo, C., Shaw, P., Logie, E., Privat-Maldonado, A., Dewilde, S., Smits, E., Berghe, W. V., & Bogaerts, A. (2021). Physical plasma-derived oxidants sensitize pancreatic cancer cells to ferroptotic cell death. *Free Radical Biology & Medicine, 166*, 187–200. doi:10.1016/j.freeradbiomed.2021.02.026 PMID:33636332

Laroussi, M., Bekeschus, S., Keidar, M., Bogaerts, A., Fridman, A., Lu, X., Ostrikov, K., Hori, M., Stapelmann, K., Miller, V., Reuter, S., Laux, C., Mesbah, A., Walsh, J., Jiang, C., Thagard, S. M., Tanaka, H., Liu, D., Yan, D., & Yusupov, M. (2021). Low-temperature plasma for biology, hygiene, and medicine: Perspective and roadmap. *IEEE Transactions on Radiation and Plasma Medical Sciences, 6*(2), 127–157. doi:10.1109/TRPMS.2021.3135118

Lin, A., Razzokov, J., Verswyvel, H., Privat-Maldonado, A., De Backer, J., Yusupov, M., Cardenas De La Hoz, E., Ponsaerts, P., Smits, E., & Bogaerts, A. (2021). Oxidation of innate immune checkpoint CD47 on cancer cells with non-thermal plasma. *Cancers (Basel), 13*(3), 579. doi:10.3390/cancers13030579 PMID:33540720

Liu, L., Liu, R., Liu, Y., Li, G., Chen, Q., Liu, X., & Ma, S. (2021). Cystine-glutamate antiporter xCT as a therapeutic target for cancer. *Cell Biochemistry and Function, 39*(2), 174–179. doi:10.1002/cbf.3581 PMID:32749001

Liu, M., Tolg, C., & Turley, E. (2019). Dissecting the dual nature of hyaluronan in the tumor microenvironment. *Frontiers in Immunology, 10*, 947. doi:10.3389/fimmu.2019.00947 PMID:31134064

Lo, M., Wang, Y. Z., & Gout, P. W. (2008). The x cystine/glutamate antiporter: A potential target for therapy of cancer and other diseases. *Journal of Cellular Physiology, 215*(3), 593–602. doi:10.1002/jcp.21366 PMID:18181196

Marengo, B. (2016). *Redox homeostasis and cellular antioxidant systems: crucial players in cancer growth and therapy.* Oxidative medicine and cellular longevity.

Misra, S., Hascall, V. C., Markwald, R. R., & Ghatak, S. (2015). Interactions between hyaluronan and its receptors (CD44, RHAMM) regulate the activities of inflammation and cancer. *Frontiers in Immunology, 6*, 201. doi:10.3389/fimmu.2015.00201 PMID:25999946

Papadopoulos, M. C., & Saadoun, S. (2015). *Key roles of aquaporins in tumor biology.* Biochimica et Biophysica Acta (BBA)-. *Biochimica et Biophysica Acta. Biomembranes, 1848*(10), 2576–2583. doi:10.1016/j.bbamem.2014.09.001

Radi, R. (2018). Oxygen radicals, nitric oxide, and peroxynitrite: Redox pathways in molecular medicine. *Proceedings of the National Academy of Sciences of the United States of America, 115*(23), 5839–5848. doi:10.1073/pnas.1804932115 PMID:29802228

Razzokov, J., Fazliev, S., Erkinova, D., Mamatkulov, S., & Chen, Z. (2022). Understanding the effect of nitrosylation on dynamics of human epidermal growth factor: A μs simulation study. *Journal of Physics. D, Applied Physics, 55*(47), 475201. doi:10.1088/1361-6463/ac9148

Razzokov, J., Fazliev, S., Yusupov, M., Sharipov, A., Ruziev, Z., & Mamatkulov, S. (2021). Effect of mutation and disulfide bond formation on the catalytic site of monomeric cytoglobin: A molecular level insight. *Plasma Medicine, 11*(4), 41–51. doi:10.1615/PlasmaMed.2021041420

Shaw, P., Kumar, N., Privat-Maldonado, A., Smits, E., & Bogaerts, A. (2021). Cold atmospheric plasma increases temozolomide sensitivity of three-dimensional glioblastoma spheroids via oxidative stress-mediated DNA damage. *Cancers (Basel), 13*(8), 1780. doi:10.3390/cancers13081780 PMID:33917880

Smith, L. J., Kahraman, A., & Thornton, J. M. (2010). Heme proteins—Diversity in structural characteristics, function, and folding. *Proteins, 78*(10), 2349–2368. doi:10.1002/prot.22747 PMID:20544970

Stadtman, E., & Levine, R. (2003). Free radical-mediated oxidation of free amino acids and amino acid residues in proteins. *Amino Acids, 25*(3-4), 207–218. doi:10.1007/s00726-003-0011-2 PMID:14661084

Stadtman, E. R. (2006). Protein oxidation and aging. *Free Radical Research, 40*(12), 1250–1258. doi:10.1080/10715760600918142 PMID:17090414

Stoscheck, C. M., & King, L. E. Jr. (1986). Role of epidermal growth factor in carcinogenesis. *Cancer Research, 46*(3), 1030–1037. PMID:3002608

Surowsky, B., Fischer, A., Schlueter, O., & Knorr, D. (2013). Cold plasma effects on enzyme activity in a model food system. *Innovative Food Science & Emerging Technologies, 19*, 146–152. doi:10.1016/j.ifset.2013.04.002

Takai, E., Kitamura, T., Kuwabara, J., Ikawa, S., Yoshizawa, S., Shiraki, K., Kawasaki, H., Arakawa, R., & Kitano, K. (2014). Chemical modification of amino acids by atmospheric-pressure cold plasma in aqueous solution. *Journal of Physics. D, Applied Physics, 47*(28), 285403. doi:10.1088/0022-3727/47/28/285403

Thapa, R. & Wilson, G. (2016). *The importance of CD44 as a stem cell biomarker and therapeutic target in cancer.* Stem cells international.

Trachootham, D., Alexandre, J., & Huang, P. (2009). Targeting cancer cells by ROS-mediated mechanisms: A radical therapeutic approach? *Nature Reviews. Drug Discovery, 8*(7), 579–591. doi:10.1038/nrd2803 PMID:19478820

Vandamme, M., Robert, E., Lerondel, S., Sarron, V., Ries, D., Dozias, S., Sobilo, J., Gosset, D., Kieda, C., Legrain, B., Pouvesle, J.-M., & Pape, A. L. (2012). ROS implication in a new antitumor strategy based on non-thermal plasma. *International Journal of Cancer, 130*(9), 2185–2194. doi:10.1002/ijc.26252 PMID:21702038

Wee, P., & Wang, Z. (2017). Epidermal growth factor receptor cell proliferation signaling pathways. *Cancers (Basel), 9*(5), 52. doi:10.3390/cancers9050052 PMID:28513565

Welz, C., Emmert, S., Canis, M., Becker, S., Baumeister, P., Shimizu, T., Morfill, G. E., Harréus, U., & Zimmermann, J. L. (2015). Cold atmospheric plasma: A promising complementary therapy for squamous head and neck cancer. *PLoS One, 10*(11), e0141827. doi:10.1371/journal.pone.0141827 PMID:26588072

Xu, G., & Chance, M. R. (2007). Hydroxyl radical-mediated modification of proteins as probes for structural proteomics. *Chemical Reviews, 107*(8), 3514–3543. doi:10.1021/cr0682047 PMID:17683160

Yan, D., Sherman, J. H., & Keidar, M. (2017). Cold atmospheric plasma, a novel promising anti-cancer treatment modality. *Oncotarget*, *8*(9), 15977–15995. doi:10.18632/oncotarget.13304 PMID:27845910

Yan, D., Talbot, A., Nourmohammadi, N., Sherman, J. H., Cheng, X., & Keidar, M. (2015). Toward understanding the selective anticancer capacity of cold atmospheric plasma—A model based on aquaporins. *Biointerphases*, *10*(4), 040801. doi:10.1116/1.4938020 PMID:26700469

Yusupov, M. (2019). Transport of reactive oxygen and nitrogen species across aquaporin: A molecular level picture. Oxidative Medicine and Cellular Longevity, vol. 2019, Article ID 2930504, 11 pages

Yusupov, M., Lackmann, J.-W., Razzokov, J., Kumar, S., Stapelmann, K., & Bogaerts, A. (2018). Impact of plasma oxidation on structural features of human epidermal growth factor. *Plasma Processes and Polymers*, *15*(8), 1800022. doi:10.1002/ppap.201800022

Yusupov, M., Privat-Maldonado, A., Cordeiro, R. M., Verswyvel, H., Shaw, P., Razzokov, J., Smits, E., & Bogaerts, A. (2021). Oxidative damage to hyaluronan–CD44 interactions as an underlying mechanism of action of oxidative stress-inducing cancer therapy. *Redox Biology*, *43*, 101968. doi:10.1016/j.redox.2021.101968 PMID:33895486

Zou, Z., Chang, H., Li, H., & Wang, S. (2017). Induction of reactive oxygen species: An emerging approach for cancer therapy. *Apoptosis*, *22*(11), 1321–1335. doi:10.1007/s10495-017-1424-9 PMID:28936716

ABBREVIATIONS

AQPs: Aquaporins
CD44: Cluster Of Differentiation 44
CYGB: Cytoglobin
FEP: Free Energy Profile
GPX4: Glutathione Peroxidase 4
GSH: Glutathione
HA: Hyaluronan
hEGF: Human Epidermal Growth Factor
MD: Molecular Dynamics
NTP: Non-Thermal Plasma
RNS: Reactive Nitrogen Species
RONS: Reactive Oxygen and Nitrogen Species
ROS: Reactive Oxygen Species
SIRPα: Signal-Regulatory Protein Alpha

Section 2

Transport Properties of Plasmas

Chapter 5
Spacecraft Charging in Non-Maxwellian Plasmas at GEO Altitudes

Shahid Ali
https://orcid.org/0000-0003-1027-1317
National Centre for Physics, Pakistan

ABSTRACT

Spacecraft charging phenomenon at GEO altitudes is revisited by highlighting various spacecraft missions, spacecraft orbits, surface charge mitigating techniques, and particle distribution models. A comparative analysis is presented to study the onset of negative charging for spacecraft with single Maxwellian, Kappa, and q-nonextensive distributions. Since the plasma constitutes a mixture of two clouds having cold and hot electrons with their distinct thermal energies and densities, therefore it is important to investigate surface charging with double extensive and non-extensive distributions as well as analyzing the current-balance equations (CBEs) both analytically and numerically. The Whittaker function integral is employed to solve the power-law integrals. The q-distributed electrons show more pronounced energy tails to obey Tsallis statistics having nonstationary equilibrium states. Non-extensivity effect strongly modifies the CBE for spacecraft charging, supporting better in the limit $q<1$ in contrast to Maxwell and κ distributions. Numerically critical and anticritical thresholds are also identified to understand the onset of negative charging of various space-grade materials.

INTRODUCTION

Plenty of spacecraft and satellites orbit around the Earth for their special purposes, including the weather satellites, communication satellites, navigation satellites, scientific satellites, etc. From smartphones to broadcast and weather forecasting, satellites directly influence our daily lives and provide access to various space environments for scientific discoveries. The direct exposure of plasma disturbances to spacecraft generates charging or discharging on its surfaces and excess of charge affects the functionality of onboard electronic devices. This may in turn lead to incorrect analyses and conclusions about space

DOI: 10.4018/979-8-3693-0904-9.ch005

plasma environments. The surface charging also reduces the lifetime and performance of spacecraft & spaceships.

Space is a vibrant and complex realm, full of particles, radiations, and objects spanning a wide range of sizes (from very small size to very large size). The Earth is positioned at third in planets having an average distance of 150 million km from the sun. The solar wind as streams of charged particles moving with a speed ranging from 300-to-700 km per second compresses the Earth's magnetosphere via the pressure force during the dayside and stretches it on the night side. The interaction of solar wind results in the formation of asymmetrical Earth's geomagnetic field lines and an energetic hot plasma is eventually propelled towards the Earth from magnetotail, reaching geomagnetic altitudes at midnight. Hence space plasma environments can affect satellites and spacecraft as well as the climate on Earth.

Spacecraft and satellites can be charged by a number of charging combinations at GEO altitudes (Harris, 2003) including the (i) ambient electrons and ions, (ii) ambient electrons and ions with photoelectrons, (iii) ambient electrons and ions with secondary electrons, and (iv) ambient electrons and ions with both photo- and secondary electrons. The level of charging depends on the energy of plasma particles that are often described by the particle distribution functions. The use of sophisticated designs and quality materials has prevented spacecraft against the electrostatic discharges.

Spacecraft charging primarily depends on the charging mechanisms, including the plasma bombardment and photoelectric effect (see Fig.1). If a spacecraft is subjected to the plasma bombardment in eclipse, it will acquire a negative charge or potential (around 1kV) on its surface (Harris, 2003). Such a negative potential is frequently observed in many regions of Van Allen radiation belts and in magnetotail environment.

Figure 1. Surface charging mechanisms that account for plasma bombardment and photoelectric effects

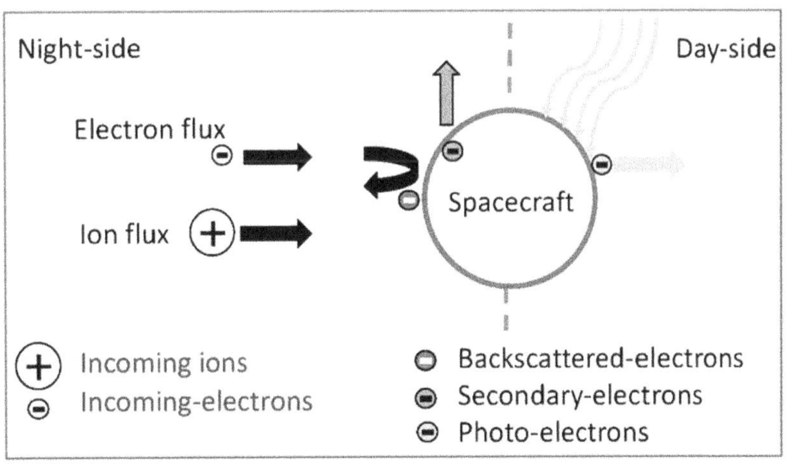

On the other hand, if solar radiations interact with the surface of spacecraft in sunlit side, it would lead to trigger a photoelectric effect and in turn releases photoelectrons from its surface. This produces a net positive charge about 2-3V potential on the spacecraft's surface (Harris, 2003). Thus, a spacecraft has negative potential in shady region of the Earth due to plasma bombardment and maintains a positive potential in sunny regions owing to the photoelectric effect.

SPACECRAFT'S ORBITS AROUND THE EARTH

These are the specified paths or orbits on which spacecraft and satellites orbit around the Earth, can be classified into the following three main categories (see Fig. 2a):

Figure 2. Schematic diagrams for (a) spacecraft Earth orbits and (b) GEO orbits orientation

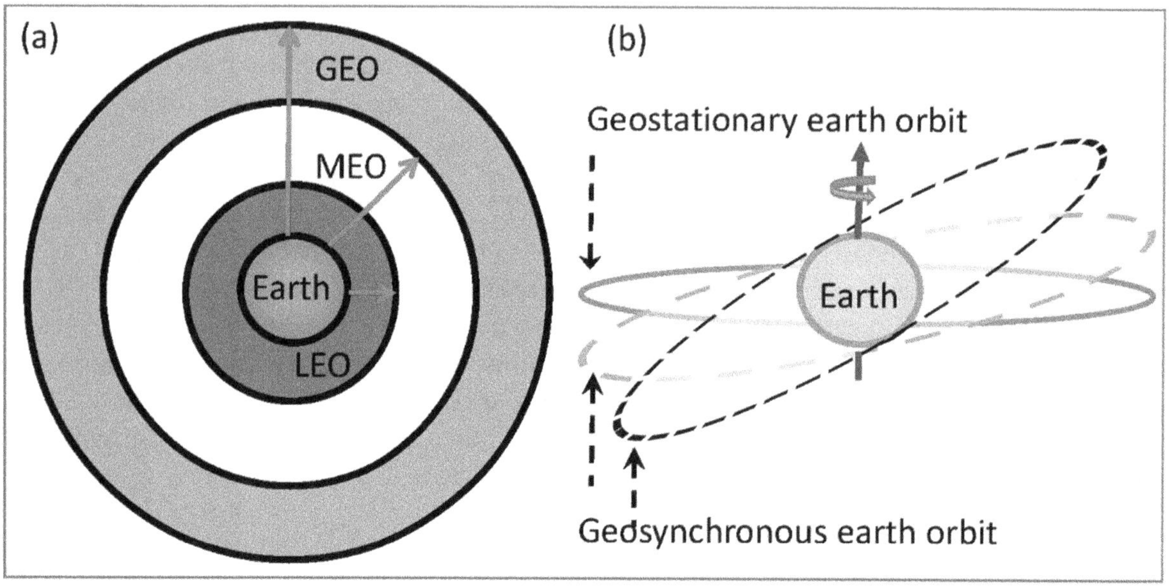

a. **Low-Earth Orbit (LEO)** is the most closer orbit to the Earth's surface, typically ranges from 160 km to 2,000 km in altitude. In LEO orbits, the satellites relatively complete a revolution around the Earth in short time and can be utilized for the Earth's observation, tele-communications and scientific missions.

b. **Medium Earth Orbit (MEO)** is located at intermediate altitudes usually between 2,000 km and 35,786 km above the Earth's surface. One of the best examples, are navigation satellites that are positioned in MEO and used in the Global Positioning System (GPS).

c. **Geosynchronous Earth Orbit (GEO)** is situated at an altitude approximately 36,000km from the Earth's surface. The GEO stationary satellites orbit around the Earth with the same rotational speed as that of Earth, allowing them to remain fixed relative to specific point on Earth's surface (see Fig. 2b). This unique characteristic makes them ideal for communication and weather satellites. It is also important to note that GEO spacecraft experiences unique conditions due to their high altitudes, for example, the number density (n_0 ~ 0.1-to-1cm^{-3}) is often low and thermal energy (~ keV) is high. The electron flux is comparatively two orders of magnitude higher than those of ion flux (Lai & Della-Rose, 2001).

SPACE-GRADE MATERIALS

Space-grade materials are specifically designed, tested, and certified before using, because these are materials that face harsh and unforgiving space environments. The latter contain extreme temperatures, powerful radiations and showers of micrometeoroids. In space, the plasma temperatures vary from hundreds of degrees Celsius in direct sunlight to hundreds of degrees below freezing in the shady areas. Multi-layer insulation materials can be used to protect satellites from the extreme temperatures and thermal gradients. Space-grade materials play a central role not only in minimizing charging effects on the spacecraft surface but also in construction and operation of satellites, space telescopes, space-based systems, etc. They have the ability to resist degradation and to maintain their properties during the missions. The vacuum can also cause outgassing and degradation of materials. Thus, for protecting spacecraft and satellites, some specialized wires, connectors, and circuit boards are designed to withstand space conditions and environments.

AN OVERVIEW OF SPACE MISSIONS AND SURFACE CHARGING

More than thousands of spacecraft and satellites orbit around the Earth, primarily operating to explore solar system including planets, moons, comets, and asteroids. Satellites often follow specific orbits for special purposes, for instance, to forecast weather conditions, controlling and positioning of aircrafts and ships, tele communication, etc. In early 1950s-to-1960s, many spacecraft missions were launched to carry out investigations involving space exploration and observed frequently accumulation of electrical charge on spacecraft surfaces. High energetic particles are found in the Earth's radiation belts that interact with spacecraft to emit secondary electrons from its surface building up an electric current. On the other hand, Vanguard and Explorer satellites have identified several charging issues. In the next decade, Astronauts were sent to the Moon by the Apollo program and who took the charging phenomenon as challenging task. In turn, many engineers and scientists have utilized the conductive coatings and grounding straps to minimize spacecraft surface charging. In 1970s-to-1980s, Skylab and Space Shuttle have also encountered spacecraft charging and mitigated these effects by using special tiles and coatings. In deep space missions, numerous investigations have been carried out by Voyager 1 and 2, to study charging effects of spacecraft during its interaction with harsh environments of Saturn, Jupiter and their larger moons. Later, Galileo and magnetospheric missions (Kivelson et al. 1997) have provided significant insights into space plasmas as regard to charging and magnetic configurations. In recent years, International Space Station has successfully been able to observe the Earth from space and to address charging phenomenon using specific materials and electrical grounding on the spacecraft surfaces (Clark et al. 2016). All above missions have adopted various strategies to mitigate charging effect on the spacecraft surfaces and provided a deep understanding of spacecraft-plasma interaction.

About two decades ago, charging mechanisms through spacecraft-plasma interaction have received a considerable attention of space plasma community due to adversarial effects (Garrett and Whittlesey 2000) to onboard delicate electronic devices. The concept of electric probe was implemented to study spacecraft charging phenomenon, where spacecraft acts as floating probe in the dynamical plasma environments (Langmuir et al. 1924). Matéo-Vélez et al. (2016) have utilized the sixteen years old Los Alamos National Laboratory (LANL) data and compared it with Polar Orbiting Environmental Satellites (POES) data to highlight impact of plasma environments on the surface charging due to electrons having energy in the range from 10-to-50keV. They provided design guidelines for spacecraft by taking into account the worst-case conditions of sunlight and eclipse. The existence of induced electrostatic potential difference is always a severe threat to space vehicles, which can be minimized substantially by knowing the characteristics of space-grade materials (Yu et al. 2022), the background plasma dynamics, designing and spinning of spacecraft. However, the existence of high energetic electrons (Ali et al. 2020, Pervaiz et al. 2023) significantly influence surface charging and dielectric materials of spacecraft can lead to internal charging of spacecraft (Hasting and Garrett 2024, Deforest 1972).

The charging of spacecraft is a very complex mechanism that can be categorized into two types, the surface charging due to electron-ion bombardment and surface charging caused by photoelectric effect. In particular, Garrett (Garrett 1981) has classified the study of spacecraft charging into four phases, where the first phase corresponds to pre-space age having probe theory (Johnson & Meadows 1955) and charge accumulates on the rockets in ionosphere, eventually ending with the successful launch of "Sputnik" in 1957. The second phase focuses on the distribution of charged particles and their concentration on the space assets in ionospheric plasmas (Gringauz 1958), whereas third phase begins with an accurate measurement of spacecraft charging, analysis of electric field induction and Earth's geomagnetic field rotation. In this context, some efforts have been made to study interactions of rapidly moving systems in atmosphere (Chopra, 1961) and equilibrium charge potential on a body in upper atmosphere and interplanetary space (Whipple 1965). The fourth phase highlights the fundamentals of space plasma dynamics and spacecraft charging. In particular, Deforest (Deforest 1972) describes spacecraft charging at GEO altitudes with the analysis from SCATHA P78-2 spacecraft (Garrett, 2000) at the end.

In 1980, Rubin et al. (1980) studied electron threshold temperature to determine the dysfunctionality of the equipment below a threshold. The latter varies from material to material and strongly depends on the surface material constants relating to secondary and backscattered electron coefficients. It is found that threshold temperature (Garrett 1981) essentially occurs for specific material and configuration with an electrostatic potential (Lai and Della-Rose 2001) in the range from 1.5-to-2.5eV and temperature 400eV-to-3000eV. Lai and Della-Rose (2001) have presented the charging of spacecraft at high voltages and highlighted the importance of equilibrium number density, temperature, space weather conditions, and geophysical regions. In particular, Lai and Tautz (2006) presented a statistical study and theoretically found the existence of critical temperature (in a range from 1-to-2keV) at GEO altitudes for potential below -200V. They also compared their study with observed data of geo satellites. After analyzing a large amount of data, there exists a linear trend between the spacecraft potential and ambient electron temperature at potential below -5kV. This has led to an intercept at finite temperature and well-agreed to critical temperature of theoretical prediction. In the sunlit side, spacecraft surfaces are mostly charged positive because of direct exposure to sun's radiations to giving rise to emission of photo-electrons. Conversely, in the shaded region of the Earth, a negative charge is developed on spacecraft surface due to plasma bombardment.

How To Mitigate Spacecraft Surface Charging?

When spacecraft and satellites pass through the GEO environment, they intercept solar wind electron and ions, solar radiations, Earth's magnetic fields, and cosmic rays. As a result, harsh space environments substantially damage the surfaces of satellites, as well as dysfunction the electronic devices and circuits within it. Some of the following techniques that can be useful for reducing adversarial effects of the spacecraft charging are as under:

a. **Conductive Coatings** The surfaces of spacecraft and satellites are commonly coated with Indium Tin Oxide and Aluminum in order to prevent them from electric charge accumulation.
b. **Tethering** is a process used to tie up one spacecraft with another object (parent spacecraft), to provide an electric path for the charge dissipation. Consequently, tethers maintain a balance between the electric charge on spacecraft and surrounding environment.
c. **Active Discharge Systems** can be installed on spacecraft to control accumulated electric charge on its surface. In such systems, ion emitters and corona brushes are used to discharge spacecraft and to maintain electrical neutrality.
d. **Potential Control** provides same electrical potential to various parts of spacecraft, making them electrically connected. This may also prevent the buildup of charge on spacecraft's surface.
e. **Dielectric Materials** are used during the manufacturing of spacecraft's components that do not to allow electric conduction and thus help in minimizing charging effects on its surface.
f. **Electron Emission Devices** are installed in spacecraft emitting low-energy electrons to neutralize the charges on its surface. Hollow cathodes and plasma contactors are the common examples.
g. **Orientation Control** by orientation of spacecraft, one can effectively reduce charging effect on its surface. This means that one has to make spacecraft to be less exposed to solar radiation and charged particles.
h. **Material Selection** matters a lot in way to reduce charging impact. Therefore, materials of less susceptibility are preferred, which have attributes in minimizing the charging effects on spacecraft surface.
i. **Proper Solar Panel Designs** also help us to minimize charging effects. The panels have large surface area and their electrical bonding to spacecraft's structure can be addressed by using the conductive coatings.
j. **Monitoring and Mitigation Algorithms** the real-time monitoring systems and algorithms are used to detect charging events and trigger mitigation automatically. This proactive approach is helpful to prevent damages to spacecraft systems.
k. **Space Weather Forecasting:** Forecasting of space weather conditions and geomagnetic storms play a crucial role and the risk of spacecraft charging can be reduced if preventive measures are taken during the periods of heightened space weather activity.

PARTICLE DISTRIBUTION MODELS

Maxwellian and Kappa Distributions

In standard situations, the plasma particles obey the well-known "Maxwellian (extensive) distribution" provided particle velocities and energies are relatively low (~10keV or less) in the plasma system (Harris 2003). The plasma particles exchange their momenta and energies through inter-particle collisions and eventually gain a thermal equilibrium. The Maxwellian distribution represents the probability of particles per unit volume, strictly based on the Maxwell-Boltzmann statistics. Since this distribution is achieved applied to collisions dominated plasmas, therefore it characterizes large collisional cross-sections. However, the use of Maxwellian distribution function in all circumstances is not a good choice especially when an observed data corresponds to space plasma environments. The power-law (in velocity space) distribution that exhibits high-energy tails (Vasyliunas, 1968) in the distribution curves, is the only choice to be adopted for adequate modelling of high energetic plasma particles (>10keV). The enhancement in energies can caused by the wave-particle interaction, plasma turbulence, and external forces acting on the plasma systems (e.g., solar wind, ionosphere, magnetosphere, interstellar medium, etc.). Thus, power-law distributions play an essential role in the situations where particle nonthermality effects exist due to smaller collisional cross-sections and consequently collisional effects become no longer effective. This necessitates the use of non-Maxwellian distributions in space plasmas. Figure 3 represents the normalized distribution functions [including the Maxwell and Kappa (Vasyliunas, 1968) distributions] against the normalized particle speed for varying the Kappa index κ. It is examined that high energetic tails appear at low values of κ. As the Kappa values increase, the plasma system departs from superthermal to thermal one and eventually approach to the Maxwellian distribution at $\kappa \rightarrow \infty$.

Figure 3. High energy tails are associated with Kappa distribution at low-values of Kappa index and distribution curve approaches to the Maxwellian one at $\kappa \rightarrow \infty$

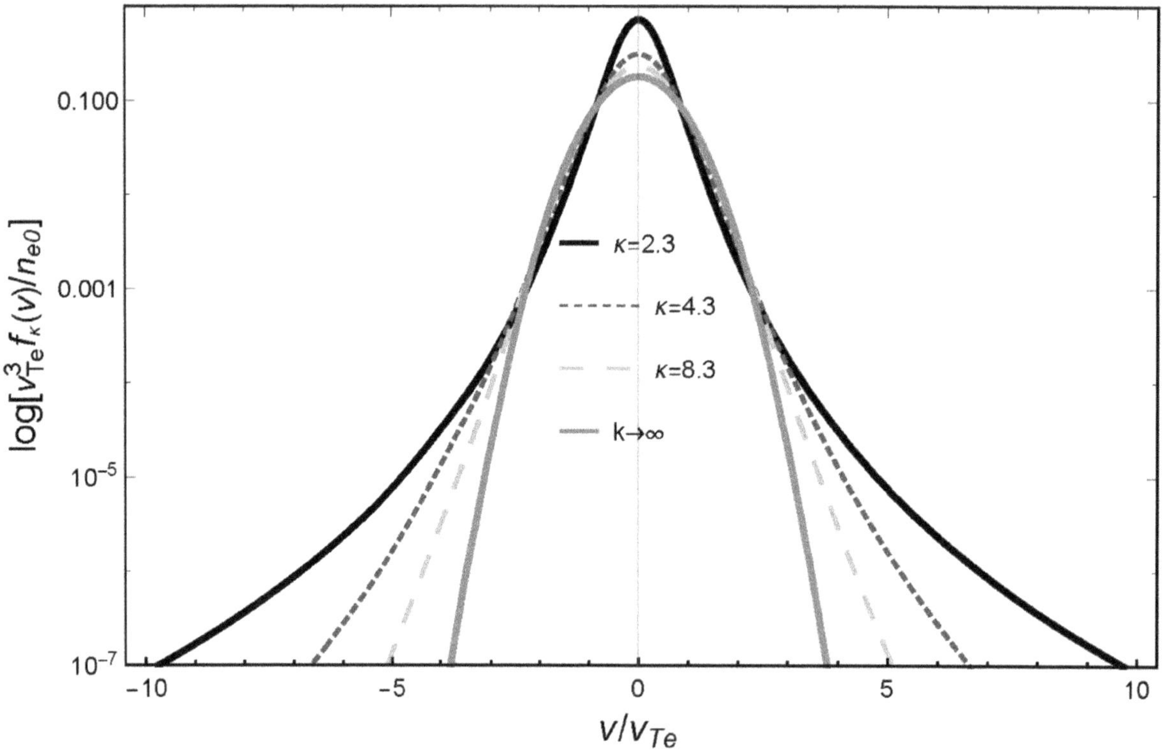

The Maxwellian and Kappa distributions (Vasyliunas, 1968) in 3D forms can be, respectively, represented by

$$f_M(v) = \frac{n_{e0}}{\pi^{3/2} v_{Te}^3} \exp\left(-\frac{v^2}{v_{Te}^2}\right), \tag{1}$$

and

$$f_{\kappa}(v) = \frac{n_{e0}\Gamma(\kappa+1)}{\pi^{3/2}v_{Te}^{3}\Gamma(\kappa-1/2)(\kappa-3/2)^{3/2}}\left(1+\frac{v^2}{(\kappa-\frac{3}{2})v_{Tc}^2}\right)^{-\kappa-1}. \tag{2}$$

Note that the Maxwell distribution describes particle velocities in terms of exponential law at thermal equilibrium, while the Kappa distribution being a power-law distribution in velocity space indicates the departure from the Maxwell distribution. We may convert (1) and (2) into corresponding energy distributions by taking into account the substitutions such as $\varepsilon=mv^2/2$, $vdv=d\varepsilon/m$ and $f(v)\mathbf{dv}=f(\varepsilon)d\varepsilon$, respectively, yielding the following results (Ali et al. 2020)

$$f_M(\varepsilon)d\varepsilon = \frac{2n_{e0}\varepsilon^{1/2}}{\sqrt{\pi}}\left(\frac{1}{T_e}\right)^{3/2}\exp\left(-\frac{\varepsilon}{T_e}\right)d\varepsilon, \tag{3}$$

and

$$f_{\kappa}(\varepsilon)d\varepsilon = \frac{2n_{e0}\Gamma(\kappa+1)\varepsilon^{1/2}}{\sqrt{\pi}\Gamma(\kappa-1/2)\{(\kappa-3/2)T_e\}^{3/2}}\left\{1+\frac{\varepsilon}{(\kappa-\frac{3}{2})T_e}\right\}^{-\kappa-1} d\varepsilon \tag{4}$$

where n_{e0} denotes an equilibrium electron number density, $v_{Te}\left[=\left(2T_e/m_e\right)^{1/2}\right]$ is the electron thermal speed, T_e the electron temperature in energy units. The symbols Γ and κ indicate the gamma function and superthermality (spectral) index, respectively. The Kappa index pinpoints the deviating effect from a thermal equilibrium and satisfies the condition $\kappa > 3/2$ in our numerical investigation for making sure that particle effective speed is assumed to be realistic in the present model. In the past decades, a great deal of interest was developed to study spacecraft-plasma interaction with non-Maxwellian distributions. In particular, Ginet et al. (2013) have introduced new models to investigate high energetic trapped particles and space plasma environments. Later, Plis et al. (2017) and Cooper et al. (2017) have discussed degradation of materials by exposing them to an electron energy of 90keV and observed significant changes in optical absorption of the materials. Several theoretical and computational models were considered to obtain threshold conditions for negative charging of spacecraft with different non-Maxwellian distributions (Lai 2009, Lai 2012, Harris 2003, Ali et al. 2020). In addition, Meyer-Vernet (1999) have also utilized the simplest kinetic approach for investigating the electrostatic potential in a solar wind plasma to account for the Maxwellian and Kappa distributed particles.

Non-Extensive Distribution Function

Plasma being an ionized state of matter exhibits the behavior of charged particles, depending on both the extensive (Maxwellian) and non-extensive statistics and properties. Non-extensive properties (Boon and

Tsallis 2005) have their vital role in describing dynamics of plasmas in fusion research, astrophysics, and plasma technologies. In an Euclidean space-time environment, thermodynamic extensivity (or additivity) holds and total entropy of the compound system is equivalent to the sum of individual entropies, namely, $S_T(A+B)=S_1(A)+S_2(B)$, where $S_T(A+B)$ is the total entropy of the combined system with individual entropies $S_1(A)$ and $S_2(B)$. However, in a non-Euclidean space-time environment, the entropy is treated as non-additive quantity relying on the Tsallis statistics (Tsallis, 1988) rather than the Maxwell-Boltzmann statistics. Therefore, pseudo-additivity (or non-extensivity) demands the total entropy of independent subsystems A and B as $S_T(A+B)=S_1(A)+S_2(B)+(1-q)S_1(A)S_2(B)$, where q represents the strength of non-extensivity. It is worth to mention that extensive properties scale linearly with the size or amount of the system, whereas non-extensive properties are affected by the size (or extent) and hence do not represent linear scaling behavior. Extensive and non-extensive properties of the particles have a significant role in thermodynamics and statistical physics (Du 2004a, 2004b). In recent years (Ali et al. 2020, Fareeha et al. 2023), spacecraft charging and its numerical analyses have been carried out with both extensive and nonextensive electron distribution functions. The latter exhibits more pronounced energy tails in comparison with Kappa distribution. It is also examined that q-distribution triggers the charging onset more earlier (at low thermal energies) than the extensive and Kappa distributions.

Figure 4. Non-extensive distribution is plotted against the normalized particle speed to show the behavior of non-extensive particles in the limits (a) q<1 (left panel) and (b) q>1 (right panel)

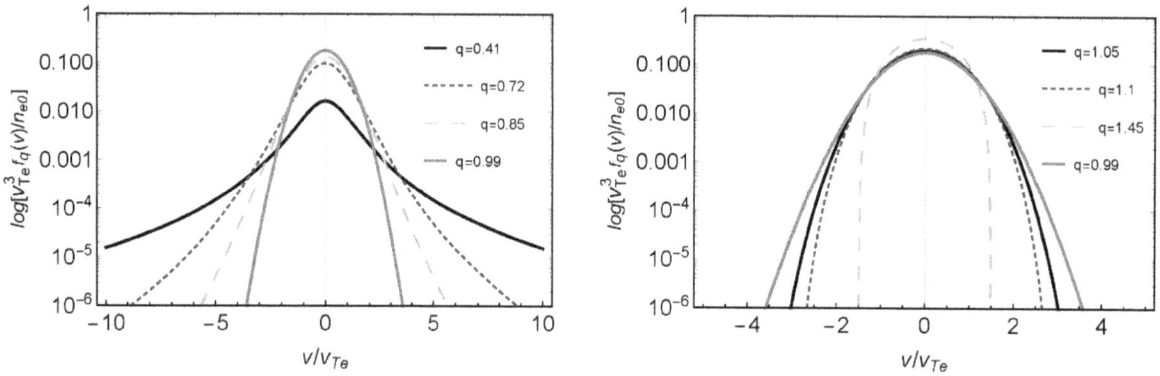

Thus, q-distribution (Boon and Tsallis 2005, Tsallis 1988) is utilized for calculating particle probabilities of more pronounced energy tails (Livadiotis and McComas, 2009), to account long range interactions. Recently, spacecraft charging (Ali et al. 2020) has been studied by incorporating the following 3D q-distribution function purely relying on the Tsallis statistics:

$$f_q(v) = C_q\left(1-(q-1)\frac{v^2}{v_{T_e}^2}\right)^{1/q-1} \tag{5}$$

with

$$C_q = \frac{n_{e0}\Gamma\left(\frac{1}{1-q}\right)\left(\frac{1-q}{\pi v_{Te}^2}\right)^{3/2}}{\Gamma\left(\frac{1}{1-q}-\frac{3}{2}\right)} \text{ for -1<}q\text{<1}$$

and

$$C_q = \frac{n_{e0}\Gamma\left(\frac{1}{q-1}+\frac{3}{2}\right)\left(3q-1\right)\left(\frac{q-1}{\pi v_{Te}^2}\right)^{3/2}}{2\Gamma\left(\frac{1}{q-1}\right)} \text{ for } q\geq 1.$$

The conversion of Eq. (5) can be made to obtain the *q-nonextensive energy distribution* as

$$f_q(\varepsilon)d\varepsilon = \frac{2n_0\Gamma\left(\frac{1}{1-q}\right)\left(\frac{1-q}{T}\right)^{3/2}}{\sqrt{\pi}\Gamma\left(\frac{1}{1-q}-\frac{3}{2}\right)}\varepsilon^{1/2}d\varepsilon\left\{1-(q-1)\frac{\varepsilon}{T}\right\}^{\frac{1}{q-1}}, \tag{6}$$

for -1<q<1 and

$$f_q(\varepsilon)d\varepsilon = \frac{n_0\Gamma\left(\frac{1}{q-1}+\frac{3}{2}\right)(3q-1)\left(\frac{q-1}{T}\right)^{3/2}}{\sqrt{\pi}\Gamma\left(\frac{1}{q-1}\right)}\varepsilon^{1/2}d\varepsilon\left\{1-(q-1)\frac{\varepsilon}{T}\right\}^{\frac{1}{q-1}} \tag{7}$$

for $q\geq 1$.

Observe that non-extensive behavior of the plasma particles is only relevant in the limit $q\neq 1$ and if q tends to unity, the non-extensive q-distribution exactly approaches to the Maxwellian distribution. Figure (4) displays the impact of normalized q-distributed function against the normalized speed with varying q-parameter. It may be noticed that more pronounced energy tails in appear the distribution curves when q<1 is taken into account. Here high energy tails exist at low probabilities and high probabilities are suppressed. On the other hand, the limit $q>1$ suppresses the lower probabilities and modification of high probabilities (see Figure 4b) occurs.

CURRENT BALANCE EQUATION FOR SPACECRAFT CHARGING

When a GEO spacecraft or satellite enters into the outer magnetosphere, it interacts with relatively cool electrons and ions, because the plasma conditions correspond to quiet period (Pandya et al. 2022) and favor the single particle distribution. Since ions being the heavier species, their effects often neglected (Lai & Della-Rose, 2001) in the charging of spacecraft for larger mass as compared to lighter electrons that carry larger amount of electron flux. The present model is valid in the eclipse (shaded region behind the Earth) where photoelectron currents are ignored. Now assuming a normal incidence of the electron flows to the spacecraft's surface, we can easily determine spacecraft charging by the current-balance equation due to incoming and outgoing electrons.

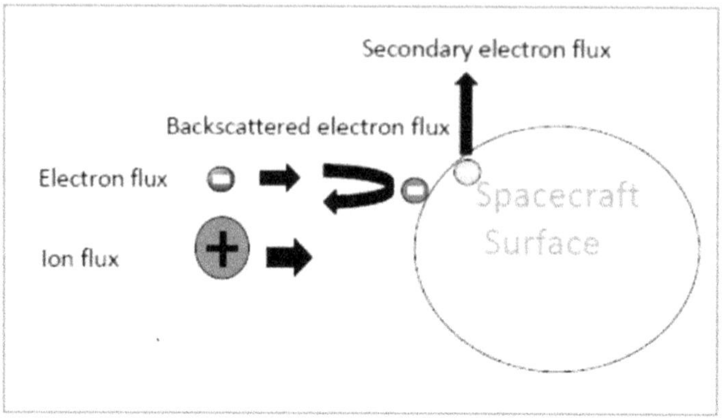

In the perspective of incoming electrons, we must consider a uniform electron current density flux (i.e., the rate of flow of charge per unit area having a uniform incident flux on the surface area "A"). On the other hand, the outgoing electrons contribute through the yields of secondary and backscattered electron emissions, often described by the respective coefficients $\delta(\varepsilon)$ and $\eta(\varepsilon)$ playing an effective role in the onset of negative charging.

CBE With Single Maxwellian df

The current-balance equation (CBE) for a single Maxwellian plasma can be governed by the incoming and outgoing electron current density fluxes (Lai 2012), as

$$J_{in} = J_{out} \equiv J_{se} + J_{be}$$
$$\int_0^\infty \varepsilon^{1/2} f_M(\varepsilon) d\varepsilon = \int_0^\infty \varepsilon^{1/2} f_M(\varepsilon)[\delta(\varepsilon) + \eta(\varepsilon)] d\varepsilon, \tag{8}$$

with secondary and backscattered electron coefficients (Sanders and Inouye, 1978; Prokopenko and Laframboise 1980), respectively, as

$$\delta(\varepsilon) = c\left\{\exp\left(-\frac{\varepsilon}{a}\right) - \exp\left(-\frac{\varepsilon}{b}\right)\right\}, \tag{9}$$

and

$$\eta(\varepsilon) = A - B\exp(-\varepsilon C), \tag{10}$$

where J_{in}, J_{se}, and J_{be} represent the current densities of incident electrons, secondary electrons and backscattered electrons. Generally, the electron current density flux is defined in terms of particle distribution function $[f_M(v_j)]$ by the relation $J_j = Q_j \int v_j f_M(v_j) dv_j^3$ with Q_j being the charge of the jth species ($j=in$ for incident electrons, $j=se$ for secondary electrons and $j=be$ for backscattered electrons). It is worth to mention that for calculating current densities, we have converted the velocity distribution function $f_M(v_j)\, dv_j^3$ into the energy variable. As a result, one experiences the contribution of primary (Maxwellian) electrons on the LHS of (8) in the form of energy ε. The contribution of outgoing (Maxwellian) electrons appears on the RHS of (8) containing the coefficients of both secondary and backscattered electrons, which are strongly dependent on the energy and surface material constants (Lai, 2011). However, these coefficients do not rely on the electron temperature and density parameters. The secondary electron yield is involved in small alphabets $a(=4.3E_{max})$, $b(=0.367E_{max})$, and $c(=1.37\delta_{max})$ specifying the surface material constants with maximum energy ε_{max} and maximum yield δ_{max} Moreover, in the backscattered yield, we have the capital alphabets, i.e., A, B, and C, which refer to the surface material constants and are significantly depended on atomic number Z of the materials.

Equation (8) can be written in the following form

$$\frac{\int_0^\infty d\varepsilon \varepsilon^{1/2} f_M(\varepsilon)[\delta(\varepsilon)+\eta(\varepsilon)]}{\int_0^\infty d\varepsilon \varepsilon^{1/2} f_M(\varepsilon)} = 1. \tag{11}$$

This is the CBE in the form of average, and more precisely by absorbing the factor $\varepsilon^{1/2}$ within the distribution function, we may eventually express (11) in a shorthand notation, as

$$\langle \delta(\varepsilon) + \eta(\varepsilon) \rangle = 1 \tag{12}$$

Inserting Eqs. (1), (9) and (10) into (8), the integrals can be solved to finally obtain the following result (Harris 2003, Ali et al. 2020)

$$c\left\{\left(1+\frac{T}{a}\right)^{-2} - \left(1+\frac{T}{b}\right)^{-2}\right\} + A - \frac{B}{(1+TC)^2} - 1 = 0. \tag{13}$$

This is the well-known sixth-order current-balance equation in terms of electron temperature and is firstly numerically solved by Harris (2003) for spacecraft charging due to Maxwellian electron currents. We have noticed that the first four roots are unphysical and therefore neglected, whereas the remaining two roots are physical and lead to anti-critical (T_A*) and critical (T_C*) temperatures. For non-activated Copper-Beryllium (CuBe) materials, these roots in terms of temperature are turned out to be as Te(= –3.924, –1.687, –1.559, –0.211, 0.0196, and 1.341) keV. However, anti-critical and critical (threshold condition) temperatures (i.e., T_A*=0.0196keV and T_C*=1.341keV) not only vary from material to material but also strongly rely on the particle energy distributions. Equation (13) is only applicable in the eclipse or night-side of the Earth and gives rise to the threshold condition above which the negative charging onset is triggered for spacecraft surface. Also the normalized net current flux initially reaches to a maximum value at low electron energy and then gradually decreases with increasing electron thermal energies.

Figure 5. Threshold conditions exist for the onset of negative charging at GEO altitude Maxwellian plasma
Source: Harris (2003)

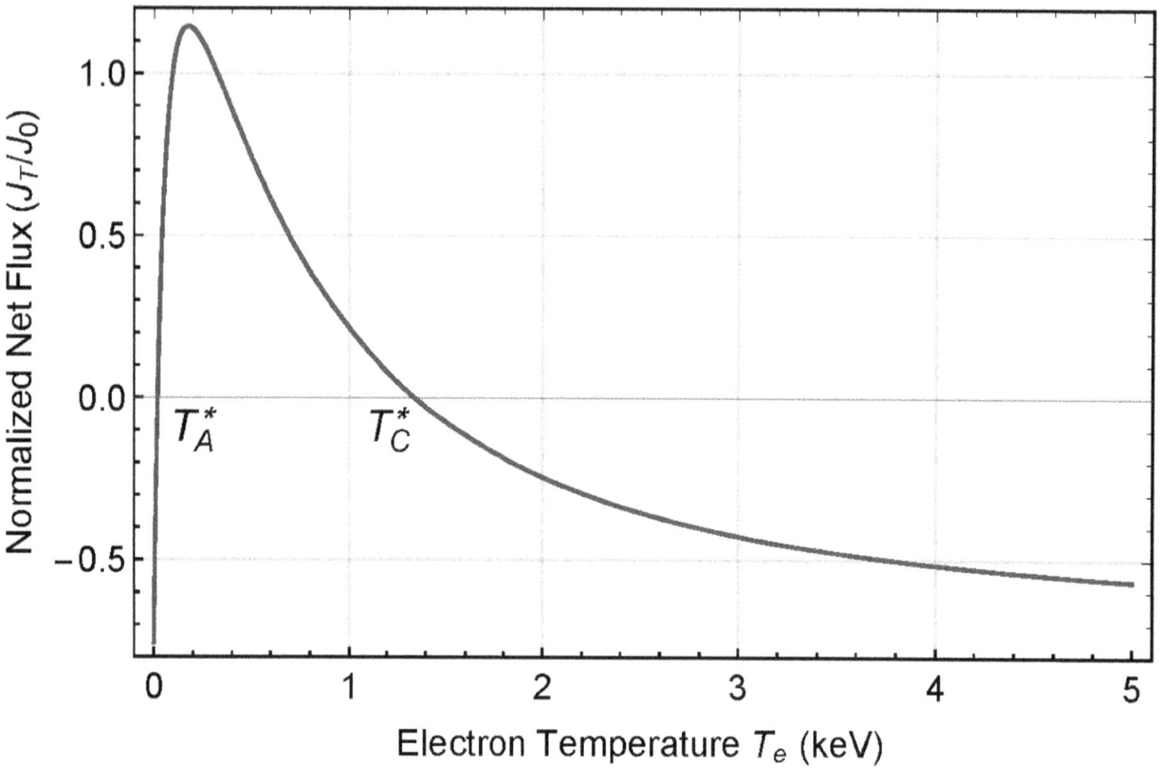

CBE With Single Kappa-df

The Kappa distribution is also known as the power-law distribution, is normally used for studying superthermal (accelerated) particles. It generalizes the Maxwell-Boltzmann distribution function and fits better based on the empirical data coming from space plasma environments. Kappa distribution function (Vasyliunas, 1968) exhibits high energy tails which cannot be captured by the usual Maxwellian distribution. The current-balance equation thus becomes modified by the presence of Kappa-distributed electrons and hence Eq. (8) takes the following form

$$\int_0^\infty \left(1+\frac{\varepsilon}{\left(\kappa-\frac{3}{2}\right)T_e}\right)^{-\kappa-1} \varepsilon d\varepsilon = c\int_0^\infty \exp\left(-\frac{\varepsilon}{a}\right)\left(1+\frac{\varepsilon}{\left(\kappa-\frac{3}{2}\right)T_e}\right)^{-\kappa-1} \varepsilon d\varepsilon$$
$$-c\int_0^\infty \exp\left(-\frac{\varepsilon}{b}\right)\left(1+\frac{\varepsilon}{\left(\kappa-\frac{3}{2}\right)T_e}\right)^{-\kappa-1} \varepsilon d\varepsilon$$
$$-B\int_0^\infty \exp(-\varepsilon C)\left(1+\frac{\varepsilon}{\left(\kappa-\frac{3}{2}\right)T_e}\right)^{-\kappa-1} \varepsilon d\varepsilon$$
$$+A\int_0^\infty \left(1+\frac{\varepsilon}{\left(\kappa-\frac{3}{2}\right)T_e}\right)^{-\kappa-1} \varepsilon d\varepsilon \tag{14}$$

It is worth expressing that electron equilibrium density, electron temperature and other constants are cancelled out as common quantities form both sides of Eq. (14). Using the Whittaker function integrals, we may easily integrate over the energy variable to obtain the following result (Harris, 2003)

$$1=\kappa(\kappa-1)\left\{cU_1\left(2\mid 2-\kappa\mid\left(\kappa-\frac{3}{2}\right)\frac{T_e}{a}\right)-cU_2\left(2\mid 2-\kappa\mid\left(\kappa-\frac{3}{2}\right)\frac{T_e}{b}\right)\right.$$
$$\left.-BU_3\left(2\mid 2-\kappa\mid\left(\kappa-\frac{3}{2}\right)T_eC\right)\right\}+A, \tag{15}$$

where U_1, U_2, and U_3 are the Whittaker functions that arise from the power-law integrals. Numerical analysis (Harris, 2003) has shown that current-balance equation based on the Whittaker function approach reduces more quickly to the Maxwellian case owing to very little chance of error to round off in numerical calculations. It is found that Eq. (15) agrees numerically if we compare it with the following current-balance equation obtained by hypergeometric function (Harris, 2003)

$$1=\kappa(\kappa-1)\left(\frac{2\kappa-3}{2}T_e\right)^{-\kappa-1} ca^{1-\kappa}U\left(\kappa+1,\kappa,(2\kappa-3)\frac{T_e}{2a}\right)-cb^{1-\kappa}$$
$$\times\ U\left(\kappa+1,\kappa,(2\kappa-3)\frac{T_e}{2b}\right)-BC^{1-\kappa}U\left(\kappa+1,\kappa,(2\kappa-3)\frac{T_eC}{2}\right)+A \tag{16}$$

It may be noticed that Kappa-distributed curves approach to the Maxwellian curve (solid blue curve) by increasing the value of electron superthermality index κ = 1.54, 1.64, 2, and Maxwllian. At high superthermality index κ (= 1.54), we obtain anti-critical temperature T_A*=0.105keV at very low electron thermal energy while critical temperature (T_C*=11.445keV) at higher electron thermal energy above which the onset of negative charging begins. As long as the Kappa index increases the critical temperature shifts towards the lower electron thermal energies and eventually approaching the Maxwellian curve. It is clear that the magnitude of the normalized net current flux increases with increasing

Kappa-values. Harris (2003) has revealed that space observations at GEO altitudes support the onset for negative currents caused by the Maxwellian electrons as compared to superthermal electrons modelled by the Kappa distribution function.

Figure 6. The current-balance equation (15) is plotted to study normalized net flux density as a function of electron thermal energy with different values of Kappa index

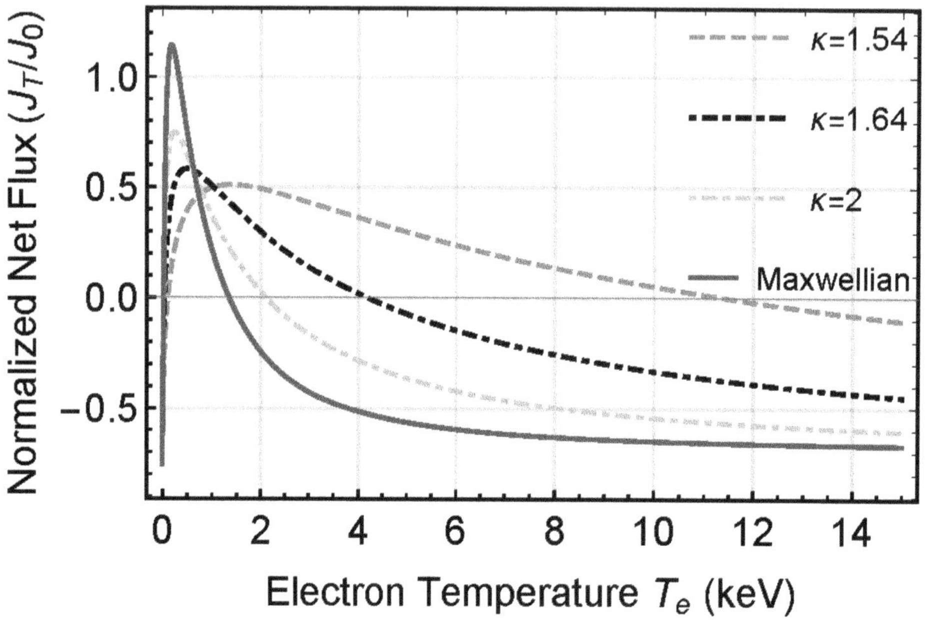

CBE With Single Non-Extensive q-df

Superthermal electrons with even more pronounced energy tails can be described by a generalized nonextensive (power-law) q-distribution, which is convertible to non-extensive Kappa distribution (Livadiotis and McComas, 2009). The latter is completely different from the standard Kappa distribution (Vasyliunas, 1968). For $q\neq1$ a non-extensive distribution is accounted for describing non-equilibrium stationary states and showing high energetic tails at low probabilities in the distribution function. For $q\to1$ non-extensive distribution exactly coincides with the Maxwellian distribution. Proceeding as before, the current-balance equation with a non-extensive distribution function can be solved within the framework of Whittaker function approach, as

$$1=\frac{2q^2-q}{(1-q)^2}\left\{cU_1\left(2\middle|3+\frac{1}{q-1}\middle|\frac{1}{1-q}\frac{T_e}{a}\right)-cU_2\left(2\middle|3+\frac{1}{q-1}\middle|\frac{1}{1-q}\frac{T_e}{b}\right)\right.$$
$$\left.-BU_3\left(2\middle|3+\frac{1}{q-1}\middle|\frac{1}{1-q}CT_e\right)\right\}+A \tag{17}$$

This is a modified current-balance equation due to non-extensive q-distribution function, which can be used to determine the critical temperature for triggering onset of spacecraft charging. Here U_1, U_2, and U_3 denote the Whittaker functions arising from the power-law integrals. The current-balance expression (17) approaches to the Maxwellian (extensive) case in the limit $q \to 1$ and for $q \neq 1$, the critical temperature appears at low electron thermal energy due to Tsallis statistics to identify the nature of systems involving the long-range interactions.

Sub- and Superextensivity Limits

In the super-extensive limit $q<1$, the q-nonextensive distribution function can be expressed into a non-extensive Kappa distribution function (Livadiotis, & McComas, 2009; Du, 2004a; 2004b) by taking into account $q = 1 - \frac{1}{\kappa}$. Consequently, the current-balance equation (17) yields the following result

$$(2 - 3\kappa + \kappa^2)\,(cU_1 - cU_2 - BU_3) + A - 1 = 0, \tag{18}$$

where

$$U_1(a_1|c_1|\chi) = U_1\left(2|3-\kappa|\tfrac{\kappa T_e}{a}\right),\ U_2(a_1|c_1|\chi) = U_2\left(2|3-\kappa|\tfrac{\kappa T_e}{b}\right),$$

and $U_3(a_1|c_1|\chi)=U_2(2|3-\kappa|\kappa T_e C)$. Similarly, for sub-extensive (Lima et al., 2000; Leubner, 2004) limit q>1, we need to substitute $q = 1 + \frac{1}{\kappa}$ in Eq. (17) to obtain

$$(2 + 3\kappa + \kappa^2)\,(cU_1 - cU_2 - BU_3) + A - 1 = 0, \tag{19}$$

with

$$U_1(a_1|c_1|\chi) = U_1\left(2|3+\kappa|\tfrac{-\kappa T_e}{a}\right),\ U_2(a_1|c_1|\chi) = U_2\left(2|3+\kappa|\tfrac{-\kappa T_e}{b}\right),$$

and $U_3(a_1|c_1|\chi)=U_3(2|3+\kappa|-\kappa T_e C)$. Thus, Eq. (17) reduces to non-extensive Kappa distribution in the limits of super-extensivity and sub-extensivity. It may be noticed that super-extensivity system [via Eq. (18)] contains an opposite sign to spectral index as compared to subextensivity system [via Eq. (19)].

Solid blue curve represents the standard Maxwellian electrons whereas dashed red, dot-dashed black and dot-dashed green curves exhibit non-extensivity effect when $q \neq 1$. The current initially increases and reaches to maximum at low electron temperature and then gradually decreases with increasing thermal energy. The onset of negative charging (T_C^*=0.351keV) occurs at low thermal energy as compared to the Maxwellian case at T_C^*=1.341keV. In this model, we have carried out spacecraft charging investigations within the limit 0.33 < q < 1 in order to have the finite energy (Verheest, 2013), which is important for introducing concepts of pressure/temperature in the context of fluid description for Tsallis particles.

Figure 7. Normalized CBE (17) is plotted against the electron thermal energy for different values of non-extensive parameter q

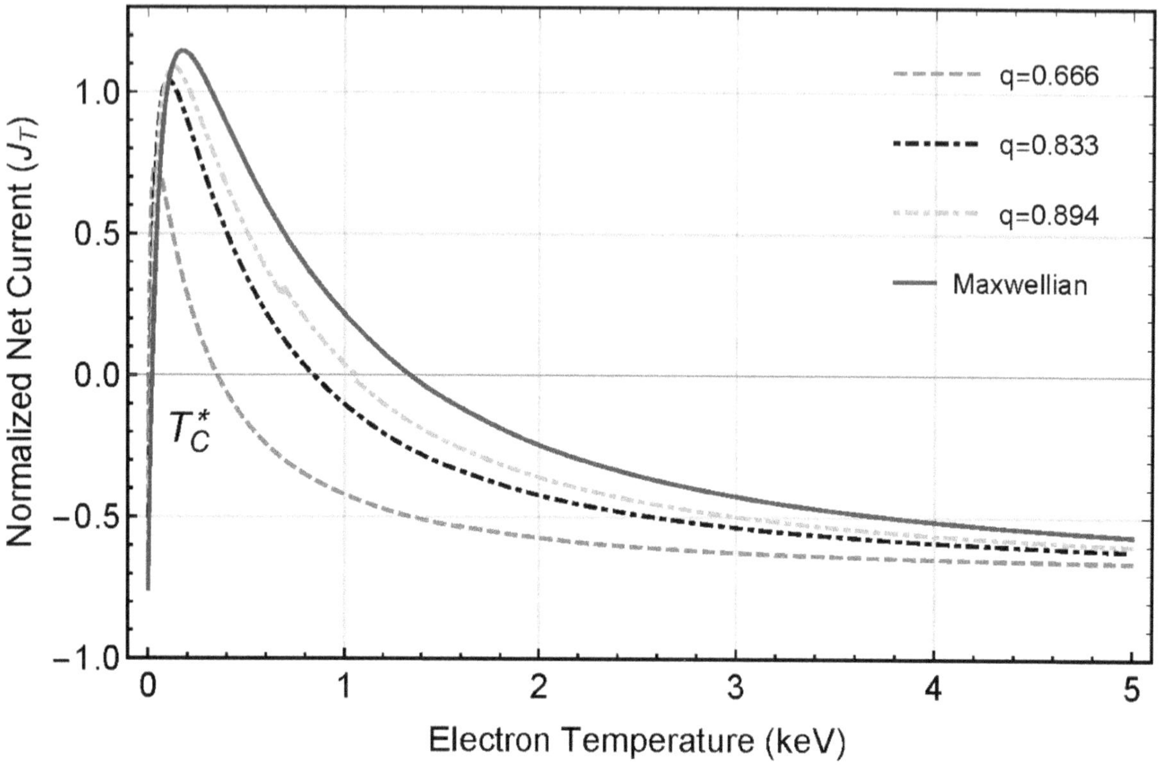

CBE With Double Maxwellian dfs

At GEO altitudes, the plasma varies in time especially during the quiet period of sun. It can be treated as cold to follow a single distribution function (f_1) with electron equilibrium number density and temperature (n_1, T_1). However, the arrival of a relatively hot plasma cloud in such an environment leads to a new particle distribution function (f_2) for electrons with different number density and temperature (n_2, T_2) for electrons. As a result, the plasma is mixed and exhibits two groups of electrons (Lai, 2012; Huang et al., 2015; 2017) at different energies (T_{01},T_{02}) and number densities (n_{01},n_{02}). In such a scenario, the plasma can be described by the sum of two Maxwellian distribution functions for electrons, given by $f_M(\varepsilon)=f_{M1}(\varepsilon)+f_{M2}(\varepsilon)$. This model also neglects the contribution of ion currents due to larger inertia as compared to electrons. Taking into consideration the normal incidence of electron flux, the modified current-balance equation (8) then yields the following form (Lai, 2012):

$$\int_0^\infty \varepsilon^{1/2} f_{M1}(\varepsilon) d\varepsilon = -\int_0^\infty \varepsilon^{1/2} f_{M2}(\varepsilon) d\varepsilon + \{\delta(\varepsilon)+\eta(\varepsilon)\} \times \left\{\int_0^\infty \varepsilon^{1/2} f_{M1}(\varepsilon) d\varepsilon + \int_0^\infty \varepsilon^{1/2} f_{M2}(\varepsilon) d\varepsilon\right\}, \tag{20}$$

where two distribution functions are defined by the Maxwellian energy functions, respectively, as

$$f_{M1}(\varepsilon) = \frac{2n_{01}\varepsilon^{1/2}}{\sqrt{\pi}}\left(\frac{1}{T_{01}}\right)^{3/2}\exp\left(\frac{-\varepsilon}{T_{01}}\right) \text{ and } f_{M2}(\varepsilon) = \frac{2n_{02}\varepsilon^{1/2}}{\sqrt{\pi}}\left(\frac{1}{T_{02}}\right)^{3/2}\exp\left(\frac{-\varepsilon}{T_{02}}\right)$$

Note that (n_{01},T_{01}) and (n_{02},T_{02}) are the first and second type of electron densities and temperatures, respectively. It is assumed that the initial (first electron) population is relatively cold in comparison with the second one, i.e., $T_{01}<T_{02}$ and the surface of spacecraft is assumed to be uncharged initially. Absorbing the factor $\varepsilon^{1/2}$ within the distribution function, Eq. (20) can be expressed in terms of average (Lai, 2012), as

$$1 = \frac{\sum_{i=1,2}\int_0^{\infty} f_{Mi}(\varepsilon)d\varepsilon\left\{\delta(\varepsilon)+\eta(\varepsilon)\right\}}{\sum_{i=1,2}\int_0^{\infty} f_{Mi}(\varepsilon)d\varepsilon} \equiv \left\langle \delta(\varepsilon)+\eta(\varepsilon)\right\rangle_i \tag{21}$$

Here $i=1$ identifies the single Maxwelian distribution for modelling cold electrons and $i=2$ corresponds to the second Maxwelian distribution for studying hot electrons. Now substituting the expressions of $f_{M1}(\varepsilon)$, $f_{M2}(\varepsilon)$, $\delta(\varepsilon)$ and $\eta(\varepsilon)$ into Eq. (20), and after solving integrals, we eventually obtain (Huang et al. 2015)

$$1 = \frac{\alpha\sqrt{T_{01}}\left\{\xi_\delta(T_{01})+\xi_\eta(T_{01})\right\}+\sqrt{T_{02}}\left\{\xi_\delta(T_{02})+\xi_\eta(T_{02})\right\}}{\alpha\sqrt{T_{01}}+\sqrt{T_{02}}}, \tag{22}$$

with $\alpha(=n_{01}/n_{02})$ the cold-to-hot electron density ratio, whereas $\xi_\delta(T_{01,02})= c(1+T_{01,02}/a)^{-2} - c(1+T_{01,02}/b)^{-2}$ and $\xi_\eta(T_{01,02})= A-B(1+CT_{01,02})^{-2}$ show the contributions of secondary and backscattered electrons involving the cold and hot electron temperatures $T_{01,02}$. In order to consider a single electron distribution, we take $T_{02}=0$, and so the current-balance equation (22) immediately reduces to

$$1 = \xi_\delta(T_{01})+\xi_\eta(T_{01}) \equiv c\left(1+\frac{T_{01}}{a}\right)^{-2} - c\left(1+\frac{T_{01}}{b}\right)^{-2} + A - \frac{B}{(1+CT_{01})^2}. \tag{23}$$

For the sake of simplicity in numerical analysis, we have taken into account $n_{01}=n_1$, $n_{02}=n_2$, $T_{01}=T_1$, and $T_{02}=T_2$. Equation (23) exactly coincides with the previous result (Harris 2003) and Eq. (13) above.

We have examined that the magnitude of normalized current flux density decreases in two populations of Maxwellian electrons as long as thermal energy of hot electrons increases. This variation results in triggering the onset for negative charging of spacecraft and satellites at low thermal energies as can be seen from Fig. 8a. However, an opposite behavior for the surface charging is observed by changing the cold-to-hot electron density ratio (see Fig. 8b) and consequently the threshold points shifts to higher values of electron thermal energies.

Figure 8. The CBE (22) that accounts for two group of Maxwellian electrons is solved by showing the normalized net flux density as a function of T_1 with varying (a) the hot electron thermal energy T_2 at fixed α=3 and (b) the cold-to-hot electron density ratio $\alpha(=n1/_{n}2)$ at fixed $T2=_{3}$

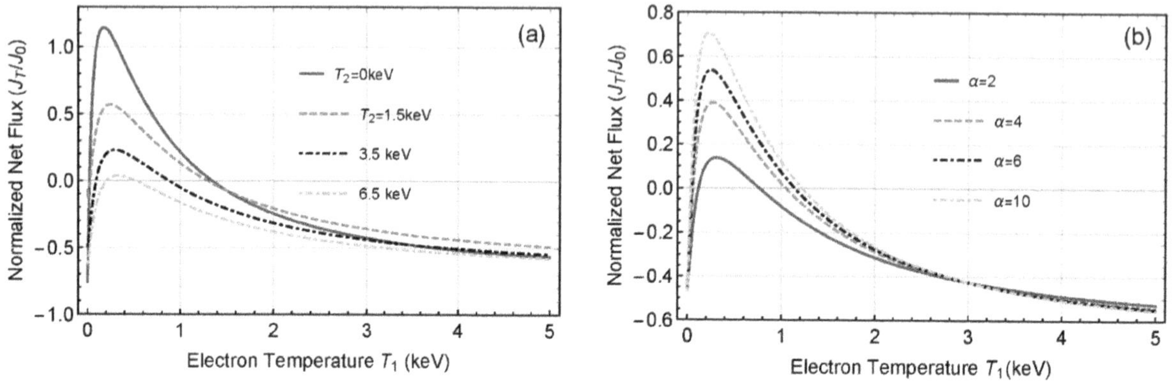

Figure 9. The CBE (25) modified with a q-distribution is solved for showing normalized net flux density as a function of T_1 with varying (a) hot electron thermal energy T_2 at fixed α=3 and q=0.8 and (b) cold-to-hot electron density ratio $\alpha(=n_1/n_2)$ at fixed T_2=10 and q=0.8

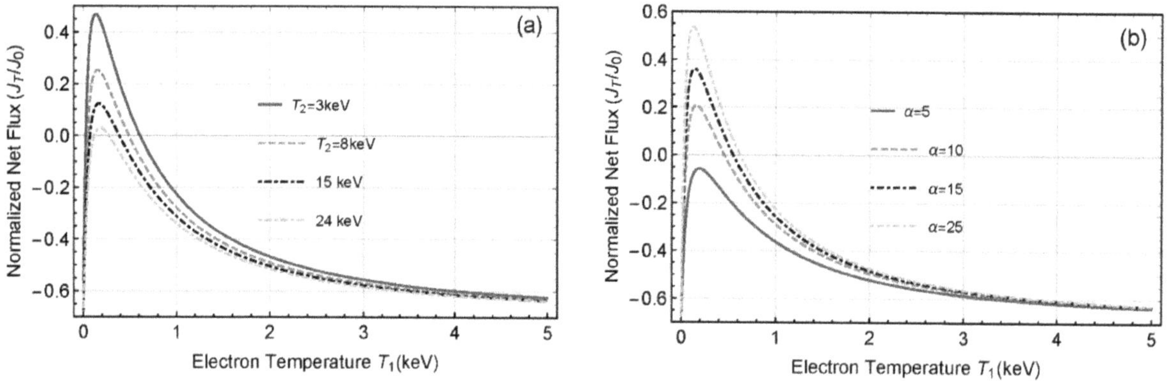

Table 1 indicates threshold conditions for the onset of negative charging of common space-grade materials with variation of entropic index q in the presence of two non-extensive electron distributions.

CBE With Double Non-Extensive dfs

In the presence of double q-distributions (Javed et al. 2023), the current-balance equation can be determined by incoming and outgoing electron currents on the spacecraft surface (which is initially uncharged), is given by

Table 1.

Threshold conditions for different space-grade materials at fixed T_2 = 15keV and α = 10						
Double q-distribution with various q-indices	Non-activated CuBe	Activated CuBe	Gold	Aluminium oxide	SiO2 (quartz)	Glass
0.8	TA = 0.070 TC = 0.385	TA = 0.022 TC = 1.784	TA = 0.213 TC = 0.906	TA = 0.0821 TC = 0.330	TA = 0.096 TC = 0.500	TA =0.089 TC = 0.397
0.85	TA = 0.070 TC = 0.521	TA = 0.024 TC = 2.133	-----	TA = 0.079 TC = 0.461	TA = 0.095 TC = 0.678	TA =0.089 TC = 0.539
0.9	TA = 0.072 TC = 0.656	TA = 0.027 TC = 2.484	TA = 0.221 TC = 1.572	TA = 0.080 TC = 0.591	TA = 0.099 TC = 0.856	TA = 0.089 TC = 0.697
Single Maxwellian distribution	TA =0.019 TC = 1.340	TA = 0.098 TC = 3.648	TA = 0.114 TC = 2.911	TA =0.019 TC = 1.265	TA = 0.0290 TC = 1.704	TA = 0.024 TC = 1.444

$$\int_0^{\infty} \varepsilon^{1/2} f_{q1}(\varepsilon) d\varepsilon = -\int_0^{\infty} \varepsilon^{1/2} f_{q2}(\varepsilon) d\varepsilon + \{\delta(\varepsilon) + \eta(\varepsilon)\} \times \left\{ \int_0^{\infty} \varepsilon^{1/2} f_{q1}(\varepsilon) d\varepsilon + \int_0^{\infty} \varepsilon^{1/2} f_{q2}(\varepsilon) d\varepsilon \right\}, \tag{24}$$

where

$$f_{q1}(\varepsilon) = \frac{2n_{01}\Gamma\left(\frac{1}{1-q}\right)\left(\frac{1-q}{T_{01}}\right)^{3/2} \varepsilon^{1/2}}{\sqrt{\pi}\Gamma\left(\frac{1}{1-q} - \frac{3}{2}\right)} \left\{1 - (q-1)\frac{\varepsilon}{T_{01}}\right\}^{\frac{1}{q-1}}$$

and

$$f_{q2}(\varepsilon) = \frac{2n_{02}\Gamma\left(\frac{1}{1-q}\right)\left(\frac{1-q}{T_{02}}\right)^{3/2} \varepsilon^{1/2}}{\sqrt{\pi}\Gamma\left(\frac{1}{1-q} - \frac{3}{2}\right)} \left\{1 - (q-1)\frac{\varepsilon}{T_{02}}\right\}^{\frac{1}{q-1}}.$$

Substituting the values of $f_{q1}(\varepsilon)$, $f_{q2}(\varepsilon)$, $\delta(\varepsilon)$ and $\eta(\varepsilon)$ into (24) and performing integrations over energy parameter ε, we finally come up with this form

$$1 = \frac{\frac{2q^2 - q}{(1-q)^2}\left\{\alpha\sqrt{T_{01}}\left(\xi_\delta + \xi_\eta\right) + \sqrt{T_{02}}\left(\xi_\delta' + \xi_\eta'\right)\right\}}{\alpha\sqrt{T_{01}} + \sqrt{T_{02}}} \tag{25}$$

where

$$\xi_\delta = c(U_1 - U_2),\ \xi_\delta' = c(U_1' - U_2'),\ \xi_\eta = A\frac{(1-q)^2}{2q^2 - q} - BU_3, \text{and } \xi_\eta' = A\frac{(1-q)^2}{2q^2 - q} - BU_3'.,$$

with Whittaker functions

$$U_1 = U_1\left(2\middle|3 + \frac{1}{q-1}\middle|\frac{T_{01}}{(1-q)a}\right),\ U_2 = U_2\left(2\middle|3 + \frac{1}{q-1}\middle|\frac{T_{01}}{(1-q)b}\right),\ U_3 = U_3\left(2\middle|3 + \frac{1}{q-1}\middle|\frac{CT_{01}}{1-q}\right)$$

and

$$U_1' = U_1'\left(2\middle|3 + \frac{1}{q-1}\middle|\frac{T_{02}}{(1-q)a}\right),\ U_2' = U_2'\left(2\middle|3 + \frac{1}{q-1}\middle|\frac{T_{02}}{(1-q)b}\right),\ U_3' = U_3'\left(2\middle|3 + \frac{1}{q-1}\middle|\frac{CT_{02}}{1-q}\right).$$

To present our results in a more simplest way, one needs to take into account $n_{01}=n_1$, $n_{02}=n_2$, $T_{01}=T_1$, and $T_{02}=T_2$ in Eq. (25) for verifying the CBE due to a single q-distribution using $T_2=0$, can be given as

$$1 = \frac{2q^2 - q}{(1-q)^2}\left\{cU_1 - cU_2 - BU_3\right\} + A. \tag{26}$$

This exactly corresponds to the same expression that is also given in Eq. (17).

Figure 9 displays the impact of hot electron temperature (T_2) and cold-to-hot electron density ratio (α) on the profiles of normalized current flux density accounting for two group of q-nonextensive electrons. As long as the thermal energy of hot electrons increases, the magnitude of the net current decreases. This variation triggers an onset for negative charging of spacecraft and satellites at more less thermal energies as shown in Fig. 9a. However, an opposite behavior for the surface charging is observed by changing the cold-to-hot electron density ratio (see Fig. 9b). It is worth mentioning that the onset of negative charging is significantly affected by the variation of nonextensive parameter q and surface material constants, can be seen from Table 1.

CONCLUSION

To conclude, we have revisited the spacecraft surface charging phenomena due to plasma bombardment at GEO altitudes and highlighted the significance of spacecraft missions, measurements and particle distribution models. A literature survey is made to understand various spacecraft charging mechanisms and to mitigate surface charging effects. Apart from a comparative analysis of negative charging through the current-balance equations for low and high energy particles, we have also considered a plasma that constitutes a mixture of two clouds with distinct thermal energies and electron distributions, the framework of double Maxwellian (extensive) distributions is utilized to analyze the current-balance equations on the spacecraft's surfaces both analytically and numerically. It has been found that an onset for negative charging occurs which is significantly dependent on spacecraft surface material properties, the electron density ratio $\alpha(=n_1/n_2)$, and corresponding thermal energies. The study is also extended to model high energetic electrons by using the power-law distribution functions as an alternative scheme for spacecraft charging. This has presented more realistic results to determine the onset of spacecraft charging with aid of modified current-balance equations. Furthermore, numerical calculations are carried out to identify critical and anti-critical thresholds for charging onset using various space-grade materials and non-extensivity parameter.

REFERENCES

Boon, J. P., & Tsallis, C. (2005). Special issue overview Nonextensive statistical mechanics: new trends, new perspectives. *Europhysics News, 36*(6), 185-186.

Chopra, K. P. (1961). Interactions of rapidly moving bodies in terrestrial atmosphere. *Reviews of Modern Physics*, *33*(2), 153–189. doi:10.1103/RevModPhys.33.153

Clark, G., Cohen, I., Westlake, J. H., Andrews, G. B., Brandt, P., Gold, R. E., Gkioulidou, M. A., Hacala, R., Haggerty, D., Hill, M. E., Ho, G. C., Jaskulek, S. E., Kollmann, P., Mauk, B. H., McNutt, R. L. Jr, Mitchell, D. G., Nelson, K. S., Paranicas, C., Paschalidis, N., & Schlemm, C. E. (2016). The "Puck" energetic charged particle detector: Design, heritage, and advancements. *Journal of Geophysical Research. Space Physics*, *121*(8), 7900–7913. doi:10.1002/2016JA022579 PMID:27867799

Cooper, R., Cowardin, H., Engelhar, D., Plis, E., & Hoffman, R. (2017, September). Space Weathering Experiments on Spacecraft Materials. In *Advanced Maui Optical and Space Surveillance Technologies Conference (AMOS)* (No. JSC-CN-39072).

DeForest, S. E. (1972). Spacecraft charging at synchronous orbit. *Journal of Geophysical Research*, *77*(4), 651–659. doi:10.1029/JA077i004p00651

Du, J. (2004a). Nonextensivity in nonequilibrium plasma systems with Coulombian long-range interactions. *Physics Letters. [Part A]*, *329*(4–5), 262–267. doi:10.1016/j.physleta.2004.07.010

Du, J. (2004b). The nonextensive parameter and Tsallis distribution for self-gravitating systems. [Europhysics Letters]. *Europhysics Letters*, *67*(6), 893–899. doi:10.1209/epl/i2004-10145-2

Garrett, H. B. (1981). The charging of spacecraft surfaces. *Reviews of Geophysics, 19*(4), 577–616. doi:10.1029/RG019i004p00577

Garrett, H. B., & Hoffman, A. R. (2000). Comparison of spacecraft charging environments at the Earth, Jupiter, and Saturn. *IEEE Transactions on Plasma Science, 28*(6), 2048–2057. doi:10.1109/27.902232

Garrett, H. B., & Whittlesey, A. C. (2000). Spacecraft charging, an update. *IEEE Transactions on Plasma Science, 28*(6), 2017–2028. doi:10.1109/27.902229

Ginet, G. P., O'Brien, T. P., Huston, S. L., Johnston, W. R., Guild, T. B., Friedel, R., & Su, Y. J. (2014). AE9, AP9 and SPM: New models for specifying the trapped energetic particle and space plasma environment. *The Van Allen Probes Mission*, 579-615.

Gringauz, K. I. (1958, June). Rocket measurements of electron concentration in the ionosphere by means of an ultrashort wavelength dispersion interferometer. In *Doklady Akad* (Vol. 120). Nauk SSSR.

Harris, J. T. (2003). *Spacecraft charging at geosynchronous altitudes: Current balance and critical temperature in a non-Maxwellian plasma*, [Theses Master of Science, Air Force Institute of Technology]. https://scholar.afit.edu/etd/4166

Hastings, D., & Garrett, H. (2004). *Spacecraft-environment interactions*. Cambridge University Press.

Huang, J., Liu, G., & Jiang, L. (2015). Threshold for spacecraft charging in double-Maxwellian plasma. *Journal of Geophysical Research. Space Physics, 120*(8), 6301–6308. doi:10.1002/2015JA021173

Huang, J., Liu, G., Jiang, L., & Yang, Y. (2017). Theory for geosynchronous spacecraft charging index. *Space Weather, 15*(9), 1203–1211. doi:10.1002/2017SW001670

Javed, S., Rubab, N., Zaheer, S., Poedts, S., & Jaffer, G. (2023). Numerical calculations of charging threshold at GEO altitudes with two temperature non-extensive electrons. *Space Weather, 21*(10), e2022SW003412.

Johnson, C. Y., & Meadows, E. B. (1955). First investigation of ambient positive-ion composition to 219 km by rocket-borne spectrometer. *Journal of Geophysical Research, 60*(2), 193–203. doi:10.1029/JZ060i002p00193

Kivelson, M. G., Khurana, K. K., Russell, C. T., & Walker, R. J. (1997). Intermittent short-duration magnetic field anomalies in the Io torus: Evidence for plasma interchange? *Geophysical Research Letters, 24*(17), 2127–2130. doi:10.1029/97GL02202

Lai, S. T. (2009). Dependence of electron flux on electron temperature in spacecraft charging. *Journal of Applied Physics, 105*(9), 094912. doi:10.1063/1.3125517

Lai, S. T. (2011). Novel Ideas of Spacecraft Charging Mitigation. *IEEE Transactions on Plasma Science, 40*(2), 402–409. doi:10.1109/TPS.2011.2176755

Lai, S. T. (2012). *Fundamentals of spacecraft charging: spacecraft interactions with space plasmas*. Princeton University Press. doi:10.1515/9781400839094

Lai, S. T., & Della-Rose, D. J. (2001). Spacecraft charging at geosynchronous altitudes: New evidence of existence of critical temperature. *Journal of Spacecraft and Rockets, 38*(6), 922–928. doi:10.2514/2.3764

Lai, S. T., & Tautz, M. (2006). High-level spacecraft charging in eclipse at geosynchronous altitudes: A statistical study. *Journal of Geophysical Research. Space Physics, 111*(A9).

Langmuir, I., & Mott-Smith, H. (1924). Langmuir probe technique. *General Electric Review, 27*, 449–455.

Leubner, M. P. (2004). Core halo distribution functions: A natural equilibrium state in generalized thermostatistics. *The Astrophysical Journal, 604*(1), 469–478. doi:10.1086/381867

Lima, J., Silva, R., & Plastino, A. (2001). Nonextensive thermostatistics and the *h* theorem. *Physical Review Letters, 86*(14), 2938–2941. doi:10.1103/PhysRevLett.86.2938 PMID:11290077

Livadiotis, G., & McComas, D. (2009). Beyond kappa distributions: Exploiting Tsallis statistical mechanics in space plasmas. *Journal of Geophysical Research, 114*(A11), A11105. doi:10.1029/2009JA014352

Matéo-Vélez, J. C., Ganuschkina, N., Meredith, N., Sicard-Piet, A., Maget, V., Payan, D., & Dubyagin, S. (2016, April). *From GEO/LEO environment data to the numerical estimation of spacecraft surface charging at MEO.* 14th Spacecraft Charging and Technology Conference, Noordwijk, NL. http://old. esaconferencebureau. com/2016-events/14sctc/proceedings

Meyer-Vernet, N. (1999). How does the solar wind blow? A simple kinetic model. *European Journal of Physics, 20*(3), 167–176. doi:10.1088/0143-0807/20/3/006

Pandya, A., Kothari, N., & Alad, R. (2022). A comparative study of body potential transient profile of geosynchronous orbit spacecraft in single and double Maxwellian plasma models. *International Journal of Numerical Modelling, 35*(3), e2968. doi:10.1002/jnm.2968

Pervaiz, F., Ali, S., Ali, M., & Lai, S. T. (2023). Spacecraft Charging Due To Energetic Electrons and Ions at Geosynchronous Altitudes. *Journal of Geophysical Research: Space Physics, 128* (1), e2022JA030642.

Ali, S., Rubab, N., Nawaz, K., & Sarwar, S. (2020). A comparative study of spacecraft charging onset at geosynchronous altitudes: A nonextensive particle approach. *Journal of Geophysical Research: Space Physics, 125* (9), e2020JA028107.

Plis, E., Engelhart, D. P., Barton, D., Cooper, R., Ferguson, D., & Hoffmann, R. (2017). Degradation of polyimide under exposure to 90 keV electrons. *physica status solidi (b), 254*(7), 1600819.

Prokopenko, S. M., & Laframboise, J. G. L. (1980). High voltage differential charging of geostationary spacecraft. *Journal of Geophysical Research, 85*(A8, no. A8), 4125–4131. doi:10.1029/JA085iA08p04125

Rubin, A. G., & Henry, B. (1980). Spacecraft charging on ATS-5. Geophysics Laboratory, Air Force Systems Command, United States Air Force.

Sanders, N. L., & Inouye, G. T. (1978). Secondary emission effects on spacecraft charging: energy distribution consideration. In Spacecraft Charging Technology. Air Force Geophysics Laboratory, Hanscom AFB, MA.

Tsallis, C. (1988). Possible generalization of Boltzmann-Gibbs statistics. *Journal of Statistical Physics, 52*(1–2), 479–487. doi:10.1007/BF01016429

Vasyliunas, V. M. (1968). Low-energy electrons on the day side of the magnetosphere. *Journal of Geophysical Research, 73*(23), 7519–7523. doi:10.1029/JA073i023p07519

Verheest, F. (2013). Ambiguities in the Tsallis description of non-thermal plasma species. *Journal of Plasma Physics, 79*. doi:10.1017/S0022377813001049

Whipple Jr, E. C. (1965). *The equilibrium electric potential of a body in the upper atmosphere and in interplanetary space* (No. X-615-65-296).

Yu, Y., Su, S., Cao, J., Jordanova, V. K., & Denton, M. H. (2022). Improved Boundary Conditions for Coupled Geospace Models: An Application in Modeling Spacecraft Surface Charging Environment. *Space Weather, 20*(9), e2022SW003178.

Chapter 6
Role of Electric Field on Diffusion Coefficients for Weakly Coupled Plasma

Aamir Shahzad
Modeling and Simulation Laboratory, Department of Physics, Government College University Faisalabad (GCUF), Faisalabad, Pakistan

Muhammad Kashif
Modeling and Simulation Laboratory, Department of Physics, Government College University Faisalabad (GCUF), Faisalabad, Pakistan

Rabia Waris
https://orcid.org/0009-0007-3249-2328
Modeling and Simulation Laboratory, Department of Physics, Government College University Faisalabad (GCUF), Faisalabad, Pakistan

Alina Manzoor
Modeling and Simulation Laboratory, Department of Physics, Government College University Faisalabad (GCUF), Faisalabad, Pakistan

Maogang He
Key Laboratory of Thermo-Fluid Science and Engineering, Ministry of Education (MOE), Xi'an Jiaotong University, Xi'an, China

ABSTRACT

Equilibrium molecular dynamics (EMD) has been used to investigate diffusion coefficient (D) of 3D weakly coupled complex dusty plasma (WCCDP) by varying applied electric field (E). Velocity auto correlation function technique has been used to compute diffusion coefficient of dust particles. Impact of varying electric field has been tested for complete series of plasma parameters (Γ, κ) in a canonical ensemble. EMD simulation have been used to compute D for fixed number of particles ($N = 500$) at different values of plasma parameters ($0.1 \leq \Gamma \leq 0.9$ and ($1 \leq \kappa \leq 3$). Computed results showed that value of D decreases by increasing Γ and increases by increasing E and κ. Simulation results also exposed that result obtained through Green-Kubo has same drift as previous results. But this work has some advancement due to $\bar{E}$.

DOI: 10.4018/979-8-3693-0904-9.ch006

1. INTRODUCTION

Thermophysical properties of complex systems are dynamically investigated by simulation and in laboratories to study the complex system a microscopic level. It has played a fundamental role in the field of science and technology. Thermophysical is a combination of thermodynamic and transport properties. Thermodynamic defines the equilibrium condition of complex systems such as heat, temperature, enthalpy, entropy, pressure and density. Whereas, transport properties described the energy, mass and heat flow of complex systems. It includes diffusion, shear viscosity and thermal conductivity. In this chapter, our motivation is to explore the effect of electric field on diffusion coefficient in 3-D weakly coupled dusty plasma by using molecular dynamics (MD) simulations. First, it defines the basics of plasma and dusty plasmas (DPs) in physics. The existence of DPs in nature, laboratories and industries is explained in detail. The significance of DPs is discussed by their possible applications in industry, laboratory, space and tokomak. The general theory and mechanism of DPs, its main parameters and characteristics are discussed. Next is the introduction of computer simulation techniques used in this study, and a historical overview of molecular dynamics (MD) simulations and the model and method are explained in detail. The purpose and motivations of the presented work are to be obtained from simulation data of diffusive transport properties in the presence of varying electric field at numerous DPs' parameters, together with previous work that has been done to date. It provides necessary problems, justifying the current work performed with the contribution for this chapter.

1.1 Plasma

Plasma is the fourth state of matter it was firstly introduced by Sir William Crooks in 1879. In the universe, 99% of physical matter is in the plasma state and the rest part of the world is only about 1% (Chen, 1984). In space, the most visible matter is in the plasma state; the sun and stars are the main examples of plasma in our universe. Plasma is created artificially in a lab and used for a variety of technical applications, including fusion energy research, display, fluorescent lighting, and more. The Irving Langmuir, an American physicist, defines plasma first time as "plasma is a quasi-neutral gas of the charged particles that show collective behaviors." He received Noble prize in 1927 for introducing "plasma" first time. Here quasi neutrality means "the gas becomes electrically neutral when the number of positive charge density is equal to number of ions.(ni $\approx$ ne)" and Collective behavior means that charged particle collides with each other due to electric field and coulomb potential. . Plasma is used in over daily life fields such as laser, sterilizing of medical instruments, lightening, intense power beams, water purification planet and many more (Fortov & Morfill, 2009).

1.2 Dusty Plasma

Dusty plasmas are also known as complex plasmas that contain particulates of condensed matter. Lyman Spitzer along with Hannes Alfven described that dust in the universe was not just an interruption to optical observation, but that it was an essential component of the universe. One of the most interesting events in the field of dusty plasmas occurred in early 1980 during the Voyager 2 flyby of planet Saturn. In 2005, Cassini spacecraft took new and improved images of spokes with detail that would provide a better understanding of their origin. In 1992, the European spacecraft Ulysses flew by the planet Jupiter and detected the dust particles and measured their masses and impact speed. Dust particles sizes ranging from few nanometers to micrometers, and typically more massive than that of plasma ions, electrons, and neutrons. When dust particles immersed in the plasma, they attain a high electric charge (mostly negative charge) which makes dusty plasmas interesting and technological important in the area of applied plasma physics. Dust particles are found in the large abundant in planetary plasmas, cosmic plasmas, plasmas in the laboratory, and plasmas near earth (Gregori, 2012). As a matter of fact, one may consider that except in hottest regions of fusion plasmas, where dust particles would not survive, most are known as dusty plasmas in sense that some dust particles might be present. To understand this fact, here are two cases in which: there are just few secluded (non-interacting) dust particles, with the goal that they have nearly nothing if any impact on the plasma, and there are countless number of dust particles in the plasma so that their existence really changes properties and behavior of plasma. Dust particles are charged by their interactions with plasma particles yet do not change plasma in any manifest way. In other case, charge dust particle is a component of plasma; have to face collective interactions that plasma becomes an ionized gas from a neutral gas and in this case it is characterized as "dusty plasmas." Currently, dusty plasma becomes more interesting field of research due to increase the pollution in environment. It is a huge occurrence in magneto and ionosphere of earth, atmosphere of stars, sun, solar wind, galaxies cosmic radiations, planetary rings comet tails and nebulae components.

1.3 Plasma Parameters

There are two basic dimensionless parameters which are used for investigation of transport coefficients in 2-D and 3-D dusty plasma systems, and which are liable for mass transfer and phase state in plasma systems (Vaulina et al., 2008). These dimensionless parameters are known as Coulomb coupling (Γ) parameter and Screening parameter (κ).

1.3.1 Debye Screening Strength

The Debye length is an important physical parameter in plasma. It provides the distance over which the influence of the electric field of an individual charged particle is felt by other charged particles (such as ions) inside the plasma. Mathematically it is the ratio of interparticle distance to the Debye length as $\kappa = \frac{a_{ws}}{\lambda_D}$ here "a_{ws}" is the Wigner-Seitz (WS) radius and "λ_D" is the Debye length.

1.3.2 Coulomb Coupling

Charged dust grains can be either weakly or strongly depending on the strength of the Coulomb coupling parameter (Thomas et al., 1994). Coulomb coupling factor has relationship as $\Gamma = \left(\frac{Q^2}{4\pi\varepsilon_0}\right)\left(\frac{1}{a_{ws}k_B T}\right)$,

Where k_B is Boltzmann's constant and T is system temperature.

1.4 Types of Dusty Plasma

On the basis of Coulomb coupling (Γ) parameter, dusty plasmas are classified into two classes: one is called weakly coupled (ideal) complex dusty plasmas and other is called strongly coupled (non-ideal) complex dusty plasmas. These are discussed below.

1.4.1 Weakly Coupled Complex Dusty Plasmas (WCCDPs)

The coupling parameter can also be denoted such as Γ=P.E/K.E, it is ratio of potential to kinetic energy of dusty particles. When dust charged particles have average kinetic energy much larger than the average potential energy, then that plasma is known as weakly coupled complex dusty plasma (WCCDPs) or ideal plasma (Boulos et al., 1994). The WCCDPs have a high temperature but low density and value of Coulomb coupling parameter less than 1 (mathematically $\Gamma < 1$). In case of WCCDPs, any structural form does not exist. WCCDPs is considered mostly exist as ionized gases.

1.4.2 Strongly Coupled Complex Dusty Plasmas (SCCDPs)

The SCCDPs is speedily emerging filed from last three decades. It is very significant to astrophysical plasma and quickly progressing in laboratory experiments. In this type of dusty plasma average thermal energy of charged particles are much smaller than average potential energy and mathematically $\Gamma \geq 1$ it's also called non ideal plasma (Vaulina et al., 2010). In case of SCCDPs, it has high density and low temperature and can be specified in any structural form.

1.5 Dust Particle in Plasma

Dust in atmosphere and in whole universe is in different sizes and shapes. Generally, it is in solid form but also presents in liquid and gaseous forms. If dust particle exists with plasma, it forms dusty plasma (complex dusty plasma). Dusty plasma also known as complex plasma has electrons, ions, neutral and dust particles. . Dust is present everywhere in the space, such as interplanetary dust. These charged dust particles change properties of plasma by electric and magnetic fields. The value of Coulomb coupling parameter between dust particles is high due to the massive dust charge, which leads to liquid and solid phase of dust system at room temperature. For study phase transitions and structural properties of solids, Yukawa liquids and plasma crystals are applicable in these systems (Piel, 2010). Plasma with dust particles can be termed as either "dust in plasma" or "dusty plasma" depending on organization of numerous characteristic lengths and radius between interacting particles (r_d and λ_D). If Debye length of dust particles is less than interparticle distance then it is called dust in plasma. In this case, there are no dust particles in plasma sphere. If Debye length of dust particles is greater than interparticle distance then it is called dusty plasma. It shows that dust particles are present in plasma sphere. Dust particle in plasma attains charge and is affected by electric and magnetic fields. Temperature on the dust particle is almost 10 K, and values of electric potential varies from 1 to 10 V is mostly negatively potential. It can be grown or inserted in low temperature plasma at low degree of ionization in the laboratory. It becomes a central constituent of dusty plasma for transports of energy, momentum and mass. In recent years, dusty plasmas have opened up an entirely new field of research of science and technology by investigating transport properties (thermal conductivity, shear viscosity, and diffusion coefficient) of dusty plasmas in the laboratory.

a) **Charge on dust particle**

The value of charge ranges from (zero to thousands of electronic charge) totally depending on the shape and size of the dusty particles. A lot of mechanisms are adopted to produce charge on a dust particle. If all mechanisms are considered at once, then the measurement of the equilibrium charge condition on a dust particle becomes very difficult. Mostly, the charge on dust particles is negative in low-temperature laboratory plasma (Rodriguez, 2018). When an electron from background plasma strikes surface of dust particle and is attached and lost from plasma. Because electrons are more moveable than ions, the surface dust grain collects electrons, attracts ions, and repels electrons until the state of immobility is attained. This charge is accountable for a long lifetime of particles and confinement in plasma

b) **Size of dust particle**

Dust particles are much larger than electrons and ions, and its size varies from few millimeters to nanometer (10^{-3}m to 10^{-9} m). A typically used particle may have a diameter of 3.50±0.05 μm and a mass ~3×10^{-11} kg. Such particles are named mono diffuse (Shahzad & He, 2013). It has any shape, mostly spherical, and also easily observed in laboratory without any microscope. It can be made by conducting or dielectric material.

c) **Nature of dust particle:**

Dust particles that interact with plasma particles such as electrons and ions on their surface become an interacting particle. In dusty plasma energy, momentum and mass are transferred through dust particle. Dust particle may be in liquid, solid and gaseous forms.

1.6 Forces on Dust Particles in Plasma

The knowledge about several forces acting on dust particles in plasma is required for an understanding of their dynamics and transport.

a) **Force of gravity**

Dust particle is subject to force of gravity that is proportional to its mass, but under microgravity state, it is snubbed. $F_g = mg = 4/3\ \pi b^3 \rho g$, where g is acceleration due to gravity and ρ is mass density of particle (Kikuchi, 2013).

b) **Electric force**

In some situation, there is an electric field E in plasma, electric force acting on dust particles of charge q is F_e=qE .electric field is smaller in bulk of plasma while it is larger in sheaths next to plasma wall boundary.

c) **Neutral drag force**

This force is produced from collisions with neutral gas atoms or molecules, and is proportional to neutral pressure in vacuum chamber. F_n is given as $F_n = Nm_n v^2 n \pi b^2$, where N is neutral density, m_n is mass of neutral atoms (or molecules), and v_n is average relative velocity (Shukla & Mamun, 2015).

1.7 Applications of Dusty Plasma

The study of dusty plasmas had become motivating for research of laboratory plasma when formation of dust and dust trapping was examined during plasma etching of silicon wafers and to limit deposition ratio when powder formation in plasma-enhanced CVD was acknowledged.In industrial applications, dust particles produced disturbing effects in plasma and this contamination has shattering effects on fabricated circuits. Other applications, such as surface processing make use of dust particles that have spread in plasma. For example, growth of carbon-based nanostructures on surface used for electronic devices such as sensors, silicon-based films which have used in solar cells, and flat-panel displays illustrates an enhanced performance of nanoparticles produced in plasma through chemical reactions, are inserted into film. It has used to grow or modify existing materials. Primarily, dust has not considered a useful scientific significance in plasma. It has simply considered an unwanted pollutant in plasma. To reduce negative effects of dust particles in plasma leads to advance of material science. The nanoparticles have considered as building blocks of nanotechnology in plasma liberations. There are many advantages of dust particles in dusty plasmas such that nano crystalline silicon particles have used to enhance efficiency of silicon solar cells, which have developed in silane plasmas (Merlino, 2006). In hydrocarbon plasmas such as methane or acetylene, carbon-based nanostructures have developed to manage thin carbon films, which lead to materials high hardness, wear-resistance, and chemical dullness. Dust particles are also used in ceramic industry for sintering and in fabrication of hard coatings, and also used in optical devices. Dust particles decrease performance and the yield of many electronic devices. By using various techniques of plasma, like techniques of cold plasma widely used to vary the surface of fabrics properly. Different techniques exhibit different and excellent positive effects in dying, printing and shrinking of wool properly. It have major importance in energy productions, intense beam of laser, in development of fusion reactions, high power radiation source, lightening, sterilizations of various surgical and medical tools, purification of water plants, as well as for controlling radiations etc. X-rays and different types of radiations which are emitted through plasma sources have many applications in daily life.

1.8 Molecular Dynamics Simulation

MD simulation is a powerful method that can be used to solve many physical problems in atomic material research. MD simulation is handled usually all microscopic information and molecular methods have proven to be productive in applied research. Plasmas and complex liquids have various uses, ranging from semiconductor chips, colloids, thin films, and electrochemistry to biochemical films and other important areas where simulation play an important role. It also plays an important role in all advanced fields, such as textile science, engineering physics, plasma physics, astronomy, life sciences or organic sciences, and chemical industry. Computer simulation has become gradually significant in identifying complex systems. Using faster and more sophisticated computer systems, it can be studied composition and behavior of large complex systems (Shahzad & He, 2014). In the 1950s, Alder Wainwright and

Rahman used first MD simulations for liquid argon. MD simulation has two basic kinds which rely on properties so far which we can be working to calculate: one is EMD simulations (EMDS) and other one is NEMD simulations (NEMDS). In present work, EMDS is applied to investigate effect of electric field on diffusion coefficient in weakly coupled complex plasma at different dusty plasma parameters.

1.9 Diffusion Coefficient

Diffusion coefficient (*D*) is one of main transport property of dusty plasma. During diffusion process, particles drift from high concentrations to low concentrations. Diffusion rate can be controlled by variations of exterior parameters such as temperature, concentration, external electromagnetic forces (electric and magnetic). Investigations of diffusion have different purposes in liquids, such as exploring dynamical properties and understanding microscopic phenomena. For DPs research, the investigations of diffusion motion for applying external fields such as electric and magnetic become advance topic in current research. Different types of diffusion motion exist in dusty plasma systems depends on temperature and forces. Diffusion also plays an important role in discovering dynamical properties (structure, waves, and instabilities) of many biological, physical, and chemical systems. Diffusion motion of dust particles in CDPs referred as one of main research topics, an essential consequence of understanding the transport and dynamical properties. It is one of main property to find loss of energy (stopping power) in DPs. A lot of work has been made to understand diffusion motion in DPs. Molecular dynamics simulations were performed and investigate velocity autocorrelation function for 3D WCCDPs (Shahzad et al., 2013). Langevin dynamics simulation for 2D SCCDPs (Dzhumagulova et al., 2014) and diffusion coefficient (Dzhumagulova et al., 2012). For the investigations of self-diffusion coefficient MD simulations were performed for 3D CDPs (Begum & Das, 2016), one-component plasma (Daligault, 2012) and ionic mixtures (Daligault, 2012).

1.10 Scope and Objective of Chapter

The MD simulation methods have widely been used in the field of DPs to compute numerous physical properties of different physical systems. Till 2021, the MD simulation was widely used to calculate the thermo physical, structural and waves, and instability only for Yukawa potential. Develop an NEMD simulation technique to tune the effect of electric field on diffusion coefficient of 3D WC-CDPs for different series of plasma parameters. Three different varieties of EEF are used to understand the behavior of diffusion coefficient in 3D Yukawa liquids with variations of electric field. The presented developed numerical model can be used to understand nature of diffusion in gas-like, liquid-like and solid-like regimes of dusty plasmas with variations of the electric field in different directions.

2. NUMERICAL MODEL AND SIMULATION

The pair interactions between charged dust particles in plasma are computed through Yukawa potential, and this interaction is known as pure isotropic and generally repulsive. Numerous physical phenomena widely examined through Yukawa potential, which is a good approximation for screened charge potential and makes it possible to designate many experimental results. Such isotropic potential scattering is helpful for qualitative analytical and numerical estimations of dynamical, structural properties of DPs.

Moreover, interparticle interactions in 3D DPs are calculated to explore and understand various vital methods, such as phase transition (liquid-solid) and related perilous phenomena (Kompaneets et al., 2009).

This potential is used in an experimental study of plasmas under microgravity and space conditions. This interactions model is also used for physical systems such as colloidal suspension, astrophysics, aerodynamics, hydrodynamics, materials for energy generations, polymer physics, biomedicine, biological and chemical systems, etc. It has benefits for the reduced computational cost compared to another many body potential (Shahzad & He, 2012). In the absence of an external electric field, interaction between dust particles is pure isotropic and computed through Yukawa potential as:

$$\Phi(r) = \frac{Q_d^2}{4\pi\epsilon_0} \frac{e^{-r/\lambda_D}}{r} \tag{1}$$

Here Q_d is charge on dust particles, r is interparticle distance and λ_D is Debye length of plasmas system, ϵ_0 is permittivity of free space.

The total force that acts on dust particles is sum of ion, electron drag force, and electric force, which induced drift velocity $\mathbf{u} = \mathbf{E}/mv$, where m represents mass and v is collision frequency. Drift velocity of particle increases with increasing strength of EEF as $\mathbf{E}\ \alpha\ \mathbf{u}_d$ (Shakoori, He, Shahzad et al, 2022). The homogenous EEF for z-axis can be written as:

$$\mathbf{E}_z = (0,0,\mathbf{E}_z) = eE(\mathbf{r}_z\cos\theta) \tag{2}$$

Here E represents strength of an EEF.It is considered that there is no electric force considerably affected along the x-axis and y-axis. θ is the angle that correspond to direction of EEF; thus, interaction between dust particle in DPs are calculated for case of EEF as:

$$\Phi(\mathbf{r},\theta) = \frac{Q_d^2}{4\pi\epsilon_0} \frac{e^{-r/\lambda_D}}{r} - eE(\mathbf{r}_z \cos\theta) \tag{3}$$

The first term represents isotropic interaction of dust particles, and second term is anisotropic interaction in presence of an EEF. EEF is normalized by Debye length for EEF as $E^* = e\lambda_D E_z / T_d$ (Sukhinin et al., 2017).The 3D trajectories of interacting charged dust particles with charge Q_d and mass m under influence of homogeneous EEF can be written as:

$$m\mathbf{a} = \mathbf{F}_y + Q_d E^*(\mathbf{m}_z \cos\theta) \tag{4}$$

Where $\mathbf{F}_y$ is Yukawa interaction force $\mathbf{F}_y = -\partial\phi(\mathrm{r})/\partial\mathrm{r}$ and acceleration on i^{th} particle ($\mathrm{a}\boldsymbol{i} = \mathrm{Fi}/\mathrm{m}i_j$ and $\mathbf{m}z$ is unit vectors along corresponding axis of cubic simulation box. The intensity of normalized external electric field E* classifies nature of anisotropy. Three dimensionless parameters (Coulomb coupling, Debye screening, electric field) are used to characterize the DP systems. Coulomb coupling parameter, “ $= \mathrm{Q}_d^2 / (3\grave{\mathrm{A}}\epsilon_0 a_{ws} k_B \mathrm{T}_d)$ here a_{ws}, k_B and T_d are average Wigner–Seitz radius (inter-particle distance), Boltzmann's constant, and absolute systems temperature. Debye screening strength, $\kappa = aw_a/\lambda D$, for 3D case Wigner–Seitz given as a_{ws}=$(1/\pi n)^{1/2}$, n is the equilibrium dust density (nd = N/V). Plasma frequency for 3D system is $\omega_{pd} = \left(Q_d^2 / 2\pi\epsilon_0 m a_{ws}^3\right)^{1/2}$, m *is* accounted as mass of dust particle.

3. PARAMETERS AND SIMULATION TECHNIQUE

The simulations start from face-centered cubic (FCC) lattice configurations of Yukawa identical particles having randomly initial velocities. The $N = 500$ particles are used for whole calculation at different κ and Γ. The periodic boundary conditions for each direction are imposed on simulation box to simulate infinite system and to avoid image convection. The dimension of a cubic simulation box with edge length is L_x/ L_y (Shahzad et al., 2020).The standard Ewald summation method calculates the long-range interaction between the dust particles. In Ewald summation technique, the interactional potential is divided into two parts: real space and reciprocal space. Both are rapidly converging. The Ewald convergence parameter value ($\alpha = 5.6/La_{ws}$) is used. The interaction and forces are calculated into two parts: repulsive interaction through Yukawa potential and forces induced due to EEF also computed (Shakoori, He, & Shahzad, 2022). The Equation of motion is integrated using Predictor corrector integration method. The isotropic interactions are considered in absence of an electric field and anisotropic interaction accounted for different cases of E. When an electric field is substituted on, many physical phenomena occur; depending on nature of electric field.

Appropriate input factors are selected for numerical scheme's steadiness and accuracy. These input parameters, such as system temperature ($T = 1/\Gamma$), screening strength (κ), external electric field (E_i), number of particles (N), simulation time step and total run time, are selected to provide Yukawa system and diffusion coefficient. The time step for this simulation technique is given as $\mathrm{d}t = 0.005/\omega_p$ that help us to evaluate sufficient data for long simulation run. The reported simulations are performed between $5.0\times10^5/\omega_p$ and $3.0\times10^5/\omega_p$ time units. In this paper, EMD method is used for Yukawa systems to approximate diffusion over a wide range of plasma coupling ($\Gamma = 0.1$-0.9), screening ($\kappa = 1$-3), and number of particles ($N = 500$).

4. COMPUTATIONAL RESULTS

The task of this work is to investigate effect of EEF on diffusion coefficients for 3D WCCDPs computed through EMD simulations within an appropriate statistical limit. Figure 1 demonstrates computational results obtained using EMD simulations for weakly coupled plasma system to evaluate diffusion coefficient. EMD simulation are performed with various values of Γ for κ 1-3 at E= 0.4, 0.6, 0.8. . It is analyzed that obtained data sets of plasma D are usually close to each other within a statistical limit, though these data are over predicted as expected. EMD outcomes show that D increases with increasing E for each state of plasma (κ, Γ). Further, it is observed that D slightly increases at low Γ (nearly noni deal state of plasma), and at higher Γ electric field significantly affects the plasma D. EMD outcomes also revealed that value of D is increases with increasing κ. The EMD results obtained in current work clearly explain that total electric potential taking into account (sum of Yukawa and potential induced due to EEF), increasing trends of plasma for 3D WCCDPs and well-defined plasma states dependence is predicted. We advance two possible reasons for increasing trends of plasma diffusion coefficients with an increasing E that may arise from dust interaction in Yukawa systems, and they explain the results. First, the EMD method needs simulations for multiple values of E in order to establish linear regime. The applied EEF to diffusion coefficients affect the interactions of dust particles that cause to "tune" the interaction between the charged dust particles in Yukawa systems. Moreover, particles are polarized under action of an electric field; therefore, dipole-dipole coupling is induced. It is expected that the drift velocity of charged dust particles increased with increasing applied EEF. Electric field can change the interaction between the particles as well as it can cause an additional force. The flow of charged dust particles induced in different phases of WCCDPs due to EEF is one reason for increasing diffusion coefficients. The present simulations are consistent with CPs (Begum & Das, 2016), ionic liquids (Daligault, 2012) and simple liquids (Daligault, 2012)

Figure 1. Variation of Diffusion confident's w.r.t Coulomb coupling for three Screening parameters with (a) E = 0.4, (b) E = 0.6 and (c) E = 0.8

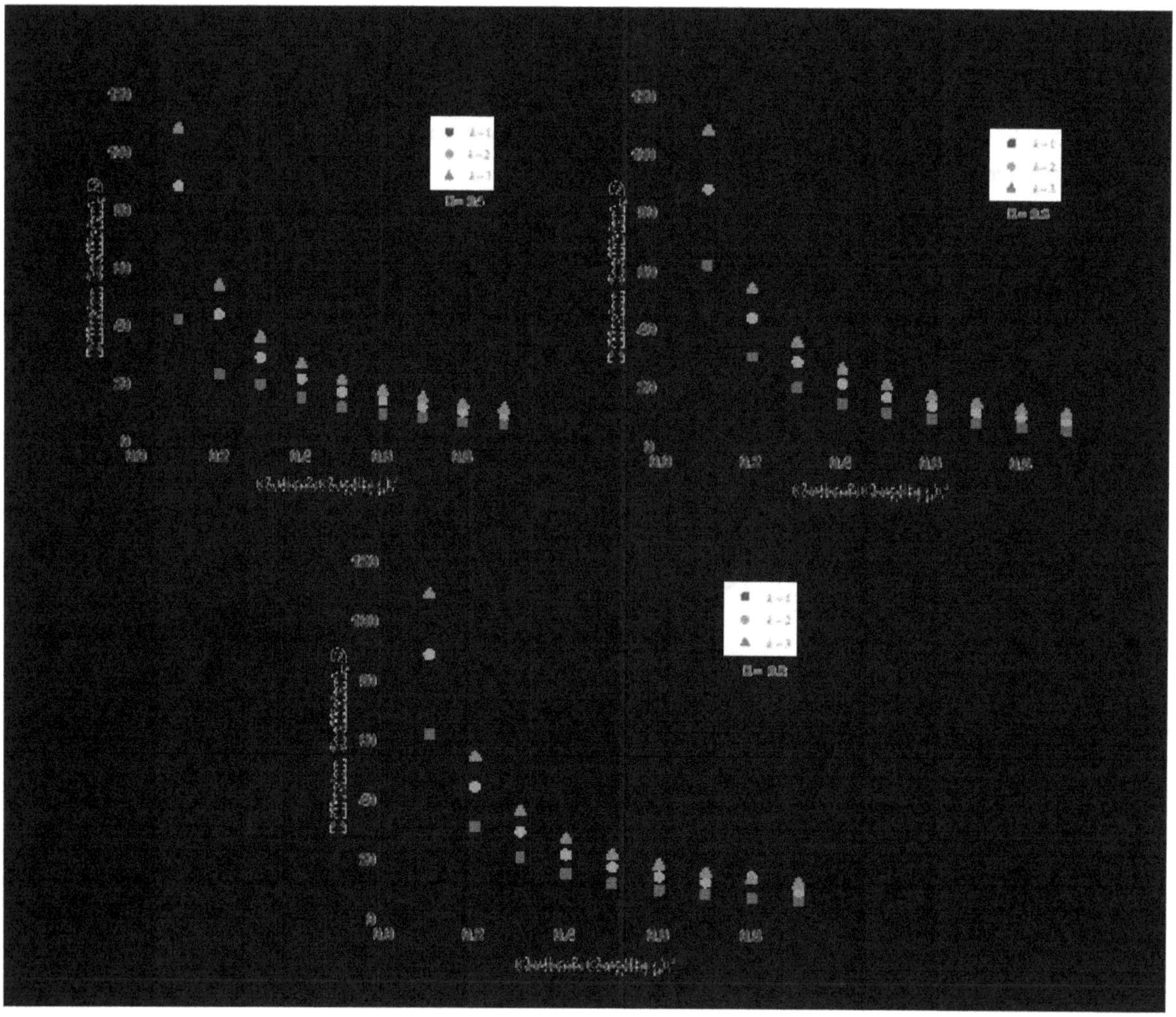

5. SUMMARY

A New algorithm of EMD simulations method is used to compute the diffusion coefficient for 3D complex plasma at varying EEF. The diffusion coefficient is calculated for constant electric fields (E = 0.4, 0.6, 0.8), and a number of particles (N = 500) at κ = 1- 4 and Γ = 0.1-0.9. The obtained results are compared and discussed theoretically for 3D systems. These comparisons and discussions show that the algorithms of an electric field of different cases with Yukawa potential have precision and reliability. These comparisons authenticated a new numerical model that can be used for electric field variations. It is concluded that the suggested numerical model to tune the diffusion coefficient from EMD yields consistent results, and this model is entirely accurate. This model has advantages in studying the behavior of diffusion coefficient ain the presence of electric field for 3D WCCDPs and other physical systems.

ACKNOWLEDGMENT

This work was sponsored by the National Research Program for Universities (NRPU) with (No. 20-15251/ NRPU/R&D/HEC/2021) for utilization of computational/ manpower resources.

REFERENCES

Begum, M., & Das, N. (2016). Self-diffusion as a criterion for melting of dust crystal in the presence of magnetic field. *The European Physical Journal Plus*, *131*(2), 1–12. doi:10.1140/epjp/i2016-16046-2

Boulos, M. I., Fauchais, P., & Pfender, E. (1994). *Thermal plasmas: fundamentals and applications*. No Title. doi:10.1007/978-1-4899-1337-1

Chen, F. F. (1984). *Introduction to plasma physics and controlled fusion* (Vol. 1). Plenum press. doi:10.1007/978-1-4757-5595-4

Daligault, J. (2012). Practical model for the self-diffusion coefficient in Yukawa one-component plasmas. *Physical Review. E*, *86*(4), 047401. doi:10.1103/PhysRevE.86.047401 PMID:23214710

Daligault, J. (2012). Diffusion in ionic mixtures across coupling regimes. *Physical Review Letters*, *108*(22), 225004. doi:10.1103/PhysRevLett.108.225004 PMID:23003608

Dzhumagulova, K. N., Masheeva, R. U., Ramazanov, T. S., & Donkó, Z. (2014). Effect of magnetic field on the velocity autocorrelation and the caging of particles in two-dimensional Yukawa liquids. *Physical Review. E*, *89*(3), 033104. doi:10.1103/PhysRevE.89.033104 PMID:24730953

Dzhumagulova, K. N., Ramazanov, T. S., & Masheeva, R. U. (2012). Velocity autocorrelation functions and diffusion coefficient of dusty component in complex plasmas. *Contributions to Plasma Physics*, *52*(3), 182–185. doi:10.1002/ctpp.201100070

Fortov, V. E., & Morfill, G. E. (2009). *Complex and dusty plasmas: from laboratory to space*. CRC Press.

Gregori, G. (2012). *Molecular dynamics simulations of the equilibrium dynamics of non-ideal plasmas* [Doctoral dissertation, University of Oxford].

Kikuchi, H. (2013). *Electrohydrodynamics in dusty and dirty plasmas: gravito-electrodynamics and EHD* (Vol. 258). Springer Science & Business Media.

Kompaneets, R., Morfill, G. E., & Ivlev, A. V. (2009). Design of new binary interaction classes in complex plasmas. *Physics of Plasmas*, *16*(4), 043705. doi:10.1063/1.3112703

Merlino, R. L. (2006). Dusty plasmas and applications in space and industry. *Plasma Physics Applied*, *81*, 73–110.

Piel, A. (2010). An introduction to laboratory, space, and fusion plasmas. *Plasma Physics*.

Rodriguez, I. J. (2018). *Some Assembly Required: computational simulations of dusty plasma*.

Shahzad, A., & He, M. (2014). *Computer Simulation of Complex Plasmas: Molecular Modeling and Elementary Processes in Complex Plasmas*. Scholars' Press.

Shahzad, A., & He, M. G. (2012). Thermal Conductivity of Three-Dimensional Yukawa Liquids (Dusty Plasmas). *Contributions to Plasma Physics*, *52*(8), 667–675. doi:10.1002/ctpp.201200002

Shahzad, A., & He, M. G. (2013, July). Interaction contributions in thermal conductivity of three-dimensional complex liquids. In. AIP Conference Proceedings: Vol. 1547. *No. 1* (pp. 173–180). American Institute of Physics. doi:10.1063/1.4816866

Shahzad, A., He, M. G., & He, K. (2013). Diffusion motion of two-dimensional weakly coupled complex (dusty) plasmas. *Physica Scripta*, *87*(3), 035501. doi:10.1088/0031-8949/87/03/035501

Shahzad, A., Khan, M. Q., Shakoori, M. A., He, M., & Feng, Y. (2020). Thermal Conductivity of Dusty Plasmas through Molecular Dynamics Simulations. *Thermophysical Properties of Complex Materials, 13*.

Shakoori, M. A., He, M., & Shahzad, A. (2022). Diffusion coefficients of dusty plasmas in electric field. *The European Physical Journal D*, *76*(11), 227. doi:10.1140/epjd/s10053-022-00553-w

Shakoori, M. A., He, M., Shahzad, A., & Khan, M. (2022). Tuning the structure and transport properties of complex plasmas using electric field. *Physica Scripta*, *98*(1), 015608. doi:10.1088/1402-4896/aca6b2

Shukla, P. K., & Mamun, A. A. (2015). *Introduction to dusty plasma physics*. CRC press. doi:10.1201/9781420034103

Sukhinin, G. I., Fedoseev, A. V., Salnikov, M. V., Rostom, A., Vasiliev, M. M., & Petrov, O. F. (2017). Plasma anisotropy around a dust particle placed in an external electric field. *Physical Review. E*, *95*(6), 063207. doi:10.1103/PhysRevE.95.063207 PMID:28709289

Thomas, H., Morfill, G. E., Demmel, V., Goree, J., Feuerbacher, B., & Möhlmann, D. (1994). Plasma crystal: Coulomb crystallization in a dusty plasma. *Physical Review Letters*, *73*(5), 652–655. doi:10.1103/PhysRevLett.73.652 PMID:10057503

Vaulina, O. S., Adamovich, X. G., Petrov, O. F., & Fortov, V. E. (2008). Evolution of the mass-transfer processes in nonideal dissipative systems II: Experiments in dusty plasma. *Physical Review. E*, *77*(6), 066404. doi:10.1103/PhysRevE.77.066404 PMID:18643380

Vaulina, O. S., Lisin, E. A., Gavrikov, A. V., Petrov, O. F., & Fortov, V. E. (2010). Analysis of pair interparticle interaction in nonideal dissipative systems. *Journal of Experimental and Theoretical Physics*, *110*(4), 662–674. doi:10.1134/S1063776110040138

ABBREVIATIONS

CDP: Complex Dusty Plasma
DPs: Dusty Plasmas
DSFs: Diffusion Coefficient
EEF: External Electric Field
EMD: Equilibrium Molecular Dynamics
NEMD: Non Equilibrium Molecular Dynamics
WCCDP: Weakly Coupled Complex Dusty Plasma
κ: Debye Screening Strength

Chapter 7
A Mesoscopic Analysis for Diffusion Transport Phenomena

Soraya Trabelsi
Laboratory of Mathematical Modeling and Multi-scale Simulation for Physics and Engineering (2MSiPI), Faculty of Science of Tunis, University of Tunis El Manar, Tunisia & Faculty of Science, Bizerte University of Carthage, Tunisia

Ezeddine Sediki
Laboratory of Mathematical Modeling and Multi-scale Simulation for Physics and Engineering (2MSiPI), Faculty of Science of Tunis, University of Tunis El Manar, Tunisia

ABSTRACT

A mesoscopic study of linear irreversible transport phenomena is proposed. This chapter takes a phenomenological and statistical approach to non-equilibrium phenomena. At thermodynamic equilibrium, the intensive physical quantities of a system are uniform in space and time. These quantities can be defined at any point and at any time, and are referred to as local thermodynamic equilibrium. Otherwise, the system is out of thermodynamic equilibrium. This is the case for all irreversible phenomena, which are generally induced by an external input of energy and/or matter to the system. In this chapter, we will be focusing on the phenomenon of transport, which is a key process in non-equilibrium physics. There are various transport phenomena. Each is characterized by macroscopic properties. A microscopic approach is taken to study the transport phenomena. However, we are particularly interested in the phenomenon of particle diffusion and of the thermal diffusion.

1. INTRODUCTION

In the last years, understanding the transport phenomena become one of the most interesting issue for its physical significance of an advance in industrial technology.

Transport equations for heat and mass can be investigated on three scales. The first scale is the macroscopic scale for which partial differential equations (PDE) are used such as Navier-Stokes equation (N.S). These equations are hard to solve. In fact, due to the non-linearity terms and complex geometry, it is difficult to achieve the analytical solution. However, through computational fluid dynamic method, such

DOI: 10.4018/979-8-3693-0904-9.ch007

as the finite volume method (FVM), the ordinate discrete method (ODM), the finite difference method (FDM), finite element method (FEM) and others, the PDE are changed to a system of algebraic equations that are solved by an iterative procedure until reaching the convergence of the results. The second scale is a microscopic scale. The microscopic equation of motion of particle is governed by the Hamilton's equation. Therefore, with systems of a great number of particles we have to identify the position and the velocity of every particle to describe the motion of the each single particle. This task is difficult so we have to use several variables in order to describe the system properties. To the contrary, mesoscopic scale is an association among the microscopic and macroscopic scales. The LBM is a statistical model. It is based on the mesoscopic approach that provides its origin from kinetic theory (Liao & Jen, 2011; Mohamad, 2011; Sukop, 2006; Wolf-Gladrow, 2004).

In the last years, several works have been developed in the LBM, the method has received prodigious attention as an alternative and enhancing numerical scheme for simulating complex fluid flows.

The property of the assembly of particles is described by a distribution function. For that reason, LBM it has recognized to be an efficient method for simulating complex flow and it is becoming more current in the field of the numerical conventional method.

In the present chapter, a statistical study of transport phenomenon particularly molecular mass transport phenomenon and thermal diffusion transport phenomenon is devoted for which the microscopic approach is used to evaluate macroscopic quantities. The breakdown of this chapter is as follows:

In sections 2 and 3, the physical background of diffusion involving a microscopic analysis of diffusion phenomenon is presented. Next, the fundamental of LBM is detailed in section 4. Section 5 is dedicated to the application of LBM to energy and mass transport equation solution by using LBM.

Figure 1. Microscopic, mesoscopic, and macroscopic scales of fluid flows

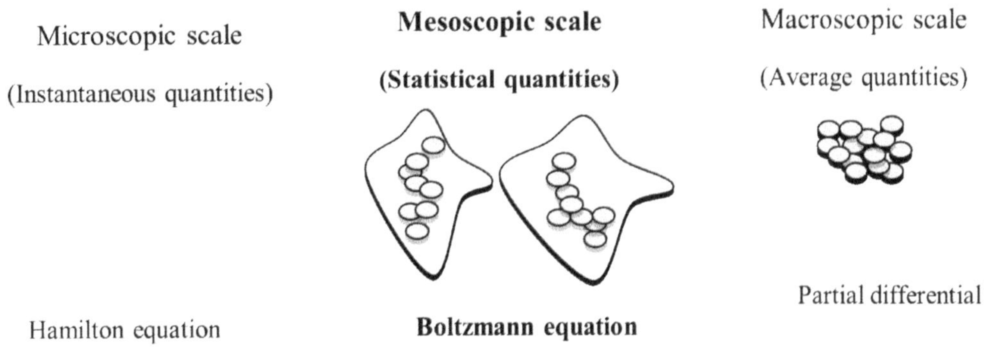

2. MASS TRANSPORT PHENOMENON: PARTICLE DIFFUSION IN PLASMA

Transport processes in plasma physics result from particle and energy transport. The first diffusion phenomenon corresponds to the movement of particles from higher density region to lower density region due to the internal energy of the plasma medium. Fick's law governs this phenomenon. The second is the convection phenomenon, which corresponds to the movement of particles into different regions of the plasma. This is due to the external forces applied to the system.

In this section, a special attention is devoted to the particle diffusion. We present the Fick's law and we explain the mechanisms of diffusion particles. The behavior of the diffusion coefficient, which is the main transport property, is exposed throw some current works in literature. Then the governing equation of article diffusion is established. In the end of this section, we provided a microscopic analysis of the particle diffusion phenomena.

2.1. Fick's Law

A. Fick, in 1855, has carried out an experimental study revealing that a system, presenting a gradient of particle density, evolves towards uniformity of particle distribution in the medium.

The Fick's law is given by:

$$\boldsymbol{J}_n = -D\overrightarrow{gradn_v}\left(\vec{r},t\right) \tag{1}$$

Where $\boldsymbol{J}_n$ is the current density vector. D is the diffusion coefficient or diffusivity always positive. It is expressed by m^2s^{-1}.

The diffusion coefficient depends on the temperature and pressure of the diffusing medium. The diffusivity also depends on the nature of the diffusing particles and the medium in which diffusion takes place.

The negative sign (-) means that the particle flow is directed towards low concentrations. This means that particle diffusion occurs from regions of high particle concentration to regions of low concentration.

The Fick's law is phenomenological, but not universal. It is valid only when the number of particles does not vary abruptly over short distances. This law has its limits for very large and sufficiently time-varying particle density gradients.

2.2. Fundamental Transport Property for Particle Diffusion

The diffusion coefficient is an important transport property, crucial and fundamental to describe mass transfer processes in many fields such as chemical, biological, geological, physics and engineering.

Indeed, this coefficient dependent on particle size and other properties of the diffusing material and particularly on temperature. Diffusion coefficients of one substance into the other are commonly determined experimentally and presented in reference tables. We summarized in table 1 some self-diffusion coefficients and in tables 2 and 3 those of binary diffusion in gaseous and liquids medium at temperature T=273K and pressure p=0.1MPa. These results are selected from A-TO-Z guide to thermodynamics, Heat & Mass Transffer, and Fluids Engineering.

Table 1. Self-diffusion coefficient D_A of some gases

Gas	$D_A(cm^2s^{-1})$	Gas	$D_A(cm^2s^{-1})$
H_2	1.604	CO_2	0.106
H_e	1.386	H_2O	0.176
Ar	0.157	CH_4	0.188
O_2	0.192	NH_3	0.192

Tables 2. Binary diffusion coefficient D_{AB} for gaseous

Diffusing gas *A*	Medium *B*	*P.* (mmH_2O)	*T(K)*	$D_{AB}(cm^2s^{-1})$
H_2	H_e	760	298	1.64
Ar	H_e	751	288	0.703
N_2	CO_2	760	288	0.158
H_2O	*Air*	760	273	0.219
H_2	N_2	760	288	0.743
H_2O	CO_2	760	273	0.146

Tables 3. Binary diffusion coefficient D_{AB} for liquids medium

Gas	Liquid	*T(K)*	$D_{AB}10^{-5}(cm^2s^{-1})$
Air	H_2O	293	2.5
CO_2	H_2O	298	1.92
H_2	H_2O	298	4.50
NH_3	H_2O	285	1.64
CO_2	C_2H_6O	298	3.42
CO_2	C_7H_{16}	298	6.03

The diffusion coefficient depends on various factors and particularly its behavior changed in a certain temperature range. Previously, numerical and experimental methods have been evaluated to study the behavior of diffusion coefficient and its dependency on temperature (Cussler, 2009; Marinakis & Samios, 2005; Meier et al., 2005; Zabaloy et al., 2006; Zhang et al., 2011).

Recently, Lapuerta, M. et al. (Lapuerta et al., 2014) presented an experimental analysis of the effect of temperature on the diffusion coefficient for various alcohols. They used nine different equations among those proposed in the literature to estimate the diffusion coefficient for binary gaseous mixtures alcohol-air, alcohol. The two former equations (the original one from Chapman-Enskog and the Wilke-Lee equation), were expressed as a function of the collision integral, which accounts for the molecular interactions. The resulting diffusion coefficients are plotted against temperature for five alcohols. Tsimpanogiannis, et al. (2019) presented a detailed overview of classical molecular simulation studies examining the self-diffusion coefficient of water. They discussed different aspects that affect the diffusion process, including the molecular models, the system-size effects, the temperature and pressure conditions and the type of confinement. Sean McGivern et al. (2021) have used a reverse-flow gas chromatography technique to determine the diffusion coefficients. These coefficients have been measured in N_2 for the homologous series of refrigerant-related (fluoro) methanes: methane (CH_4), fluoro-methane (CH_3F), di-fluoro-methane (CH_2F_2), and tri-fluoro-methane (CHF_3) over a wide range of temperatures (300–550) *K*. Amooghin et al. (2020) have studied a time-dependent 2-D axisymmetric model of a multilayer hollow fiber composite membrane for gas separation. The effect of temperature on the diffusion coefficients of gases in the membrane for various gases such as H_2, O_2, N_2, CH_4, CO_2 and He is investigated. They found that the diffusion coefficient of all gases increases with the increasing temperature. Xiao Zhao, and Hui Jin (2020) have employed molecular dynamics simulations for the calculation of self-diffusion coefficients of H_2, CH_4, CO, O_2 and CO_2 in supercritical water over a wide range of temperature (673–973*K*) and pressure (250–280*atm*). Results show that self-diffusion coefficient of H_2 is always the largest while self-diffusion coefficient of CO_2 is always the smallest. Self-diffusion coefficients of CH_4, CO and O_2 are close to each other. The effect of temperature is that higher temperature will increase diffusion coefficient. Others recent interesting studies are found in references (Koirala et al., 2022; Michalis et al., 2016; Zhao et al., 2021).

2.3. Balance Equation of Particle Diffusion

Let us consider a container with rigid adiabatic walls divided into two equal partitions of volume and each partition contains a number of particles and are separated by a piston that can move. The first case of same number of particles in each partition is shown in figure 2(a). It is the case of uniform density. In each one of the two partitions, the number of particles moving to the right side is equal to the number of particles moving to the left side. This is due to the isotropic distribution velocity. The average flow velocity is zero.

In the case of figure 2 (b), the number of particles is different at each partition. The left partition have more particles than the right one. Therefore, the density is non-uniform. In this case, a diffusive current flux appear from left to right partition when the piston is removed. This result is due to the thermal motion of particles, which is associated to their internal energy.

Figure 2. (a) System with uniform density and (b) non-uniform density

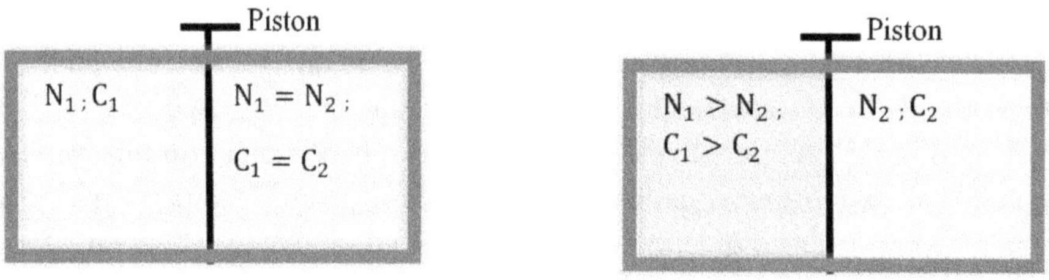

The local balance equation is given by:

$$\frac{\partial n_v}{\partial t} = -div\boldsymbol{J}_n + \sigma_n \tag{2}$$

Substituting the particle flux with that given by Fick's law, we obtain:

$$\frac{\partial n_v}{\partial t} = Ddiv\left[\nabla n_v\right] + \sigma_n \tag{3}$$

The particle diffusion equation is written as follows:

$$\frac{\partial n_v}{\partial t} = D\Delta n_v + \sigma_n \tag{4}$$

At steady state $\frac{\partial n_v}{\partial t} = 0$; the diffusion equation becomes Poisson's equation $\Delta n_v = -\frac{\sigma_n}{D}$.
In the absence of particle, production σ_n=0 is Laplace's equation Δ_{nv}=0.

The diffusion phenomenon is irreversible. In fact, the particle diffusion equation changes by reversing the time from t to $(-t)$. This irreversibility phenomenon is explained by the Fick's law.

The general form of the particle diffusion equation, in the absence of particle production, is expressed as:

$$\frac{\partial n_v}{\partial t} - D\Delta n_v = 0 \tag{6}$$

2.4 Diffusion Characteristic Length

The characteristic diffusion length, L, is related to the characteristic period τ via the diffusion coefficient D.

Indeed, according to the particle diffusion equation without production term, in order of magnitude, the following relationship can be written:

$$\frac{\partial^2 n_v(x,t)}{\partial x^2} \cong \frac{n_v}{L^2} \tag{7}$$

It follows:

$$\frac{1}{D}\frac{\partial n_v(x,t)}{\partial t} \cong \frac{n_v}{D\tau} = \frac{n_v}{L^2} \tag{8}$$

So that:

$$L^2 = D\tau \text{ and } L = \sqrt{D\tau} \tag{9}$$

Over a period of time τ, particle diffusion takes place over a characteristic length L.

2.5 Microscopic Analysis of Particle Diffusion

Particle diffusion is the result of non-homogeneous particle agitations. At the microscopic scale, the particle diffusion phenomenon can be analyzed. For that purpose, we consider an enclosure made up of two compartments, C_1 and C_2, separated by a partition and containing the same ideal gas at the same temperature, but with different numbers of particles. Let N_1 and N_2 the number of particles in the compartments C_1 and C_2 respectively. The density is non-uniform, we have: $n_{v_1} > n_{v_2}$

Then, the molecular agitation is greater in C_1 than in C_2.

By removing the partition between C_1 and C_2, molecular agitation progresses in the direction of uniformity, which implies a double flow of particles:

- The first particles flow is from C_1 to C_2.
- The second particles flow is from C_2 to C_1.

Figure 3. A control volume

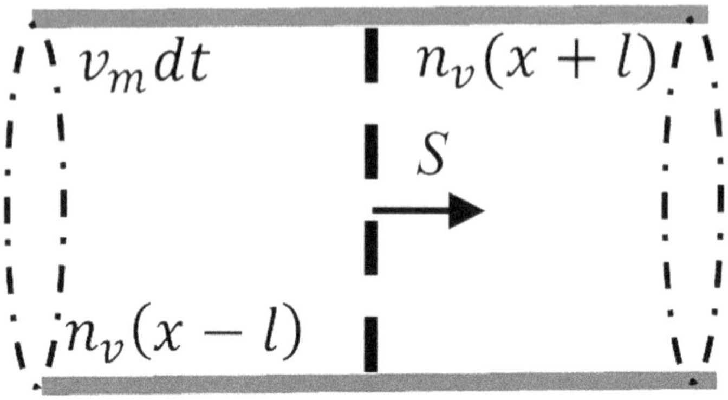

However, the current particle from C_1 to C_2 is more intense because the number of collisions in C_1 is greater than the number of collisions in C_2. Consequently, a diffusion current density from C_2 to C_2 is occurred.

We consider the one-dimensional case, during time dt and along the x direction, there are $\frac{n_v}{6}$ molecules per unit volume moving in the direction $x>0$ or ($x<0$) due to isotropy.

As shown in figure 3, the particles inside the volume V_1 are those confined in the cylinder of which the height is equal to $v_m dt$, centered at the position (x–l) and of a cross-section S. The number of these particles is given by $\frac{n_v(x-l)}{6} v_m S dt$. Similarly, in the volume V_2, the number of particles is equal to $\frac{n_v(x+l)}{6} v_m S dt$. Where l is the mean free path.

One can express that by the expression:

$$J_{n,x} = \frac{dN}{dt} = \frac{v_m}{6}\left[n_v(x-l) - n_v(x+l)\right] \tag{10}$$

One can simplify this formula to:

$$J_{n,x} \cong -\frac{v_m l}{3}\frac{\partial n_v}{\partial x} \tag{11}$$

Using Fick's law for the one-dimensional case $J_{n,x} = -D\frac{\partial n_v}{\partial x}$, and by identification, we finally acquired the formula of the diffusion coefficient for particle diffusion phenomena:

$$D = \frac{v_m l}{3} \tag{12}$$

This relationship shows that when the mean free path l is small, particle diffusion is low because of the great number of collisions. The particle flux is therefore low. Similarly, at low velocities, particles move slowly, which restraints their diffusion.

3. ENERGY TRANSPORT PHENOMENON. THERMAL DIFFUSION

3.1 Fourier's Law

Joseph Fourier carried out an experimental study (Grattan-Guinness, 2005) in order to compare thermal diffusion in several metallic materials. It consisted of introducing different identical metal rods (Ion, Aluminum, Zinc, Copper …etc.) into a boiling water bath. Observing the temperature at same points on the different stems showed that the temperature increases over time but more or less quickly from one stem to another. The existence of temperature gradient in a material leads to the existence of heat that flow in direction of the decreasing temperature. There is no movement of matter or work received; the heat is transferred from one point of the stem to another point by conduction.

Fourier's law states that, "Any system with a uniform temperature has no heat transfer to its environment. Heat transfer is only possible when there is a temperature gradient in the system".

Mathematically, this translates into the following relationship:

$$\boldsymbol{J}_{th} = -\lambda \nabla \mathrm{T} \tag{13}$$

λ is a positive constant and represents the thermal conduction coefficient or conductivity.

The minus sign in this relationship indicates the orientation of the thermal current towards low temperatures.

This phenomenological law was established by Fourier in 1815 and expresses the proportionality between the thermal volume current $\boldsymbol{J}_{th}(\boldsymbol{r},t)$ and the temperature gradient $T(\boldsymbol{r},t)$.

Fourier's law is valid for temperature differences that do not vary abruptly over short distances. Otherwise, very intense diffusion currents appear, and convection transfer is taken into account.

3.2 Fundamental Transport Property for Thermal Diffusion

The thermal diffusivity is one of the main transport properties of heat. This physical quantity delivers an indication of the rate at which heat energy is transported through a medium by conduction phenomena. The two important properties related to conduction heat transfer are thermal conductivity and thermal diffusivity. The relation between thermal conductivity and thermal diffusivity is given by:

$$\kappa = \frac{\lambda}{\rho c_v} \tag{14}$$

Where κ is the thermal diffusivity.

However, the role played by diffusivity is important in thermal diffusion phenomenon. A special attention is devoted in this sub-section to thermal diffusivity and to the proposed methods of measurements. Several methods of thermal diffusivity measurements in various materials are found in literature (Salazar, 2003; Sun & Venart, 2005).

Recently, Marín et al. (2013) proposed a technique for thermal diffusivity measurement in fluids. It is based on the Angstrom method, but with excitation of thermal waves by electromagnetic energy absorption and pyroelectric detection. They found that the method could be suitable for characterization of non-homogeneous materials.

More recently, Huangfei et al. (2019) measured simultaneously the thermal diffusivity and thermal conductivity of four natural granitoid samples at high pressures (up to $1.5GPa$) and temperatures (up to $988K$) in a multi-anvil apparatus using the transient plane-source method. They examined factors influencing thermal transport properties. Owczarek and Baryłka (2019) determined the thermal diffusivity of the material through methods based on the measurement of the temperature profile in the wall. This method take into account the layers of material close to the surface. It works by measuring the temperature in the element's volume and assuming that the obtained results meet the Fourier equation. Chen et al. (2022) measured the thermal diffusivity of ethanol by the dynamic light scattering method in the temperature ranging from (293 to 654)K and along six isobaric lines including 0.1, 2.5, 5, 6.2, 7.5, and 10 MPa. They analyzed the variation of thermal diffusivity with the temperature and pressure. In addition, they made the comparison between the experimental data and the calculated values from thermal conductivity. Sun et al. (2022) presented a review of thermo-physical properties of natural gas hydrates. they summarized the progress of laboratory measurements and models of NGH thermo-physical parameters, such as phase equilibrium, thermal conductivity, dissociation enthalpy, thermal diffusivity, and specific heat capacity. They discussed the advantages and disadvantages of measurement methods, numerical models, typical experimental data, and characteristics. White et al. (2022) detailed three novel methods for measuring both the thermal conductivity and the thermal diffusivity to within 5.6% which are specifically designed for bodies like lithium-ion batteries. They recommended these novel methods to represent the future for thermal characterization of lithium-ion batteries. Many others works are given in references (Athanasaki et al., 2023; Nikitin et al., 2022; Zhang et al., 2023).

3.3 The Mechanisms of Thermal Transport Phenomenon

The existence of temperature difference in the medium leads to a heat transfer across a system boundary by the mechanism of intermolecular interactions. The heat is transferred due to particle collision and lattice vibration as illustrated in figure 4.

Diffusion occurs when the particles collide randomly and spread out. They move from a region of high concentration to a region of lower concentration. The diffusion process in the medium increases with the increases of temperature because the when temperature increases (warm particles) the kinetic energy of particles increases.

Diffusion occurs in liquids and gases the particles move everywhere and collide with each other randomly (figure 3(a)). However, in a gaseous substance particles move faster than those in a liquid substance and are more spread out. Consequently, diffusion process is faster through gases than through liquids.

In solids, when there is a temperature difference, the particles are bond together intensely but they vibrate closely to their site. The warm side of the solid performs more energetic particle movements (figure 3(b)). Particularly, for metals free electrons can move throw the solid and more the temperature rises more they move.

Figure 4. (a) Particle collision and (b) lattice vibration

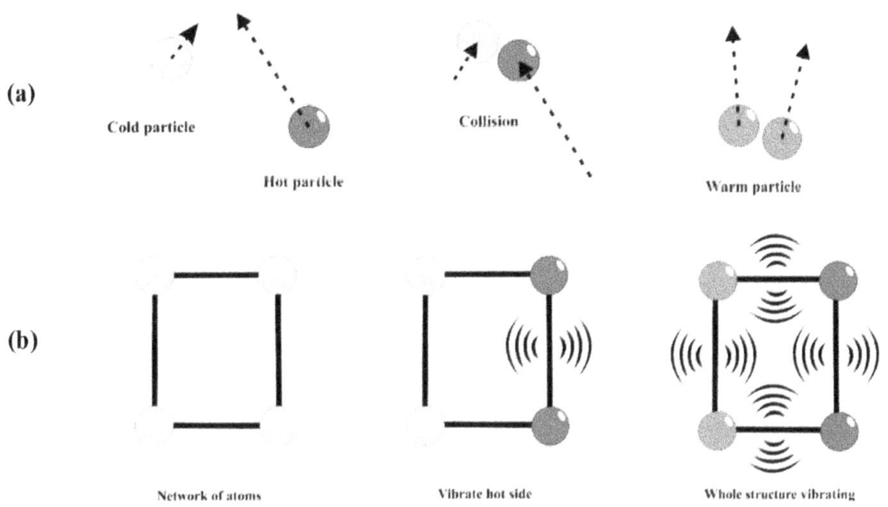

3.4 Balance Equation of Thermal Diffusion

The balance equation of thermal diffusion is given by:

$$\frac{\partial(\rho u)}{\partial t} = -div\boldsymbol{J}_{th} + \sigma_p \tag{15}$$

Then, we have $d(\rho u)=\rho c_v dT$; this leads to:

$$\rho c_v \frac{\partial T}{\partial t} = -div\boldsymbol{J}_{th} + \sigma_p \tag{16}$$

In the absence of energy production σp_0; eq.(16) becomes:

$$\rho c_v \frac{\partial T}{\partial t} + div\boldsymbol{J}_{th} = 0 \tag{17}$$

The following equation is obtained:

$$-\lambda div(\nabla T) + \rho c_v \frac{\partial T}{\partial t} = 0 \tag{18}$$

The thermal diffusion equation, also known as the heat equation, takes the general form:

$$\frac{\partial T}{\partial t} - \frac{\lambda}{\rho c_v}\Delta T = 0 \tag{19}$$

The diffusion equation is not invariant to time reversal.

During a characteristic time τ, the diffusion phenomenon takes place over a characteristic distance L as:

$$\frac{T}{\tau} = \frac{\lambda}{\rho c_v}\frac{T^2}{L^2} \tag{20}$$

So that,

$$\mathrm{T} = \frac{L^2}{\kappa} \text{ or } L = \sqrt{\kappa.\tau} \tag{21}$$

At the end,

$$\kappa = \frac{\lambda}{\rho c_v} \tag{22}$$

κ designates the thermal diffusion coefficient or diffusivity.

3.5 Microscopic Analysis of Thermal Diffusion

We reason in the same way as for particle diffusion phenomena. Consider an enclosure made up of two compartments, C_1 and C_2, containing particles of the same density, but with different thermal agitation such us $U_1 > U_2$. A similar study to that detailed for analyzing particle diffusion at the microscopic scale lead to the following relation:

$$\boldsymbol{J}_{th} \cong -\frac{\rho c_v v_m l}{3}\frac{\partial T}{\partial x} \tag{23}$$

By identification with Fourier's law:

$$\boldsymbol{J}_{th} = -\frac{\lambda \partial T}{\partial x} \tag{24}$$

After some calculation step, one can obtain:

$$\lambda = \frac{\rho c_v v_m l}{3} = \rho c_v D \tag{25}$$

Finally

$$D = \frac{v_m l}{3} \tag{26}$$

4. RELATION BETWEEN THE BOLTZMANN EQUATION AND THE DIFFUSION EQUATIONS

The general form of the corresponding governing equations (6) and (19) can be expressed by:

$$\frac{\partial \Phi}{\partial t} - \Lambda \Delta \Phi = 0 \tag{27}$$

The Governing equation and the physical parameters of transport Φ and Λ are summarized in table 4.

The Boltzmann equation is consequently a transport equation of the distribution function *f*.

To solve the proposed governing equation, lattice Boltzmann method (LBM), which is a statistical method based on a mesoscopic approach, is proposed. This method has reached an advanced state over the last three decades and become a strong alternative for simulating complex problems and particularly for understanding fluid flow at the microscopic scale.

Table 4. Transport phenomena: Governing equations and properties transport.

	Particle diffusion	**Thermal diffusion**
Equation of diffusion	$\frac{\partial n_v}{\partial t} - D\Delta n_v = 0$	$\frac{\partial T}{\partial t} - \frac{\lambda}{\rho c_v}'' T = 0$
Variable of transport (Φ)	n_v	T
Transport coefficient (Λ)	D	$\frac{\lambda}{\rho c_v}$

5. NUMERICAL METHOD

Computational fluid dynamic methods (CFD), used for diffusion transport problems, such as finite volume method (FVM), finite element method (FEM), discrete ordinates method (DOM) and others need large computational time and storage volume and cannot deal with complex geometry.

The lattice Boltzmann method (LBM) is an alternative to the conventional methods. It takes its origin from the kinetic theory of gases. It is based on the determination of a single function, representing the particle distribution, through which macroscopic quantities are expressed. This method reproduces the fluid behavior using a mesoscopic approach based on statistical analysis. It provides access to integro-differential equations based on a mathematical development attributed to Chapman-Enskog. In this section, an exhaustive study of the lattice Boltzmann method is devoted.

5.1. The Boltzmann Equation

The Boltzmann equation (BE) is obtained from a particle conservation balance. In the absence of collisions between molecules, the number of molecules before and after the perturbation is conserved within the control volume. The balance is written:

$$f\left(\boldsymbol{r}+\boldsymbol{c}dt,\boldsymbol{c}+\frac{\boldsymbol{F}}{m}dt,t+dt\right)d\boldsymbol{r}d\boldsymbol{c}-f\left(\boldsymbol{r},\boldsymbol{c},t\right)d\boldsymbol{r}d\boldsymbol{c}=0 \tag{28}$$

Where $\boldsymbol{F}$ is an external force vector applied to the system and m is the particle mass.

The BE is a continuous equation that describes the evolution of a non-equilibrium thermodynamic system using statistical study.

When the collisions between molecules are taken into account, we diverge from the statistical equilibrium and the balance equation becomes:

$$f\left(\boldsymbol{r}+\boldsymbol{c}dt,\boldsymbol{c}+\frac{\boldsymbol{F}}{m}dt,t+dt\right)d\boldsymbol{r}d\boldsymbol{c}-f\left(\boldsymbol{r},\boldsymbol{c},t\right)d\boldsymbol{r}d\boldsymbol{c}=\Omega\left(f\right)d\boldsymbol{r}d\boldsymbol{c}dt \tag{29}$$

In the general case, the Boltzmann equation can be written as follows:

$$\frac{\partial f}{\partial t}+\boldsymbol{c}.\nabla f+\frac{\boldsymbol{F}}{m}\nabla_{c}f=\Omega\left(f\right) \tag{30}$$

In the case of the absence of external force, the BE is written in the simplest form:

$$\frac{\partial f}{\partial t}+\boldsymbol{c}.\nabla f=\Omega\left(f\right) \tag{31}$$

Macroscopic quantities are calculated by integration over the velocity space throw the following expressions:

$$\rho\left(\boldsymbol{r},t\right)=\int f\left(\boldsymbol{r},\boldsymbol{c},t\right)d\boldsymbol{c} \tag{32}$$

$$\rho\left(\boldsymbol{r},t\right)\boldsymbol{u}\left(\boldsymbol{r},t\right)=\int \boldsymbol{c}f\left(\boldsymbol{r},\boldsymbol{c},t\right)d\boldsymbol{c} \tag{33}$$

$$\rho(\boldsymbol{r},t)e(\boldsymbol{r},t)=\frac{1}{2}\int(\boldsymbol{c}-\boldsymbol{u})^2 f(\boldsymbol{r},\boldsymbol{c},t)\,d\boldsymbol{c} \tag{34}$$

$\rho(r,t)$ is macroscopic mass density, u the macroscopic velocity vector and c are velocity vectors due to thermal agitation.

The equilibrium distribution function is defined by:

$$f^{eq}=\frac{\rho}{(2\pi rT)^{\frac{D}{2}}}\exp\left(-\frac{c^2}{2rT}\right)exp\left(\frac{\boldsymbol{c}.\boldsymbol{u}}{rT}-\frac{u^2}{2rT}\right) \tag{35}$$

Where $r=\frac{R}{M}$ and D is the space dimension.

5.2. From Boltzmann Equation to Lattice Boltzmann Method

The discretized lattice Boltzmann equation, under BGK approximation, is written as follows:

$$\frac{\partial f_k}{\partial t}+\boldsymbol{c}_k.\nabla f_k(\boldsymbol{r},t)=\frac{1}{\tau_f}\left(f_k^{eq}(\boldsymbol{r},t)-f_k(\boldsymbol{r},t)\right) \tag{36}$$

Where f_k is the discrete transport quantity distribution function, τ_f is the single relaxation time.

The second member of eq.(36) defined by $\Omega=\frac{1}{\tau_f}\left(f_k^{eq}(\boldsymbol{r},t)-f_k(\boldsymbol{r},t)\right)$ is the collision operator, under BGK approximation, it characterizes the effect of collisions between particles.

Equation (36) is the most current kinetic model and alternates Navier-Stokes equation in Computational fluid dynamic simulations.

For the time step Δt and space step $\Delta \boldsymbol{r}_k = \boldsymbol{c}_k \bullet t$, the fully discretized equation is given by:

$$f_k(\boldsymbol{r}+\boldsymbol{c}_k\Delta t,t+\Delta t)=f_k(\boldsymbol{r},t)+\frac{\Delta t}{\tau}\left(f_k(\boldsymbol{r},t)-f_k^{eq}(\boldsymbol{r},t)\right) \tag{37}$$

f_k^{eq} designates the local equilibrium distribution function.

The discretized equation is solved in two steps:

- Collision step:

$$f_k(\boldsymbol{r},t+\Delta t)=(1-\omega)f_k(\boldsymbol{r},t)+\omega f_k^{eq}(\boldsymbol{r},t) \tag{38}$$

- Streaming step:

$$f_k\left(\boldsymbol{r}+\boldsymbol{c}_k\Delta t, t+\Delta t\right)=f_k\left(\boldsymbol{r}, t+\Delta t\right) \tag{39}$$

The macroscopic quantities defined above () are calculated by summing over all the distribution functions of the lattice arrangements *DnQm* selected, and are written as follows:

$$\rho\left(\boldsymbol{r},t\right)=\sum f_k\left(\boldsymbol{r},t\right) \tag{40}$$

$$\rho\left(\boldsymbol{r},t\right)\boldsymbol{u}\left(\boldsymbol{r},t\right)=\sum \boldsymbol{c}_k f_k\left(\boldsymbol{r},t\right) \tag{41}$$

$$\rho\left(\boldsymbol{r},t\right)e\left(\boldsymbol{r},t\right)=\frac{1}{2}\sum \boldsymbol{c}_k^2 f_k\left(\boldsymbol{r},t\right) \tag{42}$$

Chapman (1916) and Enskog (1917) developed individually the Chapman-Enskog method (CE) for the continuum Boltzmann equation, under hypothesis of Maxwell equilibrium distribution function. For low Mach number, the equilibrium distribution function is mainly an expansion of the Maxwell's distribution function. Therefore, CE method allows calculating transport coefficients for macroscopic quantities from the microscopic information of the gas. In this work the CE method is applied to derive the heat transfer equations from the Boltzmann equation in a form of a series in Knudsen number, denoted by ε. The series approximation yields the Navier–Stokes–Fourier equations at first order. Others references on CE expansion are found in literature (Sinkevich & Semenov, 2003; Struchtrup, 2005).

A Taylor expansion in time and space is firstly used and leads to the following equation:

$$\left(\frac{\partial}{\partial t}+\boldsymbol{c}_k.\nabla\right)f_k+\frac{\Delta t}{2}\left(\frac{\partial}{\partial t}+\boldsymbol{c}_k.\nabla\right)^2 f_k+O\left(\Delta t^2\right)=-\frac{1}{\tau_f}\left(f_k-f_k^{eq}\right) \tag{43}$$

The discrete distribution function f_k is written in a form of a series in Knudsen number:

$$f_k=f_k^{(0)}+\varepsilon f_k^{(1)}+\varepsilon^2 f_k^{(2)}+O\left(\varepsilon^3\right) \tag{44}$$

Where ε is a small dimensionless parameter that corresponds to the relaxation time in the case of BGK approximation. It is proportional to the Knudsen number and is the ratio between the mean free path and the characteristic length.

Consequently, the non-linearity of the collision operator is covered.

5.3. LBM Schemes

The lattice arrangements *DnQm* used in LBM are illustrated in table 5. Where *n* denotes the number of dimensions of the problem and *m* is the number of particle velocities vectors.

The nodes in different positions regarding to the central node have suitable weighting factors w_k, which depend on the distance from central node with maximum value. The weighing factors are used in the calculation of equilibrium distribution function f_k^{eq} .

Figure 5. Particle- D3Q15

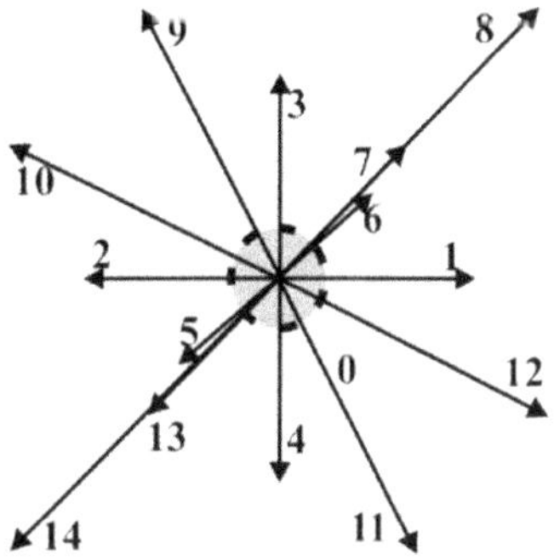

The normalization condition for the weights is: $\sum_k w_k = 1$

The moments of the distribution functions are determined by:

$$\int (c) c^n = \sum_k w_k c_k^n \tag{45}$$

(c) is the Maxwell function.

For 2-D configuration, the Maxwell function and the equilibrium distribution function are respectively:

$$(c) = \frac{1}{(2\pi rT)^{\frac{2}{2}}} exp\left(-\frac{c^2}{2rT}\right) \tag{46}$$

$$f_k^{eq} = \rho w_k \left(1 + \frac{\boldsymbol{c_k}.\boldsymbol{u}}{rT} - \frac{u^2}{2rT} + \frac{(\boldsymbol{c_k}.\boldsymbol{u})^2}{2(rT)^2} + \Theta(u^3)\right) \tag{47}$$

Table 5 summarizes various lattice models with corresponding discrete velocity c_k, weighting factors w_k and distribution functions $f_k(w_k)$.

Table 5. Lattice models DnQm with corresponding discrete velocity c_k, weighting factors w_k and distribution functions $f_k(w_k)$

DnQm model	Discrete velocity c_k	Weights w_k	$f_k(w_k)$
D1Q2	±1	±1/2	$f_{1,2}(1/2)$
D1Q3	0, ±1	4/6, 16	$f_0(4/6), f_{1,2}(1/6)$
D1Q5	0, ±1, ±2	6/12, 2/12, 1/12	$f_0(6/12), f_{1,2}(2/12), f_{3,4}(1/12)$
D2Q5	(0;0), (±1;0), (0, ±1)	4/6, 1/12	$f_0(4/6), f_{1,2,3,4}(1/12)$
D2Q9	(0;0), (±1;0), (0; ±1), (±1; ±1)	4/9, 1/9, 1/36	$f_0(4/9), f_{1\to4}(1/9), f_{5\to8}(1/36)$
D3Q15	(0;0;0), (±1;0;0), (0; ±1;0), (0;0; ±1), (±1; ±1; ±1)	2/9, 1/9, 1/72	$f_0(16/72), f_{1\to6}(8/72), f_{7\to14}(1/72)$
D3Q19	(0;0;0), (±1;0;0), (0; ±1;0), (0;0; ±1), (±1; ±1;0), (±1;0; ±1), (0;±1;±1)	1/3, 1/18, 1/36	$f_0(12/36), f_{1\to6}(2/36), f_{7\to18}(1/36)$

6. ENERGY AND MASS TRANSPORT EQUATION SOLUTION BY USING LBM

In this section, the phenomenon of heat and mass transfer of humid air through a building wall are proposed as an application of transport problem simulated with the mesoscopic lattice Boltzmann method.

6.1. Physical Problems

The configuration of the geometry is a wall of thickness *e* and height *h*. It is exposed to high temperatures and concentrations (T_w,C_w) on the west face which is placed at the position (x=0,y) whereas the other faces are maintained at equal temperature and concentration such that:

$T_E=T_N=T_S<T_W$ and $C_E=C_N=C_S<C_W$.

The building wall is isotropic porous medium transfer, viscous dissipation and the work of pressure forces are negligible. The thermophysical properties of humid air are assumed to be constant. Mach number condition (Ma<1).

Figure 6. Schematic geometry and boundaries conditions

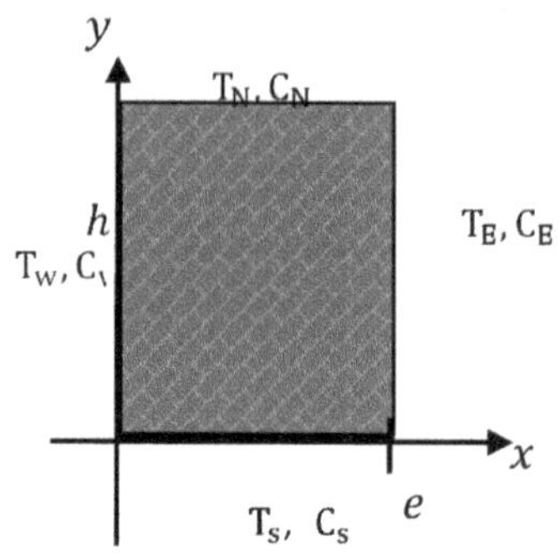

6.2. Temperature and Concentration Distributions

The computations are performed using the Lattice Boltzmann Method with D2Q9 scheme.

The thermo-physical properties are:

$D=10^{-7}m^2s^{-1}$, $K_s=0.52Wm^{-1}K^{-1}$, $K_f=0.026Wm^{-1}K^{-1}$.

The thermal capacities of the porous medium and the fluid are respectively:

$(\rho c_p)_s=1, 512.10^{-6}$ Kg/$s^2.mK$, $(\rho c_p)_f=1, 31.10^3$ Kg/$s^2.mK$.

The west face at position (x=0, y) is at temperature T_W=30°C and a concentration C_W=15g/m^3 whereas the other faces are at temperature 10°C and concentration 4.8g/m^3.

Temperature and concentration fields, inside a porous medium, obtained by solving the balance equations with the Lattice Boltzmann Method, with D2Q9 model, inside a porous medium are illustrated in figures 6 and 7 respectively.

Figure 7. Evolution of temperature inside the porous media from high (a) to small (b) porosity

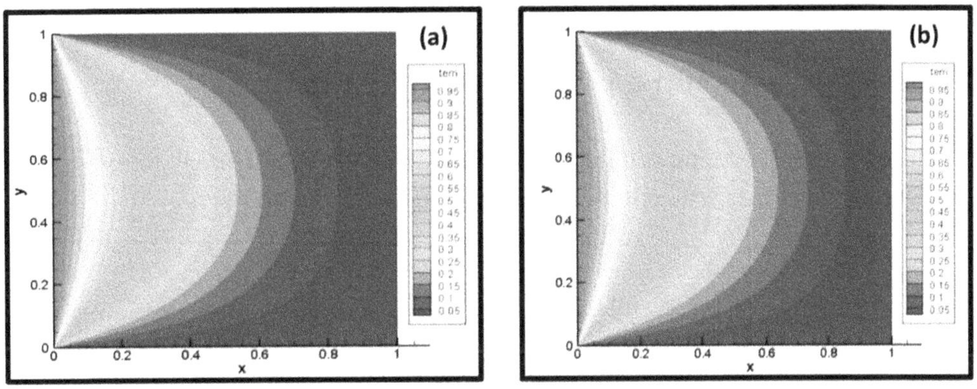

Figure 8. Evolution of concentration inside the porous media from high (a) to small (b) porosity

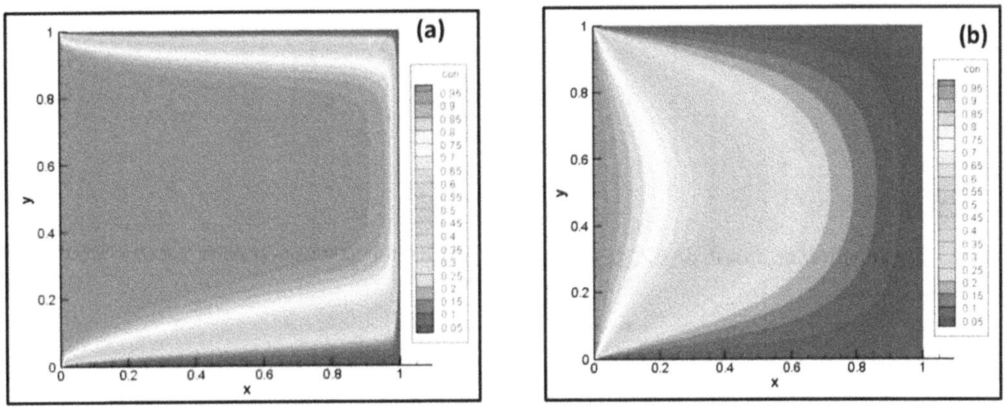

6.3. Effect of Temperature on Mass Transport Distributions

The effect of temperature on mass transport distributions inside porous media is illustrated in figure 8. Calculations are performed with Dufour number *Du*=0 and various Soret number (*Sr*=0; *Sr*=0.1 and *Sr*=0.2).

Figure 9. Temperature effect on mass transport distributions inside porous media for Sr=0 (a); Sr=0.1 (b) and Sr=0.2 (c)

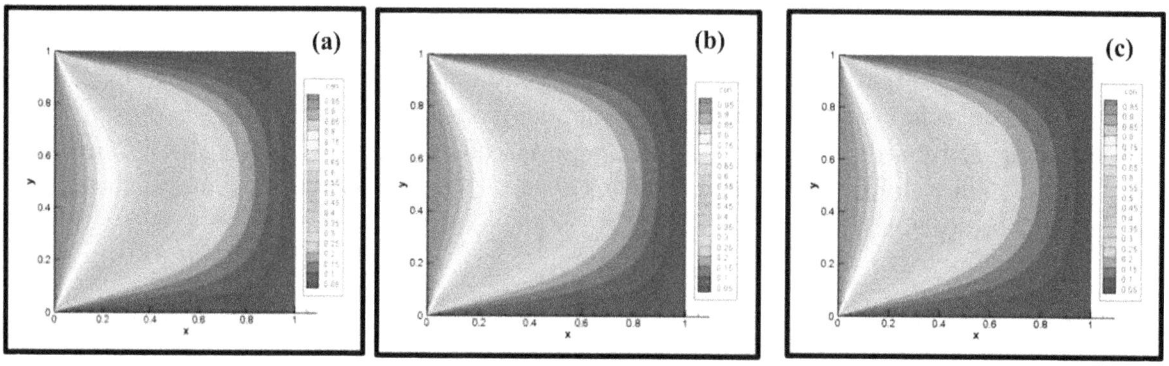

6.4. Effect of Mass Transport on Temperature Distributions

The effect of mass transport on temperature distributions is observed in figure 9. Results are carried out with various Dufour number (*Du*=0, *Du*=0.1, *Du*=0.2) and for Soret number *Sr*=0.

Figure 10. Mass transport effect on temperature distributions inside porous media for Sr=0 (a); Sr=0.1 (b) and Sr=0.2 (c)

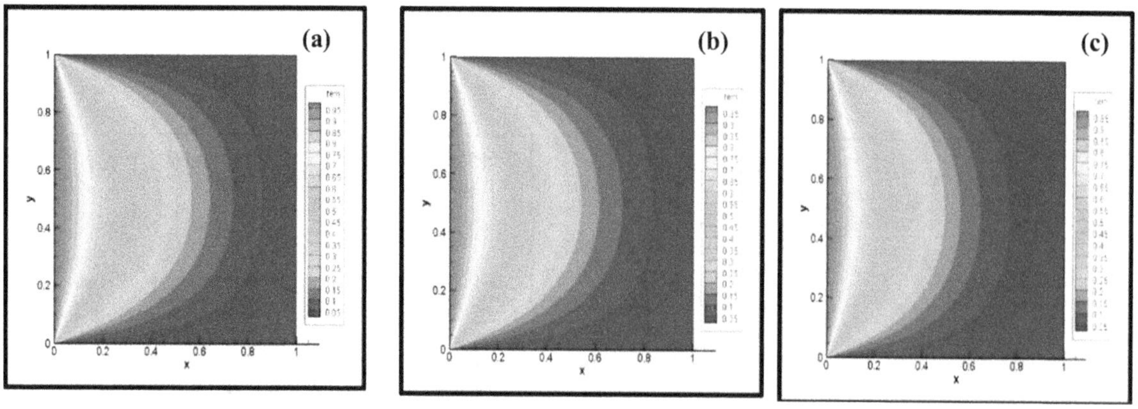

7. CONCLUSION

The present chapter offer a comprehensive study of a mesoscopic thermodynamic analysis for diffusion transport phenomena particularly mass and thermal transport phenomena. The advanced works based on the examination of the behavior of transport properties throw numerical and experimental methods are presented. Several computational fluid dynamic methods (CFD) are used for diffusion transport problems. These methods need large computational time and storage volume. Therefore, a special attention is devoted in this chapter to the lattice Boltzmann method (LBM). It is a statistical method based on a mesoscopic approach. The LBM has reached an advanced state over the last three decades and become a strong alternative to the conventional Navier–Stokes equations for simulating complex problems and particularly for understanding fluid flow and heat transfer at the microscopic scale. LBM has employed over a large industrial, academic and engineering applications. the fundamental of LBM is detailed in this work. As a final point, the phenomenon of heat and mass transfer of humid air through a building wall are proposed as an application of transport problem simulated with the mesoscopic lattice Boltzmann method.

Nomenclature

J_n	Flux Density Vector For Particle Transport	c_k	Velocity Vector in k-Direction
n_v	Volume Number Density	c	Speed of Light
D	Particle Diffusivity	t	Time
D_A	Self-Diffusion Coefficient	c	Lattice Velocity Vector
D_{AB}	Binary-Diffusion Coefficient	w	Weight Factor for LBM
T	Temperature	Sr	Soret Number
p	Pressure	Du	Dufour Number
L	Characteristic Diffusion Length		**Greek Symbols**
l	Mean Free Path	κ	Diffusivity
J_{th}	Flux Vector for Heat Transport	λ	Thermal Conductivity
c_v	Heat Capacity at Constant Volume	σn	Particle Production
f_k	Distribution Function	τ	Characteristic Period
f_k^{eq}	Equilibrium Distribution Function (Dimensionless)	ρ	Density
F	External Force Vector	Ω	Collision Term

REFERENCES

Athanasaki, G., Jayakumar, A., & Kannan, A. M. (2023). Gas diffusion layers for PEM fuel cells: Materials, properties and manufacturing–A review. *International Journal of Hydrogen Energy*, *48*(6), 2294–2313. doi:10.1016/j.ijhydene.2022.10.058

Chapman, S. (1916). On the law of distribution of molecular velocities, and on the theory of viscosity and thermal conduction, in a non-uniform simple monotomic gas. *Phil. Trans. R. Soc. A*, *216*, 279–348.

Chen, J., Xing, H., Zhan, T., Chen, L., He, M., & Zhang, Y. (2022). Measurement of thermal diffusivity of ethanol from (293 to 564) K and up to 10 MPa in vicinity of the critical point. *Fluid Phase Equilibria, 552*, 113276. doi:10.1016/j.fluid.2021.113276

Cussler, E. L. (2009). *Diffusion: mass transfer in fluid systems*. Cambridge university press. doi:10.1017/CBO9780511805134

Ebadi Amooghin, A., Mirrezaei, S., Sanaeepur, H., & Moftakhari Sharifzadeh, M. M. (2020). Gas permeation modeling through a multilayer hollow fiber composite membrane. *Journal of Membrane Science and Research, 6*(1), 125–134.

Enskog, D. (1917). *Kinetische Theorie der Vorgänge in mässig verdünnten Gasen*, [Doctoral Dissertation, Upsala].

Fu, H., Zhang, B., Ge, J., Xiong, Z., Zhai, S., Shan, S., & Li, H. (2019). Thermal diffusivity and thermal conductivity of granitoids at 283–988 K and 0.3–1.5 GPa. *The American Mineralogist, 104*(11), 1533–1545. doi:10.2138/am-2019-7099

Grattan-Guinness, I. (2005). Joseph Fourier, Théorie analytique de la chaleur (1822). In *Landmark Writings in Western Mathematics 1640-1940* (pp. 354–365). Elsevier Science. doi:10.1016/B978-044450871-3/50107-8

Koirala, R. P., Dawanse, S., & Pantha, N. (2022). Diffusion of glucose in water: A molecular dynamics study. *Journal of Molecular Liquids, 345*, 117826. doi:10.1016/j.molliq.2021.117826

Lapuerta, M., Hernández, J. P., & Agudelo, J. R. (2014). An equation for the estimation of alcohol-air diffusion coefficients for modelling evaporation losses in fuel systems. *Applied Thermal Engineering, 73*(1), 539–548. doi:10.1016/j.applthermaleng.2014.08.009

Liao, Q., & Jen, T. C. (2011). Application of Lattice Boltzmann method in fluid flow and heat transfer. *Computational Fluid Dynamics Technologies and Applications*, 29-68.

Marín, E., Hernández-Rosales, E., Mansanares, A. M., Ivanov, R., Rojas-Trigos, J. B., & Calderón, A. (2013). A method for thermal diffusivity measurement in fluids. *The Review of Scientific Instruments, 84*(10), 104903. doi:10.1063/1.4824191 PMID:24182147

Marinakis, S., & Samios, J. (2005). The temperature and density dependence of fluid xenon self-diffusion coefficients: A comparison between experimental, theoretical and molecular dynamics results. *The Journal of Supercritical Fluids, 34*(1), 81–89. doi:10.1016/j.supflu.2004.10.002

McGivern, W. S., & Manion, J. A. (2021). Binary diffusion coefficients for methane and fluoromethanes in nitrogen. *Journal of Chemical & Engineering Data, 66*(8), 3047-3056.

Meier, K., Laesecke, A., & Kabelac, S. (2005). Transport coefficients of the Lennard-Jones model fluid. III. Bulk viscosity. *The Journal of Chemical Physics, 122*(1), 014513. doi:10.1063/1.1828040 PMID:15638680

Michalis, V. K., Moultos, O. A., Tsimpanogiannis, I. N., & Economou, I. G. (2016). Molecular dynamics simulations of the diffusion coefficients of light n-alkanes in water over a wide range of temperature and pressure. *Fluid Phase Equilibria, 407*, 236–242. doi:10.1016/j.fluid.2015.05.050

Mohamad, A. A. (2011). *Lattice Boltzmann Method* (Vol. 70). Springer. doi:10.1007/978-0-85729-455-5

Nikitin, E. D., Popov, A. P., Bogatishcheva, N. S., & Faizullin, M. Z. (2022). Critical temperatures, pressures, heat capacities, and thermal diffusivities of C3 to C6 dinitriles. *Journal of Chemical & Engineering Data, 67*(4), 836-845.

Owczarck, M., & Baryłka, A. (2019). Determining the thermal diffusivity of the material based on the measurement of the temperature profile in the wall. *Rynek energii.*

Salazar, A. (2003). On thermal diffusivity. *European Journal of Physics, 24*(4), 351–358. doi:10.1088/0143-0807/24/4/353

Sinkevich, O. A., & Semenov, A. M. (2003). Solution of the Boltzmann equation by expanding the distribution function with several time and coordinate scales in the Enskog series in Knudsen parameter. *Technical Physics, 48*(10), 1221–1225. doi:10.1134/1.1620111

Struchtrup, H. (2005). Macroscopic transport equations for rarefied gas flows: approximation methods in kinetic theory. In *Interaction of Mechanics and Mathematics*. Springer-Verlag.

Sukop, M. C. (2006). *DT Thorne, Jr.* Lattice Boltzmann Modeling Lattice Boltzmann Modeling.

Sun, L., & Venart, J. E. S. (2005). Thermal conductivity, thermal diffusivity, and heat capacity of gaseous Argon and nitrogen. *International Journal of Thermophysics, 26*(2), 325–372. doi:10.1007/s10765-005-4502-0

Sun, S., Gu, L., Yang, Z., Lin, H., & Li, Y. (2022). Thermophysical properties of natural gas hydrates: A review. *Natural Gas Industry B, 9*(3), 246–263. doi:10.1016/j.ngib.2022.04.003

Tsimpanogiannis, I. N., Moultos, O. A., Franco, L. F., Spera, M. B. D. M., Erdős, M., & Economou, I. G. (2019). Self-diffusion coefficient of bulk and confined water: A critical review of classical molecular simulation studies. *Molecular Simulation, 45*(4-5), 425–453. doi:10.1080/08927022.2018.1511903

White, G., Hales, A., Patel, Y., & Offer, G. (2022). Novel methods for measuring the thermal diffusivity and the thermal conductivity of a lithium-ion battery. *Applied Thermal Engineering, 212*, 118573. doi:10.1016/j.applthermaleng.2022.118573

Wolf-Gladrow, D. A. (2004). *Lattice-gas cellular automata and lattice Boltzmann models: an introduction*. Springer.

Zabaloy, M. S., Vasquez, V. R., & Macedo, E. A. (2006). Description of self-diffusion coefficients of gases, liquids and fluids at high pressure based on molecular simulation data. *Fluid Phase Equilibria, 242*(1), 43–56. doi:10.1016/j.fluid.2005.12.031

Zhang, C., Chen, Y., Yang, L., & Shi, M. (2011). Self-diffusion for Lennard-Jones fluid confined in a nanoscale space. *International Journal of Heat and Mass Transfer, 54*(21-22), 4770–4773. doi:10.1016/j.ijheatmasstransfer.2011.05.032

Zhang, Z., Xu, L., Qu, T., Lei, M., Lin, Z. K., Ouyang, X., & Huang, J. (2023). Diffusion metamaterials. *Nature Reviews. Physics, 5*(4), 218–235. doi:10.1038/s42254-023-00565-4

Zhao, X., & Jin, H. (2020). Correlation for self-diffusion coefficients of H2, CH4, CO, O2 and CO2 in supercritical water from molecular dynamics simulation. *Applied Thermal Engineering*, *171*, 114941. doi:10.1016/j.applthermaleng.2020.114941

Zhao, X., Luo, T., & Jin, H. (2021). A predictive model for self-, Maxwell-Stefan, and Fick diffusion coefficients of binary supercritical water mixtures. *Journal of Molecular Liquids*, *324*, 114735. doi:10.1016/j.molliq.2020.114735

Chapter 8
Interaction of Beam-Driven Electron-Acoustic Solitons in Auroral Region

Rabia Jahangir
National Centre for Physics, Pakistan

Shahid Ali
https://orcid.org/0000-0003-1027-1317
National Centre for Physics, Pakistan

ABSTRACT

The interaction of beam driven solitons is studied in an unmagnetized plasma, comprising (r,q) distributed hot electrons, cold inertial electrons, warm electron beam and static ions. For this purpose, fluid equations are solved to obtain four distinct roots of linear dispersion relation which vary with beam speed. For small-amplitude limit, the standard Korteweg-de Vries (KdV) equation is derived that admits soliton solution and critical condition gives rise to modified KdV equation (mKdV). The overtaking interaction of KdV and mKdV solitons is studied by Hirota method, while the head-on collision (HOC) by the Poincare-Lighthill-Kuo (PLK) technique. The analytical expressions of phase shifts are examined for the plasma parameters, namely density, temperature, and velocity of beam electrons. In contrast to Hirota method, the PLK method provides the linear sum of the solitons at the point of interaction. However, despite this limitation, PLK technique is only analytical technique available to study HOC. The relevance of this model is interesting to understand tripolar electric field structures in Earth's Auroral region, which may be associated to the interaction of EASWs in space plasmas.

1. INTRODUCTION

In this chapter, we take into account the *hydrodynamic* or *fluid* model for plasma system, which consists of a mixture of electrons and ions along with energetic electron beam particles, intermediated by the electromagnetic field. For this model the separate set of equations for density, momentum and energy

DOI: 10.4018/979-8-3693-0904-9.ch008

conservation for all the constituent particles is utilized. The electron waves are produced on the time scale of electron motion, so are high frequency waves as compared to the ion waves. We particularly study the nonlinear wave interaction of electron acoustic waves in this chapter driven by electron beam particles. The nonlinear wave interaction is a significant phenomenon as it plays a vital role for the transport of energy and momentum in a plasma system.

1.1 Electron Acoustic Waves

The Electron Acoustic Waves (EAWs) are the high frequency electrostatic waves found in various regions of space. They have been detected by many satellite missions, including S3-3 and THEMIS in auroral region (Temerin et al., 1982; Angelopoulos et al., 2008), POLAR in polar cusp (Gary & Tokar, 1985), Viking and FAST satellites in polar cusp boundary layer (Bostrom et al., 1989; Ergun et al., 1998; Pottelette et al., 1999), GEOTAIL in plasma sheet boundary layer (Matsumoto et al., 1994), CLUSTER in Earth's magnetosphere/magnetopause (Pickett et al., 2005; Graham et al., 2015), etc. The broadband electrostatic noise (BEN) has been observed in bow shock, magnetotail, auroral region, polar cusp and other regions of the Earth's magnetosphere (Singh & Lakhina, 2001). In BEN, there is a large asymmetric bipolar electric field component parallel to the magnetic field, which is found in the range from a few *mV/m* to 100*mV/m*. The large amplitude electric field component has successfully been interpreted in terms of EAWs in various regions of magnetosphere (Dubouloz et al., 1991; Lakhina et al., 2011; Dillard et al., 2018). The existence of these waves has been reported by the Magnetospheric Multi-scale (MMS) during asymmetric magnetic reconnection at Earth magnetopause (Ergun et al., 2016).

The pre-requisite for the generation of stable EAWs is the presence of two temperature electron populations in a plasma system. The cold electrons provide the necessary inertia and the thermal pressure of hot electrons furnish the essential restoring force for the wave. The quasi-neutrality is ensured by the presence of ions which act stationary for these high frequency waves. The phase velocity of the EAW lies intermediate between the thermal speed of hot electrons and the thermal speed of cold electrons. The EAWs were first predicted by Fried and Gould (1961) while studying the dispersion relation of linearized longitudinal waves in hot, collisionless, unmagnetized plasma. They anticipated the presence of this mode in plasmas, however, unlike Langmuir mode and ion acoustic waves, this mode is heavily Landau damped. Later, Watanabe established that this mode becomes less weakly damped in the presence of cold and hot electrons. Gary and Tokar (1985) presented the necessary conditions on plasma parameters for the formation of EAWs. presented the necessary conditions on plasma parameters for the formation of EAWs. They established that the temperature of cold electrons (T_c) must be much lower than the temperature of hot electrons (T_h), and the minimum percentage of the hot electrons density (n_h) must be 20 percent of the total electron density ($n_e=n_c+n_h$), mathematically equivalent to $T_c \leq 10T_h$ and $0.2<n_c/n_e<0.8$, respectively.

The presence of two temperature electrons has been detected in terrestrial and planetary magnetospheres as well as the laser-plasma corona (Bezzerides et al., 1978; Mozer et al., 1979; Lakhina et al., 2018), which allows the investigation of EAWs in these regimes. The existence of two electron temperatures is also justified for the regions where there is diffusion of particles emerging from different regions of space. One such frequent situation occurs when the energetic particles of the solar wind interact with various regions of magnetosphere. The electron-electron collisions in such space plasmas are low, so that the high frequency mode appears before the electrons attain the thermal equilibrium. The magnetic reconnection observed by MMS at the Earth's magnetopause is assumed to be due to the mixture of hot magnetospheric particles and the warm, dense magnetosheath plasma particles (Rufai et al., 2021).

1.2 Beam Electrons

The drifting electrons aligned to magnetic field, termed as beam electrons, coming from the solar wind have been detected in outer layer of Earth's magnetosphere (Tokar & Gary, 1984; Ogilvie et al., 1984; Pottelette et al., 1990; Tsurutani et al., 1998). These beam electrons act as the main source of energy to excite EAWs in a two temperature electron plasma and alter the parametric regimes for the propagation and formation of nonlinear waves. Initially, Schriver and Ashour-Abdalla (1989) interpreted BEN in the distant magnetotail by the combination of cold electron beams with the instabilities in a thermal plasma. The energetic electron beam bursts with the hot electrons (100–200eV) and the electron beam (50–100eV) have been observed in the auroral region of Earth's magnetosphere (Dubouloz et al., 1991), assumed to be strongly related to BEN emission.

The observation of BEN in plasma sheet boundary layer (PSBL) by GEOTAIL satellite has been interpreted in terms of series of solitary waves termed as Electrostatic Solitary Waves (ESW) (Matsumoto et al., 1994). Based upon particle simulation, the nonlinear dynamics of electron beam were proposed to play role for generation of these nonlinear structures. Later, the GEOTAIL observations have provided significant evidences for the electron beam source and instabilities corresponding to the reconnection region at dayside magnetopause boundary, which were linked to the formation of electrostatic waves in this region (Matsumoto et al., 2003). The MMS space mission has observed large negative potential electrostatic solitary waves driven by interaction between magnetosheath and magnetospheric plasma population having cold electron beam (Holmes et al., 2018).

The EAWs, driven by streaming electron beam, have been studied by few researchers for various plasma models (Berthomier et al., 2000; Taibany, 2005; Elwakil et al., 2007, Pickett et al., 2004; Singh et al., 2011). In particular, Berthomier et al. (2000) have studied the EA waves with beam electrons in a plasma system and noticed that the non-zero streaming velocity of beam electrons generates new class of EAWs. They mentioned a critical condition in terms of temperature and density for the beam electrons in which electron density holes appeared for cold electrons. The study was later extended to include higher order nonlinearity with electron beam and hot electron vortex distribution, strongly modifying the wave amplitudes and widths of solitary pulses (Taibany, 2005). Singh et al. (2016) explained the EAW characteristics in a magnetized plasma with nonthermal electrons in the presence of electron beam and applied the model for numerical analysis of auroral field lines observed by Viking satellite.

1.3 Non-Maxwellian Distributions

The thermal equilibrium in a system is attained by the frequent collision among its particles which cause the exchange of momentum and kinetic energy. The particles in thermal equilibrium follow the Maxwellian distribution given as

$$f(u) = n_0 \left(\frac{m}{2\pi k_B T} \right)^{3/2} \exp\left[\frac{e\varphi}{k_B T} \right], \tag{1}$$

where m and T are the mass and temperature of particle, n_0 is the number density at equilibrium and φ is the electrostatic potential. Most of the phenomena in plasma physics may be explained by assuming the Maxwellian distribution for the constituent charged particles. However, certain observational data of the satellite in various regions of space show that the distribution functions vary significantly from the Maxwellian distribution (Abid et al., 2017). One of the reasons of these nonthermal distributions is the presence of solar wind in space which acts as the constant source of nonthermality. The other reason being that due to less collisions in space, the particles may not achieve thermal equilibrium on a time scale of observation of wave. The elongated tails of distribution function observed by OGO 1 and OGO 2 satellite missions were first explained by Vasyliunas (1968) using an empirical formula in terms of spectral index κ, termed as kappa distribution. Furthermore, the flat top for low energy particles have also been observed by satellite missions in electron velocity distributions for terrestrial bow shock and magnetotail (Feldman et al., 1983; Masood et al., 2006; Masood et al., 2008; Asano et al., 2008).

Figure 1. Comparison of Maxwellian and (r,q)-distributions

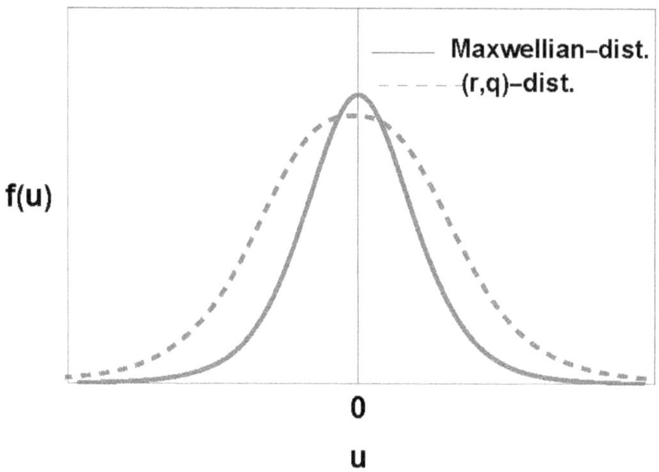

Qureshi et al. (2004) described the modified distribution functions with elongated tails and flat top by double spectral indices, coined as (r,q)-distribution. The distribution function is defined as

$$f_{(r,q)}(v)=\Theta\left[1+\frac{1}{q-1}\left(\frac{v^2-2e\varphi/m_e}{\Lambda v_{th}^2}\right)^{r+1}\right]^{-q}, \tag{2}$$

where

$$\Theta=\frac{3\Gamma[q](q-1)^{-3/(2+2r)}}{4\pi\Lambda^{3/2}v_{th}^3\Gamma\left[q-\frac{3}{2+2r}\right]\Gamma\left[1+\frac{3}{2+2r}\right]}, \tag{3}$$

and

$$\Lambda = \frac{3(q-1)^{-1/(1+r)}\Gamma\left[q-\frac{3}{2+2r}\right]\Gamma\left[\frac{3}{2+2r}\right]}{2\Gamma\left[q-\frac{5}{2+2r}\right]\Gamma\left[\frac{5}{2+2r}\right]}. \tag{4}$$

Here, $v_{th} = \sqrt{2T_e / m_e}$ is the electron thermal velocity and $\Gamma[-]$ is the gamma function. In this double spectral distribution, the index "q" describes the elongated tails of the curve due to superthermality of highly energetic particles and "r" called the flatness parameter, alters the hump of Maxwellian distribution to a flat top due to low energetic particles. It would be beneficial to mention here that (r,q) distribution has an immense advantage over the other distribution functions as it can impersonate almost all the distribution functions by using different choices of r and q. In particular, the (r,q) tends to Maxwellian distribution for $(r\rightarrow 0, q\rightarrow\infty)$ and kappa distribution for $(r\rightarrow 0, q\rightarrow\kappa+1)$. Moreover, the conditions q>1 and q(r+1)>5/2 need to be fulfilled for recovering physically meaningful results. The comparison of Maxwellian and (r,q) distribution is illustrated in Fig (1).

The effects of double spectral indices (r,q) have been studied by many authors to explain various phenomena in plasma systems. Qureshi et al. (2005) first described the Landau damping of Langmiur and ion acoustic waves in plasma for (r,q)-distribution. The phenomenon of terrestrial lion roars has been explained by using (r,q) distribution (Qureshi et al., 2014) which showed a good agreement of theoretical results with data. Shah et al. (2018) studied the nonlinear EAW structures in planetary magnetospheres to explore the effect of spectral indices on solitary waves. Furthermore, there has been a considerable amount of literature to study the linear and nonlinear modes in plasmas that highlights importance of using (r,q)-distribution specifically in space and dusty plasmas (Qureshi et al., 2019; Ullah et al., 2020; Naeem et al., 2020; Shohaib et al., 2022; Masood et al., 2022).

1.4 Solitary Wave Interaction

Solitons are the unique nonlinear coherent structures that retain their shape during the propagation as well as after the interaction with other solitons. These properties are significant for the redistribution or the transfer of energy and momentum and comprehending nonlinear wave interaction in a system. In the standard Korteweg de Vries (KdV) equation the quadratic nonlinearity balances the dispersion for the emergence of soliton. The soliton appears as hump or dip in the electric potential of a charged particle in a plasma system. There is ample amount of literature for the study of nonlinear waves based on KdV equation for various physical systems of interest where solitary waves are generated (Korteweg & de-Vries, 1895; Kakutani et al., 1968; Jeffery, 1973). Furthermore, in a multi-component plasma there is a certain critical point, for which the basic quadratic nonlinearity of KdV vanishes and we must consider the higher orders. For higher orders, one obtains the cubic nonlinearity in place of quadratic nonlinearity which balances dispersion to produce stable structures. The nPDE containing cubic nonlinearity is termed as modified Korteweg de Vries (mKdV) equation. However, in the vicinity of the critical point both the quadratic and cubic nonlinearities contribute to give the soliton solution for the Gardner equation. Many investigators have explored these KdV type equations and their other variants in the higher dimensions to study the nonlinear wave propagation (Tamang et al., 2020; Mannan & Mamun, 2011; Masud et al., 2012; Nawaz et al., 2022).

The type of interaction of two solitons depends mainly on the direction of propagation of solitons. The unique feature of the generation of phase shift is the constant factor for all types of interactions of solitons. The phase shift appears as a delay (negative phase shift) or advancement (positive phase shift) for the emergence of solitons from the point of interaction. The phase shift represents the memory of the soliton interaction for each soliton after interaction.

The solitons travelling in the same direction may experience overtaking interaction when the taller faster soliton overtakes the shorter slower soliton as shown in Figure (2). At the point of overtaking interaction, the amplitude of two solitons do not merely sum up, but they interact in a nonlinear fashion. Both KdV and mKdV equations are integrable which means that they possess infinite conservation laws and their N-soliton solutions can be attained. However, the difference of interaction is that the mKdV equation also allows the study of peculiar interaction of the opposite polarity solitons (Verheest & Hereman, 2019), which is not possible for KdV case. The two soliton solutions of nonlinear partial differential equations (nPDEs) to study overtaking collision may be comprehended by a number of methods. A few of them include inverse scattering transform (Ablowitz & Clarkson, 1991), Bäcklund transform (Miura1 et al., 968), Darboux transformation (Matveev & Salle, 1991), Hirota's bilinear formalism (Hirota, 1971; 2004), etc. The benefit of utilizing the Hirota's method is that it is a rather simple method which converts the nPDE to a solvable ordinary differential equation.

Figure 2. Overtaking interaction of two solitons

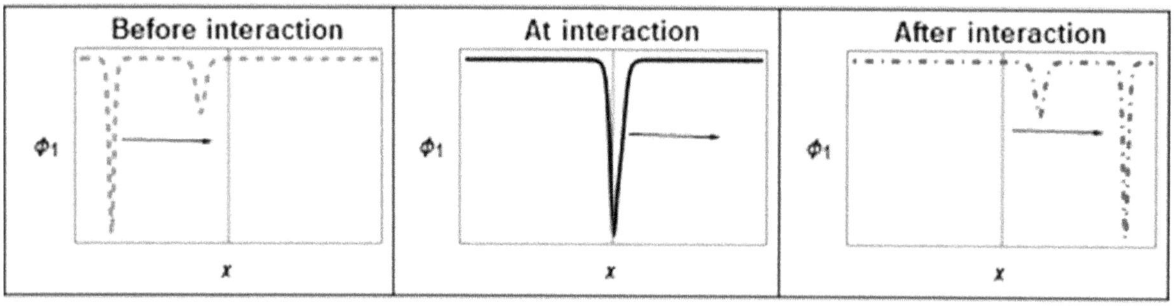

In ingenious Hirota's formalism, the dependent variable function transformation is introduced to convert nPDE to a bilinear form. Furthermore, the bilinear operators, named as Hirota operators then convert the bilinear equation into a linear algebraic equation. The functions are then expanded in terms of a small parameter, where each order of small parameter give the multi-soliton solution. The description of the method would become clear in next sections where it is applied to Korteweg-de Vries (KdV) and modified KdV equations. Hirota (1971) first used the method for KdV equation, but it was later applied to large classes of nonlinear evolution equations. The only condition of application of method is that the equations should be integrable. Therefore, Hirota method has been employed efficiently now to evaluate the multi-soliton solutions of a number of nPDEs including Korteweg-de Vries (KdV) (Jahangir et al., 2020), Kadomtsev Petviashvili (KP) (Jahangir et al., 2016), modified KP (Shohaib et al, 2022), Zakharov Kuznetsov (ZK) (Yousaf et al., 2021), Gardner (Nawaz et al., 2022) equations and many more in various unmagnetized and magnetized plasma systems. It is important to note that in all above mentioned equations, the Gardner equation is the partially integrable equation, however, Hirota method successfully demonstrates the two soliton interaction for Gardner equation as well.

Another possible type of soliton interaction is the head on collision of solitons when the counter propagating solitons move towards one another and interact, termed as head on collision (HOC) of solitons. Unlike overtaking interaction, the HOC may take place between equi-amplitude solitons besides the taller and shorter solitons. The HOC of solitons is studied analytically by extended Poincare-Lighthill-Kuo (PLK) technique (Poincare, 1880; Lighthill, 1949; Kuo, 1953). The PLK method has been used to study the HOC of solitons in several physical systems, such as, shallow water waves (Maxworthy, 1976), nonlinear transmission lines (Tsuboi, 1989), Rayleigh Bénard convecting fluid (Huang et al., 1993), blood solitary waves (Xue, 2004), Bose-Einstein condensates (Li & Duan, 2008), one dimensional bead chains (Wang et al., 2018), and many more. The HOC of ion-acoustic, magnetoacoustic and dust/(dust ion) acoustic solitons in planar and nonplanar geometry has been investigated theoretically for various plasma models using PLK technique (Huang & Velarde, 1996; Verheest et al., 2012; Jahangir et al., 2019; Hasan et al., 2019; Jahangir et al., 2023).

Figure 3. Head on collision of two solitons

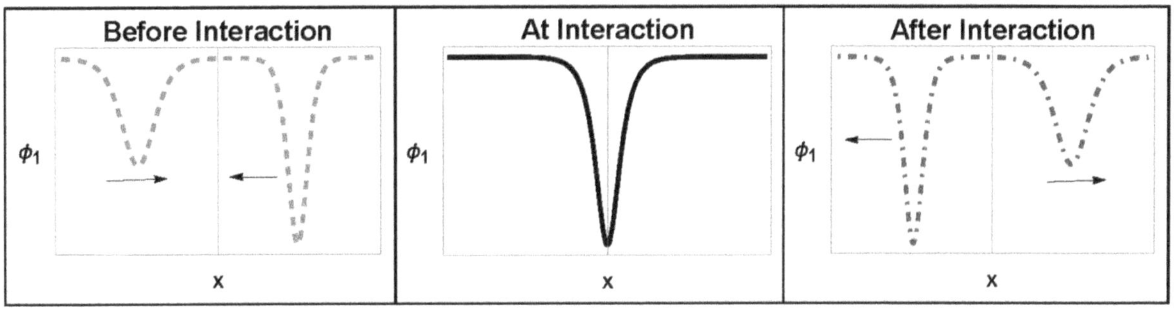

The HOC of nonlinear structures has been studied experimentally for water waves (Maxworthy, 1976) and strongly coupled dusty plasma (Harvey et al., 2010). The results indicated that the superposition of waves does not hold at the point of interaction and additionally the solitons suffer a time delay during their interaction. The analytical technique of PLK gives the analytical expressions of phase shifts, however, the limitation of this technique is that amplitudes of interacting solitons simply add up linearly at the point of interaction. However, it is important to realize that despite this limitation of the PLK technique, it is so far the only available analytical technique to study the HOC of solitons. Most recently, the Particle in cell (PIC) simulation has been introduced to study the HOC by calculating the phase shift and by studying the nonlinear superposition of solitons accurately during the interaction (Hosseini2019).

We study the two soliton interaction of EAWs in an unmagnetized, collisionless, four component plasma system, in this chapter, comprising (r,q) distributed hot electrons, cold inertial electrons, electron beam and stationary ions. The model is analyzed numerically in the coming sections for broadband electrostatic noise (BEN) in the auroral region and then compared with the observational data.

2. MATHEMATICAL MODELLING OF ELECTRON ACOUSTIC WAVES

The mathematical modeling of a plasma system using fluid dynamics is based upon the use of conservation laws for the constituent particles along with certain approximations. Here, we assume a collisionless, homogeneous, unmagnetized plasma system comprising cold dynamical electrons, (r,q) distributed inertialess hot electrons, static ions along with the warm beam electrons. The nomenclature of cold electrons is sometimes used for electrons with temperature T_c=0 case, which ensures the simplification in model, however, we have used it for electrons with low temperature, satisfying the basic condition $T_h \geq 10T_c$ for generation of EAWs. Therefore, the words of hot and cold electrons are essentially utilized to differentiate the temperatures of the two electron populations. Furthermore, the streaming electrons are assumed to have temperature intermediate between cold and hot electrons based upon the data of Viking satellite in auroral region (Dubouloz et al., 1993). The propagation of EAWs is governed by the following normalized continuity equation (mass conservation) and equation of motion (momentum conservation), respectively

$$\frac{\partial n_j}{\partial t} + \frac{\partial n_j v_j}{\partial x} = 0, \tag{5}$$

$$\frac{\partial v_j}{\partial t} + v_j \frac{\partial v_j}{\partial x} = \frac{e}{m_j}\frac{\partial \varphi}{\partial x} - \frac{1}{m_j n_j}\frac{\partial P_j}{\partial x}, \tag{6}$$

where $j=c,b$ for cold and beam electrons and $P_j \equiv n_j T_j$ is the equation of state for pressure. The intermediating electric field, for electrostatic case ($E= -\partial\varphi/\partial x$), is taken in to account by Poisson's equation as

$$\frac{\partial^2 \varphi}{\partial x^2} = 4\pi e\left(n_h + n_c + n_b - n_{i0}\right). \tag{7}$$

Here, $n_{h,c,b}$ are the number densities of hot, cold and beam electrons, φ is the electrostatic potential and $v_{c,b}$ are the velocities of cold and beam electrons.

The hot electrons are governed by the double index generalized (r,q)-distribution. Integrating the distribution function over the velocity space gives the number density of generalized (r,q) distributed hot electrons

$$n_h = n_{h0}\left(1 + \sum_{i=1}^{\infty} \mu_i \varphi^i\right), \tag{8}$$

where μ_i denote the expansion coefficients, given as

$$\mu_1 = \frac{\Gamma[q-R']\Gamma[R'](q-1)^{-2R'}}{2\Lambda\Gamma[3R']\Gamma[q-3R']},\ \mu_2 = \frac{-\Gamma[q+R']\Gamma[-R'](q-1)^{-4R'}}{8\Lambda^2\Gamma[3R']\Gamma[q-3R']},\ \mu_3 = \frac{\Gamma[q+3R']\Gamma[-3R'](q-1)^{-6R'}}{16\Lambda^3\Gamma[3R']\Gamma[q-3R']}, \tag{9}$$

and so on, with $R' = 1/2(1+r)$.

2.1 Dispersion Properties of Beam Driven EAWs

The linear dispersion relation is significant and fundamental relation for understanding the dispersion and wave properties in a plasma system. To establish the dispersion relation for the plasma system, we linearize the model equations (5)-(7) and apply the Fourier analysis by assuming the perturbations of the form exp $i(kx-\omega t)$. After many algebraic steps, we attain

$$\mu_1 + k^2\lambda_{dh}^2 - \frac{1}{\left(\frac{\omega^2}{k^2C_e^2} - \frac{n_{h0}}{n_{c0}}\frac{T_c}{T_h}\right)} - \frac{n_{b0}/n_{c0}}{\left(\frac{(\omega - kv_{b0})^2}{k^2C_e^2} - \frac{n_{h0}}{n_{c0}}\frac{T_b}{T_h}\right)} = 0, \tag{10}$$

where electron speed $C_e=\omega_{pc}\lambda_{dh}=(T_h/\alpha m_e)^{1/2}$, $\alpha=n_{h0}/n_{c0}$, Debye length $\lambda_{dh}=(T_h/4\pi e^2 n_{ho})^{1/2}$ and plasma frequency $\omega_{pc}=(4\pi e^2 n_{c0}/m_e)^{1/2}$, respectively. Here, μ_1 denotes the contribution due to (r,q) distribution and $k^2\lambda_{dh}^2$ is the dispersive term which appears due to the Poisson's equation. This term vanishes in the long wavelength limit or if the quasi-neutrality condition is considered instead of Poisson's equation.

The linear dispersion relation may be analyzed for the plasma parameters of auroral region of BEN observed by Viking satellite. For this region, cold, hot and beam electron densities are, respectively, $n_{c0}=0.5cm^{-3}$, $n_{h0}=1.5\text{–}2.0cm^{-3}$, $n_{b0}=1.0cm^{-3}$ and the corresponding temperatures are $T_h=250eV$, $T_c=2eV$, and $T_b=50eV$, respectively. For the non Maxwellian distribution, we choose $q=2.5$ and $r=0.4$. The four roots of ω from the dispersion relation Equation (10) for beam driven EAWs are plotted in Figure (4). Initially, for smaller values of beam velocity v_{b0} ($v_{b0}\leq 0.2$), we obtain four distinct roots for EAWs as depicted in Figure 4(a). Here, two of the roots exist due to the beam electrons (negative and positive roots depicting forward and backward propagation), while remaining two are obtained for the cold electrons (forward propagating and backward propagating). These roots may be termed as beam electron acoustic (BEA) and cold electron acoustic (CEA) modes. For higher values of beam velocity v_{b0} ($0.2<v_{b0}<0.27$), the two BEA modes become forward propagating (the fast and slow modes with reference to the beam velocity), as shown in Figure 4(b).

The remaining two modes are the forward and backward propagating CEA modes. Figure Fig 4(c) shows the case of a wide range of beam velocities v_{b0}, i.e., $0.27\leq \varepsilon \leq 2.8$ where only two modes exist which correspond to the BEA modes. For this particular case, the other two roots of CEA modes become complex which correspond to the two stream instability. For larger super-sonic beam velocity (e.g., $v_{b0}\geq 2.8$), we again get three forward and one backward propagating mode, where two forward propagating modes correspond to the BEA modes (fast and slow) and the third one is the CEA mode, whereas the backward propagating mode corresponding to the cold electrons. Furthermore, the validity of different regions of these modes corresponding to various values of the beam velocity v_{b0}, plotted in Figure 4 depends upon the remaining plasma parameters and change if the fixed set of remaining parameters are altered.

Figure 4. The plot of linear dispersion relation for various values of beam speed

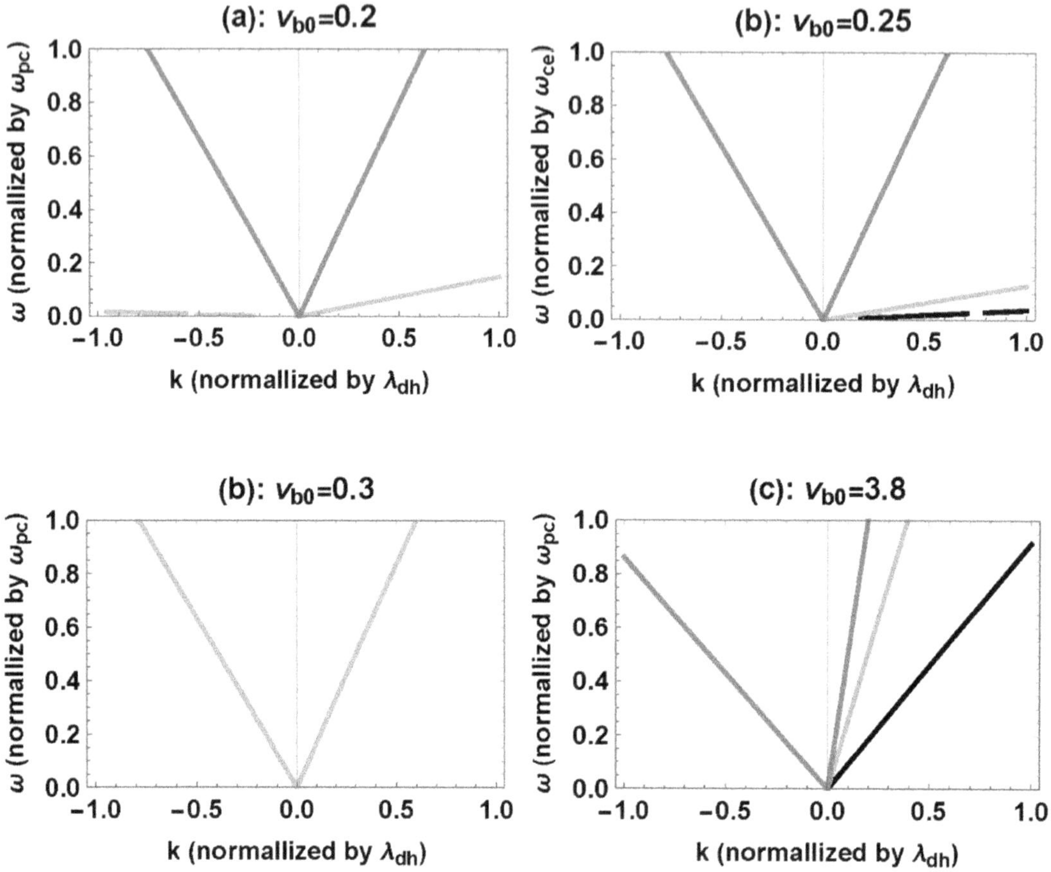

2.2 Dispersion Properties of EAWs

The dispersion relation simplifies to the usual dispersion relation of EAWs in the absence of electron beam as

$$\frac{\omega^2}{k^2} = \omega_{pc}^2 \lambda_{dh}^2 \left(\frac{n_{h0}}{n_{c0}} \frac{T_c}{T_h} + \frac{1}{\mu_1 + k^2 \lambda_{dh}^2} \right). \tag{11}$$

We get two roots of ω for cold electron acoustic mode termed as electron acoustic mode, one positive and one negative depicting the forward and backward propagating modes, respectively. The first term represents the thermal motion effect of cold electrons which becomes negligible when the cold electron temperature is very low. In this particular case, it is important to examine that the dispersion relation of EAW still varies with the ratio of hot-to-cold electrons through the electron speed $C_e = \omega_{pc} \lambda_{dh}$. Here, $\mu_1 \rightarrow 1$ for the Maxwellian distribution and $k^2 \lambda_{dh}^2$ vanishes in long wavelength limit.

Figure 5. The linear dispersion relation in the absence of beam electrons

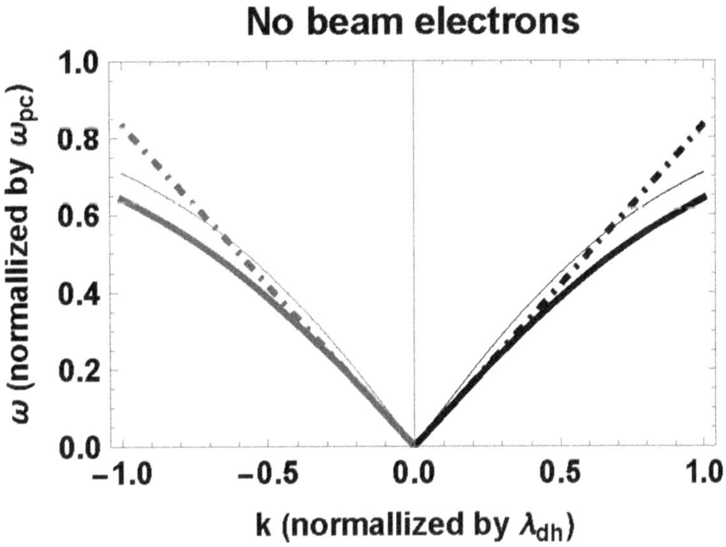

The linear dispersion relation for the stable EAWs in the absence of beam electrons is compared for various cases in Figure (5). Each mode on right and left show the forward propagating (positive) and backward propagating mode. The comparison of linear dispersion relation with (r=0.4, q=2.5) distributed (thick solid curves) and Maxwellian distributed (thin solid curves) hot electrons shows that the frequency of Maxwellian distributed electrons is greater than the (r,q) distributed hot electrons. Furthermore, the frequency curve becomes a straight line for long wavelength limit (dashed curves).

3. WEAKLY NONLINEAR ELECTRON ACOUSTIC WAVE

The nonlinear wave propagation is studied by transforming the set of model equations to a nonlinear evolution equation. We first normalize/scale the model equations by dividing the variables by the largest system parameters or by their equilibrium values. The normalization is utilized in plasma physics to achieve the dimensionless parameters, which aids in the analysis and comparison of various plasma systems. Therefore, following normalization scheme is utilized for model equations (5-7)

$$n_{h,c,b} \to n_{h0,c0,b0}, v_{c,b} \to C_e, \varphi \to T_h / e, x \to \lambda_{dh}, \text{ and } t \to \omega_{pc}^{-1}, \tag{12}$$

This normalization of speed, space and time is also obvious from the dispersion relation which simplifies it. Then the normalized model equations are given as

$$\frac{\partial n_j}{\partial t} + \frac{\partial}{\partial x} n_j v_j = 0, \tag{13}$$

$$\frac{\partial v_j}{\partial t}+v_j\frac{\partial v_j}{\partial x}=\alpha\frac{\partial \varphi}{\partial x}-\frac{\alpha\sigma_j}{n_j}\frac{\partial n_j}{\partial x}, \tag{14}$$

and

$$\frac{\partial^2\varphi}{\partial x^2}=\frac{1}{\alpha}n_h+n_c+\frac{1}{\beta}n_b-\left(1+\frac{1}{\alpha}+\frac{1}{\beta}\right). \tag{15}$$

Note that the ratios are defined as

$$\alpha=\frac{n_{h0}}{n_{c0}},\beta=\frac{n_{h0}}{n_{b0}},\sigma_c=\frac{T_c}{T_h}\text{ and }\sigma_b=\frac{T_b}{T_h}. \tag{16}$$

The small but finite amplitude propagation of an EAW in the presence of beam electrons along x-axis is studied by the reductive perturbation method (RPM). For RPM, the space-time coordinates are stretched and the dependent variables n_j, v_j, and φ are expanded (Washimi & Tanuiti, 1966). The stretching and expansion is mainly derived from the linear dispersion relation and the presence or absence of magnetic field for the plasma system. For detailed derivation, we refer the readers to (Infeld & Rowland, 1990).

3.1 Standard Korteweg-de Vries (KdV) Equation

For standard KdV equation of an unmagnetized plasma in one dimension, the required stretchings and expansions in terms of a small parameter ε, are defined as

$$\xi=\varepsilon^{1/2}\left(x-\lambda t\right),\text{ and }\tau=\varepsilon^{3/2}t, \tag{17}$$

and

$$R_j\left(x,t\right)=R_{j0}+\sum\nolimits_{i=1}^{\infty}\varepsilon^{i+\frac{1}{2}}R_{ji}\left(x,t\right). \tag{18}$$

Here, ε determines the amplitude of nonlinearity ($0<\varepsilon\leq 1$) and $R_j(x,t)=[n_j,v_j,\varphi]$ represent various physical variables, where $j=c,b$ are for cold and beam electrons with $R_{j0}=[1,v_{j0},0]$. Note that the cold electrons are assumed to be stationary initially, so $v_{c0}=0$, whereas the beam electrons are assumed to be streaming with an initial constant speed v_{b0}. Employing the stretching and expansion from Eqs. (17) and (18) into the model Eqs. (13-15), we obtain the equations for different orders of ε. The lowest most order of ε terms (i.e., $\varepsilon^{3/2}$-order) give expressions for first order perturbed quantities as

$$n_{j1} = \frac{-\alpha\varphi_1}{\left(\lambda - v_{j0}\right)^2 - \alpha\sigma_j},\tag{19}$$

$$v_{j1} = \frac{-\alpha\left(\lambda - v_{j0}\right)\varphi_1}{\left(\lambda - v_{j0}\right)^2 - \alpha\sigma_j},\tag{20}$$

and the dispersion relation

$$\mu_1 - \frac{1}{\lambda^2 - \alpha\sigma_c} - \frac{\alpha}{\beta\left(\left(\lambda - v_{b0}\right)^2 - \alpha\sigma_b\right)} = 0.\tag{21}$$

Note that this normalized dispersion relation has four distinct roots, and is similar to the linear dispersion relation of Equation (10) in the long wavelength limit. The dispersion relation also agrees well with the previous result for non-thermal distribution and dynamical ions of Singh et al. (2016) for static ions and (r,q) distributed electrons of our case. This dispersion relation reduces to $\lambda = \sqrt{\alpha\sigma_c + 1/\mu_1}$ in the absence of beam electrons.

For higher-order terms (i.e., $\varepsilon^{5/2}$-order), we obtain the equations in terms of the first and second order perturbed quantities. The set of higher order equations is used to obtain the following standard Korteweg de Vries (KdV) equation

$$\partial_\tau\varphi_1 + P\varphi_1\partial_\xi\varphi_1 + Q\partial_{\xi,\xi,\xi}\varphi_1 = 0.\tag{22}$$

The quadratic nonlinearity coefficient P and dispersive coefficient Q are defined as

$$P = \frac{X_2}{X_1}, \text{ and } Q = \frac{1}{X_1},\tag{23}$$

where

$$X_1 = \frac{2\lambda}{(\lambda^2 - \alpha\sigma_c)^2} + \frac{2\alpha\left(\lambda - v_{b0}\right)}{\beta((\lambda - v_{b0})^2 - \alpha\sigma_b)^2},\tag{24}$$

and

$$X_2 = \frac{\alpha\left(-3\lambda^2 + \alpha\sigma_c\right)}{(\lambda^2 - \alpha\sigma_c)^3} + \frac{\alpha^2(-3\left(\lambda - v_{b0}\right)^2 + \alpha\sigma_b)}{\beta((\lambda - v_{b0})^2 - \alpha\sigma_b)^3} - 2\mu_2.\tag{25}$$

Therefore, the beam, hot, and cold electrons parameters affect the nonlinearity and dispersion whereas the (r,q) distribution of hot electrons only alters the nonlinearity coefficient through μ_2. The obtained KdV equation has the soliton solution of the shape of a hump or a dip in the potential of charged particle in a plasma system.

3.2 Modified KdV Equation

The quadratic nonlinearity coefficient P carry both the positive and negative competing terms in X_2, so may become zero to give a critical condition for $\beta(=n_{h0}/n_{b0})$ as

$$\beta_c = \frac{\alpha(-3(\lambda - v_{b0})^2 + \alpha\sigma_b)(\lambda^2 - \alpha\sigma_c)^3}{((\lambda - v_{b0})^2 - \alpha\sigma_b)^3(2\mu_2(\lambda^2 - \alpha\sigma_c)^3 - \alpha(-3\lambda^2 + \alpha\sigma_c))}. \tag{26}$$

This critical value of density ratio β depends upon the temperatures and densities of cold, hot and beam electrons, beam velocity v_{b0} and the distribution of hot electrons. At this critical point, the quadratic nonlinearity P vanishes and higher order nonlinearity has to be considered for stable coherent structure. Therefore, the stretchings of space-time coordinates are altered to obtain higher order nonlinearity which give the modified Korteweg de Vries (mKdV) equation for given critical value.

The stretching then becomes (Abdikian et al., 2021)

$$\xi = \varepsilon(x - \lambda t), \text{ and } \tau = \varepsilon^3 t. \tag{27}$$

This stretching is used for the normalized model equations gives the same first order perturbed quantities and dispersion relation as given by Eqs. (19-20) and (21). For higher order (ε^3-order), we obtain

$$n_{j2} = \left(\frac{-\alpha}{(\lambda - v_{j0})^2 - \alpha\sigma_j}\right)\varphi_2 + \frac{\alpha^2(3(\lambda - v_{j0})^2 - \alpha\sigma_j)}{2((\lambda - v_{j0})^2 - \alpha\sigma_j)^3}\varphi_1^2, \tag{28}$$

$$v_{j2} = \left(\frac{-\alpha(\lambda - v_{j0})}{(\lambda - v_{j0})^2 - \alpha\sigma_j}\right)\varphi_2 + \frac{\alpha^2(\lambda - v_{j0})((\lambda - v_{j0})^2 + \alpha\sigma_j)}{2((\lambda - v_{j0})^2 - \alpha\sigma_j)^3}\varphi_1^2, \tag{29}$$

and

$$\mu_1\varphi_2 + \frac{1}{\alpha}n_{c2} + \frac{1}{\beta}n_{b2} + \mu_2\varphi_1^2 = 0. \tag{30}$$

The set of above equations can be solved simultaneously to obtain $X_2\varphi_1\partial\varphi_1/\partial\xi=0$, which shows that quadratic nonlinearity coefficient vanishes as $\varphi_1\neq 0$. For further higher orders (ε^4-order) are simplified to

$$\left(\lambda - v_{j0}\right)\partial_\xi n_{j3} - \partial_\xi v_{j3} = \partial_\tau n_{j1} + \partial_\xi\left(v_{j2}n_{j1}\right) + \partial_\xi\left(v_{j1}n_{j2}\right), -\left(\lambda - v_{j0}\right)\partial_\xi v_{j3} - \alpha\partial_\xi\varphi_3 + \alpha\sigma_j\partial_\xi n_{j3}$$
$$= -\partial_\tau v_{j1} - \partial_\xi\left(v_{j2}v_{j1}\right) + \alpha\sigma_j\partial_\xi\left(n_{j1}n_{j2}\right) - \alpha\sigma_j n_{j1}^2\partial_\xi n_{j1}, \frac{1}{\alpha}n_{c3} + \frac{1}{\alpha}n_{b3} + \mu_1\varphi_3 = \partial_{\xi,\xi}\varphi_1 - 2\mu_2\varphi_1\varphi_2 - \mu_3\varphi_1^3. \tag{31}$$

Solving these equations simultaneously along with the expressions of first and second order perturbed quantities lead to the following modified Korteweg de Vries (mKdV) equation with cubic nonlinearity is obtained

$$\partial_\tau\Psi + R\Psi^2\partial_\xi\Psi + S\partial_{\xi,\xi,\xi}\Psi = 0, \tag{32}$$

where, $\Psi=\varphi_1$ and

$$R = \frac{3Y_2}{Y_1}, \text{ and } S = \frac{1}{Y_1}, \tag{33}$$

are, respectively, the coefficients of cubic nonlinearity and dispersion with

$$Y_1 = 2\alpha\left[\frac{\alpha\lambda}{(\lambda^2 - \alpha\sigma_c)^2} + \frac{\left(\lambda - v_{b0}\right)}{\beta((\lambda - v_{b0})^2 - \alpha\sigma_b)^2}\right], \tag{34}$$

and

$$Y_2 = \frac{\alpha^2\left(5\lambda^4 - \frac{4}{3}\lambda^2\alpha\sigma_c + \frac{1}{3}\alpha^2\sigma_c^2\right)}{2(\lambda^2 - \alpha\sigma_c)^5} + \frac{\alpha^3(5(\lambda - v_{b0})^4 - \frac{4}{3}\left(\lambda - v_{b0})^2\alpha\sigma_b + \frac{1}{3}\alpha^2\sigma_b^2\right)}{2\beta((\lambda - v_{b0})^2 - \alpha\sigma_b)^5} - \mu_3. \tag{35}$$

Both KdV and mKdV are the integrable nPDEs which provide the solitary solution. The single soliton solutions of KdV and mKdV equations are proportional to sec $h^2(\xi)$ and sec h(ξ), respectively, thus the difference of the nonlinearity portraits itself qualitatively in terms of difference of width of resultant solitons as shown in Figure 6. The multi-soliton solutions of the KdV and mKdV equations for overtaking interaction are obtained by using the Hirota method in next section.

Figure 5. Difference of KdV and mKdV equation

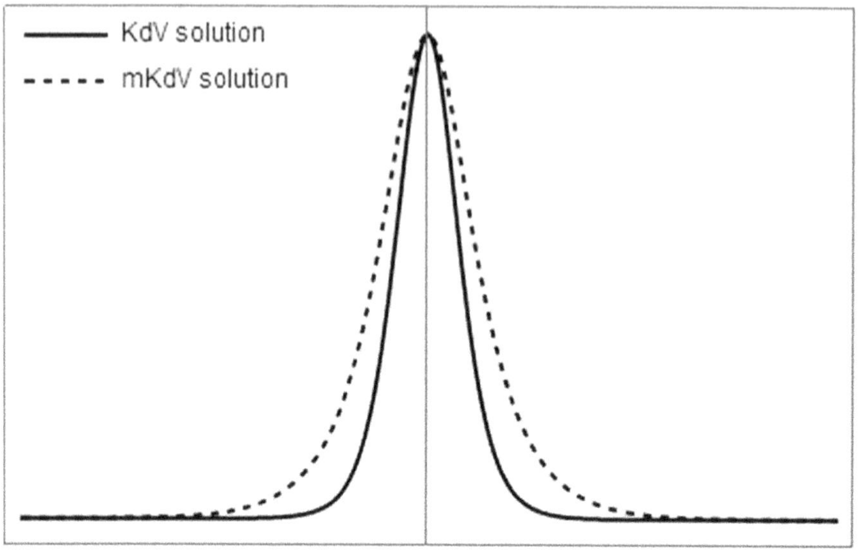

4. OVERTAKING INTERACTION OF KDV SOLITONS

The integrable nPDEs can be solved to obtain multi-soliton solutions for overtaking interaction by Hirota's direct method (Hirota, 1971; 2004) The standard KdV equation (22) can be solved using this method by assuming a dependent variable transformation of φ_1 as

$$\varphi_1(\xi,\tau)=\frac{12Q}{P}\frac{\partial^2\left[\log f(\xi,\tau)\right]}{\partial\xi^2}. \tag{36}$$

Inspiration or the prediction of this transformation is sometimes based on performing the Painleve integrability test (Ablowitz & Clarkson, 1991). This transformation then translates the KdV Equation (22) to the bilinear form as

$$ff_{\xi,\tau}-f_\xi f_\tau+3Q(f_{\xi,\xi})^2-4Qf_\xi f_{\xi,\xi,\xi}+Qff_{\xi,\xi,\xi,\xi}=0. \tag{37}$$

It may be noted that the transformation takes nonlinearity to a scaling factor in the amplitude of transformation and omits it from the bilinear form of equation. The bilinear equation is only generated for the solitonic solutions. The bilinear equation can be simplified to the ordinary differential equation, named Hirota equation, by the aid of Hirota derivatives as

$$D_\xi\left(D_\tau+BD_\xi^3\right)(f.f)=0. \tag{38}$$

Here, Hirota operators are same as the usual Leibnitz derivatives of products except that they operate anti-symmetrically on the pair of functions. The Hirota derivatives/operators D_ξ, D_τ are defined as

$$D_\xi^l D_\tau^m (f.f) = \left(\frac{\partial}{\partial \xi} - \frac{\partial}{\partial \xi'}\right)^l \left(\frac{\partial}{\partial \tau} - \frac{\partial}{\partial \tau'}\right)^m \left(f(\xi,\tau) f(\xi',\tau')\right)\Big|_{\xi=\xi',\tau=\tau'} . \tag{39}$$

The dependent function $f(\xi,\tau)$ is expanded in terms of power series of a small parameter ε as

$$f(\xi,\tau) = 1 + \varepsilon f_1 + \varepsilon^2 f_2 + \varepsilon^3 f_3 + ... \tag{40}$$

We substitute this expansion into the Hirota bilinear Equation (37) and collect the terms with different orders of ε. Eventually, we find the multi-soliton solutions for each order of ε.

4.1 Single KdV Soliton Solution

For exact one soliton solution of KdV equation, we take the first order function $f_1 = e^\eta$. The benefit of assuming the exponential function is that the infinite expansion of f is truncated to finite terms (up to first order for single soliton case). Here $\eta = k\xi + \omega\tau + \gamma$, where k (≡ propagation vector) and γ (≡ initial phase) are arbitrary constants and ω is found by substituting expansion of $f(\xi,\tau)$ in the ε- order Hirota bilinear Equation (37), which gives $\omega = Qk^3$. Therefore the single soliton solution of KdV becomes

$$\varphi_1(\xi,\tau) = \frac{12Q}{P}\frac{\partial^2}{\partial \xi^2} log\left[1 + exp\left(k\xi - Qk^3\tau + \gamma\right)\right] \Rightarrow \varphi_1(\xi,\tau) = \frac{3Q}{P} k^2 sech^2\left(\frac{k\xi - Qk^3\tau + \gamma}{2}\right). \tag{41}$$

Figure 7. Electron acoustic soliton (a) electric potential and (b) electric field

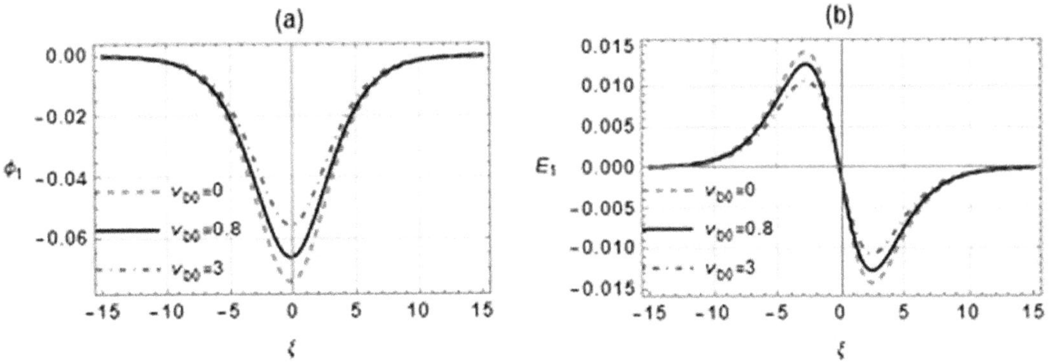

Here, ε has been assumed to be one in the dependent variable function f for simplicity and 3Qk2/P is *th*e *a*mplitude of the single soliton. The soliton structure for EAW is plotted in Figure 7 for the plasma parameters consistent with the energetic burst of beam electrons (Dubouloz et al., 1991) in the auroral region, i.e., $(n_{c0}, n_{h0}, n_{b0})=(0.5, 1.5\text{–}2.0, 1.0)cm^{-3}$, $(T_c, T_h, T_b)=(2, 250, 50)eV$ and $(r,q)=(0.4, 2.5)$. It shows a potential dip, termed as rarefactive soliton, in curve (a) for the potential of electrons for various values of beam velocity. Since most of the observed data of space plasmas is collected by satellites missions in terms of electric fields, therefore, the curves in (b) are plotted for electric field, where $E_1(\xi,\tau)= -\partial_\xi\varphi_1(\xi,\tau)$. The soliton potential then corresponds to a bipolar electric field, where the hump and dip of electric field appear in this order only for rarefactive soliton. Both of the curves show that the increase of the beam velocity reduces the amplitude of solitary waves. The amplitude of the electric field may be measured from the above curves by un-normalizing the electric field which ranges from 76mV/m to 85mV/m *for* fixed *rema*ining plasma parameters. This range of electric field agrees with observed data (Dubouloz et al., 1991) of electric field amplitude of BEN in auroral region, where the electric field varies up to 100mV/m.

4.2 Two *KdV* Soliton Solution

The exact two soliton solution is attained by considering $f_1 = e^{\eta_1} + e^{\eta_2}$, as a superposition function and $f_2 = a_{12}e^{\eta_1+\eta_2}$ as the interaction function, where $\eta_i = k_i\xi + \omega_i\tau + \gamma_i$, $(i=1,2)$. Here, k_i, γ_i are the arbitrary constants, $\omega_i = -Qk_i^3$ (evaluated by ε- order of Equation (37)) and a_{12} (evaluated by ε^2- order of Equation (37)) is the interaction parameter given as

$$a_{12} = \left(\frac{k_1 - k_2}{k_1 + k_2}\right)^2. \tag{42}$$

Then, $f(\xi,\tau) = 1 + e^{\eta_1} + e^{\eta_2} + a_{12}e^{\eta_1+\eta_2}$ and two-soliton solution turns out to be

$$\varphi_1(\xi,\tau) = \frac{12Q}{P}\frac{\partial^2}{\partial\xi^2}log\left[1 + e^{\eta_1} + e^{\eta_2} + a_{12}e^{\eta_1+\eta_2}\right], \tag{43}$$

which represents the two solitons travelling in the same direction. It is emphasized that all the physical effects of plasma parameters on the two solitons are consistent with their behavior for single soliton.

4.3 Interaction of KdV Solitons

For the solitons travelling as a train of solitons, the taller soliton may overtake the shorter soliton due to the greater speed of taller soliton, namely overtaking interaction. Therefore, one of the obvious results is that overtaking interaction cannot take place among the equi-amplitude solitons. The overtaking interaction induces a phase shift for each soliton after interaction which is proportional to the "log a_{12}" obtained by the asymptotic limit of the two soliton solution, when the solitons become well separated.

Figure 8. 3-D plot for soliton interaction of KdV rarefactive solitons

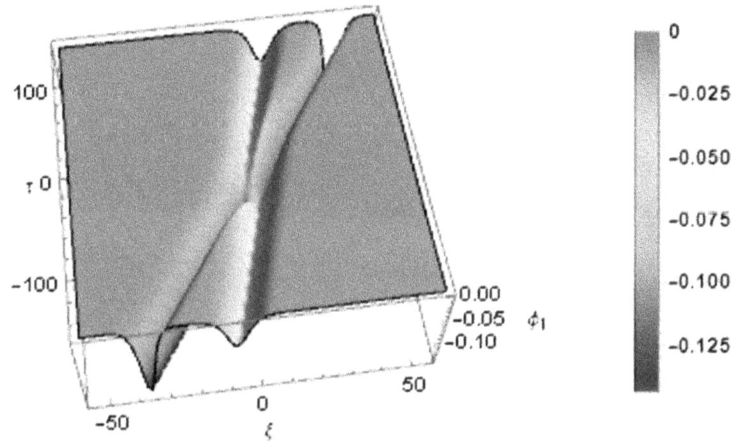

Figure 8 shows the 3D plot of overtaking interaction of electron acoustic solitons which shows the overtake of shorter rarefactive soliton by the taller rarefactive soliton. It is observed from the plot that at the point of interaction the resultant amplitude is much smaller than the sum of amplitudes of constituent solitons. This indicates that the linear superposition principle fails at the point of collision as the interaction takes place among the nonlinear waves.

Figure 9. (a) Electric potential for k_1=0.7, k_2=0.3 (b): Electric potential for k_1=0.7, k_2=0.45 (c): Electric field for k_1=0.7, k_2=0.45. Here, dashed, solid and dot-dashed curves represents the times before, at and after collision, respectively.

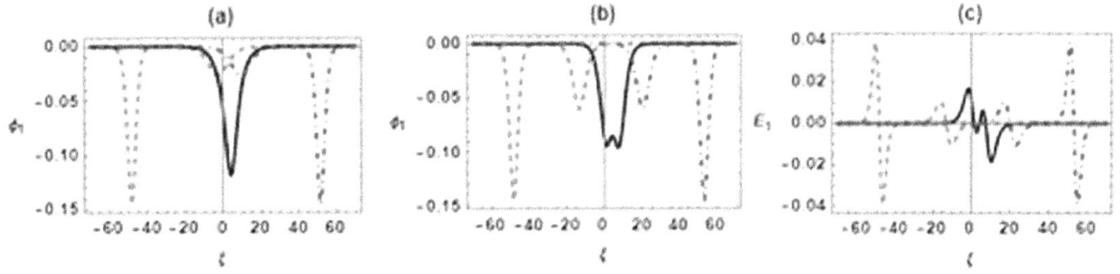

It is also important to mention that the choice of propagation vectors changes the type of composite structure generated at the point of interaction. It has been observed that the ratio of the propagation vectors k_1/k_2, where k_1 (k_2) denote the propagation vector of taller (shorter) soliton, mainly determines the shape of composite structure. Therefore, we get a single dip composite soliton if the ratio k_1/k_2 is high as shown in Figure 9(a) else we get double dips at the peak of composite rarefactive solitary structure at the interaction point, depicted in (b) of Figure (9). These two dips of solitary wave potential correspond to the tripolar electric field composite structures (shown in (c) of Figure (9)), which have been identified frequently in space plasmas (Pickett et. al., 2004). Therefore, one possible interpretation of such tripolar electric field structures is the interaction of two solitons in space.

Figure 10. Contour plots of two soliton interaction for different values of beam velocity

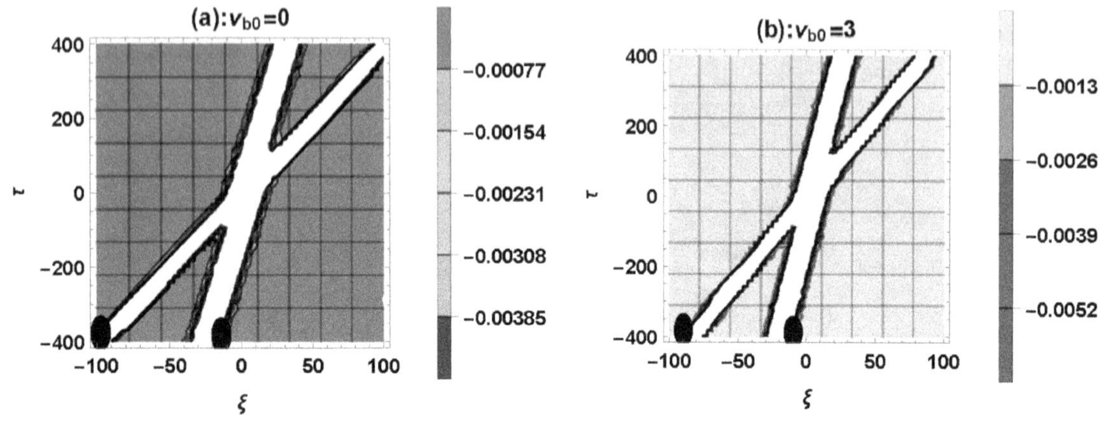

Figures 10(a-b) manifest the impact of beam velocity on the interaction regime of two KdV solitons in terms of contour plots of electric potential. The comparison of both curves show that the increase of beam velocity contracts the interaction regime from ~13km to 10km. The interaction region is determined by the distance between dots on the spatial scale in the curves and then un-normalizing by the Debye length λ_{dh}. This is in accordance with the result of single soliton, i.e., the beam velocity reduces the amplitude of single soliton, thereby, decreasing the speed of the soliton. Therefore, the spatial regime of interaction contracts or the temporal regime of interaction expands.

5. OVERTAKING INTERACTION OF mKdV SOLITONS

The mKdV is also a completely integrable equation, so we can obtain the N-soliton solutions for the mKdV equation. For mKdV Equation (32), we use a different dependent variable Hirota transformation (Wazwaz, 2010) given as

$$\Psi = \sqrt{\frac{24S}{R}}\partial_{\xi}\left[\tan^{-1}\left(\frac{\tilde{f}}{g}\right)\right], \tag{44}$$

where $\tilde{f}$ and g are the dependant variable functions. The Hirota transformation given by Equation (44) produces a pair of following bilinear equations

$$D_{\xi}^{2}\left(\tilde{f}.\tilde{f}+g.g\right)=0, \tag{45}$$

$$\left(SD_{\xi}^{3}+D_{\tau}\right)\left(\tilde{f}.g\right)=0. \tag{46}$$

The arbitrary functions $\tilde{f}$ and g may be expanded in terms of small parameter ε as

$$\tilde{f}(\xi,\tau)=\tilde{f}_0+\varepsilon\tilde{f}_1+\varepsilon^2\tilde{f}_2+\varepsilon\tilde{f}_3+....,$$

$$g(\xi,\tau)=g_0+\varepsilon g_1+\varepsilon^2 g_2+\varepsilon^3 g_3+.....$$

We assume the first order expansion functions $\tilde{f}_1=F_1e^{\eta}$ and $g_1=G_1e^{\eta}$, where $\eta=k\xi+\omega\tau+\gamma$ as in above case of KdV equation. Putting these values in Equation (45) we get the following expression for the expansion functions

$$\tilde{f}_0F_1+g_0G_1=0. \tag{47}$$

This general expression gives the conditions on expansion functions to determine whether the electron acoustic solitary wave would be compressive or rarefactive.

5.1 Single mKdV Soliton Solution

One of the possible conditions on arbitrary expansion functions given by Equation (45) is $\tilde{f}_0=G_1=0$.

Plugging this condition in Equation (46) gives the relation $\omega_1=-Sk_1^3$. For single mKdV soliton, the complete functions f and g become

$$\tilde{f}=e^{k_1(\xi-Sk_1^2\tau)},\ \text{and}\ g=1 \tag{49}$$

Then the following single compressive soliton solution for mKdV equation is attained

$$\Psi=\sqrt{\frac{24S}{R}}\partial_\xi\left[\tan^{-1}\left(e^{k_1(\xi-Sk_1^2\tau)}\right)\right].$$

For the other possible condition $g_0=F_1=0$ given by Equation (47), we get the single rarefactive soliton solution of mKdV equation of the form

$$\Psi=\sqrt{\frac{24S}{R}}\partial_\xi\left[\tan^{-1}\left(\frac{1}{e^{k_1(\xi-Sk_1^2\tau)}}\right)\right]. \tag{50}$$

The Figure (11) shows the potential and electric fields of the compressive and rarefactive single soliton solutions of mKdV equation for various values of beam speed. The compressive and rarefactive solitons are complete mirror images having the same amplitudes and widths, with only change of positive and negative polarity for case of electric potential or the change of order of hump and dip for the case of electric field. The graphs again show a decrease in the amplitude of soliton for enhanced beam velocity consistent with the KdV solitons, however, the amplitude and the spatial scales vary for the KdV and mKdV solitons. This information is significant for data physicists of space plasmas who may compare the spatial and temporal scales of various electrostatic solitary waves to determine the type of nonlinearity of nPDEs.

Figure 11. Compressive and rarefactive electric potentials and fields for mKdV soliton. Here, dashed, solid and dot-dashed curves are plotted for v_{b0}=0, 0.8 and 3, respectively.

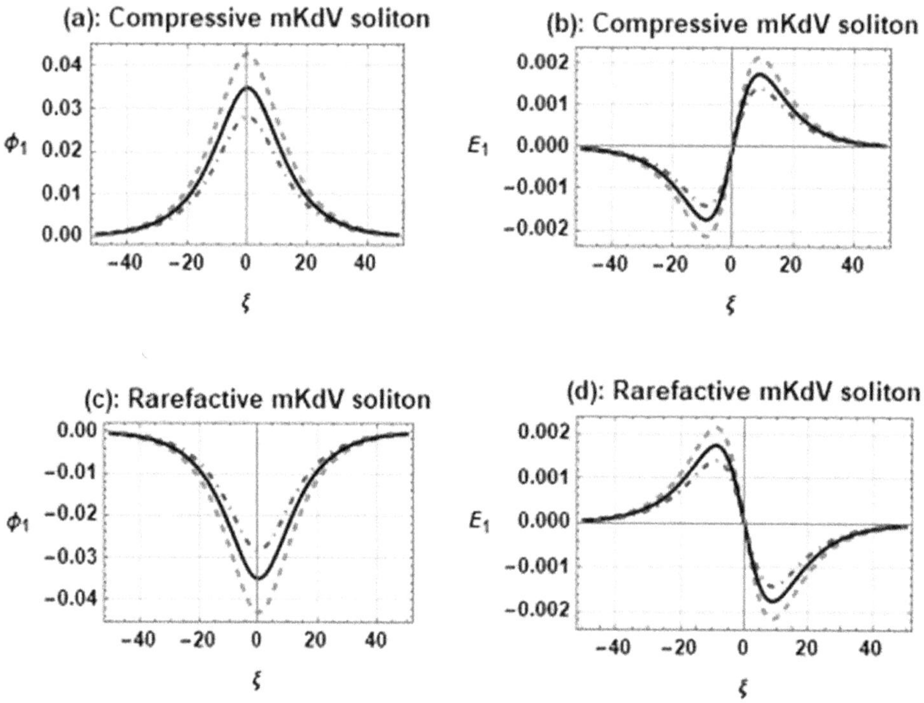

5.2 Two mKdV Soliton Solution

For attaining the two soliton solution of mKdV for compressive case given by Equation (47), we choose

$$\tilde{f}_1 = e^{\eta_1} + e^{\eta_2}, \tag{51}$$

and

$$g_2 = a_{121} e^{\eta_1+\eta_2}, \tag{52}$$

where $\eta_i = k_i\xi + \omega_i\tau + \gamma_i$ with i=1,2 and a_{121} is the interaction parameter. For the condition of compressive solitons, substituting Eqs. (51), (52) and expansion $\tilde{f}$ and g in Equation (46) give us $\omega_i = -Sk_i^3$ and $a_{121} = -((k_1 - k_2)^2 / (k_1 + k_2)^2)$. The dependent variable functions then become

$$\tilde{f} = e^{k_1(\xi - Sk_1^2\tau)} + e^{k_2(\xi - Sk_2^2\tau)},\ g = 1 + a_{121}e^{k_1(\xi - Sk_1^2\tau) + k_2(\xi - Sk_2^2\tau)}. \tag{53}$$

Substituting the functions in dependent variable transformation of Equation (44), the compressive two soliton solution for mKdV equation turns out to be

$$\Psi = \sqrt{\frac{24S}{R}}\partial_\xi\left[\tan^{-1}\left(\frac{e^{k_1(\xi - Sk_1^2\tau)} + e^{k_2(\xi - Sk_2^2\tau)}}{1 + a_{121}e^{k_1(\xi - Sk_1^2\tau) + k_2(\xi - Sk_2^2\tau)}}\right)\right]. \tag{54}$$

For the rarefactive case, we take

$$g_1 = e^{\eta_1} + e^{\eta_2},\ \text{and}\ \tilde{f}_2 = a_{121}e^{\eta_1 + \eta_2}, \tag{55}$$

which gives the rarefactive two soliton solution of mKdV equation as

$$\Psi = \sqrt{\frac{24S}{R}}\partial_\xi\left[\tan^{-1}\left(\frac{1 + a_{121}e^{k_1(\xi - Sk_1^2\tau) + k_2(\xi - Sk_2^2\tau)}}{e^{k_1(\xi - Sk_1^2\tau)} + e^{k_2(\xi - Sk_2^2\tau)}}\right)\right], \tag{56}$$

by adopting the same above procedure.

5.3 Interaction of mKdV Solitons

The overtaking interaction of mKdV solitons is studied by the two-soliton solutions derived in the above section.

The possibility of both compressive and rarefactive two-soliton solutions for mKdV equation attributes to the peculiar types of interaction of compressive-rarefactive and rarefactive-compressive mKdV solitons besides the interaction of compressive-compressive and rarefactive-rarefactive solitons. The curves of compressive-compressive and rarefactive-rarefactive soliton interactions show the nonlinear superposition of waves along with phase shift which is the similar behavior qualitatively as that of the KdV solitons, as shown in Figs. (12) (a-b). However, the interaction of a taller compressive (hump) soliton with a shorter rarefactive (dip) soliton shown in (c) of Figure (12) depicts a completely different shape of composite structure at the point of interaction. Now, the taller compressive soliton seems to pass through the shorter rarefactive soliton by splitting the shorter soliton into two dips. Similarly, when a taller dip soliton overtakes a shorter hump soliton, a single taller dip and two shorter humps seem to be produced at the point of interaction as shown in (d) of Figure (12).

Figure 12. Types of two soliton interaction of mKdV solitons. Here, dashed, solid and dot-dashed curves show the time before, at, and after interaction, respectively.

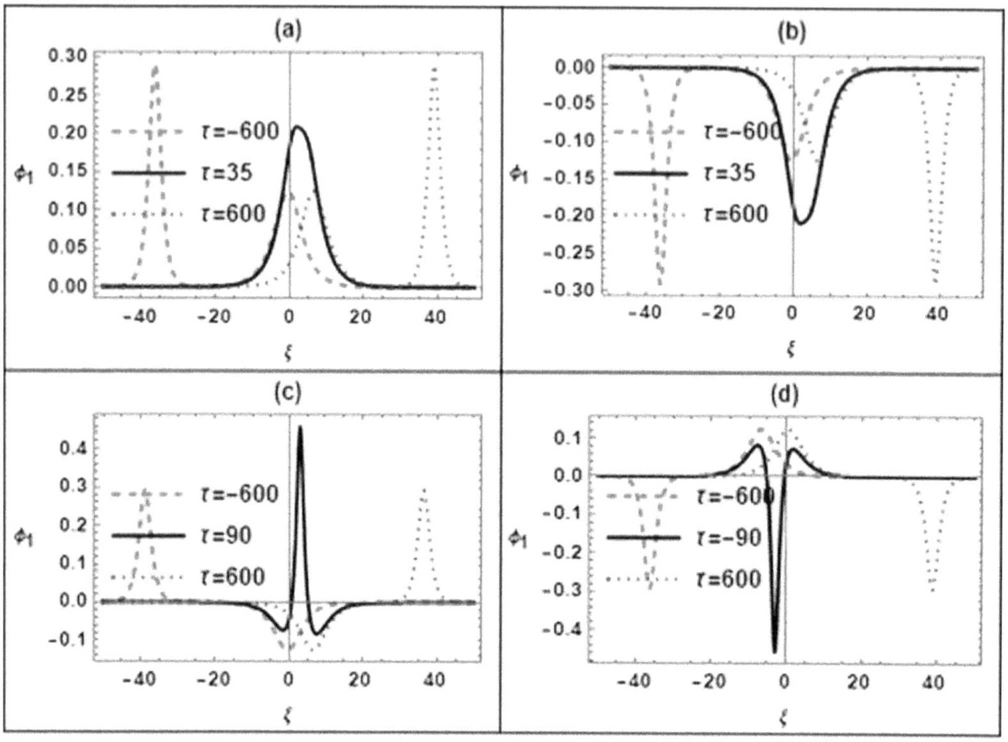

6. HEAD ON INTERACTION OF SOLITONS

The head on interaction of solitons is studied by altogether different technique known as extended Poincare-Lighthill-Kuo (PLK) technique. The space-time coordinates are stretched in a different manner to study the interaction of two EASWs (Su, 1980), using the extended PLK method. Therefore,

$$\xi = \varepsilon\left(x - \Lambda t\right) + \varepsilon^2 P_0\left(\eta, \tau\right) + \ldots\ldots.,$$
$$\eta = \varepsilon\left(x + \Lambda t\right) + \varepsilon^2 Q_0\left(\xi, \tau\right) + \ldots\ldots.,$$

$$\tau = \varepsilon^3 t, \quad (57)$$

where Λ is the normalized phase velocity of the EAWs. Here, two stretched coordinates ξ and η are introduced for right and left propagating solitons with the opposite signs of velocities. The functions P_0 and Q_0 represent the first order coefficients of phase shift, to be determined later. It may be noted that P_0 is the function of η and Q_0 is the function of ξ only. The phase functions actually determine the phase shift and the phase shift induced in a soliton after interaction is due to the other soliton, so the P_0, Q_0 are functions of other coordinates only. We may define new compact operators for simplicity of the form

$$X=\left(\frac{\partial}{\partial\eta}+\frac{\partial}{\partial\xi}\right),T=\Lambda\left(\frac{\partial}{\partial\eta}-\frac{\partial}{\partial\xi}\right), \tag{58}$$

$$X'=\left(\frac{\partial P_0}{\partial\eta}+\frac{\partial P_0}{\partial\xi}\right)\frac{\partial}{\partial\xi}+\left(\frac{\partial Q_0}{\partial\eta}+\frac{\partial Q_0}{\partial\xi}\right)\frac{\partial}{\partial\eta}, \tag{59}$$

$$T'=\Lambda\left(\left(\frac{\partial P_0}{\partial\eta}-\frac{\partial P_0}{\partial\xi}\right)\frac{\partial}{\partial\xi}+\left(\frac{\partial Q_0}{\partial\eta}-\frac{\partial Q_0}{\partial\xi}\right)\frac{\partial}{\partial\eta}\right). \tag{60}$$

Using stretching and new operators of Eqs. (plk) and (operators), we have the following space-time operators

$$\begin{aligned}\frac{\partial}{\partial x}&=\varepsilon X+\varepsilon^3X'+\ldots,\\ \frac{\partial}{\partial t}&=\varepsilon T+\varepsilon^3T'+\varepsilon^3\frac{\partial}{\partial\tau}+\ldots.\end{aligned} \tag{61}$$

The dependent variables n_j, v_j and φ are expanded about their equilibrium value as follows

$$R_j(x,t)=R_{j0}+\sum_{i=1}^{\infty}\varepsilon^{i+1}R_{ji}(x,t). \tag{62}$$

Here, $R_j(x,t)=[n_j,v_j,\varphi]$, and $R_{j0}=[1,v_{j0},0]$. It should be noted that we may ignore the ε -order terms in the expansion as they give the redundant equations only. Using stretchings and expansions into normalized governing equations and collecting the lowest order of ε, we get

$$(T+v_{j0}X)n_{j1}+Xv_{j1}=0,$$

$$\left(T+v_{j0}X\right)v_{j1}+\alpha\sigma_jXn_{j1}-\alpha X\varphi_1=0,$$

$$\mu_1\varphi_1+\frac{n_{c1}}{\alpha}+\frac{n_{b1}}{\beta}=0. \tag{63}$$

Solving the equations simultaneously gives

$$(\left(T+v_{j0}X\right)^2-\alpha\sigma_bX^2)\left(\mu_1\left(T^2-\alpha\sigma_cX^2\right)-X^2\right)\varphi_1-\frac{\alpha}{\beta}\left(T^2-\alpha\sigma_cX^2\right)X^2\varphi_1=0. \tag{64}$$

Substituting the values of operators provide us with the following complex general equation

$$C_1\partial_{\xi,\xi,\xi,\xi}\varphi_1 + C_2\partial_{\eta,\eta,\eta,\eta}\varphi_1 + C_3\partial_{\xi,\xi,\eta,\eta}\varphi_1 + C_4\partial_{\xi,\xi,\xi,\eta}\varphi_1 + C_5\partial_{\xi,\eta,\eta,\eta}\varphi_1 = 0, \tag{65}$$

where all the coefficients C_i's are independent and do not provide any separability condition. Therefore, we consider the case of EAWs without beam electrons for simplicity. In the absence of beam electrons, the first order equations give

$$\left(\alpha\sigma_c - \Lambda^2 + \frac{1}{\mu_1}\right)\left(\partial_{\xi,\xi} + \partial_{\eta,\eta}\right)\varphi_1 + 2\left(\alpha\sigma_c + \Lambda^2 + \frac{1}{\mu_1}\right)\partial_{\xi,\eta}\varphi_1. \tag{66}$$

Then the solvability condition gives the phase velocity

$$\Lambda = \sqrt{\frac{1}{\mu_1} + \alpha\sigma_c}, \tag{67}$$

and the other part of equation is $4\Lambda^2\partial_{\xi,\eta}\varphi_1 = 0$, which ensures the separability of the potential and allows it to be considered as

$$\varphi_1 = \chi_1(\xi,\tau) + \chi_2(\eta,\tau), \tag{68}$$

The next order equations of ε give the same above results in terms of φ_2, so can be ignored. For further higher order, we obtain

$$\begin{aligned} -2\Lambda^3\varphi_3 &= \int\left(\partial_\tau\chi_1 - L\chi_1\partial_\xi\chi_1 + M\partial_{\xi,\xi,\xi}\chi_1\right)d\eta - \int\left(\partial_\tau\chi_2 + L\chi_2\partial_\eta\chi_2 - M\partial_{\eta,\eta,\eta}\chi_2\right)d\xi \\ &+ \iint\left[N\partial_\eta P_0 + O\chi_2\right]\partial_{\xi,\xi}\chi_1 d\eta d\xi + \iint\left[N\partial_\xi Q_0 + O\chi_1\right]\partial_{\eta,\eta}\chi_2 d\eta d\xi. \end{aligned} \tag{69}$$

Here, the phase coefficients are defined to be

$$L = \frac{3\alpha}{2\Lambda} + \frac{\alpha^2\sigma_c\mu_1}{\Lambda} + \frac{\mu_2}{\mu_1^2\Lambda}, M = \frac{1}{2\mu_1^2\Lambda}, \tag{70}$$

$$N = \frac{2\alpha\sigma_c}{\Lambda} + \frac{3}{2\mu_1\Lambda}, O = \frac{\alpha^2\sigma_c\mu_1}{\Lambda} + \frac{\alpha}{2\Lambda}. \tag{71}$$

The coefficients of nonlinearity L and dispersion M are same as the one obtained for KdV Equation (22) in the absence of beam electrons. The nonlinearity coefficient in the absence of beam electrons contain no more competing terms, so can't be zero and we cannot attain the pair of mKdV equations. Most importantly, the PLK method allows us to obtain a couple of nonlinear equations for left and right propagating solitons, while Hirota method applies to a single nPDE. The first integral in Equation (69) is such that the integrand is independent of η, so it becomes the secular term. Similar is the case with second integral. Therefore, the first two integrals must be annulled to avoid secularities. The next two integrals are not secular at this order, but they tend to be secular for higher order [Su1980], therefore, must be taken to be zero. Then we get a pair of KdV equations

$$\partial_\tau \chi_1 - L\chi_1 \partial_\xi \chi_1 + M\partial_{\xi,\xi,\xi}\chi_1 = 0, \tag{72}$$

$$\partial_\tau \chi_2 + L\chi_2 \partial_\eta \chi_2 - M\partial_{\eta,\eta,\eta}\chi_2 = 0. \tag{73}$$

with the solutions

$$\chi_1(\xi,\tau) = -\frac{3v_\xi}{L}\sec h^2\left[\sqrt{\frac{v_\xi}{4M}}\left(\xi - v_\xi\tau\right)\right], \tag{74}$$

$$\chi_2(\eta,\tau) = -\frac{3v_\eta}{L}\sec h^2\left[\sqrt{\frac{v_\eta}{4M}}\left(\eta + v_\eta\tau\right)\right], \tag{75}$$

which represent the two counter propagating solitons moving towards one another. The most important result for HOC is given by Equation (68), which may be interpreted that unlike the overtaking interaction, the two soliton potentials simply add up linearly for PLK method, which contradicts with the experimental observations. However, the advantage of PLK method is the evaluation of analytical expressions of trajectories and phase shifts.

6.1 Phase Shifts After Collision

The phase shifts after head on interaction may be studied analytically with the aid of phase coefficients in PLK method. From the last two integrals of Equation (69) we obtain the following expressions

$$P_0 = \frac{O}{N}\int_0^\eta \chi_2(\eta,\tau)\,d\eta,\;\; Q_0 = \frac{O}{N}\int_0^\xi \chi_1(\xi,\tau)\,d\xi. \tag{76}$$

Substituting the soliton solutions and integrating these expression yield,

$$P_0 = -\frac{6}{L}\frac{O}{N}\sqrt{Mv_\eta}\,\tan h\left[\sqrt{\frac{4v_\eta N}{O}}\left(\eta + v_\eta\tau\right) + 1\right], \tag{77}$$

$$Q_0 = -\frac{6}{L}\frac{O}{N}\sqrt{Mv_\xi}\,\tan h\left[\sqrt{\frac{4v_\xi N}{O}}\left(\xi - v_\xi\tau\right)-1\right]. \tag{78}$$

To ensure that the solitons are well separated before collision, we assume them to be asymptotically far away from each other, therefore, at the initial time $t\rightarrow\infty$, the soliton χ_1 is considered at ξ=0, η=-∞ and χ_2 is initially at η=0, ξ=∞, respectively. After the collision, the phase shift is evaluated when solitons have passed through each other and are again separated. Therefore, at latter time t→∞, χ_1 moves towards right so that it is at ξ=0, η=∞ and χ_2 propagates towards left and reaches at η=0, ξ=-∞.

Figure 13. Phase shift variation with various parameters

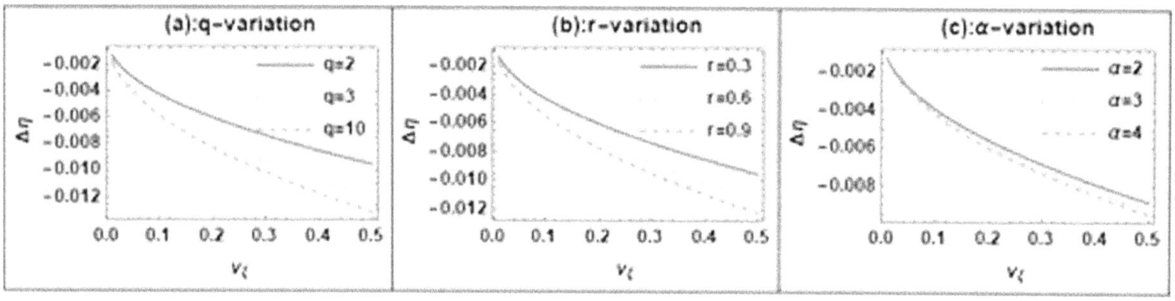

Phase shifts after collision induced for the solitons χ_1 and χ_2 are given by the following relations

$$\Delta_\xi = -12\varepsilon^2 \frac{N}{LO}\sqrt{Mv_\eta}, \tag{79}$$

$$\Delta_\eta = 12\varepsilon^2 \frac{N}{LO}\sqrt{Mv_\xi}. \tag{80}$$

Thus, the phase shift introduced due to the collision for one soliton depends upon the velocity of the other soliton and all plasma parameters through the coefficients *L, M, N,* and *O*. We get the positive phase shift for both solitons in their directions of propagation for this plasma system. The negative (positive) phase shifts after collision usually refer to slowing down (speeding up) of the resultant solitons on average than the solitons without interaction.

SUMMARY

In conclusion, we have studied the impact of electron beam on various types of interaction for Electron Acoustic waves. The roots of linear dispersion relation are changed completely for the beam driven EAWs. Both overtaking and HOC have been studied and compared for this mode. The results have been numerically analyzed for the energetic bursts of beam electrons in BEN belonging to auroral region of

Earth's magnetosphere. Besides the evaluation of analytical expressions of phase shifts, the limitation of PLK to study HOC is emphasized.

REFERENCES

Abdikian, A., Vasheghani Farahani, S., & Hussain, S. (2021). The characteristics of daughter waves emerging from colliding solitary waves in astrophysical plasma media. M*onthly Notices of the Royal Astronomical Society, 506(*1), 997–1006. doi:10.1093/mnras/stab1781

Abid, A. A., Khan, M. Z., Lu, Q., & Yap, S. L. (2017). A generalized AZ-non-Maxwellian velocity distribution function for space plasmas. *Physics of Plasmas*, *24*(3), 033702. doi:10.1063/1.4977447

Ablowitz, M. J., Ablowitz, M. A., Clarkson, P. A., & Clarkson, P. A. (1991). *Solitons, nonlinear evolution equations and inverse scattering*. Cambridge university press. doi:10.1017/CBO9780511623998

Ali, S., Masood, W., Rizvi, H., Jahangir, R., & Mirza, A. M. (2021). Contribution of the generalized (r, q) distributed electrons in the formation of nonlinear ion acoustic waves in upper ionospheric plasmas. *AIP Advances*, *11*(12), 125020. doi:10.1063/5.0075007

Angelopouulos, V. (2008). *Space Science Reviews*, *141*, 5. doi:10.1007/s11214-008-9336-1

Asano, Y., Nakamura, R., Shinohara, I., Fujimoto, M., Takada, T., Baumjohann, W., Owen, C. J., Fazakerley, A. N., Runov, A., Nagai, T., Lucek, E. A., & Rème, H. (2008). Electron flat-top distributions around the magnetic reconnection region. *Journal of Geophysical Research*, *113*(A1), 2007JA012461. doi:10.1029/2007JA012461

Berthomier, M., Pottelette, R., Malingre, M., & Khotyaintsev, Y. (2000). Electron-acoustic solitons in an electron-beam plasma system. *Physics of Plasmas*, *7*(7), 2987–2994. doi:10.1063/1.874150

Bezzerides, B., Forslund, D. W., & Lindman, E. L. (1978). Existence of rarefaction shocks in a laser-plasma corona. *The Physics of Fluids*, *21*(12), 2179–2185. doi:10.1063/1.862176

Bostrom, R., Holback, B., Holmgren, G., & Koskinen, H. (1989). Solitary structures in the magnetospheric plasma observed by Viking. *Physica Scripta*, *39*(6), 782–786. doi:10.1088/0031-8949/39/6/022

Dillard, C. S., Vasko, I. Y., Mozer, F. S., Agapitov, O. V., & Bonnell, J. W. (2018). Electron-acoustic solitary waves in the Earth's inner magnetosphere. *Physics of Plasmas*, *25*(2), 022905. doi:10.1063/1.5007907

Dubouloz, N., Pottelette, R., Malingre, M., & Treumann, R. A. (1991). Generation of broadband electrostatic noise by electron acoustic solitons. *Geophysical Research Letters*, *18*(2), 155–158. doi:10.1029/90GL02677

Dubouloz, N., Treumann, R. A., Pottelette, R., & Malingre, M. (1993). Turbulence generated by a gas of electron acoustic solitons. *Journal of Geophysical Research*, *98*(A10), 17415–17422. doi:10.1029/93JA01611

El-Taibany, W. F. (2005). Electron-acoustic solitary waves and double layers with an electron beam and phase space electron vortices in space plasmas. *Journal of Geophysical Research*, *110*(A1), 2004JA010525. Advance online publication. doi:10.1029/2004JA010525

Elwakil, S. A., Zahran, M. A., & El-Shewy, E. K. (2007). Nonlinear electron-acoustic solitary waves in a relativistic electron-beam plasma system with non-thermal electrons. *Physica Scripta*, *75*(6), 803–808. doi:10.1088/0031-8949/75/6/010

Ergun, R. E., Carlson, C. W., McFadden, J. P., Mozer, F. S., Delory, G. T., Peria, W., Chaston, C. C., Temerin, M., Roth, I., Muschietti, L., Elphic, R., Strangeway, R., Pfaff, R., Cattell, C. A., Klumpar, D., Shelley, E., Peterson, W., Moebius, E., & Kistler, L. (1998). FAST satellite observations of large-amplitude solitary structures. *Geophysical Research Letters*, *25*(12), 2041–2044. doi:10.1029/98GL00636

Ergun, R. E., Holmes, J. C., Goodrich, K. A., Wilder, F. D., Stawarz, J. E., Eriksson, S., Newman, D. L., Schwartz, S. J., Goldman, M. V., Sturner, A. P., Malaspina, D. M., Usanova, M. E., Torbert, R. B., Argall, M., Lindqvist, P.-A., Khotyaintsev, Y., Burch, J. L., Strangeway, R. J., Russell, C. T., & André, M. (2016). Magnetospheric Multiscale observations of large-amplitude, parallel, electrostatic waves associated with magnetic reconnection at the magnetopause. *Geophysical Research Letters*, *43*(11), 5626–5634. doi:10.1002/2016GL068992

Feldman, W. C., Anderson, R. C., Bame, S. J., Gary, S. P., Gosling, J. T., McComas, D. J., Thomsen, M. F., Paschmann, G., & Hoppe, M. M. (1983). Electron velocity distributions near the Earth's bow shock. *Journal of Geophysical Research*, *88*(A1), 96–110. doi:10.1029/JA088iA01p00096

Fried, B. D., & Gould, R. W. (1961). Longitudinal ion oscillations in a hot plasma. *The Physics of Fluids*, *4*(1), 139–147. doi:10.1063/1.1706174

Gary, S. P., & Tokar, R. L. (1985). The electron-acoustic mode. *The Physics of Fluids*, *28*(8), 2439–2441. doi:10.1063/1.865250

Graham, D. B., Khotyaintsev, Yu. V., Vaivads, A., & Andre, M. (2015). Electrostatic solitary waves with distinct speeds associated with asymmetric reconnection. *Geophysical Research Letters*, *42*(2), 215224. doi:10.1002/2014GL062538

Harvey, P., Durniak, C., Samsonov, D., & Morfill, G. (2010). Soliton interaction in a complex plasma. *Physical Review E: Statistical, Nonlinear, and Soft Matter Physics*, *81*(5), 057401. doi:10.1103/PhysRevE.81.057401 PMID:20866360

Hasan, U., Masood, W., Jahangir, R., & Mirza, A. M. (2019). Head-on collision of nonlinear electrostatic shock waves in a relativistically degenerate plasma. *Physica Scripta*, *95*(1), 015601. doi:10.1088/1402-4896/ab20e9

Hirota, R. (1971). Exact solution of the Korteweg-de Vries equation for multiple collisions of solitons. *Physical Review Letters*, *27*(18), 1192–1194. doi:10.1103/PhysRevLett.27.1192

Hirota, R. (2004). *The Direct Method in Soliton Theory*. Cambridge University Press. doi:10.1017/CBO9780511543043

Holmes, J. C., Ergun, R. E., Newman, D. L., Wilder, F. D., Sturner, A. P., Goodrich, K. A., Torbert, R. B., Giles, B. L., Strangeway, R. J., & Burch, J. L. (2018). Negative potential solitary structures in the magnetosheath with large parallel width. *Journal of Geophysical Research. Space Physics*, *123*(1), 132–145. doi:10.1002/2017JA024890

Hosseini Jenab, S. M., & Brodin, G. (2019). Head-on collision of nonlinear solitary solutions to Vlasov-Poisson equations. *Physics of Plasmas*, *26*(2), 022303. doi:10.1063/1.5078865

Huang, G., Lou, S., & Xu, Z. (1993). Phys. Rev. E Stat. *Phys. Plasmas Fluids Relat. Interdiscip*, *47*, R3830. doi:10.1103/PhysRevE.47.R3830

Huang, G., & Velarde, M. G. (1996). Head-on collision of two concentric cylindrical ion acoustic solitary waves. *Physical Review E: Statistical Physics, Plasmas, Fluids, and Related Interdisciplinary Topics*, *53*(3), 2988–2991. doi:10.1103/PhysRevE.53.2988 PMID:9964593

Infeld, E., & Rowlands, G. (1990). Nonlinear Waves, Solitons and Chaos. Cambridge University Press

Jahangir, R., Ali, S., Mirza, A. M., & Masood, W. (2019). Head on interaction of magnetoacoustic solitons in a spin-1/2 dense plasma with geometrical effects. *Physica Scripta*, *94*(12), 125602. doi:10.1088/1402-4896/ab3499

Jahangir, R., & Masood, W. (2020). Interaction of electron acoustic waves in the presence of superthermal electrons in terrestrial magnetosphere. *Physics of Plasmas*, *27*(4), 042105. doi:10.1063/1.5143400

Jahangir, R., Masood, W., & Rizvi, H. (2023). Face to face interaction of dust acoustic solitons with trapped ions. *Chaos, Solitons, and Fractals*, *174*, 113694. doi:10.1016/j.chaos.2023.113694

Jahangir, R., Masood, W., Siddiq, M., & Batool, N. (2016). Interaction of solitons for obliquely propagating magnetoacoustic waves in stellar atmosphere. *Physics of Plasmas*, *23*(12), 122301. doi:10.1063/1.4971219

Jeffrey, A. (1973). The role of Korteweg de Vries Equation in Plasma Physics. Jeffrey, A. Role of the Korteweg-de Vries equation in Plasma Physics. *Quarterly Journal of the Royal Astronomical Society*, *14*, 183.

Kakutani, T., Ono, H., Taniuti, T., & Wei, C.-C. (1968). Taniuti, Wei CC, Reductive perturbation method in non-linear wave propagation II: Application to hydromagnetic waves in a cold plasma. *Journal of the Physical Society of Japan*, *24*(5), 1159–1166. doi:10.1143/JPSJ.24.1159

Korteweg, D. J., & Vries, G. D. (1895). On the change of form of long waves advancing in a rectangular channel, and on a new type of long stationary waves. *The London, Edinburgh and Dublin Philosophical Magazine and Journal of Science*, *39*(240), 422–443. doi:10.1080/14786449508620739

Kuo, Y. (1953). On the Flow of an Incompressible Viscous Fluid Past a Flat Plate at Moderate Reynolds Numbers. *Journal of Mathematics and Physics*, *32*(1-4), 83–101. doi:10.1002/sapm195332183

Lakhina, G. S., Singh, S. V., Kakad, A. P., & Pickett, J. S. (2011). Generation of electrostatic solitary waves in the plasma sheet boundary layer. *Journal of Geophysical Research*, *116*(A10), n/a. doi:10.1029/2011JA016700

Lakhina, G. S., Singh, S. V., Rubia, R., & Sreeraj, T. (2018). A review of nonlinear fluid models for ion-and electron-acoustic solitons and double layers: Application to weak double layers and electrostatic solitary waves in the solar wind and the lunar wake. *Physics of Plasmas*, *25*(8), 080501. doi:10.1063/1.5033498

Li, S., & Duan, W. (2008). Interactions between two solitons with an arbitrary collision direction in two-dimensional Bose-Einstein condensates. *The European Physical Journal B*, *62*(4), 485–491. doi:10.1140/epjb/e2008-00192-y

Lighthill, M. J. (1949). A Technique for rendering approximate solutions to physical problems uniformly valid. *The London, Edinburgh and Dublin Philosophical Magazine and Journal of Science*, *40*(311), 1179–1201. doi:10.1080/14786444908561410

Mannan, A., & Mamun, A. A. (2011). Nonplanar dust-acoustic Gardner solitons in a four-component dusty plasma. *Physical Review. E*, *84*(2), 026408. doi:10.1103/PhysRevE.84.026408 PMID:21929121

Masood, W., Hamid, N., Ullah, S., Aman-ur-Rahman, Shah, H. A., Alyousef, H. A., & El-Tantawy, S. A. (2022). Formation of acoustic nonlinear structures in non-Maxwellian trapping plasmas. *Physics of Fluids*, *34*(5), 053113. doi:10.1063/5.0090923

Masood, W., & Schwartz, S. J. (2008). Observations of the development of electron temperature anisotropies in Earth's magnetosheath. *Journal of Geophysical Research*, *113*(A1), 2007JA012715. doi:10.1029/2007JA012715

Masood, W., Schwartz, S. J., Maksimovic, M., & Fazakerley, A. N. (2006). Electron velocity distribution and lion roars in the magnetosheath. *Annales Geophysicae*, *24*(6), 1725–1735. doi:10.5194/angeo-24-1725-2006

Masud M., Asaduzzaman M., & Mamun, A.A. (2012). Dust-ion-acoustic Gardner solitons in a dusty plasma with bi-Maxwellian electrons. *Physics of Plasmas, 19*(10), 103706. doi:10.1063/1.4753922

Matsumoto, H., Kojima, H., Miyatake, T., Omura, Y., Okada, M., Nagano, I., & Tsutsui, M. (1994). Electrostatic solitary waves (ESW) in the magnetotail: BEN wave forms observed by GEOTAIL. *Geophysical Research Letters*, *21*(25), 2915–2918. doi:10.1029/94GL01284

Matveev, V. B., & Salle, M. A. (1991). *Darboux Transformations and Solitons*. Springer. doi:10.1007/978-3-662-00922-2

Maxworthy, T. (1976). Experiments on the collision between two solitary waves. *Journal of Fluid Mechanics*, *76*(1), 177–186. doi:10.1017/S0022112076003194

Miura, R. M., Gardner, C. S., & Kruskal, M. D. (1968). Korteweg-de vries equation and generalizations. II. Existence of conservation laws and constants of motion. *Journal of Mathematical Physics*, *9*(8), 1204–1209. doi:10.1063/1.1664701

Mozer, F. S., Cattell, C. A., Temerin, M., Torbert, R. B., Von Glinski, S., Woldorff, M., & Wygant, J. (1979). The dc and ac electric field, plasma density, plasma temperature, and field-aligned current experiments on the S3-3 satellite. *Journal of Geophysical Research*, *84*(A10), 5875–5884. doi:10.1029/JA084iA10p05875

Naeem, I., Masood, W., & Mirza, A. M. (2020). Dipolar and Kelvin-Stuart's cat's eyes vortices in magnetoplasmas with non-Maxwellian electron distribution. *Astrophysics and Space Science*, *365*(3), 1–1. doi:10.1007/s10509-020-03759-9

Nawaz, H., Jahangir, R., Masood, W., & Siddiq, M. (2022). Cubic nonlinearity driven dust ion acoustic solitons with superthermal two-temperature electrons in the Saturn's magnetosphere. *Physica Scripta*, *97*(7), 07560110. doi:10.1088/1402-4896/ac735e

Ogilvie, K. W., Fitzenreiter, R. J., & Scudder, J. D. (1984). Observations of electron beams in the low-latitude boundary layer. *Journal of Geophysical Research*, *89*(A12), 10723 10732. doi:10.1029/JA089iA12p10723

Pickett, J. S., & Chen, L.-JKahler, S. WSantolik, OGoldstein, M. LLavraud, BDecreau, P. MKessel, RLucek, ELakhina, G. STsurutani, B. TGurnett, D. ACornilleau-Wehrlin, NFazakerley, AReme, HBalogh. A. (2005). On the generation of solitary waves observed by Cluster in the near-Earth magnetosheath. *Nonlinear Processes in Geophysics.*, *12*(2), 181-193. doi:10.5194/npg-12-181-2005

Pickett, J. S., Kahler, S. W., Chen, L. J., Huff, R. L., Santolik, O., Khotyaintsev, Y., Décréau, P. M., Winningham, D., Frahm, R., Goldstein, M. L., Lakhina, G. S., Tsurutani, B. T., Lavraud, B., Gurnett, D. A., André, M., Fazakerley, A., Balogh, A., & Rème, H. (2004). Solitary waves observed in the auroral zone: The Cluster multi-spacecraft perspective. *Nonlinear Processes in Geophysics*, *11*(2), 183–196. doi:10.5194/npg-11-183-2004

Pottelette, R., Ergun, R. E., Treumann, R. A., Berthomier, M., Carlson, C. W., McFadden, J. P., & Roth, I. (1999). Modulated electron-acoustic waves in auroral density cavities: FAST observations. *Geophysical Research Letters*, *26*(16), 2629–2632. doi:10.1029/1999GL900462

Pottelette, R., Malingre, M., Dubouloz, N., Aparicio, B., Lundin, R., Holmgren, G., & Marklund, G. (1990). High-frequency waves in the cusp/cleft regions. *Journal of Geophysical Research*, *95*(A5), 5957–5971. doi:10.1029/JA095iA05p05957

Qureshi, M. N., Nasir, W., Bruno, R., & Masood, W. (2019). Whistler instability based on observed flat-top two-component electron distributions in the Earth's magnetosphere. *Monthly Notices of the Royal Astronomical Society*, *488*(1), 954–964. doi:10.1093/mnras/stz1702

Qureshi, M. N., Nasir, W., Masood, W., Yoon, P. H., Shah, H. A., & Schwartz, S. J. (2014). Terrestrial lion roars and non-Maxwellian distribution. *Journal of Geophysical Research. Space Physics*, *119*(12), 10–059. doi:10.1002/2014JA020476

Qureshi, M. N., Shah, H. A., Murtaza, G., Schwartz, S. J., & Mahmood, F. (2004). Parallel propagating electromagnetic modes with the generalized (r, q) distribution function. *Physics of Plasmas*, *11*(8), 3819–3829. doi:10.1063/1.1688329

Qureshi, M. N., Shi, J. K., & Ma, S. Z. (2005). Landau damping in space plasmas with generalized (r,q) distribution function. *Physics of Plasmas*, *12*(122902), 122902. doi:10.1063/1.2139504

Rufai, O. R., Khazanov, G. V., & Singh, S. V. (2021). Finite amplitude electron-acoustic waves in the electron diffusion region. *Results in Physics*, *24*, 104041. doi:10.1016/j.rinp.2021.104041

Shah, K. H., Qureshi, M. N., Masood, W., & Shah, H. A. (2018). Electron acoustic nonlinear structures in planetary magnetospheres. *Physics of Plasmas*, *25*(4), 042303. doi:10.1063/1.5026186

Shohaib, M., Masood, W., Jahangir, R., Siddiq, M., Alkhateeb, S.A., & El-Tantawy, S.A. (2021). On the dynamics of nonlinear propagation and interaction of the modified KP solitons in multicomponent complex plasmas. *Journal of Ocean Engineering and Science.* doi:10.1016/j.joes.2021.10.005

Shohaib, M., Masood, W., Jahangir, R., Siddiq, M., & Rizvi, H. (2022). Interaction of dust ion acoustic solitons with cubicnonlinearity in a magnetized dusty plasma with (r, q) distributed electrons. *Contributions to Plasma Physics, e202100134*(3), e202100134. doi:10.1002/ctpp.202100134

Shriver, D., & Ashour-Abdalla, M. (2003). 1989. Geophys. Res. Lett. 16, 899. Matsumoto H, Deng XH, Kojima H, and Anderson RR, Observation of Electrostatic Solitary Waves associated withreconnection on the dayside magnetopause boundary. *GEOPHYSICAL RESEARCH LETTERS, VOL., 30*(6), 1326. doi:10.1029/2002GL016319

Singh, S. V., Devanandhan, S., Lakhina, G. S., & Bharuthram, R. (2016). Electron acoustic solitary waves in a magnetized plasma with nonthermal electrons and an electron beam. *Physics of Plasmas, 23*(8), 082310. doi:10.1063/1.4961961

Singh, S. V., & Lakhina, G. S. (2001). Generation of electron-acoustic waves in the magnetosphere. *Planetary and Space Science, 49*(1), 107–114. doi:10.1016/S0032-0633(00)00126-4

Singh, S. V., Lakhina, G. S., Bharuthram, R., & Pillay, S. R. (2011). Electrostatic solitary structures in presence of non-thermal electrons and a warm electron beam on the auroral field lines. *Physics of Plasmas, 18*(12), 122306. doi:10.1063/1.3671955

Su, C. H., & Mirie, R. M. (1980). On head-on collisions between two solitary waves. *Journal of Fluid Mechanics, 98*(3), 509. doi:10.1017/S0022112080000262

Tamang, J., & Saha, A. (2020). Bifurcations of small-amplitude supernonlinear waves of the mKdV and modified Gardner equations in a three-component electron-ion plasma. *Physics of Plasmas, 27*(1), 012105. doi:10.1063/1.5115821

Temerin, M., Cerny, K., & Lotko, W. (1982). W. and F. S. Mozer. *Physical Review Letters, 48*, 1175. doi:10.1103/PhysRevLett.48.1175

Tokar, R. L., & Gary, S. P. (1984). Electrostatic hiss and the beam driven electron acoustic instability in the dayside polar cusp. *Geophysical Research Letters, 11*(12), 1180–1183. doi:10.1029/GL011i012p01180

Tsuboi, T. (1989). Phase shift in the collision of two solitons propagating in a nonlinear transmission line. *Physical Review A: General Physics, 40*(5), 2753–2755. doi:10.1103/PhysRevA.40.2753

Tsurutani, B. T., Arballo, J. K., Lakhina, G. S., Ho, C. M., Buti, B., Pickett, J. S., & Gurnett, D. A. (1998). Plasma waves in the dayside polar cap boundary layer: Bipolar and monopolar electric pulses and whistler mode waves. *Geophysical Research Letters, 25*(22), 4117–4120. doi:10.1029/1998GL900114

Ullah, S., Masood, W., & Siddiq, M. (2020). Electron acoustic envelope solitons in non-Maxwellian plasmas. *The European Physical Journal D, 74*(2), 1–9. doi:10.1140/epjd/e2019-100589-1

Vasyliunas, V. M. (1968). A survey of low-energy electrons in the evening sector of the magnetosphere with OGO 1 and OGO 3. *Journal of Geophysical Research, 73*(9), 2839–2884. doi:10.1029/JA073i009p02839

Verheest, F., Hellberg, M. A., & Hereman, W. A. (2012). Head-on collisions of electrostatic solitons in nonthermal plasmas. *Physical Review E: Statistical, Nonlinear, and Soft Matter Physics*, *86*(3), 036402. doi:10.1103/PhysRevE.86.036402 PMID:23031029

Verheest, F., & Hereman, W. A. (2019). Collisions of acoustic solitons and their electric fields in plasmas at critical compositions. *Journal of Plasma Physics*, *85*(1), 905850106. doi:10.1017/S0022377818001368

Wang, F.-G., Yang, Y.-Y., Han, J.-F., & Duan, W.-S. (2018). Head-on collision between two solitary waves in a one-dimensional bead chain. *Chinese Physics B*, *27*(4), 044501. doi:10.1088/1674-1056/27/4/044501

Washimi, H., & Tanuiti, T. (1966). Propagation of ion acoustic solitary waves of small amplitude. *Physical Review Letters*, *17*(19), 996–998. doi:10.1103/PhysRevLett.17.996

Watanabe, K., & Taniuti, T. (1977). Electron-acoustic mode in a plasma of two-temperature electrons. *Journal of the Physical Society of Japan*, *43*(5), 1819–1820. doi:10.1143/JPSJ.43.1819

Wazwaz A M. (2010). *Partial differential equations and solitary waves theory*. Verlag, Berlin, Heidelberg: Springer.

Xue, J.-K. (2004). Head-on collision of the blood solitary waves. *Physics Letters. [Part A]*, *331*(6), 409–413. doi:10.1016/j.physleta.2004.09.029

Yousaf Khattak, M., Masood, W., Jahangir, R., & Siddiq, M. (2021). Interaction of ion acoustic solitons for Zakharov Kuznetsov equation in relativistically degenerate quantum magnetoplasmas. *Waves in Random and Complex Media*, 1–7. doi:10.1080/17455030.2021.1968536

Chapter 9
Numerical Approach to Thermal Conductivity of Electrorheological Complex (Dusty) Plasmas

Muhammad Asif Shakoori
School of Materials Science and Physics, China University of Mining and Technology, China

Misbah Khan
Department of Refrigeration and Cryogenic Engineering, School of Energy and Power Engineering, Xi'an Jiaotong University, China

Haipeng Li
https://orcid.org/0000-0002-3244-383X
School of Materials Science and Physics, China University of Mining and Technology, China

Maogang He
Key Laboratory of Thermo-Fluid Science and Engineering of Ministry of Education, Xi'an Jiaotong University, China

Aamir Shahzad
Modeling and Simulation Laboratory, Department of Physics, Government College University Faisalabad (GCUF), Faisalabad, Pakistan

Syed Ali Raza
School of Electrical Engineering, Xi'an Jiaotong University, China

ABSTRACT

Electrorheological complex (dusty) plasmas (ER-CDPs) are type of plasmas with rheological behavior under external electric field (EEF). In this chapter, the nonequilibrium molecular dynamics simulations method is used to tune thermal conductivity for two-dimensional (2D) complex plasma liquids using EEF in different directions. Anisotropic thermal conductivity is investigated in three different cases, (i) E_x(x-axis), (ii) E_y(y-axis), and (iii) E_{xy}(xy-axis) and with constant external perturbation force ($P_x = 0.02$). The thermal conductivity under the influence of EEF is different in different directions. Obtained results are compared and discussed with previous known theoretical, simulation, and experimental data for 2D systems in constant EEF. The appropriate normalized thermal conductivity with Einstein frequency at constant EEF follows universal temperature scaling law. These comparisons and discussions show that algorithms of EEF of different cases with Yukawa potential have accuracy and consistency. These comparisons validated a new numerical model that can be used for variations of EEF along with different system sizes and plasma parameters.

DOI: 10.4018/979-8-3693-0904-9.ch009

1. INTRODUCTION

In this chapter, our focus is to explore the electrorheological complex (dusty) plasmas (ER-CDPs) by investigating thermal conductivity for two-dimensional (2D) systems using molecular dynamics (MD) simulations. First, it describes the basics of plasmas, CDPs, and ER plasma physics. The existence of CDPs in nature, laboratories and industries is explained in detail. Further, reviewed the CDPs concepts in the sense of basics with a historical overview and current stage of knowledge. The importance of ER-CDPs are discussed by their possible applications in industry, laboratory and space. The general theory and mechanism of ER-CDPs, its main parameters and characteristics are discussed. Next is the introduction of computer simulation techniques used in this study, and a historical overview of molecular dynamics (MD) simulations and the model and method are explained in detail. The purpose and motivations of the presented work are to be obtained from simulation data of thermal transport properties at numerous CDPs' parameters, together with previous work that has been done to date. It provides necessary problems, justifying the current work performed with the contribution for this chapter.

1.1. Plasma

Plasma physics, a fascinating and complex branch of physics, explores the behavior and properties of plasma, often called the fourth state of matter. Unlike solids, liquids, or gases, plasmas are ionized gases where electrons and ions coexist in a high-energy state, generating a complex interplay of electromagnetic forces. This field of study explores the behavior of plasmas in different contexts, ranging from the burning temperatures of stars and astrophysical phenomena to the controlled environments of fusion reactors. Understanding plasma physics is essential for advancing fusion energy research, aiming to harness nuclear fusion, the same process that powers the sun, as a clean and virtually limitless energy source. Additionally, plasma physics plays a fundamental role in astrophysics, explaining phenomena like solar systems and the dynamics of interstellar and intergalactic mediums. Moreover, it finds applications in technologies such as plasma TVs, spacecraft propulsion systems, and cutting-edge materials processing techniques. The study of plasma physics continues pushing scientific knowledge's boundaries, offering insights into the fundamental nature of matter and energy while shaping the future of sustainable energy solutions and technological innovations (Fortov & Morfill, 2009; Shahzad, et al., 2018; Shukla, 2001).

1.2. Review Concepts of CDPs

Dust is present everywhere in space and on Earth. The CDPs are a composition of four elements: electrons, ions, neutral atoms and dust particles (grains). In nature, CDPs are ubiquitous in astrophysical situations like comets, interplanetary, interstellar clouds, or the rings of Giant planets, liquids in the white dwarf, interior of supernova cores, neutron stars, etc., (Fortov & Morfill, 2009; Shukla, 2001). The coulomb crystal of dust particles was observed in (Chu & Lin, 1994). After that, research in this direction increased drastically, the dust particles can be grown in a laboratory through a chemical reaction and sputtering.

Nowadays, research on CDPs has become an exciting field in the sciences and technologies. The most thrilling event in the area of CDPs occurred in 1980 for the planet Saturn mission. Mendis (1997) observed the bright comet in a distant ancestor, an excellent cosmic laboratory for investigating the CDPs and their dynamical and physical consequences. The other appearances of CDPs are the zodical lights, the origin nebula, the noctilucent cloud etc., at the laboratory level; CDPs are an available terrestrial laboratory in the remote past. Lyman Spitzer and Hannes Alfven recognized that dust in the universe was not merely a hindrance to a visual opinion but an essential component of the universe. Dust particle images took the shape of spokes rotating around the Saturn ring and the last surveys designated that these spokes are fine dust material. In 2005, Cassini spacecraft made a new and improved observation of spokes with a feature that would provide an enhanced consideration of their source. In 1992, the European spacecraft Ulysses flew by the planet Jupiter, detected the dust particles and measured their masses and impact speed. Again, in 1995, NASA spacecraft Galileo perceived the origin of dust streams around Jupiter. The enormous current interest in CDPs started in the mid-1980s, with a laboratory investigation of the dynamical structure factor of dusty plasmas. The field of CDPs has been overgrown since the 1990s. Dust charging, interaction forces, wave phenomena and phase transitions were comprehensively investigated (Chu & Lin, 1994; Fortov & Morfill, 2009; Merlino, 2021; Shahzad, et al., 2018; Shukla, 2001). CDPs in the laboratory showed new physics, like the formation of 2D (Chu & Lin, 1994) and 3D plasma crystals (Thomas et al., 1994) or spherical Yukawa balls (Arp et al., 2004). The high attractively of this field of investigation lies in the increased transparency of the dust clouds and the slow motion of the dust particles, which can be traced with fast video cameras. This is one of the rare occasions where plasma phenomena can be studied by simultaneously observing the many-particle system at the "atomic level". The current situation (2000-23) of dusty plasma is stable in laboratory conditions and is very significant in industries, science, technology, medical science, and energy sectors. Different characteristics of dust particles are investigated via theoretical, computer simulation and experimental techniques.

1.3. Electrorheological CDPs

Under the influence of an external electric field (EEF), the CDPs show ER characteristics that are the same as conventional ER fluids (Ivlev et al., 2008). ER-CDPs are unlike traditional CDPs plasmas, is governed by the interplay of electric force, which result from the charged particles interacting with EEF, which arise from the collisions and interactions between the suspended particles. The basic phenomena exhibited by ER-CDPs are rooted in the response of these charged particles to force originate from applied EEF. When subjected to an EEF, the particles in ER-CDPs experience forces that can cause them to self-organize into various structures such as chains, clusters, or crystals. This phenomenon is known as particle ordering or crystallization and is a fundamental aspect of ER-CDPs behavior. The strength and nature of the applied field can significantly influence the arrangement and stability of these structures. Another key phenomenon observed in ER-CDPs is their remarkable change in rheological properties under the influence of EEF. Rheology refers to the study of the flow and deformation of matter, and ER-CDPs exhibit dramatic alterations in their viscosity and flow behavior in response to EEF (Dietz et al., 2021; Ivlev et al., 2008; Joshi et al., 2023; Kana et al., 2020; Kompaneets et al., 2009).

CDPs have ER characteristics and behaviors same as conventional ER fluids, was first time observed in 2008 by scientist (Ivlev et al., 2008) on the board of international space station (ISS) through Plasmakristall-3 (PK-3) laboratory setting under microgravity conditions. Several authors examined the structural properties (phase transition) of CDPs under the influence of EEF using PK-3 laboratory (Ivlev et al., 2008) and PK-4 (Dietz et al., 2021; Ivlev et al., 2011; Joshi et al., 2023; Mitic et al., 2021) on the board of ISS, theoretical simulations (Brandt et al., 2009, 2010). Slowing of acoustic waves in EEF (Schwabe et al., 2020) and waves in 1D ER plasma (Rosenberg, 2015) also reported. Shakoori, et al., (2023) reported the MD simulation and reported tuning the structure and transport properties of ER-CDPs and diffusion under influence of uniaxial AC electric field (Shakoori, et al., 2021a; 2022b) uniaxial homogeneous (Shakoori, et al., 2022a; Shakoori, et al., 2021b) and biaxial homogeneous (Shakoori et al., 2022c). Sukhinin et al., (2017) used a Monte Carlo simulation of plasma polarization around dust particles in an externally applied electric field. They mentioned that due to induced dipole potential, the formation of dust particle alignment, chain (string) multi-layered structure, and coagulation of charged dust particles. A self-consistent model was developed for plasma anisotropy (string) of charged dust particles under the external electric field's action by (Sukhinin et al., 2017). The literature shows that no substantial studies were reported on the thermal conductivity of ER-CDPs.

1.3.1. Significance of ER-CDPs

ER-CDPs are exploited in various industrial and technological applications, such as in the development of smart fluids and adaptive materials with tunable mechanical properties. Additionally, ER-CDPs are of great interest in the study of fundamental physics and condensed matter phenomena. Researchers explore these systems to gain insights into the collective behavior of particles in complex environments, offering valuable information about phase transitions, self-organization, and emergent phenomena. The study of ER-CDPs continues to be a thriving area of research, with implications ranging from advanced materials engineering to our understanding of the fundamental principles governing matter at the microscopic level. The significance of ER-DPs reverberates across many scientific and technological domains, underscoring their crucial role in contemporary research. In materials science, they enable the development of novel smart materials with adjustable mechanical properties, revolutionizing industries like automotive engineering and robotics (Brandt et al., 2009, 2010; Joshi et al., 2023; Kana et al., 2020; Kompaneets et al., 2009; Rosenberg, 2015; Schwabe et al., 2020; Shakoori, et al., 2023).

1.4. Molecular Dynamics Simulations

In thermal fluids science and engineering, the experimental setups are very complex and available at a high cost, and some experiments require more time for calculations for specific investigations. In the modern age, cost-effectiveness, low time consumption, and computational power are the most prominent. The art of MD simulations became a standard computational tool for investigating and developing new models in the field of plasma physics, polymer, materials, biological and chemical sciences, and other physical systems. This scheme became a standard computation tool to investigate various physical systems' phase transition, structure, and dynamical properties. The research presented in this chapter is based on the MD technique, which is a powerful computational tool to study the thermophysical properties of physical conditions in CDPs. As computational power continues to advance, molecular dynamics simulations have become indispensable tools, driving innovation in fields ranging from drug discovery to materials science, revolutionizing our understanding of the fundamental principles governing the microscopic world (Rapaport, 2004).

1.5. Thermal Transport

Thermal transport in CDPs, a captivating area of study within the realm of plasma physics, presents a complex interplay of charged microparticles and neutral gas molecules. Understanding the mechanisms underlying thermal conductivity in these unique systems is pivotal, as it not only illuminates fundamental plasma physics but also holds implications for diverse applications. The charged particles, immersed in a neutral gas background, interact through a combination of Coulombic forces and collisions, giving rise to intricate energy transfer processes. Thermal energy is transmitted through both collective particle motion and individual particle collisions, creating a dynamic environment where heat conduction occurs alongside the complex interparticle interactions. Studying thermal transport in dusty plasmas is crucial for elucidating phenomena in astrophysical contexts, such as the formation of planetary rings and interstellar clouds, where similar charged particle interactions occur. Additionally, this knowledge finds practical applications in various technological domains, including the design of advanced spacecraft shielding and plasma-based propulsion systems. Researchers employ sophisticated numerical simulations and experimental techniques to unravel the complexities of thermal transport in dusty plasmas, bridging the gap between theoretical understanding and practical applications and paving the way for innovative developments in both fundamental science and applied technologies. The CDP's thermal conductivity (λ_0) is an intrinsic, influential, and sensitive property and is also expensive from a computational point of view (Shahzad et al., 2012b; 2014; 2015; 2018; 2020).

Understanding the thermal transport mechanisms in electrorheological fluids is crucial for optimizing their performance in various applications. Heat conduction in these fluids occurs through complex interactions between the suspended particles and the surrounding liquid molecules. The electric fields not only influence the arrangement and movement of particles but also modulate their thermal conductance, leading to intriguing phenomena like field-induced phase transitions and altered thermal conductivity. Researchers delve into detailed experimental studies and computational simulations to unravel the nuances of heat transfer in electrorheological fluids. This knowledge is invaluable in the development of advanced damping systems, adaptive shock absorbers, and precise positioning devices, where these fluids play a pivotal role. Moreover, understanding the thermal behavior of electrorheological fluids contributes to advancements in energy-efficient technologies and enhances our grasp of soft matter physics, opening doors to innovative solutions in diverse engineering applications. It determines the amount of energy transport of the CDPs and their behavior with applied external fields (electric, magnetic, thermal). The heat flow technologies and designing processes require detailed knowledge of the industrial thermal conductivity of numerous systems of interest. Consequently, the appropriate development, widely applicable, and predictive technique is persistent for calculating thermal conductivity. Thermophysical properties have numerous applications in engineering and science applications. The most practical characteristics of thermal science and engineering have moved into the group of engineering problems (Shahzad et al., 2012b; 2014; 2015; 2018; 2020). Understanding heat transport at the microscale and nanoscale has been recognized as the most important and practical.

1.5.1. Literature Review of Thermal Conductivity

The thermal conductivity of liquids/materials can be calculated using the Green-Kubo (GK) relation with equilibrium MD (EMD) and nonequilibrium MD (NEMD) simulation techniques. The calculations of transport properties through GK-EMD are generally much more expensive in terms of computational time. Furthermore, because of the comparatively large force needed to drive the mass or heat flow with a small system size, inhomogeneous systems may not exist at a precisely defined temperature or density, which is necessary if any comparison with the experiment is to be made. NEMD schemes are incredibly effective and well-developed statistical mechanics tools. The application of NEMD techniques for homogeneous molecular fluids like one component, Coulomb plasma has rapidly increased, which is the driving force behind this under the influence of external flow field for the thermal conductivity. The homogeneous

NEMD techniques are also used for 2D (Shahzad et al., 2020) and three-dimensional (3D) (Shahzad & He, 2012b) CDPs. The first time adapting the Evans (1982) algorithm for determining the thermal conductivity of the Yukawa dusty plasma liquids is what inspired the current effort (Yukawa liquids).

MD has been widely used to investigate thermal conductivity in the field of CDPs for a wide range of plasma parameters and system sizes. The researchers (Shahzad & He, 2012b) reported deep investigations in the last decades for 3D and 2D using NEMD simulation techniques for numerous prospectus (Shahzad et al., 2012b; 2014; 2015; 2018; 2020). Donkó & Hartmann (2004) employed MD simulation and computed the thermal conductivity of 2D CDP (Donkó & Hartmann, 2004), Khrapak implement the vibration model to compute the thermal conductivity of 2D CDPs (Khrapak, 2018). Currently, it is valid to compute the thermal conductivity of CDPs using MD simulation.

1.6. Scope and Objective of the Chapter

The MD simulation methods have widely been used in the field of CDPs to compute the numerous physical properties of different physical systems. Till 2021, the MD simulation was widely used to calculate the thermophysical, structural and waves, and instability only for Yukawa potential. Develop an homogeneous NEMD simulation technique to tune the thermal conductivity of 2D strongly coupled (SC)-CDPs for the different varieties of anisotropic interparticle interactions. Three different varieties of EEF are used to understand the behavior of anisotropic thermal conductivity of 2D Yukawa liquids with variations of the electric field. The presented developed numerical model can be used to provide the nature of thermal conductivity in gas-like, liquid-like and solid-like regimes of dusty plasmas with variations of the electric field of numerous intensities in different directions.

2. MODEL AND METHOD FOR MD SIMULATION

2.1. Numerical Model for Pair Interaction

The pair interactions between the interacting charged dust particles in plasma are computed through Yukawa potential (Debye-Hückel), and this interaction is known as pure isotropic and generally repulsive. Numerous physical phenomena extensively investigated Yukawa potential, which is a good approximation for screened charge potential and makes it possible to describe many experimental results. Such isotropic potential distribution is helpful for qualitative analytical and numerical estimations of dynamical, structural properties of CDPs and self-organization of screened dust particles (Chu & Lin, 1994; Sukhinin et al., 2017). However, interparticle interactions in 3D CDPs are calculated to investigate and understand the number of fundamental generic processes, such as phase transition (vapor-liquid-solid) and the related critical phenomena (Kompaneets et al., 2009).

This potential is used in an experimental study of plasmas under microgravity and space conditions. This interactions model is also used for physical systems such as colloidal suspension, astrophysics, aerodynamics, hydrodynamics, materials for energy generations, polymer physics, biomedicine, biological and chemical systems, etc. It has benefits for the reduced computational cost compared to another potential (many-body) (Shahzad & He, 2012b). In the absence of an external field, the interaction between the particle are pure isotropic and computed through Yukawa potential

$$\Phi(r) = \frac{Q_d^2}{4\pi\epsilon_0}\frac{e^{-r/\lambda_D}}{r} \quad (1)$$

where Q_d is the charge on dust particles, r is interparticle distance and λ_D is the Debye length of typical background plasmas, ϵ_0 is the permittivity of free space.

The total force that acts on dust particles is the sum of ion, electron drag force, and electric force, which induced the drift velocity $\mathbf{u} = \mathbf{E}/mv$, where m represents the mass v is collision frequency with neutral. The drift velocity of the particle increases with increasing the strength of the EEF ($\mathbf{E} \propto \mathbf{u}_d$) (Kompaneets et al., 2009; Shakoori, et al., 2023). The homogenous EEF for the case of uniaxial (x and y-axis) (Shakoori, et al., 2022a) can be written as

$$\mathbf{E}_x = (\mathbf{E}_x, 0, 0) = eE(\mathbf{r}_x\cos\theta) \text{ and } \mathbf{E}_y = (0, \mathbf{E}_y, 0) = eE(\mathbf{r}_y\cos\theta_1) \quad (2)$$

and for the case of biaxial (x-y plane) EEF can be written as

$$\mathbf{E}_{xy} = \left(\mathbf{E}_x, \mathbf{E}_y, 0\right) = eE(\mathbf{r}_x\cos\theta_1 + \mathbf{r}_y\cos\theta_2) \quad (3)$$

where E represents the strength of an EEF(Shakoori et al., 2022c). It is considered that there is no electric force significantly affected along the z-axis. The θ, θ_1 and θ_2 are the angles that correspond to the direction of the EEF; thus, the interaction between dust particle in CDPs are calculated for the case of uniaxial EEF as (Shakoori, et al., 2022a)

$$\Phi(\mathbf{r},\theta) = \frac{Q_d^2}{4\pi\epsilon_0}\frac{e^{-r/\lambda_D}}{r} - eE\left(\mathbf{r}_x\cos\theta\right) \quad (4)$$

and for the case of biaxial EEF(Shakoori et al., 2022c)

$$\Phi(\mathbf{r},\theta) = \frac{Q_d^2}{4\pi\epsilon_0}\frac{e^{-r/\lambda_D}}{r} - eE\left(\mathbf{r}_x\cos\theta_1 + \mathbf{r}_y\cos\theta_2\right) \quad (5)$$

The first term represents the isotropic interaction of dust particles, and the second term is the anisotropic interaction in the presence of an EEF. The EEF is normalized by Debye length for the uniaxial EEF as $E^* = e\lambda_D E_{xy} / T_d$ and for biaxial $E_0 = e\lambda_D E_{xy} / T_d$ (Shakoori, 2021a; 2022a; 2022c; Sukhinin et al., 2017). The 3D trajectories of interacting charged dust particles with charge Q_d and mass m under the influence of uniaxial homogeneous EEF can be written as

$$m\mathbf{a} = \mathbf{F}_y + Q_d E^*(\mathbf{m}_x\cos\theta) \quad (6)$$

and for biaxial homogeneous EEF as:

$$m\mathbf{a} = \mathbf{F}_y + Q_d\bar{E}(\mathbf{m}_x \cos\theta + \mathbf{m}_y \cos\theta) \tag{7}$$

where $\mathbf{F}_y$ is Yukawa interaction force $\mathbf{F}_y = -\partial\phi(\mathrm{r})/\partial\mathrm{r}$ and acceleration on the ith particle ($a_i = F_i/m_i$) (Shahzad & He, 2012c). The m_x, m_y and m_{xy} are the unit vectors along the corresponding axes of the cubic simulation box. Three dimensionless parameters (Coulomb coupling, Debye screening, electric field) are used to characterize the CDP systems. (1) Coulomb coupling parameter, $\Gamma = Q_d^2 / \left(3\pi\int_0 a_{ws} k_B T_d\right)$ here a_{ws}, k_B and T_d are the average Wigner–Seitz radius (inter-particle distance), Boltzmann's constant, and absolute systems temperature. (2) Debye screening strength, $\kappa = a_{ws}/\lambda_D$, for 2D case the Wigner–Seitz given as $a_{ws} = (1/\pi n)^{1/2}$, n is the equilibrium dust density ($nd = N/V$). The plasma frequency for the 2D system is $\omega_{pd} = \left(Q_d^2 / 2\pi\epsilon_0 m a_{ws}^3\right)^{1/2}$, m is accounted as a mass of dust particles (Shahzad & He, 2012a). The intensity of normalized external homogeneous electric field (E^*) identifies the nature of anisotropy.

2.2. Numerical Model for Thermal Transport

The well-known method used to calculate the transport coefficients of fluids is the GK relation. This method has widely been used to calculate the hydrodynamic transport coefficients for CDPs (Shahzad et al., 2020; Shakoori et al., 2022c; 2023). The coefficients of thermal conductivity were calculated by employing the GK relations with the standard MD simulation technique in 2D (Shahzad et al., 2020) and 3D (Shahzad & He, 2012b) CDP liquid.

$$\lambda = \frac{1}{2k_B A T^2} \int_0^\infty \langle \mathbf{J}_Q(t) \cdot \mathbf{J}_Q(t) \rangle dt \tag{8}$$

Where, A is the area, T is the system temperature, and angular brackets calculate the average value of all charged dust particles in the homogeneous NEMD simulation run. In Equation (8), $\mathbf{J}_Q$ is the microscopic heat flux vector for a 2D system measured as:

$$\mathbf{J}_Q A = \sum_{i=1}^{N} \left[E_i - \frac{1}{2} \sum_{i \neq j} \mathbf{r}_{ij} \cdot \mathbf{F}_{ij} \right] \frac{\mathbf{p}_i}{m} \tag{9}$$

Where, the $\mathbf{r}_{ij} = \mathbf{r}_i - \mathbf{r}_j$ is an interparticle distance (position vector), $\mathbf{F}_{ij}$ is the force of interaction on particle i due to particle j, $\mathbf{p}_i$ is the momentum vector of the particle i^{th} and E_i is the total energy of particle i (for $i \neq j$) can be written as:

$$E_i = \frac{\mathbf{p}_i^2}{2m} + \frac{1}{2} \sum_{i \neq j} \phi_{ij} |\mathbf{r}| \tag{10}$$

In the above Equation, the first term represents the kinetic energy and the second term is the potential energy, where $\phi_{ij}|\mathrm{r}|$ is the Yukawa potential given in Equation (1). The linear response theory or non-

Hamiltonian dynamics of the system perturbed the equations of motion proposed by Evans Gillan and this perturbation equation of motion can be written as:

$$\dot{\mathbf{r}} = \frac{\mathbf{p}_i}{m} \tag{11}$$

$$\dot{\mathbf{p}} = \sum_{j=1}^{N} \mathbf{F}_i + \mathbf{D}_i\left(\mathbf{r}_i, \mathbf{p}_i\right)\cdot \mathbf{F}_e\left(t\right) - \alpha p_i \tag{12}$$

In the above Equation (12), $\mathbf{F}_i$ represents the total interparticle force on particle i, written as $\sum_{j=1}^{N}\mathbf{F}_i = -\frac{\partial \phi_{ij}\left|\mathbf{r}\right|}{\partial r_i}$, $\mathbf{D}_i\{(\mathbf{r}_i, \mathbf{p}_i), i = 1, 2, 3, \ldots N\}$ tensor phase space variable function, where $\mathbf{r}_i$ and $\mathbf{p}_i$ are the position and momentum, respectively, vector of the ith particles in N-identical particles systems. The tensor phase space function describes the coupling of the system to the external perturbation force $\mathbf{F}_e(t)$, α is the Gaussian thermostat multiplier to keep the constant system temperature. Thermal conductivity is estimated by selecting an external perturbation force applied along the x-axis $\mathbf{F}_e(t) = \delta(0, \mathbf{P}_x)$, here, δ and $\mathbf{P}_x$ are the Dirac delta function and the external perturbation force strength, respectively.

$$\lambda = \frac{1}{AT\mathbf{P}_x}\int_0^{\infty}\left\langle \mathbf{J}_{Qx}\left(t\right)\cdot \mathbf{J}_{Qx}(0)\right\rangle dt \tag{13}$$

Where J_{Qx}, T and $\mathbf{P}_x$ represent the heat flux vector along the x-axis, temperature of the system and external perturbation force, respectively. Equation (13) is the fundamental formula for computations of the autocorrelations function of the heat energy current using external perturbation force. The coefficients of thermal conductivity (COTC) (λ) is obtained by measuring the thermal response of the pair Yukawa system under an appropriate weak external perturbation force. It can be stated here that the perturbation is the Dirac delta function; therefore, the heat flux vector response is proportional to the autocorrelation function rather than the time integral of the function (Shahzad et al., 2012b; 2018; 2020).

2.3. MD Simulation Executions and Parameters

The simulations start from face-centered cubic (FCC) lattice configurations of Yukawa identical particles having randomly initial velocities sampled from the Maxwellian distribution function concerning the required initial temperature value. The $N = 1024$ particles are used for the whole calculation at different κ and Γ. The 2D periodic boundary conditions for each direction are imposed on the simulation box to simulate the infinite system. The dimension of a cubic simulation box with edge length is L_x/L_y (Shahzad et al., 2012b; 2018; 2020). The standard Ewald summation method calculates the long-range interaction between the dust particles. In the Ewald summation technique, the interactional potential is divided into two parts: real space and reciprocal space. Both are rapidly converging. The Ewald convergence parameter value ($\alpha = 5.6/La_{ws}$) is used. With the higher κ values, the real space term alone shows enough accuracy and performance.

The interaction and forces are computed into two parts: the repulsive interaction through typical Yukawa potential given in Equation (1) and the additional interaction and forces induced due to EEF computed as explained in section 2.1 (Kompaneets et al., 2009; Shakoori et al., 2022a; Sukhinin et al., 2017). The Equation of motion is integrated using the Predictor corrector integration method. The isotropic interactions are considered in the absence of an electric field ($E_i \approx 0.0$) and anisotropic interaction accounted for the case $E_i > 0.0$. When an electric field is switched on, many physical phenomena occur; these novel regimes are not limited to fluid-solid transition but also exhibit chain, ring, sheet, vortex, etc., depending on the nature of the electric field (Shakoori, et al., 2022a; Sukhinin et al., 2017).

Appropriate input parameters are chosen for the numerical scheme's consistency and accuracy. These input parameters, such as system temperature ($T = 1/\Gamma$), screening strength (κ), external electric field (E_i), number of particles (N), simulation time step and total run time, are selected to provide Yukawa system and thermal conductivity. The external normalized thermal force or NEMD simulation method $\mathbf{F}_e{}^*{=}F_x a_{ws}/\mathbf{J}_{Qx}$. The $\mathbf{J}_{Qx}$ produced the thermal heat energy along the x-axis. MD simulations are executed 2 x $10^5/\omega_p$ time unit to record the simulation outcomes. In this paper, the NEMD method is used for Yukawa and ER-CDPs systems to estimate thermal conductivity over a wide range of plasma coupling (Γ = 10-300), screening (κ =1), and the number of particles (N = 1024).

3. MD SIMULATION RESULTS AND DISCUSSIONS

First, we primarily investigated the thermal conductivity for 2D SC-CDPs through a newly developed NEMD simulation code. The anisotropic pair potential is used to investigate the effects of normalized electric field strength (E_i) on thermal conductivity, which is directed along x, y and xy directions. The effects of input parameters on thermal conductivity are analyzed and discussed. These input parameters are Coulomb coupling (Γ) and Debye screening strength (κ). The thermal conductivity is normalized by plasma frequency (ω_p) in the absence and presence of applied normalized EEF as $\lambda = \lambda_0/\omega_p$. This normalization method of thermal conductivity for CDPs was extensively used in previous studies for 3D and 2D systems.

The currently obtained results are compared and discussed with previously known theoretical, numerical and experimental data, which are nearly the same values of plasma parameters (Γ, κ) of CDPs in nonideal gas-like, liquid-like, crystal/solid-like phases. The thermal conductivity is presented here for specific values of plasma coupling ($10 \leq \Gamma \leq 300$) in the absence of an electric field, the presence of constant and varying $0 < E_i \leq 2.0$ strength. To validate the MD simulation code, we first reproduced the thermal conductivity results using the NEMD simulations for pure isotropic potential (Yukawa), see Figure 1. These comparisons show that the current NEMD scheme provides satisfactory estimations of COTC over a specific range of plasma states (Γ, κ) than those earlier employed different numerical schemes for the SC-CDPs. For the second step, the same simulation is performed under the influence of constant EEF, which is directed along the x and y axes ($E_{x,y}$ = 0.01, 0.01). In the third step, EEF is applied on the xy-axis simultaneously for the strength of E_{xy} = 0.007. The simulation outcomes at constant EEF are also compared with our previously reported results. This comparison demonstrates that the current simulation outcomes at constant EEF are about the same as an absence of EEF. These results validated the new MD simulation code for anisotropic pair potential, which can be implemented further to compute the effects of EEF.

It is generally experienced in the present NEMD scheme that various restrictions can affect results. These restrictions include the screening strength (κ), the system temperature, electric field strengths (E_i), system frequency (ω), simulation time step (dt) and total simulation length are changed to determine how these restrictions may influence the thermal conductivity of SC-CDPs. Notably, results have been computed for the COTC of 2D SC-CDPs by employing NEMD simulations at nearly the same sets of plasma points (Γ, κ). The appropriate system size $N = 1024$ was used and compared with earlier simulation methods to demonstrate that the present simulation results of thermal conductivity are well identified with the previously available simulation results.

The appropriate normalized external electric field strength gives equilibrium values of COTC for the uniaxial and biaxial electric field ($E^*_x = E^*_y = 0.01$ and $E_{0xy} = 0.007$), which are suitable for a whole range of plasma states $\Gamma \equiv (10, 300)$ and $\kappa = 1$. It should be mentioned here that, in this chapter, we used the normalized electric field as $E_x = E^*_x$, $E_y = E^*_y$ and $E_{xy} = E_{0xy}$. The obtained data of COTC in 2D CDPs in $E_i = 0$ and for equilibrium strength of E_i have well matched with previous theoretical, simulation and experimental data of 2D and 3D CDPs. The wide spectrum of electric field strengths is used for the calculations of the steady state values of COTC $\lambda(E_i, \Gamma)$ [equilibrium values of $\lambda(E_i, \Gamma)$], where the signal-to-noise ratio is appropriate for mentioned plasma state points (Γ, κ). In the present case, it is exercised that the varying electric field strength $E_i < 0.01$ gives similar data as $E_i = 0$ and $E_i > 2.0$ provides higher values of COTC. It should be mentioned here that the ranges of applied external electric field $0 \leq E_i \leq 2.0$ are more suitable for understanding thermal conductivity behavior in 2D CDPs under the influence of various EEF.

Here, we have used and already established a relation for the interaction potential account for the NEMD scheme. It should be mentioned here that thermal conductivity calculations with the GK formula have been widely used in previous studies of 2D CDPs for SC regimes. The validations of the current numerical scheme are explained in Figure 1, where we calculated the plasma $\lambda_0(\Gamma, \kappa)$ using the NEMD method for a 2D CDPs system of different values of plasma Γ ($\equiv$10, 20, 50, 100, 200, 300), keeping the constant $\kappa = 1$ and $N = 1024$, $E_i = 0.01, 0.007$ and $\mathbf{F}_x = 0.02$. Figure 1 displayed earlier 2D numerical data computed from the NEMD of (Shahzad et al., 2015; 2020) and plasma $\lambda_0(\Gamma, \kappa = 1, E_i = 0.01, 0.007)$. Our simulations were performed twenty-four to get a suitable data set. The reported simulation data are generally in good agreement for the used range of plasma parameters (κ, Γ), covering from gas-like ($\Gamma = 10$) to solid-like ($\Gamma = 300$). Figure 1 shows that the value of the externally applied electric field, which is $E_i = 0.01$, is a near equilibrium strength. At $E_i = 0.01, 0.007$, the obtained plasma λ_0 is well matched with our recently published work (Shahzad et al., 2020) for $N = 1024$. Further details of comparison and discussion with earlier simulations and experimental and theoretical estimations are found in our work (Shahzad et al., 2020).

We also have calculated the plasma $\lambda_0(E_i)$ for the universal temperature scaling law at normalized $E_x = 0, 0.01$, $P^* = 0.02$ and constant $N = 1024$. The temperature scaling expression shows normalized plasma thermal conductivity by Einstein frequency ($\lambda^* = \lambda/nm\omega_E a_{ws}^2$) as a function of normalized scaled temperature defined as $T^* = \Gamma_m/\Gamma$, where Γ_m is a liquid-solid phase critical point and complete details associated with temperature scaling law in earlier literature (Shahzad et al., 2012b; 2012c; 2018; 2020) and given as

$$\lambda^* = \mathrm{K}_1 T^* + \frac{\mathrm{K}_2}{T^*} + \mathrm{K}_3 \tag{14}$$

Where, K_1, K_2 and K_3 are fitting parameters that can be obtained using Equation (14). The calculated data for plasma $\lambda^*(E_i)$, for $\kappa = 1$. Here, a quasi-universal temperature scaling law is also checked for the 2D CDPs plasmas $\lambda^*(E_x = 0, 0.01, T^*, \kappa = 1)$, proposed by Donkó & Hartmann, (2004). The quasi-universal and universal temperature scaling law explains that the plasmas $\lambda^*(E_x = 0, 0.01, T^*)$ is taken along the y-axis and scaled T^* along the x-axis, which are revealed in Figure 2. In order to check the accuracy and reliability of plasma $\lambda^*(E_x = 0, 0.01, T^*)$, we performed simulations at constant strength of EEF along different x-axes ($E_x = 0.01$) with constant $P^* = 0.02$.

Figure 1. Normalized thermal conductivity $\lambda_0(\Gamma, \kappa, E^*_i = 0.01, 0.007)$ versus plasma coulomb coupling Γ of 2D CDPs under the influence of constant external electric field $E_i = 0.01$, $i = x$, y and $E_{xy} = 0.007$ for N = 1024 of one value of Debye screening. Present simulation data are compared with earlier NEMD results of Shahzad et al., (2015; 2020).

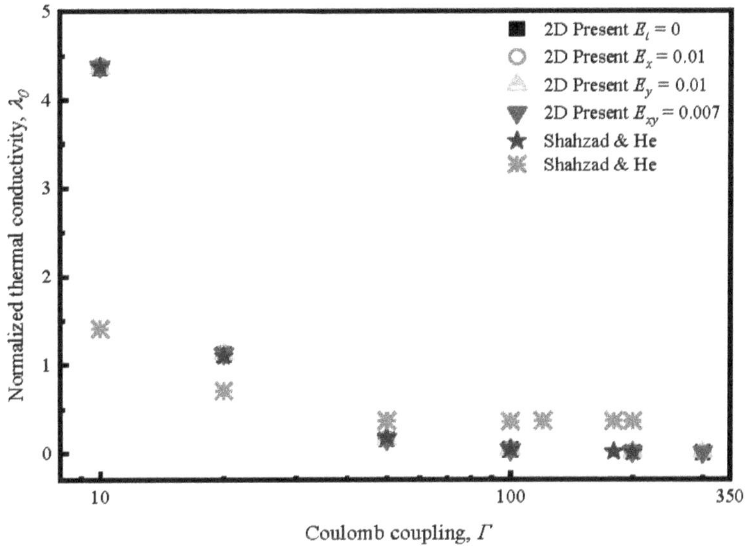

The scaled thermal conductivity is under the influence of a constant electric field for different varieties of anisotropies, thus, here presented for $\lambda^*(E_x=0.01)$ and $\lambda^*(E_y = 0.01)$. The solid red lines are the curve fitting of the functional form of Equation (14) with fitting parameters at $E_x = 0.01$ are $K_1 = 0.00889$, $K_2 = 2.2107$ and $K_3 = 0.02599$ and $E_y = 0.01$ are $K_1 = 0.00904$, $K_2 = 2.2067$ and $K_3 = 0.02647$. This curve fitting is confirmed from earlier numerical trends (Shahzad et al., 2012b; 2014; 2015; 2020) in the universal temperature functional form of Equation (14). It is observed that the solid red line is an excellent fit, as shown in Figure 2. It is to be noted here that fitting parameters (K_1, K_2, K_3) for scaled $\lambda^*(E_x)$ and $\lambda^*(E_y)$ are computed from the current NEMD simulation outcomes (λ_0) demonstrated in Figure 1. It is demonstrated here that the calculated scaled $\lambda^*(E_x = 0.01)$ and $\lambda^*(E_y = 0.01)$ explain the universal functional forms of Equation (14) and provide good fits, confirming earlier 3D and 2D DPs results.

Figure 3 shows the effects of EEF on thermal conductivity in 2D Yukawa liquids. The EEF is applied differently on the simulation box to check the thermal conductivity behavior. In the first step, E^* is applied along the x and y-axis separately as a case of uniaxial. In the second step, a biaxial (xy-axis) electric field with similar strength is used. In the third step, the effects of the electric field on COTC in different phases are also analyzed. The computed thermal conductivity is classified with respective directions of electric fields as $\lambda_0(E_x)$, $\lambda_0(E_y)$ and $\lambda_0(E_{xy})$ to differentiate the effect of different field directions.

Figure 2. thermal conductivity normalized by Einstein frequency as a function of normalized temperature for constant (a) $E_x = 0.01$ and (b) $E_y = 0.01$, $N = 1024$ at $\kappa = 1$. The solid red line was obtained from the universal temperature scaling law (Equation (14)) for 2D CDPs. The curve fitting of 2D CDPs is well matched with the NEMD simulation data displayed in Figure 1.

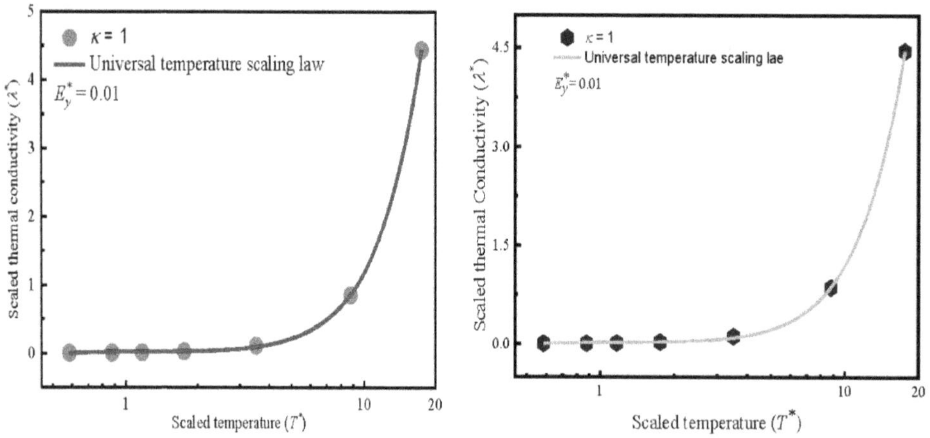

Figure 3. Variations of normalized thermal conductivity as a function of normalized EEF (E_x, E_y and E_{xy}) for $\kappa = 1$.

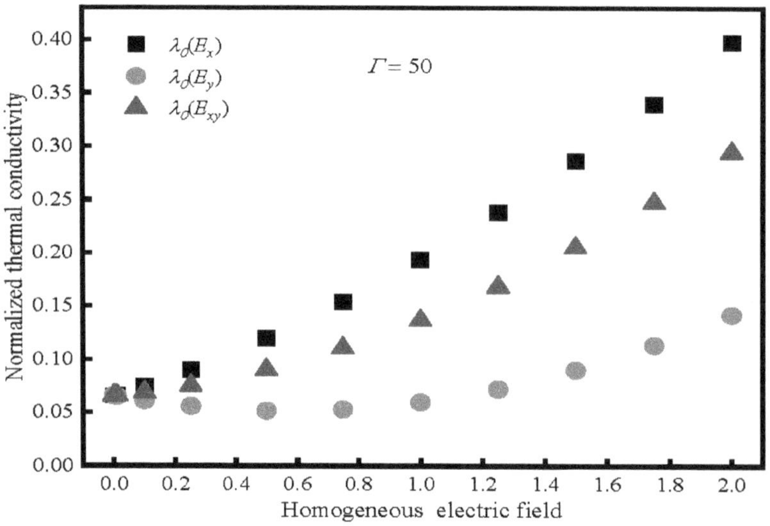

Figure 3 is a testament to the intricacies of NEMD simulations, specifically focusing on the effects of EEF on $\lambda_0(E_i)$ at $\kappa = 1$ and $\Gamma = 50$. This exploration delved into the subtle fluctuations of λ_0 concerning the incremental augmentation of E^* while keeping the system size constant at $N = 1024$. Each scenario of Γ was meticulously studied, encompassing a vast array of thirty-three simulation datasets. Rigorous validation of the employed methodology was paramount, prompting a meticulous comparison of thermal conductivity outcomes at EEF values ($E_i = 0, 0.01$) outlined in Figure 1. The range of electric field strengths tested spanned from $E_i = 0.01$ and 0.007 to 2.0, directed along the x, y, and xy-axes. Remarkably, the 2D CDPs exhibited a noteworthy surge in thermal conductivity as the intensity of applied EEF increased.

Interestingly, the direction of the electric field played a pivotal role, with the x-axis exerting a significantly more prominent influence than the y and xy directions. Intriguingly, the computed values of $\lambda_0(E_{xy})$ consistently remained lower than $\lambda_0(E_x)$ yet significantly higher than $\lambda_0(E_y)$. An interesting trend emerged concerning $\lambda_0(E_y)$; it exhibited an initial decrease with the increase of E_i. However, thermal conductivity experienced a distinctive rise beyond specific electric field thresholds. These critical points at which $\lambda_0(E_y)$ decreased and subsequently increased were intricately tied to the plasma Γ. For instance, at $\Gamma = 50$, the lower $\lambda_0(E_y)$ values were recorded at $E_i = 0.8$. Additionally, the behavior of $\lambda_0(E_x)$ stood out prominently in the analysis. With intensifying electric field strengths, $\lambda_0(E_x)$ displayed a sharp increase.

Conversely, $\lambda_0(E_y)$ exhibited an initial decline, only to rise again after reaching a specific E_y threshold, the exact value contingent on the plasma coupling (Γ). This intricate fluctuation of electric field strengths and their corresponding effects on thermal conductivities provided a profound insight into this system's complex dynamics. The in-depth examination presented in Figure 3 and its associated findings shed light on the intricate relationship between electric fields and thermal conductivities in 2D CDPs and underscored the significant influence of plasma coupling constants in shaping these interactions. This study contributes substantially to the fundamental understanding of CDP systems. It paves the way for future research endeavors in related fields, offering a rich foundation for further exploration and discoveries.

CONCLUSION

A new algorithm of the nonequilibrium MD simulations method is used to tune the thermal conductivity for 2D complex plasma liquids using the EEF in different directions. The thermal conductivity is computed for the constant electric fields ($E_i = 0, 0.01, 0.007$), perturbation force ($\mathbf{P}_x = 0.02$), and a number of particles ($N = 1024$) at $\kappa = 1$ and $\Gamma = 50$. The obtained results are compared and discussed with previous known theoretical, simulation and experimental data for 2D systems. The appropriate normalized thermal conductivity with Einstein frequency at a constant electric field follows exactly the universal temperature scaling law. These comparisons and discussions show that the algorithms of an electric field of different cases with Yukawa potential have accuracy and consistency. These comparisons validated a new numerical model that can be used for electric field variations. It is concluded that the proposed numerical model to tune the thermal conductivity from homogeneous NEMD yields consistent results, and this model is entirely accurate. This model has advantages in studying the behavior of thermal transport in electric field for 2D CDPs liquids and other physical systems.

REFERENCES

Arp, O., Block, D., Piel, A., & Melzer, A. (2004). Dust Coulomb balls: Three-dimensional plasma crystals. *Physical Review Letters*, *93*(16), 165004. doi:10.1103/PhysRevLett.93.165004 PMID:15524998

Brandt, P. C., Ivlev, A. V., & Morfill, G. E. (2009). Solid phases in electro- and magnetorheological systems. *The Journal of Chemical Physics*, *130*(20), 204513. doi:10.1063/1.3143682 PMID:19485463

Brandt, P. C., Ivlev, A. V., & Morfill, G. E. (2010). String-fluid transition in systems with aligned anisotropic interactions. *The Journal of Chemical Physics*, *132*(23), 234709. doi:10.1063/1.3442436 PMID:20572736

Chu, J. H., & Lin, I. (1994). Direct observation of Coulomb crystals and liquids in strongly coupled rf dusty plasmas. *Physical Review Letters*, *72*(25), 4009–4012. doi:10.1103/PhysRevLett.72.4009 PMID:10056356

Dietz, C., Budak, J., Kamprich, T., Kretschmer, M., & Thoma, M. H. (2021). Phase transition in electrorheological plasmas. *Contributions to Plasma Physics*, *61*(10), 1–15. doi:10.1002/ctpp.202100079

Donkó, Z., & Hartmann, P. (2004). Thermal conductivity of strongly coupled Yukawa liquids. *Physical Review. E*, *69*(1), 16405. doi:10.1103/PhysRevE.69.016405 PMID:14995723

Evans, D. J. (1982). Homogeneous NEMD algorithm for thermal conductivity—Application of non-canonical linear response theory. *Physics Letters. [Part A]*, *91*(9), 457–460. doi:10.1016/0375-9601(82)90748-4

Fortov, V. E., & Morfill, G. E. (2009). Complex and Dusty Plasmas. In *Complex and Dusty Plasmas*. CRC Press. doi:10.1201/9781420083125-c1

Ivlev, A. V., Morfill, G. E., Thomas, H. M., Räth, C., Joyce, G., Huber, P., Kompaneets, R., Fortov, V. E., Lipaev, A. M., Molotkov, V. I., Reiter, T., Turin, M., & Vinogradov, P. (2008). First observation of electrorheological plasmas. *Physical Review Letters*, *100*(9), 095003. doi:10.1103/PhysRevLett.100.095003 PMID:18352717

Ivlev, A. V., Thoma, M. H., Räth, C., Joyce, G., & Morfill, G. E. (2011). Complex plasmas in external fields: The role of non-hamiltonian interactions. *Physical Review Letters*, *106*(15), 155001. doi:10.1103/PhysRevLett.106.155001 PMID:21568565

Joshi, E., Pustylnik, M. Y., Thoma, M. H., Thomas, H. M., & Schwabe, M. (2023). Recrystallization in string-fluid complex plasmas. *Physical Review Research*, *5*(1), L012030. doi:10.1103/PhysRevResearch.5.L012030

Kana, D., Dietz, C., & Thoma, M. H. (2020). Simulation of electrorheological plasmas with superthermal ion drift. *Physics of Plasmas*, *27*(10), 103703. doi:10.1063/5.0010021

Khrapak, S. (2018). Practical formula for the shear viscosity of Yukawa fluids. *AIP Advances*, *8*(10), 105226. doi:10.1063/1.5044703

Kompaneets, R., Morfill, G. E., & Ivlev, A. V. (2009). Design of new binary interaction classes in complex plasmas. *Physics of Plasmas*, *16*(4), 043705. doi:10.1063/1.3112703

Merlino, R. (2021). Dusty plasmas: From Saturn's rings to semiconductor processing devices. *Advances in Physics: X*, *6*(1), 1873859. doi:10.1080/23746149.2021.1873859

Mitic, S., Pustylnik, M. Y., Erdle, D., Lipaev, A. M., Usachev, A. D., Zobnin, A. V., Thoma, M. H., Thomas, H. M., Petrov, O. F., Fortov, V. E., & Kononenko, O. (2021). Long-term evolution of the three-dimensional structure of string-fluid complex plasmas in the PK-4 experiment. *Physical Review. E*, *103*(6), 063212. doi:10.1103/PhysRevE.103.063212 PMID:34271636

Rapaport, D. C. (2004). *The art of molecular dynamics simulation*. Cambridge University Press. doi:10.1017/CBO9780511816581

Rosenberg, M. (2015). Waves in a 1D electrorheological dusty plasma lattice. *Journal of Plasma Physics*, *81*(4), 905810407. doi:10.1017/S0022377815000422

Schwabe, M., Khrapak, S. A., Zhdanov, S. K., Pustylnik, M. Y., Rath, C., Fink, M., Kretschmer, M., Lipaev, A. M., Molotkov, V. I., Schmitz, A. S., Thoma, M. H., Usachev, A. D., Zobnin, A. V., Padalka, G. I., Fortov, V. E., Petrov, O. F., & Thomas, H. M. (2020). Slowing of acoustic waves in electrorheological and string-fluid complex plasmas. *New Journal of Physics*, *22*(8), 083079. doi:10.1088/1367-2630/aba91b

Shahzad, A., Haider, S. I., Kashif, M., Shifa, M. S., Mu, T., & He, M. G. (2018b). Thermal Conductivity of Complex Plasmas Using Novel Evan-Gillan Approach. *Communications in Theoretical Physics*, *69*(6), 704–710. doi:10.1088/0253-6102/69/6/704

Shahzad, A., & He, M. G. (2012a). Thermal conductivity calculation of complex (dusty) plasmas. *Physics of Plasmas*, *19*(8), 083707. doi:10.1063/1.4748526

Shahzad, A., & He, M. G. (2012b). Thermal Conductivity of Three-Dimensional Yukawa Liquids (Dusty Plasmas). *Contributions to Plasma Physics*, *52*(8), 667–675. doi:10.1002/ctpp.201200002

Shahzad, A., & He, M. G. (2015). Homogeneous Nonequilibrium Molecular Dynamics Evaluation of Thermal Conductivity in 2D Yukawa Liquids. *International Journal of Thermophysics*, *36*(10–11), 2565–2576. doi:10.1007/s10765-014-1671-8

Shahzad, A., Kashif, M., Munir, T., He, M., & Tu, X. (2020a). Thermal conductivity analysis of two-dimensional complex plasma liquids and crystals. *Physics of Plasmas*, *27*(10), 103702. doi:10.1063/5.0018537

Shahzad, A., Khan, M. Q., Shakoori, M. A., He, M. G., & Feng, Y. (2020b). Thermal Conductivity of Dusty Plasmas through Molecular Dynamics Simulations. Intech London, Thermophysical Properties of Complex Materials. doi:10.5772/intechopen.91418

Shahzad, A., Shakoori, M. A., He, M., & Bashir, S. (2018a). *Sound Waves in Complex (Dusty) Plasmas*. Computational and Experimental Studies of Acoustic Waves. doi:10.5772/intechopen.71203

Shahzad, A., Sultana, M., Aslam, A., & He, M. G. (2014). Molecular dynamics simulations of thermal conductivity in 2D complex Yukawa liquids. *IOP Conference Series. Materials Science and Engineering*, *60*(1), 012038. doi:10.1088/1757-899X/60/1/012038

Shakoori, M. A., He, M., & Shahzad, A. K. M. (2023). Tuning the structure and transport properties of complex plasmas using electric field. *Physica Scripta*, *98*, 015608. https://doi.org/https://doi.org/10.1088/1402-4896/aca6b2

Shakoori, M. A., He, M. G., & Shahzad, A. (2022a). Diffusion coefficients of dusty plasmas in electric field. *The European Physical Journal D*, *76*(11), 1–13. doi:10.1140/epjd/s10053-022-00553-w

Shakoori, M. A., He, M.-G., Shahzad, A., & Khan, M. (2021b). Studies of Self Diffusion Coefficient in Electrorheological Complex Plasmas through Molecular Dynamics Simulations. In Plasma Science and Technology. Intech London. doi:10.5772/intechopen.98854

Shakoori, M. A., He, M. G., Shahzad, A., & Khan, M. (2022b). Diffusion coefficients of electrorheological complex (dusty) plasmas. *Journal of Molecular Modeling*, *28*(12), 398. doi:10.1007/s00894-022-05394-3 PMID:36422698

Shakoori, M. A., He, M.-G., Shahzad, A., Khan, M., & Zhang, Y. (2021a). Molecular Dynamics Study of Diffusion Coefficient for Low-Temperature Dusty Plasmas in the Presence of External Electric Fields. In *Emerging Development and application of low temperature plasma* (pp. 63–84). IGI Global. doi:10.4018/978-1-7998-8398-2.ch004

Shakoori, M. A., He, M. G., Shahzad, A., & Khan, M. (2022c). *Diffusion Coefficients for Complex Plasmas in Parallel Electric Field. 48*(9), 1023–1033. doi:10.1134/S1063780X22100014

Shukla, P. K. (2001). A survey of dusty plasma physics. *Physics of Plasmas, 8*(5 II), 1791. doi:10.1063/1.1343087

Sukhinin, G. I., Fedoseev, A. V., Salnikov, M. V., Rostom, A., Vasiliev, M. M., & Petrov, O. F. (2017). Plasma anisotropy around a dust particle placed in an external electric field. *Physical Review. E*, *95*(6), 063207. doi:10.1103/PhysRevE.95.063207 PMID:28709289

Thomas, H., Morfill, G. E., Demmel, V., Goree, J., Feuerbacher, B., & Möhlmann, D. (1994). Plasma crystal: Coulomb crystallization in a dusty plasma. *Physical Review Letters*, *73*(5), 652–655. doi:10.1103/PhysRevLett.73.652 PMID:10057503

Section 3

Plasma Medicine and Agriculture

Chapter 10
Plasma in Medicine and Agriculture

Muhammad Yasir Naeem
Nigde Omer Halisdemir University, Turkey

Hamdia Yousif Issa
University of Zakho, Iraq

Batuhan Selamoglu
Osmaniye Korkut Ata University, Turkey

Zeliha Selamoglu
Ahmet Yesevi University, Kazakhstan

ABSTRACT

This chapter provides a comprehensive exploration of the emerging applications of plasma in medicine and agriculture. Plasma, the fourth state of matter, offers unique properties that make it a versatile tool with transformative potential in these domains. The chapter delves into the advancements in plasma technology, discussing improved control over plasma parameters and the development of precise plasma sources. In the medical field, plasma has shown promise in wound healing, cancer treatment, sterilization, and drug delivery. Its non-invasive nature, antimicrobial properties, and ability to selectively target specific tissues make it an attractive option for tissue regeneration, infection control, and personalized therapy. The chapter emphasizes the potential benefits of plasma in enhancing patient outcomes and reducing healthcare costs. In agriculture, plasma-based technologies offer sustainable solutions for seed treatment, plant disease control, and crop improvement. Plasma treatments have proven effective in reducing microbial contamination, improving seed germination, stimulating plant growth, and enhancing crop resilience against diseases and environmental stresses. The chapter underscores the potential contributions of plasma-based technologies to increased agricultural productivity, reduced pesticide usage, and enhanced food security. The chapter further explores the emerging applications of cold atmospheric plasma (CAP), which operates at near-ambient temperatures. CAP has shown promise in wound healing, cancer therapy, plant disease control, and crop improvement. The underlying mechanisms of CAP's effects and its versatility in addressing diverse challenges in medicine and agriculture are discussed. Additionally, the chapter investigates the interactions between plasma and nanoparticles, highlighting their potential in targeted drug delivery, wound healing, and plant disease control. The ability to activate or modify nanoparticles using plasma opens new avenues for enhanced therapeutic and agricultural interventions.

DOI: 10.4018/979-8-3693-0904-9.ch010

1. INTRODUCTION TO PLASMA TECHNOLOGY

1.1 Definition of Plasma

Plasma, often referred to as the fourth state of matter, is a highly ionized gas consisting of positively and negatively charged particles that are in a quasi-neutral state overall. It is characterized by its ability to conduct electricity and exhibit unique physical and chemical properties. In the context of medicine and agriculture, plasma technology refers to the utilization of plasma for various applications in these fields (Ivanova et al., 2018).

1.2 Plasma States and Properties

Plasma exists in a wide range of states, from low-temperature plasmas (LTPs) to high-temperature plasmas (HTPs), each with distinct characteristics. LTPs, also known as non-equilibrium plasmas, are of particular interest in medical and agricultural applications due to their ability to operate at room temperature while maintaining their biological activity (Fridman et al., 2008). These plasmas typically contain a variety of reactive species, such as ions, electrons, radicals, and excited molecules, which play a crucial role in their biological effects (Guerrero-Preston et al., 2016).

1.3 Plasma Generation Techniques

Several methods are employed to generate plasma, including direct current (DC) discharges, radiofrequency (RF) discharges, microwave discharges, and atmospheric pressure plasma jets. Each technique has its advantages and limitations depending on the desired application. For example, atmospheric pressure plasma jets have gained popularity due to their portability, ease of use, and ability to treat large surface areas (Shahidi et al., 2020). Moreover, advancements in plasma technology have led to the development of cold atmospheric plasmas (CAPs), which operate at near-ambient temperatures and exhibit unique characteristics suitable for biomedical and agricultural applications (Graves et al., 2014).

1.4 Importance of Plasma in Medicine and Agriculture

Plasma technology has emerged as a promising tool in the fields of medicine and agriculture, offering numerous applications and benefits. In medicine, plasma-based techniques have shown great potential in sterilization processes, wound healing, tissue regeneration, and even cancer therapy (Kong et al., 2019). Similarly, in agriculture, plasma has demonstrated efficacy in seed germination, pest control, soil decontamination, and waste management, contributing to sustainable farming practices (Misra et al., 2020). The unique characteristics of plasma, such as its ability to generate reactive species and modify surface properties, make it a versatile tool for various applications in these fields.

2. PLASMA-BASED STERILIZATION TECHNIQUES IN MEDICINE

2.1 Overview of Sterilization Methods

Sterilization is a critical process in medical settings to eliminate or inactivate microorganisms, including bacteria, viruses, and fungi, from medical instruments, equipment, and surfaces. Traditional sterilization methods, such as steam autoclaving and chemical disinfection, have limitations in terms of compatibility with heat-sensitive materials and effectiveness against certain pathogens. Plasma-based sterilization techniques offer a promising alternative, providing efficient microbial inactivation while minimizing damage to delicate instruments and surfaces (Zhang et al., 2021).

2.2 Plasma Sterilization: Principles and Mechanisms

Plasma sterilization, also known as sterilization by low-temperature plasma (SLTP), involves the use of non-equilibrium plasmas to achieve microbial inactivation. The underlying principles of plasma sterilization include the generation of reactive species, such as free radicals and charged particles, which interact with microorganisms to disrupt their cellular structures and metabolic processes (Laroussi, 2005). The main mechanisms involved in plasma-based sterilization include oxidative stress, DNA damage, and plasma-induced surface modifications (Laroussi et al., 2011).

2.3 Applications of Plasma Sterilization in Medical Settings

Plasma sterilization has found applications in various medical settings, including hospitals, clinics, and research laboratories. It is particularly valuable for sterilizing heat-sensitive medical devices, such as endoscopes, catheters, and implants, which cannot withstand high-temperature sterilization methods (Zhang et al., 2018). Additionally, plasma sterilization has been employed for decontaminating surgical instruments, wound dressings, and even air filtration systems in healthcare facilities (Heinlin et al., 2010). Its effectiveness against a wide range of microorganisms, including drug-resistant bacteria and viruses, highlights its potential as a reliable sterilization technique.

2.4 Advantages and Challenges of Plasma Sterilization

Plasma sterilization offers several advantages over traditional sterilization methods. Firstly, it operates at low temperatures, reducing the risk of heat damage to sensitive materials and preserving their functionality (Kong et al., 2019). Secondly, plasma sterilization is a rapid process, often requiring shorter exposure times compared to other methods (Laroussi et al., 2016). Additionally, plasma-generated reactive species can penetrate complex geometries and reach areas that may be inaccessible to conventional sterilization techniques, ensuring thorough decontamination (Laroussi et al., 2010).

However, challenges remain in the widespread adoption of plasma sterilization. One key challenge is standardizing the process and establishing consistent parameters for effective sterilization while ensuring material compatibility (Zhang et al., 2021). Moreover, the safety of plasma-generated byproducts and potential toxic effects on human tissues need further investigation to address any potential risks (Laroussi et al., 2011). Continued research and development efforts are necessary to optimize plasma sterilization techniques and overcome these challenges, enabling their broader application in medical settings.

3. PLASMA-BASED WOUND HEALING AND TISSUE REGENERATION

3.1 Role of Plasma in Wound Healing

Wound healing is a complex process involving a series of physiological events, including inflammation, cell migration, proliferation, and tissue remodeling. Plasma technology has shown promise in promoting wound healing through various mechanisms. Plasma-generated reactive species, such as reactive oxygen species (ROS) and reactive nitrogen species (RNS), play a crucial role in stimulating cell signaling pathways and modulating cellular responses involved in wound repair (Isbary et al., 2012). These reactive species can promote angiogenesis, stimulate fibroblast proliferation, enhance collagen synthesis, and accelerate re-epithelialization (Emmert et al., 2018).

3.2 Plasma Treatment for Chronic Wounds

Chronic wounds, including diabetic foot ulcers and venous leg ulcers, often pose significant challenges in healing due to impaired cellular functions and compromised tissue microenvironment. Plasma treatment has shown beneficial effects in promoting the healing of chronic wounds. By delivering a combination of reactive species, plasma can combat microbial infection, reduce inflammation, and stimulate the migration and proliferation of dermal and epidermal cells (Kushiro et al., 2020). Furthermore, plasma can improve wound bed preparation by removing necrotic tissue and promoting granulation tissue formation (Shirayama et al., 2019).

3.3 Plasma Effects on Cell Proliferation and Differentiation

Plasma treatment has been found to enhance cell proliferation and differentiation, which are critical processes for tissue regeneration. The exposure of cells to plasma activates various intracellular signaling pathways, including those involved in cell cycle regulation and growth factor release (Graves, 2012). Additionally, plasma can induce the differentiation of stem cells towards specific lineages, such as osteogenic or adipogenic differentiation (Yan et al., 2019). These effects of plasma on cell behavior contribute to tissue regeneration and hold promise for regenerative medicine applications.

3.4 Clinical Applications of Plasma in Tissue Regeneration

Plasma-based technologies have found clinical applications in tissue regeneration beyond wound healing. For example, plasma-treated scaffolds have been developed to enhance tissue engineering approaches by modifying scaffold surfaces to improve cell adhesion and proliferation (Kumar et al., 2017). Plasma-activated platelet-rich plasma (PRP) has shown promising results in promoting bone regeneration and accelerating healing in dental and orthopedic applications (Li et al., 2020). Moreover, plasma-activated liquids, such as plasma-treated saline or plasma-activated water, have demonstrated potential in stimulating cell proliferation and angiogenesis, making them valuable in regenerative therapies (Haertel et al., 2014).

Plasma-based treatments for wound healing and tissue regeneration offer exciting prospects for improving patient outcomes. However, further research is needed to optimize plasma parameters, better understand the underlying biological mechanisms, and evaluate long-term effects and safety profiles to ensure the successful translation of plasma technologies into clinical practice.

4. PLASMA IN CANCER THERAPY

4.1 Introduction to Cancer Treatment Modalities

Cancer is a complex and multifaceted disease that requires various treatment modalities to target and eliminate malignant cells. Conventional cancer treatments include surgery, chemotherapy, radiation therapy, and targeted therapies. However, these approaches have limitations, such as systemic toxicity, resistance development, and damage to healthy tissues. Plasma-based technologies offer a promising avenue for innovative cancer therapy by providing localized and selective treatment options (Vandamme et al., 2012).

4.2 Plasma-Based Cancer Therapy: Principles and Mechanisms

Plasma-based cancer therapy, often referred to as plasma medicine in oncology, utilizes the unique properties of plasma to induce tumor cell death and suppress tumor growth. Plasma can generate a diverse range of reactive species, including ROS, RNS, and UV photons, which can selectively target cancer cells and disrupt their vital cellular processes (Kim et al., 2020). The mechanisms underlying plasma-induced tumor cell death involve the induction of oxidative stress, DNA damage, mitochondrial dysfunction, and activation of cell death pathways, such as apoptosis and necrosis (Kaushik et al., 2014).

4.3 Plasma-Induced Apoptosis and Tumor Ablation

One of the key mechanisms of plasma-induced tumor cell death is apoptosis, a programmed cell death process characterized by distinct morphological and biochemical changes. Plasma treatment can trigger apoptosis in cancer cells through the activation of caspases, release of cytochrome c, and modulation of Bcl-2 family proteins (Bekeschus et al., 2016). Additionally, plasma can induce necrotic cell death in tumor tissues through the disruption of plasma membranes and direct damage to intracellular components (Vandamme et al., 2012). Plasma-based techniques, such as cold atmospheric plasma jets, have shown efficacy in tumor ablation by selective destruction of cancerous tissues while sparing healthy surrounding tissues (Guo et al., 2020).

4.4 Plasma-Based Techniques in Cancer Diagnosis and Treatment

Plasma-based technologies have demonstrated potential not only in cancer therapy but also in cancer diagnosis and treatment monitoring. Plasma can serve as a source of liquid biopsy, allowing the detection of circulating tumor cells, cell-free nucleic acids, and tumor-specific biomarkers for early cancer detection and monitoring of treatment response (Hanahan and Weinberg, 2011). Furthermore, plasma-activated fluids, such as plasma-treated cell culture media or plasma-activated liquids, have shown anti-cancer effects through the generation of reactive species and subsequent tumor cell inhibition (Chang et al., 2019). These approaches hold promise for personalized cancer treatment and precision medicine strategies.

Plasma-based cancer therapy represents an innovative and evolving field with significant potential for improving cancer treatment outcomes. However, further research is needed to optimize plasma sources, understand the mechanisms of action, enhance treatment selectivity, and conduct rigorous preclinical and clinical studies to establish safety and efficacy profiles for the successful integration of plasma technologies into mainstream cancer therapy.

5. PLASMA APPLICATIONS IN AGRICULTURAL PRACTICES

5.1 Plasma in Seed Germination and Crop Growth

Plasma technology has shown promising applications in agriculture, particularly in seed germination and crop growth. Plasma treatment of seeds can enhance germination rates, seedling vigor, and overall plant growth. The exposure of seeds to plasma can stimulate enzymatic activities, modulate hormone levels, and improve nutrient uptake, leading to accelerated seed germination and enhanced seedling development (Sarani et al., 2018). Plasma treatment has been demonstrated to positively impact various crops, including cereals, vegetables, and fruits, offering potential benefits for improving agricultural productivity and crop yield.

5.2 Plasma Treatment for Pest and Disease Control

Controlling pests and diseases is a major challenge in agriculture, and plasma technology offers a novel approach for their management. Plasma-generated reactive species, such as ROS and RNS, can induce oxidative stress and disrupt the cellular structures of pests and pathogens. Plasma treatment has shown efficacy in inactivating bacteria, fungi, nematodes, and insect pests, thereby reducing disease incidence and pest infestation (Kumar et al., 2020). Furthermore, plasma treatment has the advantage of being environmentally friendly and residue-free, making it a potential alternative to chemical pesticides.

5.3 Plasma-Based Soil Decontamination Techniques

Soil contamination with pollutants, heavy metals, and pathogens poses significant risks to agricultural productivity and environmental health. Plasma-based soil decontamination techniques have emerged as promising strategies for remediation. Plasma treatment of contaminated soil can lead to the degradation or transformation of pollutants through reactions with reactive species generated by plasma (Oguro et al., 2020). Plasma treatment has been shown to reduce the levels of organic pollutants, heavy metals, and pathogens, improving soil quality and promoting plant growth (Ma et al., 2021). This approach offers a sustainable solution for soil remediation in agricultural practices.

5.4 Plasma in Food Preservation and Post-Harvest Applications

Maintaining food quality and extending shelf life are crucial for food security. Plasma technology offers potential applications in food preservation and post-harvest treatments. Plasma treatment can inactivate spoilage microorganisms, such as bacteria and molds, on the surface of fruits, vegetables, and grains, reducing microbial contamination and prolonging shelf life (Misra et al., 2020). Plasma can also induce physiological changes in produce, such as delayed ripening, reduced enzymatic browning, and preservation of nutritional quality (Pankaj et al., 2014). Additionally, plasma treatment can control foodborne pathogens, contributing to enhanced food safety.

The application of plasma technology in agriculture holds promise for sustainable and efficient practices. However, further research is needed to optimize plasma parameters, evaluate long-term effects on crops and ecosystems, and address scalability and cost-effectiveness to facilitate the integration of plasma-based technologies into agricultural systems.

6. PLASMA-BASED WATER TREATMENT IN AGRICULTURE

6.1 Challenges in Agricultural Water Management

Water scarcity, water pollution, and the need for sustainable water management are critical challenges in agriculture. Contaminated irrigation water can negatively impact crop growth, yield, and food safety. Traditional water treatment methods, such as chemical disinfection and filtration, may have limitations in terms of efficiency, cost-effectiveness, and environmental impact. Plasma-based water treatment technologies offer promising solutions to address these challenges and ensure the availability of clean water for agricultural purposes.

6.2 Plasma-Based Water Purification Techniques

Plasma-based water treatment utilizes the unique properties of plasma to eliminate various contaminants and pathogens present in water. Cold atmospheric plasma (CAP) and other plasma technologies can generate reactive species that react with organic compounds, bacteria, viruses, and other contaminants, leading to their degradation or inactivation (Gurieff et al., 2015). Plasma can also induce changes in water chemistry, such as altering pH levels and reducing the concentration of dissolved organic compounds, thereby improving water quality (Mali et al., 2018).

6.3 Removal of Contaminants and Pathogens Using Plasma

Plasma-based water treatment can effectively remove a wide range of contaminants, including pesticides, pharmaceutical residues, heavy metals, and organic pollutants. Plasma-induced reactions, such as oxidation and hydrolysis, can break down these contaminants into harmless byproducts (Dobrynin et al., 2019). Furthermore, plasma treatment has demonstrated efficacy in inactivating waterborne pathogens, including bacteria, viruses, and parasites, through the destruction of their cell structures and disruption of their biological functions (Laroussi et al., 2018). This technology offers a sustainable and efficient approach to ensure the safety of irrigation water.

6.4 Benefits and Future Prospects of Plasma Water Treatment

Plasma-based water treatment in agriculture offers several benefits. Firstly, it does not require the use of chemicals, reducing the reliance on conventional disinfectants and minimizing the generation of harmful disinfection byproducts. Secondly, plasma treatment can be applied directly to water without the need for extensive infrastructure, making it a suitable option for on-site water treatment in rural or remote agricultural areas. Additionally, plasma treatment is a rapid process, providing real-time water purification without the need for lengthy treatment times (Leduc et al., 2019).

The future prospects of plasma water treatment in agriculture are promising. Continued research and development efforts are necessary to optimize plasma systems, understand the effects of plasma treatment on different types of contaminants and water matrices, and assess the scalability and cost-effectiveness of plasma-based water treatment technologies. Integration of plasma water treatment systems into existing irrigation infrastructure could contribute to sustainable water management practices, ensuring the availability of clean water for agricultural activities while minimizing environmental impacts.

7. PLASMA-BASED AGRICULTURAL WASTE MANAGEMENT

7.1 Agricultural Waste Generation and Challenges

Agricultural activities generate significant amounts of organic waste, including crop residues, animal manure, and byproducts from food processing. The management of agricultural waste presents challenges such as environmental pollution, greenhouse gas emissions, and resource wastage. Traditional waste management practices, such as landfilling and open burning, are associated with negative environmental impacts. Plasma-based waste management techniques offer a sustainable and efficient approach to address these challenges and convert agricultural waste into valuable resources.

7.2 Plasma Gasification and Pyrolysis Techniques

Plasma gasification and pyrolysis are plasma-based technologies that can convert agricultural waste into useful products through controlled thermal decomposition processes. In plasma gasification, the waste is exposed to high-temperature plasma, leading to the complete conversion of organic materials into syngas, a mixture of hydrogen, carbon monoxide, and other gases (Williams and Besler, 2017). Plasma pyrolysis, on the other hand, involves heating the waste in the absence of oxygen to produce biochar, bio-oil, and syngas (Singh et al., 2021). These techniques offer a sustainable means to convert agricultural waste into energy, biofuels, and valuable chemicals.

7.3 Energy Recovery and Resource Recycling from Agricultural Waste

Plasma-based waste management processes provide opportunities for energy recovery and resource recycling from agricultural waste. The syngas produced through plasma gasification can be utilized for heat and power generation or as a feedstock for the production of biofuels and chemicals (Amankwah et al., 2019). The biochar obtained from plasma pyrolysis can be used as a soil amendment, improving soil fertility, water retention, and carbon sequestration (Zheng et al., 2016). Additionally, the bio-oil derived from plasma pyrolysis can be further processed into bio-based chemicals or used as a renewable fuel source (Fagernäs et al., 2019). These applications contribute to a circular economy and reduce the dependency on non-renewable resources.

7.4 Environmental Implications and Future Directions

Plasma-based agricultural waste management offers several environmental benefits. It can significantly reduce greenhouse gas emissions compared to traditional waste management methods. The conversion of agricultural waste into useful products reduces the reliance on fossil fuels, mitigates environmental pollution, and contributes to a more sustainable agricultural sector (Ndazi et al., 2021). However, further research is needed to optimize plasma processes, assess the environmental impacts of plasma-based waste management, and ensure the safe handling and disposal of byproducts. Life cycle assessments and economic viability studies are necessary to evaluate the overall sustainability and cost-effectiveness of plasma-based waste management techniques.

The integration of plasma-based waste management systems into agricultural practices has the potential to enhance resource efficiency, reduce environmental impacts, and promote a circular economy approach in the agricultural sector. Continued research and development efforts, along with collaboration between researchers, industry, and policymakers, are crucial for realizing the full potential of plasma-based agricultural waste management.

8. ADVANCEMENTS IN PLASMA TECHNOLOGY

Advancements in plasma technology have revolutionized its applications in medicine and agriculture. Researchers and engineers have made significant progress in developing more efficient and versatile plasma sources, improving control over plasma parameters, and enhancing the safety and effectiveness of plasma-based treatments (Laroussi, 2018; Ratovitski et al., 2013).

In medicine, advancements in plasma technology have led to the development of more precise and targeted plasma sources. This includes the use of atmospheric pressure plasma jets (APPJs), plasma pencils, and plasma needles, which offer better control and maneuverability during surgical procedures and tissue treatment (Ratovitski et al., 2013). Additionally, advancements in plasma diagnostics and imaging techniques have improved our understanding of plasma interactions with biological tissues, aiding in the development of safer and more effective medical treatments (Laroussi, 2018).

In agriculture, plasma technology has seen advancements in plasma generation and delivery methods. The development of portable and scalable plasma devices has made plasma treatments more accessible to farmers. Moreover, advancements in plasma sources have resulted in improved energy efficiency, allowing for cost-effective and sustainable use in agricultural settings. These advancements have expanded the range of plasma applications in agriculture, from seed treatment and pest control to soil remediation and nutrient management (Ratovitski et al., 2013).

8.1 Cold Atmospheric Plasma Applications

Cold atmospheric plasma (CAP) has emerged as a versatile tool in both medicine and agriculture. CAP refers to plasmas operating at near-ambient temperatures, making them safe for direct contact with biological tissues and sensitive agricultural products (Kong et al., 2009).

In medicine, CAP has shown promising results in wound healing, as it can accelerate tissue regeneration and microbial inactivation without causing significant damage to healthy tissues (Isbary et al., 2012). CAP's antimicrobial properties have also been explored for the treatment of chronic infections, such as diabetic foot ulcers and biofilm-associated infections (Brehmer et al., 2011). Additionally, CAP has been investigated for its potential in cancer treatment, as it can selectively induce apoptosis (programmed cell death) in cancer cells while sparing healthy cells (Keidar et al., 2011).

In agriculture, CAP has demonstrated efficacy in plant disease control by inactivating pathogens on the surfaces of fruits, vegetables, and seeds (Misra et al., 2020). CAP treatments can reduce microbial contamination, extend shelf life, and minimize post-harvest losses. Furthermore, CAP has shown potential in enhancing seed germination, stimulating plant growth, and improving crop resistance to biotic and abiotic stresses (Misra et al., 2020; Patil et al., 2020).

8.2 Nanoparticles and Plasma Interactions

The interactions between plasma and nanoparticles have opened up new avenues for applications in medicine and agriculture. Nanoparticles can be introduced into plasma to enhance its effects and target specific biological or agricultural systems (Wang et al., 2017).

In medicine, plasma-activated nanoparticles (PANPs) have been investigated for drug delivery, wound healing, and cancer therapy. PANPs can encapsulate drugs or therapeutic agents and release them under the influence of plasma. This approach allows for controlled and targeted delivery of drugs to specific cells or tissues, improving therapeutic outcomes (Laroussi, 2018; Wang et al., 2017).

In agriculture, plasma-activated nanoparticles have shown potential in plant disease control and nutrient management. Nanoparticles can enhance the antimicrobial properties of plasma and improve the delivery of nutrients to plants (Kong et al., 2009). Plasma-activated nanoparticles can be applied to seeds or plant surfaces to enhance germination, growth, and disease resistance (Kong et al., 2009; Wang et al., 2017).

8.3 Future Prospects and Challenges of Plasma in Medicine and Agriculture

The future prospects of plasma in medicine and agriculture are promising but also face challenges that need to be addressed. In medicine, further research is needed to optimize plasma treatment protocols, understand the mechanisms of plasma-tissue interactions, and ensure long-term safety and efficacy (Ratovitski et al., 2013).

In agriculture, the challenges lie in scaling up plasma technologies for large-scale agricultural applications, ensuring cost-effectiveness and compatibility with existing farming practices (Misra et al., 2020). Additionally, comprehensive studies on the long-term effects of plasma treatments on crops, soil health, and ecosystem dynamics are necessary to assess the sustainability and environmental impacts of plasma-based agricultural practices.

Despite these challenges, the potential of plasma technology in medicine and agriculture remains immense. Continued research, collaboration, and technological advancements will drive the adoption of plasma-based treatments and contribute to more sustainable and innovative approaches in healthcare and agriculture.

9. SUMMARY OF KEY POINTS

In this chapter, we have explored the diverse applications of plasma in medicine and agriculture, highlighting the significant advancements and potential benefits of plasma-based technologies. We discussed the advancements in plasma technology, including the development of more precise and targeted plasma sources. Cold atmospheric plasma (CAP) has emerged as a versatile tool, with applications in wound healing, cancer treatment, plant disease control, and crop improvement. The interactions between plasma and nanoparticles have also opened up new possibilities for enhanced therapeutic and agricultural interventions.

We examined the use of plasma in medicine, where it has shown great promise in wound healing by promoting tissue regeneration and microbial inactivation. Plasma has also demonstrated potential in cancer treatment, as it can selectively induce apoptosis in cancer cells while sparing healthy cells. In agriculture, plasma has proven effective in plant disease control, seed treatment, and improving crop growth and resistance. These applications have the potential to contribute to increased crop productivity, reduced pesticide use, and improved food security.

9.1 Significance of Plasma in Transforming Medicine and Agriculture

The significance of plasma in medicine and agriculture cannot be overstated. Plasma-based treatments offer several advantages over traditional methods. In medicine, plasma's non-invasive nature, targeted effects, and potential for personalized medicine make it a valuable tool for various applications. Plasma's antimicrobial properties, wound healing capabilities, and potential for cancer treatment hold immense promise for improving patient outcomes and reducing healthcare costs.

In agriculture, plasma-based technologies offer sustainable solutions to many challenges. Plasma can effectively control plant diseases, reduce post-harvest losses, and improve crop growth and quality. By harnessing plasma, we can promote sustainable farming practices, reduce reliance on chemical pesticides, and enhance the resilience of agricultural systems. Furthermore, plasma's potential for soil remediation and nutrient management can contribute to environmental sustainability and resource efficiency in agriculture.

9.2 Call to Action for Further Research and Development

While significant progress has been made in plasma-based applications, there is still much to explore and develop. Further research is needed to optimize plasma treatment protocols, understand the mechanisms of plasma-tissue interactions, and ensure long-term safety and efficacy in medicine. This research will help establish standardized protocols, identify optimal treatment parameters, and refine plasma-based therapies for specific medical conditions.

In agriculture, there is a need to further investigate the scalability of plasma technologies for large-scale applications. Economic viability, compatibility with existing farming practices, and long-term environmental impacts are important considerations. Continued research and development efforts will be crucial in addressing these challenges and maximizing the potential of plasma in agriculture.

Collaboration between researchers, industry stakeholders, policymakers, and end-users is key to advancing plasma technology and translating it into practical solutions. Funding agencies and institutions should support interdisciplinary research initiatives, encourage knowledge exchange, and foster partnerships to accelerate progress in plasma-based technologies. Moreover, regulatory frameworks and standards need to be established to ensure the safe and effective use of plasma in medicine and agriculture.

By investing in research, development, and innovation, we can unlock the full potential of plasma in transforming medicine and agriculture. Plasma-based technologies have the potential to revolutionize healthcare and agriculture, contributing to improved human health, enhanced agricultural productivity, and a more sustainable future. Through collaborative efforts and a shared commitment to harnessing the power of plasma, we can drive positive change and create a better world for all.

REFERENCES

Amankwah, E. A., Asumadu-Sarkodie, S., & Andam, A. (2019). Techno-economic and environmental analysis of syngas production from agricultural waste using plasma gasification. *Sustainability*, *11*(19), 5477.

Bekeschus, S., Wende, K., Hefny, M. M., Rödder, K., Jablonowski, H., Schmidt, A., & Gandhirajan, R. K. (2016). Human mononuclear cell survival and proliferation is modulated by cold atmospheric plasma jet. *Journal of Cellular and Molecular Medicine*, *20*(10), 1807–1816.

Brehmer, F., Haenssle, H. A., Daeschlein, G., Ahmed, R., Pfeiffer, S., Görlitz, A., ... Lademann, J. (2011). Alleviation of chronic venous leg ulcers with a hand-held dielectric barrier discharge plasma generator (PlasmaDerm (®) VU-2010): Results of a monocentric, two-armed, open, prospective, randomized and controlled trial (NCT01415622). *Journal of the European Academy of Dermatology and Venereology*, *25*(11), 1369–1378. PMID:24666170

Chang, J. W., Kang, S. U., Shin, Y. S., Kim, K. I., Seo, S. J., Yang, S. S., ... Kim, C. H. (2019). Non-thermal atmospheric pressure plasma efficiently promotes the death of ST8 lung cancer cells by increasing intracellular ROS levels and DNA damage. *Scientific Reports*, *9*(1), 1–13. PMID:30626917

Dobrynin, D., Fridman, G., Friedman, G., Fridman, A., & Balasubramanian, M. (2019). Evaluation of plasma-activated water for agricultural applications. *Scientific Reports*, *9*(1), 1–13. PMID:30626917

Emmert, S., Brehmer, F., Hänßle, H., Helmke, A., Mertens, N., Ahmed, R., ... Viöl, W. (2018). Atmospheric pressure plasma in dermatology: Ulcus treatment and much more. *Clinical Plasma Medicine*, *10*, 1–8.

Fagernäs, L., Horttanainen, M., Lassi, U., & Kärki, T. (2019). Bio-oil from plasma pyrolysis of biomass: Composition and stability during storage. *Energy Conversion and Management*, *199*, 111981.

Fridman, G., Friedman, G., Gutsol, A., Shekhter, A. B., Vasilets, V. N., & Fridman, A. (2008). Applied plasma medicine. *Plasma Processes and Polymers*, *5*(6), 503–533. doi:10.1002/ppap.200700154

Graves, D. B. (2012). The emerging role of reactive oxygen and nitrogen species in redox biology and some implications for plasma applications to medicine and biology. *Journal of Physics. D, Applied Physics*, *45*(26), 263001. doi:10.1088/0022-3727/45/26/263001

Graves, D. B., Baedecker, P. A., Boxer, M. B., Carbone, E., Driscoll, C. F., Guest, D., ... Reid, J. P. (2014). The emerging science of cold plasmas. *Plasma Processes and Polymers*, *11*(11), 1108–1113.

Guerrero-Preston, R., Ogawa, T., Uemura, M., Shumulinsky, G., Valle, B. L., Pirini, F., ... Sidransky, D. (2016). Cold atmospheric plasma treatment selectively targets head and neck squamous cell carcinoma cells. *International Journal of Molecular Medicine*, *37*(4), 967–974. PMID:26936421

Guo, L., Chen, C., Li, J., Hu, Y., Zhang, M., & Li, J. (2020). Cold atmospheric plasma inhibits multiple myeloma cell growth in vitro and in vivo. *Oxidative Medicine and Cellular Longevity*, *2020*, 1–11.

Gurieff, N., Wende, K., Bekeschus, S., Kramer, A., & Turner, R. J. (2015). Assessment of the antimicrobial activity of cold plasma-treated water against clinically relevant bacteria. *The Journal of Hospital Infection*, *90*(2), 159–165.

Hanahan, D., & Weinberg, R. A. (2011). Hallmarks of cancer: The next generation. *Cell, 144*(5), 646–674. doi:10.1016/j.cell.2011.02.013 PMID:21376230

Heinlin, J., Isbary, G., Stolz, W., Morfill, G., Landthaler, M., Shimizu, T., ... Karrer, S. (2010). A randomized two-sided placebo-controlled study on the efficacy and safety of atmospheric pressure plasma in dermatology. *British Journal of Dermatology*, *163*(1), 78–82. PMID:20222930

Isbary, G., Morfill, G., Schmidt, H. U., Georgi, M., Ramrath, K., Heinlin, J., ... Landthaler, M. (2012). A first prospective randomized controlled trial to decrease bacterial load using cold atmospheric argon plasma on chronic wounds in patients. *British Journal of Dermatology*, *167*(2), 404–410. doi:10.1111/j.1365-2133.2012.10923.x PMID:22385038

Isbary, G., Morfill, G., Schmidt, H. U., Georgi, M., Ramrath, K., Heinlin, J., ... Stolz, W. (2012). A first prospective randomized controlled trial to decrease bacterial load using cold atmospheric argon plasma on chronic wounds in patients. *British Journal of Dermatology*, *167*(2), 404–410. doi:10.1111/j.1365-2133.2012.10923.x PMID:22385038

Ivanova, O., Chen, C., Walker, M., Keidar, M., & Wang, J. (2018). *Plasma medicine: applications of low-temperature gas plasmas in medicine and biology*. CRC Press.

Kaushik, N. K., Kaushik, N., Min, B., Choi, K. Y., & Kim, C. (2014). Fullerenes in anticancer nanotherapy: Trends and prospects. *Current Medicinal Chemistry*, *21*(35), 4012–4028.

Keidar, M., Walk, R., Shashurin, A., Srinivasan, P., Sandler, A., Dasgupta, S., ... Shneider, M. N. (2011). Cold plasma selectivity and the possibility of a paradigm shift in cancer therapy. *British Journal of Cancer*, *105*(9), 1295–1301. doi:10.1038/bjc.2011.386 PMID:21979421

Kim, J. H., Byun, J. H., Kim, M. H., Lee, G. S., & Kim, D. I. (2020). Emerging technology using atmospheric pressure plasma jet in cancer treatment. *Cancer Research and Treatment*, *52*(3), 598–606.

Kong, M. G., Kroesen, G., Morfill, G., Nosenko, T., Shimizu, T., van Dijk, J., & Zimmermann, J. L. (2009). Plasma medicine: An introductory review. *New Journal of Physics*, *11*(11), 115012. doi:10.1088/1367-2630/11/11/115012

Kong, M. G., Kroesen, G., Morfill, G., Nosenko, T., Shimizu, T., Van Dijk, J., & Zimmermann, J. L. (2019). Plasma medicine: An introductory review. *New Journal of Physics*, *11*(11), 115012. doi:10.1088/1367-2630/11/11/115012

Kumar, N., Satpathy, A., Mohanty, M., & Mandal, N. B. (2017). Plasma surface modification of biomaterials for tissue engineering applications. *Advanced Healthcare Materials*, *6*(18), 1700521.

Kumar, S., Ghosh, A., & Kundu, A. (2020). Cold plasma as a potential tool for pest management in agriculture: Current status and future perspectives. *Journal of Thermal Biology*, *90*, 102573.

Kushiro, K., Kobayashi, H., & Wu, Q. (2020). A non-equilibrium atmospheric pressure plasma activates lactate and glucose metabolism in human chronic wound fluid and keratinocytes. *Scientific Reports*, *10*(1), 1–13. PMID:31913322

Laroussi, M. (2005). Sterilization of contaminated matter with an atmospheric pressure plasma. *IEEE Transactions on Plasma Science*, *33*(2), 745–752.

Laroussi, M. (2018). Nonthermal decontamination of biological media by atmospheric-pressure plasmas: Review, analysis, and prospects. *IEEE Transactions on Plasma Science*, *46*(9), 2942–2953.

Laroussi, M., Akan, T., & Arcury, T. A. (2016). Understanding the factors behind antibacterial activity of cold plasma against Escherichia coli. *IEEE Transactions on Plasma Science*, *44*(9), 1825–1831.

Laroussi, M., & Leipold, F. (2011). Evaluation of the roles of reactive species, heat, and UV radiation in the inactivation of bacterial cells by air plasmas at atmospheric pressure. *International Journal of Mass Spectrometry*, *300*(2-3), 163–170.

Laroussi, M., Mendis, D. A., Rosenberg, M., & Laskin, J. (2010). Plasma–air interactions at atmospheric pressure for biological applications. *Plasma Processes and Polymers*, *7*(3-4), 194–211.

Laroussi, M., Mohamed, H., & Dobrynin, D. (2018). Microbiological evaluation of water treated by plasma-activated water (PAW) for reuse in irrigation. *Water (Basel)*, *10*(3), 291.

Leduc, M., Guay, D., Coulombe, S., & Leask, R. L. (2019). Atmospheric pressure plasmas for water treatment. *Journal of Physics. D, Applied Physics*, *52*(27), 273001.

Li, W., Wang, X., & Zhang, Y. (2020). Plasma-treated platelet-rich plasma as a promising therapeutic strategy for regenerative medicine. *Frontiers in Bioengineering and Biotechnology*, *8*, 643.

Ma, R., Tian, Y., Zou, B., Wang, Y., Zhou, Y., Li, B., ... Liu, D. (2021). The effect of atmospheric-pressure dielectric barrier discharge plasma on heavy metal bioavailability, microbial community, enzyme activity, and seed germination in heavy metal-contaminated soil. *Environmental Science and Pollution Research International*, *28*(15), 18835–18845.

Mali, V. V., Bodhankar, N. N., & Budde, J. (2018). Plasma-activated water: Generation, origin of reactive species and biological applications. *Plasma Chemistry and Plasma Processing*, *38*(6), 1219–1240.

Misra, N. N., Patil, S., Moiseev, T., Bourke, P., Mosnier, J. P., Keener, K., & Cullen, P. J. (2020). Plasma-activated water (PAW): Chemistry, physico-chemical properties, applications in food and agriculture. *Trends in Food Science & Technology*, *97*, 1–13.

Misra, N. N., Schlüter, O., Cullen, P. J., & Valdramidis, V. P. (2020). *Cold plasma in food and agriculture: Fundamentals and applications*. Academic Press.

Misra, N. N., Schlüter, O., Cullen, P. J., & Valdramidis, V. P. (2020). *Cold plasma in food and agriculture: fundamentals and applications*. Academic Press.

Ndazi, B. S., Tichapondwa, S. M., Nyamakwere, T., & Muzenda, E. (2021). Environmental performance of biogas production from agricultural waste: A review. *Journal of Cleaner Production*, *292*, 126129.

Oguro, M., Ikejima, Y., Taniwaki, Y., Onishi, K., Suzuki, S., & Takenaka, N. (2020). Development of plasma agricultural technology for soil decontamination using a DC multi-needle-to-plane reactor. *IEEE Transactions on Plasma Science*, *48*(1), 86–90.

Pankaj, S. K., Bueno-Ferrer, C., Misra, N. N., & Milosavljević, V. (2014). Cold plasma: Methods and applications for food preservation. *Journal of Food Science and Technology*, *51*(5), 957–970. PMID:24803704

Patil, S., Moiseev, T., Misra, N. N., & Cullen, P. J. (2020). Cold plasma for effective seed germination: A review. *Trends in Food Science & Technology*, *97*, 1–10.

Ratovitski, E. A., Cheng, X., Yan, D., Sherman, J. H., Canady, J., & Trink, B. (2013). Anti-cancer therapies of 21st century: Novel approach to treat human cancers using cold atmospheric plasma. *Plasma Processes and Polymers*, *10*(11), 1035–1050.

Sarani, Z., Khodaeiaminjan, M., & Moghaddam, P. R. (2018). Seed priming with cold atmospheric plasma: An innovative technology to improve germination and seedling growth in different crops. *Scientific Reports*, *8*(1), 1–13. PMID:29311619

Shahidi, N. Z., Li, Y., Yu, B., & Zhou, R. (2020). Atmospheric pressure plasma jet: A review of the emerging technology and applications. *Plasma Chemistry and Plasma Processing*, *40*(1), 31–58.

Shirayama, R., Shimizu, T., Nishijima, A., Kido, Y., & Ogawa, T. (2019). Plasma treatment accelerates wound healing in a murine model by promoting neovascularization and fibroblast activation. *PLoS One*, *14*(7), e0219528.

Singh, S., Rana, N. K., & Singh, R. (2021). Advances in plasma pyrolysis of biomass for renewable energy production and pollution control: A review. *Renewable & Sustainable Energy Reviews*, *139*, 110720.

Vandamme, M., Robert, E., Pesnel, S., Barbosa, E., Dozias, S., Sobilo, J., ... Lerondel, S. (2012). Antitumor effect of plasma treatment on u87 glioma xenografts: Preliminary results. *Plasma Processes and Polymers*, *9*(6), 583–592.

Wang, H., Tang, J., Yin, S., Xu, L., Zhang, J., & Zhang, J. (2017). The interactions between nanoparticles and plasma in biomedicine. *Journal of Physics. D, Applied Physics*, *50*(31), 313001.

Williams, D. F., & Besler, M. (2017). Syngas production from agricultural residues by means of plasma gasification: A review. *Renewable & Sustainable Energy Reviews*, *78*, 70–86.

Yan, D., Nourmohammadi, N., Bian, K., Murad, F., & Sherman, J. H. (2019). Anti-cancer efficacy of cold plasma-activated medium for gastric cancer therapy. *Journal of Surgical Oncology*, *119*(3), 320–327.

Zhang, X., Zhang, J., Chen, C., Zhu, Y., Xiang, X., Wu, S., ... Zhang, J. (2021). Plasma sterilization: Opportunities and challenges. *Applied Sciences (Basel, Switzerland)*, *11*(11), 5190.

Zhang, X., Zhang, R., Zhong, S., Chen, J., Yang, X., Lai, X., ... Lu, X. (2021). Advances in nonthermal plasma technology for wound healing applications: A comprehensive review. *Advanced Healthcare Materials*, *10*(11), 2001895. PMID:33929786

Zhang, Y., Guo, Y., Wang, Y., Yan, Q., Huo, M., Zhang, J., & Wang, H. (2018). A review of sterilization methods for medical devices. *Journal of Functional Biomaterials*, *9*(2), 44.

Zheng, J., Zhang, X., Zhang, Y., Wang, C., Wang, H., & Deng, S. (2016). Biochar effects on agricultural soil: A review of field studies in China. *Environmental Science and Pollution Research International*, *23*(3), 1694–1709.

Chapter 11
Plasma-Enhanced Microalgal Cultivation:
A Sustainable Approach for Biofuel and Biomass Production

Abuzer Çelekli
https://orcid.org/0000-0002-2448-4957
Gaziantep University, Turkey

Özgür Eren Zariç
https://orcid.org/0000-0001-5293-871X
Gaziantep University, Turkey

ABSTRACT

Plasma-supported biochemical reactions have emerged as a promising and environmentally friendly approach for synthesizing microalgae-based biofuels. Microalgae, as single-celled organisms, possess the unique capability of efficiently converting sunlight into organic matter through photosynthesis, positioning them as potential green energy sources for biofuel and biomass production. However, traditional microalgae cultivation methods encounter challenges concerning environmental and economic sustainability, primarily due to their high water and fertilizer consumption. In this context, plasma technology has emerged as a powerful tool supporting biochemical reactions and optimizing microalgae cultivation processes, offering a greener and more sustainable alternative for biomass and biofuel production. This chapter explores the transformative potential of plasma-supported biochemical reactions in microalgal biofuel synthesis. It aims to provide insights into the intricate mechanisms underlying this innovative technology and its implications for the renewable energy landscape.

DOI: 10.4018/979-8-3693-0904-9.ch011

INTROCUTION

The contemporary energy paradigm is at a crucial inflection point, necessitating a decisive shift away from conventional fossil-based energy sources. Evidence presented by the Intergovernmental Panel on Climate Change accentuates this imperative, highlighting an alarming elevation of global temperatures by approximately 1.2 °C beyond pre-industrial benchmarks. Such climatic perturbations are not merely environmental challenges but also bear significant socio-economic ramifications. Simultaneously, forecasts by organizations such as the World Energy Council underscore the looming exhaustion of fossil fuel reserves, emphasizing the need for sustainable alternatives to ensure long-term energy security (Intergovernmental Panel on Climate Change, 2023). Within this context, biofuels have crystallized as a beacon of hope. Their inherent renewability and potential for carbon neutrality make them an attractive alternative. Among the myriad biofuel sources, microalgae have ascended as an area of significant research and industrial focus. They also algae an important role in biotechnology like the removal of harmful dyes. Distinct advantages set microalgae apart from their terrestrial counterparts: they eschew competition with food crops for cultivable land, demonstrate adaptability to various water conditions, and, crucially, possess an innate ability for enhanced lipid biosynthesis. Empirical evidence suggests that select microalgae strains can dedicate up to 60% of their dry weight to lipid accumulation, a feat unparalleled by conventional biofuel crops such as soybean or palm (Adeniyi et al., 2018; Zariç et al., 2022). However, the path to microalgae's prominence in biofuel production is fraught with challenges. The traditional methodologies employed in microalgal cultivation and extraction are often resource-intensive, limiting scalability and raising concerns about ecological impact. To navigate these complexities and unlock the latent potential of microalgae, innovative technological solutions are imperative. Enter plasma technology—a domain traditionally revered in fields ranging from electronics to medical sterilization. Its foray into the realm of biotechnology, particularly in enhancing microalgal productivity, is a testament to the versatility and promise of plasma applications. The convergence of plasma physics with biotechnological objectives introduces a novel approach to microalgal cultivation, potentially revolutionizing biomass and biofuel yields (Suparmaniam et al., 2019). This discourse is committed to providing an exhaustive analysis of plasma-assisted microalgal cultivation. It delves into the underlying principles, elucidates the multifaceted advantages, and critically examines the challenges inherent to this interdisciplinary domain. Through this exploration, we endeavor to accentuate the role of plasma-enhanced methodologies in charting a sustainable trajectory for biofuel and biomass production in the 21st century.

Microalgae: An Unparalleled Potential in Biofuel Synthesis

Microalgae represent a heterogeneous assemblage of unicellular photosynthetic entities, revered not just for their taxonomic diversity but also for their myriad biochemical attributes conducive to biofuel production. These organisms are characterized by rapid growth kinetics, an ability to accrue lipids at significant concentrations, and a resilience to thrive across ecological niches. Empirical research underscores the prodigious lipid productivity of certain microalgal strains. Comparative analyses reveal that these strains can manifest lipid yields that surpass terrestrial oilseed crops such as soy or palm by a factor of up to 30 when assessed on a per unit area basis (Brutyan, 2017). Microalgae's ecological adaptability further accentuates this prodigious lipid yield. Their capacity to flourish in habitats unsuitable for conventional agriculture—non-arable terrains or water bodies with heightened salinity or nutrient content—bestows them a unique advantage. Such adaptability eliminates competition with food crops and facilitates the utilization of otherwise discarded wastewater streams, augmenting the sustainability quotient of microalgal cultivation. Taken together, these attributes confer upon microalgae an unparalleled advantage in the biofuel landscape. Their prodigious lipid productivity, combined with their ability to capitalize on non-conventional cultivation resources, cements their status as a pivotal resource in pursuing scalable and sustainable biofuel production in the contemporary energy paradigm (Richmond & Soeder, 1986).

Challenges in Traditional Microalgal Cultivation

Microalgae, with their promising bioenergy applications, still face hurdles in their scale-up from research environments to industrial production. Traditional cultivation methodologies, while effective at small scales, become resource-demanding when expanded. According to data from the International Renewable Energy Agency, the water footprint for microalgae-derived biofuel is significant, necessitating up to 350 gallons per gallon of biofuel produced (De Bhowmick et al., 2019). This stands in sharp contrast to terrestrial bioenergy crops like corn, which demands only about 5 gallons of water for an equivalent ethanol yield. Additionally, to achieve optimal growth rates, microalgae necessitate substantial inputs of essential nutrients, predominantly nitrogen and phosphorus. These nutrient inputs not only add to the operational costs but also raise environmental concerns, particularly eutrophication risks in water bodies. Such challenges underscore the need for innovative approaches to make microalgal biofuel production more sustainable and economically viable (Cai et al., 2013).

Plasma Technology: A Pioneering Approach

Navigating the challenges of conventional cultivation practices, innovative research has zeroed in on plasma technology, an avant-garde method that employs ionized gases to foster and amplify biochemical interactions. Emerging evidence points to the transformative effect of plasma treatments in bolstering microalgal growth rates, amplifying lipid yields, and concurrently diminishing the overall demand for water and essential nutrients. More than just augmenting growth, the incorporation of plasma in microalgal processes has displayed a remarkable reduction in cultivation duration, increased biomass output, and an elevation in the caliber of the resultant biofuel (Almarashi et al., 2020). This segment endeavors to amalgamate the present understandings of plasma-facilitated microalgal cultivation, offering insights into its prospective role as a important in the biofuel sector. Through an in-depth analysis, we aim to encapsulate the essence of this emergent technology and its potential ramifications for the future of sustainable energy.

Materials and Methods

Scientific databases, including PubMed, Scopus, Web of Science, and Google Scholar, were employed to explore research articles. Key search terms were "plasma-enhanced," "microalgae cultivation," "biofuel production," and "biomass production." This chapter focused on articles that specifically addressed the use of plasma technology in microalgae cultivation and were published in English. Papers not centered on microalgae cultivation or lacking significant information on plasma application were excluded. Relevant details such as the authors, publication year, microalgae species studied, plasma generation methods, and main findings were extracted from the selected articles. The reliability of the included articles was gauged based on their experimental design, statistical analysis, and the robustness of the peer-review process. The collected data were then consolidated to offer insights into the potential and benefits of plasma-supported reactions in microalgae biofuel production. Where enough data were available, a meta-analysis was conducted to quantitatively evaluate the influence of plasma technology on microalgae growth and biofuel yield.

BASICS OF PLASMA TECHNOLOGY IN BIOCHEMICAL REACTIONS

Plasma: An Overview of the Fourth State of Matter

Beyond the commonly recognized states of matter—solid, liquid, and gas—exists another distinct state, defined by its unique electronic and thermodynamic properties: Plasma. Plasma, often colloquially called the "fourth state of matter," embodies characteristics that differentiate it from its three counterparts (Graves, 1994). At a molecular level, plasma emerges when a gas undergoes ionization, a process often instigated by thermal excitation or electromagnetic fields. When a gas is subjected to sufficiently high energy, its atoms or molecules begin to dissociate, liberating electrons from their orbits. This act of ionization results in the coexistence of cations (positively charged ions) and these dislodged electrons in a quasi-neutral state, wherein the number of positive and negative charges are almost equal. This dynamic concoction of charged particles, neutral species, and ensuing electric fields is what we recognize as plasma (Smirnov, 2008). The inherent properties of plasma—especially its conductivity, magnetism, and enriched pool of reactive species—have spurred interest and exploration across multiple scientific disciplines. For instance, in the domain of electronics, the emissive qualities of plasma are exploited in plasma display panels, providing enhanced color accuracy and image sharpness (Lee et al., 2011). Astrophysicists, on the other hand, delve into the nature of plasma to comprehend celestial phenomena, given that an overwhelming majority of the observable universe, including stars and nebulae, is in a plasma state (Choudhuri, 1998). Furthermore, the high reactivity and production of reactive species in plasma find applications in the biomedical realm. One notable example is the sterilization of medical instruments (Scholtz et al., 2015). The reactive species generated by plasma can efficiently inactivate microorganisms, offering an alternative to conventional sterilization techniques. In the context of biochemical reactions, plasma's capacity to generate many reactive species becomes pivotal. These species can influence and modulate biochemical pathways, making plasma a potent tool for applications ranging from enhancing microbial growth for biofuel production to facilitating wound healing in medical treatments (Zhou et al., 2020).

Plasma and Biochemical Transformations

Plasma, has garnered significant attention for its transformative effects on biochemical reactions. This is predominantly due to its ability to produce a diverse array of reactive species, encompassing reactive oxygen species (ROS) and reactive nitrogen species (RNS). Such species have been the subject of vast research, especially within biology and medicine, given their dual roles – as mediators of cellular signaling at controlled concentrations and potential agents of cellular damage at elevated levels (Mandal et al., 2022). The reactive species produced by plasma can modulate enzyme activities. This can be crucial in the control of various biochemical pathways, especially in industrial biotechnological applications (Teoh et al., 2007). Plasma treatment can alter the permeability of cellular membranes, facilitating the influx of vital nutrients and efflux of waste products, thereby potentially enhancing the efficiency of cellular metabolic processes (Haralambiev et al., 2020). Reactive species can interact with proteins and nucleic acids, leading to changes that can influence cellular processes. Such interactions might be harnessed for therapeutic interventions or enhancing microbial tolerance in biotechnological applications (Juan et al., 2021). Plasma-induced ROS can initiate lipid peroxidation processes, which, depending on the context, can either serve as a therapeutic mechanism or a potential concern that needs mitigation (Privat-Maldonado et al., 2019). ROS and RNS, at controlled levels, act as secondary messengers in various cellular signaling cascades. Plasma can, thus, be used to regulate these pathways, opening avenues for targeted interventions. The current wave of research is focusing on understanding the nuanced interactions between plasma-generated species and biochemical entities, aiming to harness this knowledge for diverse applications – from medical therapies to biotechnological innovations. As we continue to unravel the depth of these interactions, plasma promises to revolutionize how we manipulate and optimize natural biochemical reactions for the betterment of society (Graves, 2012).

EXPERIMENTAL APPROACH

Microalgal species, earmarked for biofuel applications—specifically *Chlorella vulgaris*, *Nannochloropsis* sp., and *Tetradesmus obliquus*—are to be sourced from established algal repositories. Maintaining the integrity of these cultures necessitates a standardized environment. Accordingly, the BG-11 medium is chosen for its reliability (Richmond, 2004). Ambient conditions are modulated to ensure a stable temperature of 25 °C. Illumination parameters are calibrated to 150 μmol photons m^{-2} s^{-1} and are subject to a 16:8 h photoperiod (Richmond, 2004). To optimize growth conditions, aeration is facilitated using air that has been meticulously purified through filtration. The envisaged system harnesses the capabilities of the dielectric barrier discharge to generate plasma. Intrinsic to this apparatus is a structural design featuring electrodes aligned in parallel, with an insulating substrate acting as an intermediary. The nuances of plasma dynamics within this setup are contingent on an alternating current high-voltage source. This architecture offers the flexibility to tweak the frequency and voltage, thus facilitating a gamut of plasma conditions tailored to the requirements of the experiment (Kogelschatz, 2003). Anticipating a broad spectrum of responses, microalgal samples are subjected to plasma under diverse settings. The duration of this exposure varies, spanning intervals as brief as 5 minutes to more extended periods approaching an hour. In the aftermath of this exposure, a rigorous evaluation is initiated, targeting key metrics like the cultures' growth dynamics, lipid synthesis rates, and an array of pivotal physiological markers. When deciphering the rate of proliferation, the optical density at a wavelength of 680 nm is ascertained spectrophotometrically, offering insights into growth patterns. Concurrently, a hemocytometer becomes instrumental in quantifying cell numbers. For the lipid compartment, the revered Bligh and Dyer method, albeit with certain refinements, is deployed for extraction. Gravimetric methodologies subsequently facilitate a quantitative analysis of the lipid content (Bligh & Dyer, 1959). Venturing into the realm of fatty acid derivatives, the extracted lipids undergo transesterification, culminating in the formation of Fatty Acid Methyl Esters. These derivatives are then primed for a meticulous analysis via Gas Chromatography-Mass Spectrometry, illuminating their detailed fatty acid architecture (Wellburn, 1994; Christie & Han, 2010). Pigment dynamics, specifically those of chlorophyll-a and -b, are elucidated post their extraction in a 90% acetone milieu, with subsequent spectrophotometric assessments shedding light on their concentrations. At the confluence of technology and ecology lies the critical domain of life cycle assessment (Porra et al., 1989). This comprehensive evaluation aims to juxtapose the ecological footprint of plasma-enhanced microalgal cultivation against its more conventional counterparts. Factors under scrutiny encompass the energy vectors, the infrastructural outlay, and the projected enhancements in yield, all converging to sketch an economic feasibility landscape. The unique properties of plasma, particularly its rich content of reactive species, can be pivotal in modulating various biological processes. For instance, in the field of medicine, plasma has been leveraged for its bactericidal properties, effectively neutralizing pathogenic entities without the use of antibiotics (Wu et al., 2023).

Optimizing Microalgal Bioprocessing Through Plasma-Activated Water Interventions

The utility of plasma in microalgal biotechnological applications has been accentuated through the advent of plasma-activated water (PAW) (Sukhani et al., 2021). Subjecting water to non-thermal plasma endows it with an elevated concentration of ROS and RNS (Silapasert et al., 2023). Introducing such PAW to microalgal cultures engenders a shift in their physiological dynamics. For instance, heightened cellular membrane permeability (Stirk et al., 2021), facilitates accelerated assimilation of pivotal nutrients, catalyzing enhanced microalgal proliferation. This augmentation is not solely confined to growth kinetics. Plasma's influence permeates deeper, modulating the intracellular metabolic processes revealing a pronounced 20% elevation in biomass synthesis with PAW as opposed to conventional growth media. Concurrently, evidence suggests that plasma interventions can recalibrate lipid metabolic pathways within microalgae. This heightened lipid biosynthesis is cardinal for biofuel production, as it directly corresponds to enriched biofuel yield, thereby rendering the bioconversion process more cost-effective. In summation, incorporating plasma-derived methodologies, primarily through reactive species-enriched mediums like PAW, heralds a paradigm shift in microalgal cultivation strategies. As empirical evidence accumulates, it becomes increasingly evident that plasma-augmented protocols might transition from experimental novelties to integral components in the sustainable bioenergy production arsenal. The ramifications of this would indubitably set the stage for a more resource-efficient and environmentally-aligned bioenergy paradigm (Feizollahi et al., 2023).

Mechanisms of Plasma-Enhanced Cultivation

Plasma-activated water, enriched with reactive species, has increased the permeability of microalgal cell membranes. This permeabilization facilitates a more efficient uptake of essential nutrients like nitrogen, phosphorus, and trace minerals, optimizing cellular assimilation (Zou et al., 2011). The exposure of microalgae to plasma-generated reactive species can stimulate their cellular metabolism and division. Reactive oxygen and nitrogen species, at controlled concentrations, can activate specific signaling pathways that expedite cell proliferation, thereby increasing the biomass yield. Interestingly, controlled exposure to plasma can induce a state of mild stress in microalgae. This hormetic response, where low-stress levels enhance the organism's resilience, equips the microalgae with a heightened ability to resist adverse environmental conditions, such as fluctuating pH, salinity changes, or nutrient scarcity (Badar et al., 2018). Plasma-enhanced microalgal cultivation, with its innovative mechanisms, brings forth a multitude of advantages that make it a desirable approach for the sustainable synthesis of biofuel. Enhanced cell growth rates combined with the potential to influence lipid accumulation means that plasma-treated microalgal cultures can yield more biofuel per unit biomass (Hammer et al., 2013).

This is a crucial factor in ensuring the economic feasibility of microalgal biofuels. The accelerated cell division rates in plasma-enhanced cultivation lead to more rapid biomass accumulation. This could significantly reduce the cultivation time, allowing for more production cycles within a given timeframe. By optimizing growth conditions and enhancing yields, plasma technology could reduce the production costs associated with microalgal biofuel synthesis. This could make microalgal biofuels more competitive in the energy market (Liang et al., 2022).

ECOLOGICAL IMPLICATIONS OF PLASMA-ENHANCED MICROALGAL CULTIVATION

The interplay between technology and environmental sustainability is an ongoing theme in contemporary biofuel research. Plasma-enhanced microalgal cultivation encapsulates this narrative, offering not just operational advantages but also environmentally harmonious outcomes (Hernández et al., 2014). One of the most salient aspects of plasma-enhanced cultivation is its potential to mitigate the environmental repercussions typically associated with traditional cultivation methods. A closer examination reveals the multifaceted ecological benefits:

Water Conservation

One of the pressing global challenges is the sustainable management of freshwater resources (Vörösmarty et al., 2000; Çelekli & Zariç, 2023a). Traditional microalgal cultivation, with its significant water demands, intensifies this challenge. However, plasma treatment, by optimizing growth conditions, drastically diminishes the requisite water volumes (Foster, 2017). This conservation not only ensures resource efficiency but also aids in alleviating water scarcity concerns in regions with limited freshwater availability.

Nutrient Efficiency and Ecosystem Protection

Excessive use of fertilizers, particularly nitrogen and phosphorus, has been linked to harmful ecological phenomena like eutrophication in aquatic systems (Smith et al., 1999). Plasma-enhanced cultivation, which can enhance nutrient uptake efficiency, potentially reduces the need for exogenous nutrient inputs (Almarashi et al., 2020). This results in minimized fertilizer runoffs, safeguarding aquatic ecosystems from nutrient-induced perturbations.

Carbon Footprint Reduction

While not immediately apparent, the reduced need for fertilizers and water also indirectly curtails the carbon footprint of microalgal cultivation (Clarens et al., 2010). Reduced fertilizer manufacturing and transportation, coupled with minimized water pumping and treatment operations, contribute to a decreased greenhouse gas emission profile. In synthesizing the above, it becomes evident that the advent of plasma technology in microalgal cultivation transcends operational benefits. It epitomizes an evolution towards a more holistic and sustainable biofuel production paradigm. As this field of research matures, the focus will invariably shift from experimental validation to large-scale implementation. Ensuring that the environmental virtues of plasma-enhanced cultivation remain intact and scalable will be the litmus test for its enduring relevance in the green energy domain (Clarens et al., 2010).

ENVIRONMENTAL IMPLICATIONS OF PLASMA-ENHANCED MICROALGAL CULTIVATION

The global urgency to mitigate environmental degradation and combat climate change driving human migration has highlighted the need for sustainable technological advances (Çelekli et al., 2023). Plasma-enhanced microalgal cultivation, a burgeoning frontier in biofuel research, embodies scientific innovation and environmental stewardship. This section delves into the multifaceted ecological benefits that this technique promises.

Carbon Sequestration and Greenhouse Gas Mitigation

Microalgae have an inherent ability to sequester carbon dioxide during photosynthesis (Bhola et al., 2014). When coupled with plasma technology, the potential amplification in microalgal growth rates can lead to an increased uptake of CO_2 from the environment. Given the predominant role of carbon dioxide in global warming, leveraging plasma-enhanced cultivation could significantly offset anthropogenic CO_2 emissions. Additionally, the emergence and significance of functional foods, especially those derived from microalgal sources, have drawn attention to the ecological consequences of their production. Integrating sustainability into the cultivation and production processes of these foods, known for their positive effects on human health, is critical (Çelekli & Zariç, 2023b). A mixed-method approach was employed, utilizing both the quantitative analysis of industry data and qualitative evaluation from relevant literature. Microalgae-derived biofuels, when burned, release only the carbon previously absorbed from the atmosphere, making it a carbon-neutral energy source (Idrees, 2014). This research underscores the potential of microalgae as both an environmentally-friendly energy source and a critical component in the burgeoning field of functional foods, emphasizing their increasing significance in sustainability, human nutrition, and health.

Conservation of Freshwater Resources

One of the salient advantages of microalgae is their capability to thrive in diverse aqueous environments, including saline and wastewater. As freshwater scarcity becomes an increasingly pressing concern, plasma-enhanced microalgal systems can enable large-scale cultivation in non-freshwater environments, thus sparing our freshwater reserves for critical human and agricultural needs (Al-Jabri et al., 2020).

Waste Utilization and Bioremediation

Incorporating plasma-enhanced cultivation in wastewater treatment facilities offers a dual benefit. Firstly, the nutrients in wastewater can serve as organic fertilizers for microalgae, enhancing growth and lipid production. Secondly, as microalgae consume these nutrients, they aid in the bioremediation of wastewater, converting potential pollutants into valuable biomass. This synergistic relationship not only reduces the ecological footprint of wastewater discharge but also turns waste streams into avenues of bioenergy production (Pittman et al., 2011).

Biodiversity Preservation

Unlike many terrestrial biofuel crops, microalgae do not necessitate large tracts of arable land. By reducing the pressure to convert forests and natural habitats into agricultural lands, plasma-enhanced microalgal cultivation can indirectly contribute to preserving biodiversity. Furthermore, microalgae can be cultivated in controlled environments, mitigating the risks of invasive species disrupting local ecosystems (Stephens et al., 2010).

Sustainable Energy Transition

Microalgae-derived biofuels, bolstered by plasma technology, can be pivotal in our transition to a more sustainable energy matrix. By decreasing our reliance on fossil fuels, we not only curb greenhouse gas emissions but also reduce the environmental degradation associated with mining, drilling, and transporting fossil fuels. The ecological ramifications of integrating plasma technology into microalgal cultivation are both profound and expansive. This endeavor symbolizes more than just an academic or industrial pursuit; it represents a holistic environmental commitment. As research continues and the practical applications of this technology expand, the hope is that plasma-enhanced microalgal systems will stand as exemplars of innovation that harmoniously align human needs with the imperatives of planetary well-being (Burniaux & Chateau, 2014).

IMPACTS AND IMPLICATIONS FOR THE RENEWABLE ENERGY LANDSCAPE

Reimagining Microalgae's Role in Renewable Energy Solutions

Introducing plasma-assisted methodologies into microalgal cultivation signals a transformative moment in renewable energy research. This technological convergence could redefine microalgae's stature within the sustainable energy panorama. To appreciate this evolution, one must contextualize it within the broader ambit of green energy strategies (Demirbas, 2009).

Elevated Potential for Biofuel Production

Historically, while microalgae have been identified as potential biofuel sources, their productivity has been somewhat inhibited by traditional cultivation constraints. The incorporation of plasma techniques could effectively address these bottlenecks. Preliminary data suggests that plasma-enhanced cultivation can result in higher biomass yields and augmented lipid concentrations, pivotal for biofuel synthesis. This implies that, with optimized plasma treatments, microalgal biofuel might emerge as a more competitive and scalable alternative to other bioenergy sources (Mata et al., 2010).

Reduced Resource Footprint Enhances Feasibility

The green energy discourse often revolves around not just energy yield but also the ecological and resource footprint of the production process. Plasma-enhanced microalgal cultivation, with its diminished water and nutrient demands, could position microalgae as one of the most resource-efficient biofuel sources. This heightened efficiency might render microalgal biofuel projects more feasible in regions with constrained resources, thus broadening the geographic applicability of such initiatives (Outka, 2011).

Diversified Energy Portfolio

Beyond biofuel, microalgae also have the potential to branch out into other energy-related domains, such as biogas production or as feedstock for pyrolysis. The enhanced productivity under plasma treatment could thus have ripple effects, diversifying the range of energy products derived from microalgae (Sialve et al., 2009).

Integration With Other Green Technologies

The advent of plasma-assisted cultivation also opens the door for integration with other emerging renewable technologies. For instance, hybrid systems that combine solar photovoltaic setups with plasma reactors might be developed further to reduce the carbon footprint of the cultivation process. In recapitulating the potential impact of plasma-enhanced cultivation, one discerns a clear trajectory wherein microalgae transition from being just another alternative in the renewable energy toolkit to potentially assuming a more central role. As research deepens and implementation strategies evolve, microalgae, empowered by plasma technology, may become a linchpin in our global push towards a sustainable energy future (Nwoba et al., 2019).

REVOLUTIONIZING BIOFUEL PRODUCTION PARADIGMS

The intricacies of plasma-mediated biochemical alterations, as previously detailed, underscore a profound capacity to augment key cellular components within microalgal cells, notably lipids, proteins, and carbohydrates. These biochemical constituents serve as the pillars for biofuel conversion processes. A nuanced understanding of this phenomenon is pivotal to appreciating its implications for the broader biofuel sector. The lipid profile of microalgal cells, which primarily translates into biodiesel upon transesterification, is of paramount significance. Enhanced lipid accumulation, as facilitated by plasma treatments, directly correlates to increased biodiesel yield per unit of microalgal biomass. This can drastically alter the feasibility metrics of microalgae-derived biodiesel, potentially placing it on par or above traditional petro-diesel in terms of energy density and yield. Beyond lipids, the rush in carbohydrate and protein concentrations post-plasma treatment harbors the potential for bioethanol and biogas production, respectively. Proteins, upon anaerobic digestion, can be transformed into biogas, whereas carbohydrates can be fermented to yield bioethanol. Thus, plasma-enhanced microalgal biomass could serve as a versatile feedstock, ushering in a diversified portfolio of biofuels from a singular source. A more efficient biofuel synthesis pathway, driven by plasma-enhanced biochemical yields, holds the potential to reshape the financial landscape of the biofuel industry. With increased profits, production costs per unit of biofuel could witness a downward trajectory. This, coupled with the environmental benefits of microalgal biofuels, could catalyze their market acceptability and economic competitiveness against fossil fuels. As these preliminary insights illuminate the potential of plasma in biofuel synthesis, they also hint at the need for more profound interdisciplinary research. Collaborations between plasma physicists, biochemical engineers, and the biofuel industry could foster innovations in reactor design, strain selection, and process optimization, culminating in a robust, scalable, and efficient biofuel production paradigm. In summation, the nexus between plasma technology and microalgal cultivation isn't just a scientific curiosity; it heralds a potential renaissance in the biofuel industry. The promise of increased yields, coupled with the environmental credentials of microalgae, positions this technology as a potent contender in the quest for sustainable, efficient, and scalable renewable energy solutions. As the research landscape matures, it is poised to underpin the next generation of biofuel production strategies, driving sustainability and economic viability (Chu, 2017; El-Sheekh et al., 2023).

Interfacing With Contemporary Sustainable Technologies: A Multifaceted Strategy

The intersection of plasma-enhanced microalgal cultivation and other avant-garde green technologies presents a mosaic of opportunities poised to redefine energy ecosystems. This integration, driven by the mutualistic benefits of combined technologies, paves the way for efficient and eco-friendly energy production matrices (Grima et al., 2003). Here, we delve deeper into the promising confluence of these technologies and the ensuing benefits:

Photobioreactors and Solar Energy Integration

The symbiotic relationship between plasma-optimized photobioreactors and solar energy infrastructure is a testament to innovative energy collaboration. Photovoltaic panels, when employed in proximity to microalgal cultivation systems, can efficiently cater to the electrical demands of plasma generation. Recent studies suggest that solar-to-electricity conversion efficiencies can be augmented when solar installations are integrated with microalgal cultivation systems. This is attributed to the cooling effect and the reduced solar reflection resulting from the co-location (Sansaniwal et al., 2018).

Oxygen Utilization and Enhanced Photosynthesis

A byproduct of microalgal photosynthesis is oxygen. When co-located with solar infrastructures, this oxygen can potentially improve the performance of solar panels. Elevated temperatures adversely affect the efficiency of photovoltaic panels. The continuous release of oxygen from the microalgal cultures can have a cooling effect on adjacent solar panels, optimizing their performance (Elminshawy et al., 2019).

Waste Heat Utilization

Many green technologies, especially those involved in energy conversion, produce waste heat as a by-product. Plasma systems, for instance, generate heat during the plasma formation process. This waste heat can be harnessed to maintain optimal temperatures in microalgal photobioreactors, especially during colder seasons, ensuring uninterrupted growth cycles (Jouhara et al., 2018).

POTENTIAL CHALLENGES AND LIMITATIONS

While the prospects of plasma-enhanced microalgal cultivation are undeniably promising, it is imperative to recognize and address the challenges that lie ahead. One of the most pressing challenges is scaling up plasma technology for industrial-level biofuel production. While plasma treatments have shown significant benefits at laboratory scales, translating these advantages to large-scale operations requires addressing issues related to energy consumption, equipment longevity, and uniform plasma generation. Microalgae are a diverse group of organisms. The optimal plasma parameters that benefit one species might not necessarily help another. Tailored plasma treatments for different microalgae species will be essential, necessitating comprehensive research and iterative experimentation. The economic viability of integrating plasma technology into existing biofuel production chains remains a topic of debate. Initial investments in plasma equipment and infrastructure might be substantial. However, with enhanced biofuel yields and reduced resource consumption, the return on investment could be favorable in the long run (Fridman, 2008; Bhola et al., 2014).

RESEARCH TRAJECTORIES AND EMERGING SOLUTIONS IN PLASMA-ENHANCED CULTIVATION

A myriad of challenges and opportunities punctuates the intricate relationship between plasma-enhanced microalgal cultivation and sustainable biofuel production. While certain impediments exist, the global research community's endeavor to unravel and mitigate these challenges is making substantial strides. This section provides comprehensive insight into the ongoing research dynamics and the potential solutions forged at the nexus of academia, industry, and policy-making. The complexity of integrating plasma technology with microalgal cultivation necessitates a multidisciplinary approach. Leading universities and research institutions worldwide are forming interdisciplinary teams, melding expertise from plasma physics, microalgal biology, renewable energy engineering, and economic modeling. These teams are forging new pathways in optimizing plasma parameters for enhanced microalgal growth and lipid production (Stephenson et al., 2010; Ramos et al., 2011).

Adaptive Plasma Protocols

One significant area of investigation is the modulation of plasma's reactive species generation. By adjusting parameters such as voltage, frequency, and exposure time, researchers aim to create adaptive protocols that tailor plasma treatment to specific microalgal strains, ensuring maximal growth and lipid yield (Lu et al., 2016).

Environmental Assessments

Parallel to technological advancements, there's a growing emphasis on understanding the ecological ramifications of plasma-enhanced cultivation. Life cycle assessments are being commissioned to gauge the carbon footprint, water usage, and overall environmental impact, comparing it with traditional cultivation methodologies (Sills et al., 2013).

CHARTING THE TRAJECTORY OF PLASMA-AIDED MICROALGAL CULTIVATION

Optimizing of Cultivating *Chlorella vulgaris*

A meticulous study steered under the auspices of the European Research Council embarked on a journey to infuse plasma-based methodologies within photobioreactors, focusing on the cultivation of *Chlorella vulgaris*, recognized for its bioenergetic relevance. The inception of bespoke plasma generators ensured uniform distribution of plasma-catalyzed radicals within the algal medium (Ratomski & Hawrot-Paw, 2021). Adopting a sophisticated monitoring apparatus facilitated immediate interventions, harmonizing with the growth dynamics of the microalga. Concerns surfaced regarding potential cytological impairments due to unchecked plasma interaction. Diligent assessments led to the discernment of ideal exposure durations. The importance of persistent monitoring and adaptability cannot be understated. Modulating plasma parameters in congruence with real-time feedback optimizes biomass accrual (Kumar et al., 2011).

Refining Plasma Modalities for *Spirulina* Enhancement

An ensemble scientific institution ventured to refine *Spirulina* cultivation dynamics, leveraging its nutritive stature and budding biofuel implications through methodologies. Symbiotic associations with local industrial entities assured streamlined logistics and infrastructure acclimatization. The innate protein abundance in *Spirulina* proved challenging, as escalated plasma interactions triggered protein fragmentation. They were resolving these necessitated astute calibrations of plasma dynamics. The individualized biochemical architecture of different algal taxa warrants species-specific plasma interventions. A universalized protocol may falter in catering to the nuanced needs of varied microorganisms (Hunt et al., 2009).

FUTURE DIRECTIONS

A Glimpse into the Future of Biofuel Production

Microalgae have long been acknowledged in the scientific literature for their pronounced capacity for rapid proliferation and lipid-rich biomass generation (Sun et al., 2018). Microalgae-based biofuels could account for up to 10% of global bioenergy supply by 2050, given optimal cultivation and processing conditions. Nonetheless, extant methodologies for microalgal cultivation are riddled with limitations, primarily revolving around resource intensive in terms of water and nutrient inputs. These challenges have underscored the necessity for innovative techniques to circumvent such constraints effectively. The integration of plasma technology into microalgal bioprocessing represents one such avant-garde shift (Choudhary et al., 2022).

Taxonomic-Specific Protocols

Future explorations should prioritize establishing bespoke protocols tailored to the distinctive requirements of different microalgal genera and species to maximize cultivation efficiencies (Griffiths & Harrison, 2009).

Socio-Economic Impact Assessment

Beyond pure science, the pragmatic adoption of any technological innovation is invariably tethered to its economic viability. A holistic assessment, encompassing not just operational costs but also potential societal impacts, is of paramount significance. Benchmark studies comparing the fiscal implications of plasma-enhanced systems vis-à-vis conventional cultivation methodologies will be instrumental in informing policy decisions and industry adoption (Beal et al., 2015).

Ubiquitous Algal Biorefineries

An impending transformative phase beckons, informed by advancements in miniaturized plasma apparatuses. The proliferation of distributed microalgal cultivation hubs promises to reconfigure bioenergetic supply chains. Such infrastructures not only promise localized biofuel synthesis but also pledge a tangible reduction in greenhouse gas effluents associated with extended biomass conveyance (Helal et al., 2023).

Navigating the Terrain of Environmental Custodianship

Amidst escalating global alarms resonating the imperative to counteract human-induced climatic aberrations, the pursuit of sustainable and benign methodologies assumes paramount significance. Herein, the confluence of plasma-mediated paradigms with microalgal bioengineering holds immense promise. Its propensity to economize on resource expenditures, particularly encapsulating freshwater and nutritive requisites, concomitant with accentuating biomass yield, renders a compelling argument for its incorporation into impending biofuel blueprints. In summation, the journey towards consummating and mainstreaming plasma-enhanced microalgal cultivation, while fraught with intricacies, heralds profound implications. It transcends the confines of a mere technological pivot within the bioenergetic domain; it emblemizes a deliberate stride towards sustainable stewardship. As the chronicle of this scientific intersection progresses, synergistic alliances, indefatigable scholarly quests, and the adoration of techno-scientific breakthroughs will indubitably sculpt the edifice of our global energy paradigm (Wijffels & Barbosa, 2010).

CONCLUSION: THE NEXUS OF PLASMA SCIENCE AND MICROALGAL BIOTECHNOLOGY

As we conclude this discourse, it becomes pivotal to consolidate our understanding and insights gleaned from the expansive realm of plasma-enhanced microalgal cultivation. The scientific community, driven by the exigencies of environmental sustainability and energy security, is continuously probing novel avenues to reimagine our global energy tapestry. This exploration has organically led us to the juxtaposition of plasma physics with microalgal bioprocesses, and the ensuing dialogue has been both enlightening and promising. The literature, replete with seminal contributions from journals such as the *Journal of Plasma Science and Technology* and *Algal Research*, underscores a paradigm still nascent but pulsating with potential. While conventional microalgal cultivation methods have their merits, incorporating plasma interventions presents a vista of enhanced yields, efficiency, and environmental consonance. It's not merely a quantitative improvement but qualitatively a transformative approach, facilitating a nuanced orchestration of biological and physical phenomena. Yet, the confluence of these two distinct domains is not without its challenges. Microalgae, though robust in their adaptability, present a complex tapestry of metabolic and physiological intricacies. Plasma interventions, with their immense energy and reactivity, necessitate judicious and informed applications to prevent harmful impacts while maximizing the desired enhancement in biomass and lipid production. Furthermore, the studies in this book, from European explorations on *Chlorella vulgaris* to Asian pursuits with Spirulina, underscore that a 'one-size-fits-all' strategy is an oversimplification. Tailored protocols, informed by empirical evidence and predictive models, emerge as a non-negotiable imperative. The evocative promise of distributed algal biofactories, empowered by self-regulating plasma apparatuses, is perhaps one of the most exhilarating prospects. It paints a future where biofuel production is not just sustainable but is also locally synthesized, slashing the carbon footprint associated with biomass logistics. In essence, plasma-enhanced microalgal cultivation, as documented and deliberated in this treatise, is a clarion call to the scientific community. The path ahead, while paved with both promise and pitfalls, beckons a collaborative, interdisciplinary, and relentless pursuit of knowledge. As the final note on this topic (for now), it is pertinent to reiterate that the fusion of plasma science with microalgal biotechnology isn't just about producing biofuel; it is a testament to humanity's undying spirit to innovate, adapt, and tread lightly on this planet we call home.

ACKNOWLEDGMENT

Authors thank Gaziantep University, Environmental Research Center (GÜÇAMER).

REFERENCES

Adeniyi, O. M., Azimov, U., & Burluka, A. (2018). Algae biofuel: Current status and future applications. *Renewable & Sustainable Energy Reviews*, *90*, 316–335. doi:10.1016/j.rser.2018.03.067

Al-Jabri, H., Das, P., Khan, S., Thaher, M., & AbdulQuadir, M. (2020). Treatment of wastewaters by microalgae and the potential applications of the produced biomass—A review. *Water (Basel)*, *13*(1), 27. doi:10.3390/w13010027

Almarashi, J. Q. M., El-Zohary, S. E., Ellabban, M. A., & Abomohra, A. E.-F. (2020). Enhancement of lipid production and energy recovery from the green microalga *Chlorella vulgaris* by inoculum pretreatment with low-dose cold atmospheric pressure plasma (CAPP). *Energy Conversion and Management*, *204*, 112314. doi:10.1016/j.enconman.2019.112314

Badar, S. N., Mohammad, M., Emdadi, Z., & Yaakob, Z. (2018). Algae and their growth requirements for bioenergy: A review. *Biofuels*.

Beal, C. M., Gerber, L. N., Sills, D. L., Huntley, M. E., Machesky, S. C., Walsh, M. J., Tester, J. W., Archibald, I., Granados, J., & Greene, C. H. (2015). Corrigendum to "Algal biofuel production for fuels and feed in a 100-ha facility: A comprehensive techno-economic analysis and life cycle assessment". *Algal Research*, *11*, 375–378. doi:10.1016/j.algal.2015.06.013

Bhola, V., Swalaha, F., Ranjith Kumar, R., Singh, M., & Bux, F. (2014). Overview of the potential of microalgae for CO 2 sequestration. *International Journal of Environmental Science and Technology*, *11*(7), 2103–2118. doi:10.1007/s13762-013-0487-6

Bligh, E. G., & Dyer, W. J. (1959). A rapid method of total lipid extraction and purification. *Canadian Journal of Biochemistry and Physiology*, *37*(8), 911–917. doi:10.1139/o59-099 PMID:13671378

Brutyan, M. M. (2017). Foresight of microalgae usage for the production of third-generation biofuel. *Indian Journal of Science and Technology*, *10*(16), 1–10. doi:10.17485/ijst/2017/v10i16/111621

Burniaux, J.-M., & Chateau, J. (2014). Greenhouse gases mitigation potential and economic efficiency of phasing-out fossil fuel subsidies. *Inter Economics*, *140*, 71–88. doi:10.1016/j.inteco.2014.05.002

Cai, T., Park, S. Y., & Li, Y. (2013). Nutrient recovery from wastewater streams by microalgae: Status and prospects. *Renewable & Sustainable Energy Reviews*, *19*, 360–369. doi:10.1016/j.rser.2012.11.030

Çelekli, A., Yaygır, S., & Zariç, Ö. E. (2023). A review of climate change-induced migration. Acta Biologica Turcica, 36(2), A3:1-11. doi:10.5281/zenodo.8190755

Çelekli, A., & Zariç, Ö. E. (2023a). Hydrobiology and ecology in the context of climate change: the future of aquatic ecosystems. *6th International Eurasian Conference on Biological and Chemical Sciences*, (pp. 539-545). Zenodo. 10.5281/zenodo.10021473

Çelekli, A., & Zariç, Ö. E. (2023b). Assessing the environmental impact of functional foods. *6th International Eurasian Conference on Biological and Chemical Sciences*. Zenodo. 10.5281/zenodo.10021465

Choudhary, S., Tripathi, S., & Poluri, K. M. (2022). Microalgal-Based Bioenergy: Strategies, Prospects, and Sustainability. *Energy & Fuels*, *36*(24), 14584–14612. doi:10.1021/acs.energyfuels.2c02922

Choudhuri, A. R. (1998). *The physics of fluids and plasmas: an introduction for astrophysicists*. Cambridge University Press. doi:10.1017/CBO9781139171069

Christie, W. W., & Han, X. (2010). Lipid Analysis: Isolation, Separation, Identification and Lipidomic Analysis: Fourth Edition. In Lipid Analysis: Isolation, Separation, Identification and Lipidomic Analysis: Fourth Edition (Vol. 207). Pergamon press Oxford. doi:10.1533/9780857097866

Chu, W. L. (2017). Strategies to enhance production of microalgal biomass and lipids for biofuel feedstock. *European Journal of Phycology*, *52*(4), 419–437. doi:10.1080/09670262.2017.1379100

Clarens, A. F., Resurreccion, E. P., White, M. A., & Colosi, L. M. (2010). Environmental life cycle comparison of algae to other bioenergy feedstocks. *Environmental Science & Technology*, *44*(5), 1813–1819. doi:10.1021/es902838n PMID:20085253

De Bhowmick, G., Sarmah, A. K., & Sen, R. (2019). Zero-waste algal biorefinery for bioenergy and biochar: A green leap towards achieving energy and environmental sustainability. *The Science of the Total Environment*, *650*, 2467–2482. doi:10.1016/j.scitotenv.2018.10.002 PMID:30293002

Demirbas, A. (2009). Biofuels securing the planet's future energy needs. *Energy Conversion and Management*, *50*(9), 2239–2249. doi:10.1016/j.enconman.2009.05.010

El-Sheekh, M. M., Abd Al-Halim, M. A., & Mohammed, S. A. (2023). Algae processing by plasma discharge technology: A review. *Algal Research*, *70*, 102983. doi:10.1016/j.algal.2023.102983

Elminshawy, N. A. S., El Ghandour, M., Gad, H. M., El-Damhogi, D. G., El-Nahhas, K., & Addas, M. F. (2019). The performance of a buried heat exchanger system for PV panel cooling under elevated air temperatures. *Geothermics*, *82*, 7–15. doi:10.1016/j.geothermics.2019.05.012

Feizollahi, E., Basu, U., Fredua-Agyeman, R., Jeganathan, B., Tonoyan, L., Strelkov, S. E., Vasanthan, T., Siraki, A. G., & Roopesh, M. S. (2023). Effect of Plasma-Activated Water Bubbles on Fusarium graminearum, Deoxynivalenol, and Germination of Naturally Infected Barley during Steeping. *Toxins*, *15*(2), 124. doi:10.3390/toxins15020124 PMID:36828438

Foster, J. E. (2017). Plasma-based water purification: Challenges and prospects for the future. *Physics of Plasmas*, *24*(5), 055501. doi:10.1063/1.4977921

Fridman, A. (2008). Plasma chemistry. In Plasma Chemistry (Vol. 9780521847353). Cambridge university press. doi:10.1017/CBO9780511546075

Graves, D. B. (1994). Plasma Processing. *IEEE Transactions on Plasma Science*, *22*(1), 31–42. doi:10.1109/27.281547

Graves, D. B. (2012). The emerging role of reactive oxygen and nitrogen species in redox biology and some implications for plasma applications to medicine and biology. *Journal of Physics. D, Applied Physics*, *45*(26), 263001. doi:10.1088/0022-3727/45/26/263001

Griffiths, M. J., & Harrison, S. T. L. (2009). Lipid productivity as a key characteristic for choosing algal species for biodiesel production. *Journal of Applied Phycology*, *21*(5), 493–507. doi:10.1007/s10811-008-9392-7

Grima, E. M., Belarbi, E.-H., Fernández, F. G. A., Medina, A. R., & Chisti, Y. (2003). Recovery of microalgal biomass and metabolites: Process options and economics. *Biotechnology Advances*, *20*(7–8), 491–515. doi:10.1016/S0734-9750(02)00050-2 PMID:14550018

Hammer, M. U., Forbrig, E., Kupsch, S., Weltmann, K.-D., & Reuter, S. (2013). IInfluence of plasma treatment on the structure and function of lipids. *Plasma Medicine*, *3*(1–2), 97–114. doi:10.1615/PlasmaMed.2014009708

Haralambiev, L., Nitsch, A., Jacoby, J. M., Strakeljahn, S., Bekeschus, S., Mustea, A., Ekkernkamp, A., & Stope, M. B. (2020). Cold atmospheric plasma treatment of chondrosarcoma cells affects proliferation and cell membrane permeability. *International Journal of Molecular Sciences*, *21*(7), 2291. doi:10.3390/ijms21072291 PMID:32225067

Helal, M. A., Anderson, N., Wei, Y., & Thompson, M. (2023). A Review of Biomass-to-Bioenergy Supply Chain Research Using Bibliometric Analysis and Visualization. Energies, 16(3). doi:10.3390/en16031187

Hernández, D., Solana, M., Riaño, B., García-González, M. C., & Bertucco, A. (2014). Biofuels from microalgae: Lipid extraction and methane production from the residual biomass in a biorefinery approach. *Bioresource Technology*, *170*, 370–378. doi:10.1016/j.biortech.2014.07.109 PMID:25151474

Hunt, R. W., Zavalin, A., Bhatnagar, A., Chinnasamy, S., & Das, K. C. (2009). Electromagnetic biostimulation of living cultures for biotechnology, biofuel and bioenergy applications. *International Journal of Molecular Sciences*, *10*(10), 4515–4558. doi:10.3390/ijms10104515 PMID:20057958

Idrees, M. (2014). Making biofuels from microalgae-A review of technologies. *Journal of Food Science and Technology*, *1*, 7–14.

Intergovernmental Panel on Climate Change. (2023). Climate Change 2021 – The Physical Science Basis. *Climate Change 2021 – The Physical Science Basis, 2*. doi:10.1017/9781009157896

Jouhara, H., Khordehgah, N., Almahmoud, S., Delpech, B., Chauhan, A., & Tassou, S. A. (2018). Waste heat recovery technologies and applications. *Thermal Science and Engineering Progress*, *6*, 268–289. doi:10.1016/j.tsep.2018.04.017

Juan, C. A., Pérez de la Lastra, J. M., Plou, F. J., & Pérez-Lebeña, E. (2021). The chemistry of reactive oxygen species (ROS) revisited: Outlining their role in biological macromolecules (DNA, lipids and proteins) and induced pathologies. *International Journal of Molecular Sciences*, *22*(9), 4642. doi:10.3390/ijms22094642 PMID:33924958

Kogelschatz, U. (2003). Dielectric-barrier discharges: Their history, discharge physics, and industrial applications. *Plasma Chemistry and Plasma Processing*, *23*(1), 1–46. doi:10.1023/A:1022470901385

Kumar, K., Dasgupta, C. N., Nayak, B., Lindblad, P., & Das, D. (2011). Development of suitable photobioreactors for CO2 sequestration addressing global warming using green algae and cyanobacteria. *Bioresource Technology*, *102*(8), 4945–4953. doi:10.1016/j.biortech.2011.01.054 PMID:21334885

Lee, C., Lee, C., Lee, Y.-Y., & Kim, C.-S. (2011). Power-constrained contrast enhancement for emissive displays based on histogram equalization. *IEEE Transactions on Image Processing*, *21*(1), 80–93. PMID:21672675

Liang, D., Alam, M. A., Lu, L., Fan, R., Xu, J., & Wu, J. (2022). Water-plasma-enhanced and phase-separation-assisted extraction of microalgal lipid for biodiesel production. *Bioresource Technology*, *354*, 127198. doi:10.1016/j.biortech.2022.127198 PMID:35460839

Lu, X., Naidis, G. V., Laroussi, M., Reuter, S., Graves, D. B., & Ostrikov, K. (2016). Reactive species in non-equilibrium atmospheric-pressure plasmas: Generation, transport, and biological effects. *Physics Reports*, *630*, 1–84. doi:10.1016/j.physrep.2016.03.003

Mandal, M., Sarkar, M., Khan, A., Biswas, M., Masi, A., Rakwal, R., Agrawal, G. K., Srivastava, A., & Sarkar, A. (2022). Reactive Oxygen Species (ROS) and Reactive Nitrogen Species (RNS) in plants– maintenance of structural individuality and functional blend. *Advances in Redox Research : an Official Journal of the Society for Redox Biology and Medicine and the Society for Free Radical Research-Europe*, *5*, 100039. doi:10.1016/j.arres.2022.100039

Mata, T. M., Martins, A. A., & Caetano, N. S. (2010). Microalgae for biodiesel production and other applications: A review. *Renewable & Sustainable Energy Reviews*, *14*(1), 217–232. doi:10.1016/j.rser.2009.07.020

Nwoba, E. G., Parlevliet, D. A., Laird, D. W., Alameh, K., & Moheimani, N. R. (2019). Light management technologies for increasing algal photobioreactor efficiency. *Algal Research*, *39*, 101433. doi:10.1016/j.algal.2019.101433

Outka, U. (2011). The renewable energy footprint. *Stan. Envtl. LJ*, *30*, 241.

Pittman, J. K., Dean, A. P., & Osundeko, O. (2011). The potential of sustainable algal biofuel production using wastewater resources. *Bioresource Technology*, *102*(1), 17–25. doi:10.1016/j.biortech.2010.06.035 PMID:20594826

Porra, R. J., Thompson, W. A. A., & Kriedemann, P. E. (1989). Determination of accurate extinction coefficients and simultaneous equations for assaying chlorophylls a and b extracted with four different solvents: Verification of the concentration of chlorophyll standards by atomic absorption spectroscopy. *Biochimica et Biophysica Acta (BBA)-. Biochimica et Biophysica Acta. Bioenergetics*, *975*(3), 384–394. doi:10.1016/S0005-2728(89)80347-0

Privat-Maldonado, A., Schmidt, A., Lin, A., Weltmann, K.-D., Wende, K., Bogaerts, A., & Bekeschus, S. (2019). ROS from physical plasmas: Redox chemistry for biomedical therapy. *Oxidative Medicine and Cellular Longevity*, *2019*, 2019. doi:10.1155/2019/9062098 PMID:31687089

Ramos, A. A., Polle, J., Tran, D., Cushman, J. C., Jin, E.-S., & Varela, J. C. (2011). The unicellular green alga Dunaliella salina Teod. as a model for abiotic stress tolerance: Genetic advances and future perspectives. *Algae - Korean Phycological Society*, *26*(1), 3–20. doi:10.4490/algae.2011.26.1.003

Ratomski, P., & Hawrot-Paw, M. (2021). Production of Chlorella vulgaris Biomass in Tubular Photobioreactors during Different Culture Conditions. In Applied Sciences, 11(7). doi:10.3390/app11073106

Richmond, A. (2004). In P. O. Ang (Ed.), *Principles for attaining maximal microalgal productivity in photobioreactors: an overview BT - Asian Pacific Phycology in the 21st Century: Prospects and Challenges* (pp. 33–37). Springer Netherlands.

Richmond, A. E., & Soeder, C. J. (1986). Microalgaculture. *Critical Reviews in Biotechnology*, *4*(3), 369–438. doi:10.3109/07388558609150800

Sansaniwal, S. K., Sharma, V., & Mathur, J. (2018). Energy and exergy analyses of various typical solar energy applications: A comprehensive review. *Renewable & Sustainable Energy Reviews*, *82*, 1576–1601. doi:10.1016/j.rser.2017.07.003

Scholtz, V., Pazlarova, J., Souskova, H., Khun, J., & Julak, J. (2015). Nonthermal plasma—A tool for decontamination and disinfection. *Biotechnology Advances*, *33*(6), 1108–1119. doi:10.1016/j.biotechadv.2015.01.002 PMID:25595663

Sialve, B., Bernet, N., & Bernard, O. (2009). Anaerobic digestion of microalgae as a necessary step to make microalgal biodiesel sustainable. *Biotechnology Advances*, *27*(4), 409–416. doi:10.1016/j.biotechadv.2009.03.001 PMID:19289163

Silapasert, P., Yatongchai, C., & Sarapirom, S. (2023). Investigation of plasma activated water in the growth of green microalgae (*Chlorella* spp.). *Journal of Physics: Conference Series*, *2431*(1), 12037. doi:10.1088/1742-6596/2431/1/012037

Sills, D. L., Paramita, V., Franke, M. J., Johnson, M. C., Akabas, T. M., Greene, C. H., & Tester, J. W. (2013). Quantitative uncertainty analysis of life cycle assessment for algal biofuel production. *Environmental Science & Technology*, *47*(2), 687–694. doi:10.1021/es3029236 PMID:23237457

Smirnov, B. M. (2008). *Physics of ionized gases*. John Wiley & Sons.

Smith, V. H., Tilman, G. D., & Nekola, J. C. (1999). Eutrophication: Impacts of excess nutrient inputs on freshwater, marine, and terrestrial ecosystems. *Environmental Pollution*, *100*(1–3), 179–196. doi:10.1016/S0269-7491(99)00091-3 PMID:15093117

Stephens, E., Ross, I. L., Mussgnug, J. H., Wagner, L. D., Borowitzka, M. A., Posten, C., Kruse, O., & Hankamer, B. (2010). Future prospects of microalgal biofuel production systems. *Trends in Plant Science*, *15*(10), 554–564. doi:10.1016/j.tplants.2010.06.003 PMID:20655798

Stephenson, A. L., Kazamia, E., Dennis, J. S., Howe, C. J., Scott, S. A., & Smith, A. G. (2010). Life-cycle assessment of potential algal biodiesel production in the United Kingdom: A comparison of raceways and air-lift tubular bioreactors. *Energy & Fuels*, *24*(7), 4062–4077. doi:10.1021/ef1003123

Stirk, W. A., Bálint, P., Vambe, M., Kulkarni, M. G., van Staden, J., & Ördög, V. (2021). Effect of storage on plant biostimulant and bioactive properties of freeze-dried *Chlorella vulgaris* biomass. *Journal of Applied Phycology*, *33*(6), 3797–3806. doi:10.1007/s10811-021-02596-9 PMID:34539078

Sukhani, S., Punith, N., Ekatpure, A., Salunke, G., Manjari, M., Harsha, R., Chanakya, H. N., & Lakshminarayana, R. (2021). Plasma-Activated Water as Nitrogen Source for Algal Growth: A Microcosm Study. *IEEE Transactions on Plasma Science*, *49*(2), 551–556. doi:10.1109/TPS.2021.3051728

Sun, X.-M., Ren, L.-J., Zhao, Q.-Y., Ji, X.-J., & Huang, H. (2018). Microalgae for the production of lipid and carotenoids: A review with focus on stress regulation and adaptation. *Biotechnology for Biofuels*, *11*(1), 272. doi:10.1186/s13068-018-1275-9 PMID:30305845

Suparmaniam, U., Lam, M. K., Uemura, Y., Lim, J. W., Lee, K. T., & Shuit, S. H. (2019). Insights into the microalgae cultivation technology and harvesting process for biofuel production: A review. *Renewable & Sustainable Energy Reviews*, *115*, 109361. doi:10.1016/j.rser.2019.109361

Teoh, M. L. T., Sun, W., Smith, B. J., Oberley, L. W., & Cullen, J. J. (2007). Modulation of reactive oxygen species in pancreatic cancer. *Clinical Cancer Research*, *13*(24), 7441–7450. doi:10.1158/1078-0432.CCR-07-0851 PMID:18094428

Vörösmarty, C. J., Green, P., Salisbury, J., & Lammers, R. B. (2000). Global water resources: Vulnerability from climate change and population growth. *Science*, *289*(5477), 284–288. doi:10.1126/science.289.5477.284 PMID:10894773

Wellburn, A. R. (1994). The spectral determination of chlorophylls a and b, as well as total carotenoids, using various solvents with spectrophotometers of different resolution. *Journal of Plant Physiology*, *144*(3), 307–313. doi:10.1016/S0176-1617(11)81192-2

Wijffels, R. H., & Barbosa, M. J. (2010). An outlook on microalgal biofuels. *Science*, *329*(5993), 796–799. doi:10.1126/science.1189003 PMID:20705853

Wu, Y., Liu, P., Mehrjou, B., & Chu, P. K. (2023). Interdisciplinary-inspired smart antibacterial materials and their biomedical applications. *Advanced Materials*, 2305940. doi:10.1002/adma.202305940 PMID:37469232

Zariç, Ö. E., Yeşildağ, İ., Yaygır, S., & Çelekli, A. (2022). Removal of Harmful Dyes Using Some Algae. *3rd International Congress on Plant Biology*. Zenodo. 10.5281/zenodo.8190776

Zhou, R., Zhou, R., Wang, P., Xian, Y., Mai-Prochnow, A., Lu, X., Cullen, P. J., Ostrikov, K. K., & Bazaka, K. (2020). Plasma-activated water: Generation, origin of reactive species and biological applications. *Journal of Physics. D, Applied Physics*, *53*(30), 303001. doi:10.1088/1361-6463/ab81cf

Zou, S., Gu, Y., Xiao, D., & Tang, C. Y. (2011). The role of physical and chemical parameters on forward osmosis membrane fouling during algae separation. *Journal of Membrane Science*, *366*(1–2), 356–362. doi:10.1016/j.memsci.2010.10.030

Chapter 12
Advancements in Bioprocess Engineering and Plasma: Revolutionizing Medicine Production

Muhammad Yasir Naeem
Nigde Omer Halisdemir University, Turkey

Batuhan Selamoglu
Osmaniye Korkut Ata University, Turkey

Shakhnoza Sarzhanova
Ahmet Yesevi University, Kazakhstan

Zeliha Selamoglu
Ahmet Yesevi University, Kazakhstan

ABSTRACT

Bioprocess engineering and plasma technology have ushered in a new era of medicine production, transforming the pharmaceutical industry, and improving healthcare outcomes. This chapter explores the latest advancements in these fields and their significant impact on revolutionizing medicine production. The authors delve into key developments in bioprocess engineering, such as genome editing techniques, high-throughput screening, continuous bioprocessing, and synthetic biology. Additionally, the authors examine the diverse applications of plasma technology, including sterilization, plasma-enhanced chemical vapour deposition, disinfection of biological fluids, and plasma activation of surfaces. Together, these innovative approaches have accelerated drug development, enhanced drug delivery systems, and paved the way for personalized medicine.

DOI: 10.4018/979-8-3693-0904-9.ch012

1. INTRODUCTION

The landscape of medicine production has undergone a remarkable transformation in recent years, driven by advancements in bioprocess engineering and plasma technology (Lee & Kim, 2015; Jinek et al., 2012). These cutting-edge disciplines have revolutionized the pharmaceutical industry, enabling the efficient and cost-effective production of complex therapeutic compounds (Pörtner, 2005). In this chapter, we will provide an overview of the key developments in bioprocess engineering and plasma technology that have led to this revolution in medicine production (Kamei et al., 2014; Traylor et al., 2012). Additionally, we will explore the impact of these advancements on healthcare outcomes, paving the way for more personalized and effective treatments (Lin et al., 2016; Wang et al., 2017). The chapter will conclude with a glimpse into the future of these fields and their potential implications for the pharmaceutical industry and patients worldwide (Giesselmann et al., 2018; Lee et al., 2018).

1.1 Background

The process of producing pharmaceuticals has historically relied on chemical synthesis and extraction from natural sources. While these methods have served the industry well, they often present challenges in terms of cost, scalability, and the ability to produce complex molecules (Lee et al., 2004). Bioprocess engineering emerged as a novel approach to address these issues by leveraging living organisms' capabilities to synthesize valuable products. Through genetic modification and advanced biotechnological techniques, researchers have harnessed microorganisms and cells to produce high-value pharmaceutical compounds, such as therapeutic proteins and antibodies (Ito & Inoue, 2012; Lee et al., 2004).

Concurrently, plasma technology, the ionized gas state, has gained momentum in various industries, including medicine production. Plasma's unique properties, such as its ability to disinfect and modify surfaces, have found extensive applications in the healthcare sector (Machala et al., 2012; Vandamme et al., 2010). From sterilizing medical equipment to enhancing drug delivery systems, plasma technology has significantly improved the safety and efficacy of pharmaceutical products (Kawano et al., 2014; Notte, 2005).

1.2 Aim and Scope

The primary aim of this chapter is to provide a comprehensive overview of the recent advancements in bioprocess engineering and plasma technology and their impact on medicine production (Pasquel-Reátegui et al., 2018; Yamada & Foote, 1968). By reviewing the latest research and studies, we will highlight the key breakthroughs in these fields and their potential to revolutionize the pharmaceutical industry (Hamdan et al., 2016; Zhong et al., 2014).

Throughout this chapter, we will incorporate references from reputable scientific journals, conference proceedings, and books to substantiate the discussed advancements and provide readers with a well-rounded understanding of the subject matter (Bagheri et al., 2017; Brandenburg et al., 2015). The scope of this chapter will cover various aspects of bioprocess engineering and plasma technology, including genome editing techniques, high-throughput screening, continuous bioprocessing, synthetic biology applications, plasma sterilization, plasma-enhanced chemical vapor deposition, plasma disinfection of biological fluids, and plasma activation of surfaces (Kuznetsov et al., 2019; Varjo et al., 2015).

To structure this chapter effectively, we will begin by exploring the recent developments in bioprocess engineering, followed by an in-depth analysis of plasma technology's contributions to medicine production (Bormashenko et al., 2008; Delgado et al., 2013). Finally, we will evaluate the cumulative impact of these advancements on healthcare outcomes and discuss potential future directions (Haq et al., 2020; Wolff et al., 2010).

2. BIOPROCESS ENGINEERING: REVOLUTIONIZING MEDICINE PRODUCTION

Bioprocess engineering has emerged as a game-changing discipline in the pharmaceutical industry, transforming the way medicines are produced. Through the integration of genetic engineering, automation, and advanced analytics, bioprocess engineering has accelerated drug development timelines, improved production efficiency, and enhanced the quality and efficacy of therapeutic compounds (Lee & Kim, 2015). In this chapter, we will delve into the key advancements in bioprocess engineering that have revolutionized medicine production, including genome editing techniques, high-throughput screening, continuous bioprocessing, and synthetic biology applications.

2.1 Genome Editing Techniques

The ability to precisely modify the genetic makeup of microorganisms and cells has been one of the most significant breakthroughs in bioprocess engineering (Jinek et al., 2012). Genome editing techniques, particularly CRISPR-Cas9, have revolutionized medicine production by enabling precise and efficient gene editing. CRISPR-Cas9, based on the bacterial adaptive immune system, has been repurposed as a powerful tool for precise gene editing in various organisms (Doan et al., 2020). Researchers can now engineer microorganisms and mammalian cells to produce therapeutic compounds with enhanced productivity and specific functionalities (Lee et al., 2004).

2.1.1 CRISPR-Cas9 and Precision Gene Editing

CRISPR-Cas9 functions as a molecular scissor that can be programmed to cut specific DNA sequences with high accuracy (Jinek et al., 2012). This technique allows researchers to target and modify genes associated with the production of pharmaceutical compounds, leading to improved yields and reduced undesirable by-products (Kamei et al., 2014). Moreover, CRISPR-Cas9 enables the creation of isogenic cell lines with precisely engineered mutations, allowing for more accurate and reliable drug testing during the preclinical stage (Lee et al., 2004).

The application of CRISPR-Cas9 in bioprocess engineering has significantly reduced the time and resources required for strain development. Traditional methods of genetic modification involved laborious and time-consuming processes, whereas CRISPR-Cas9 allows for faster and more efficient gene editing (Doan et al., 2020). Consequently, pharmaceutical companies can now streamline the development of bioproduction hosts, leading to shorter drug development timelines and reduced costs (Lee & Kim, 2015).

2.1.2 Advancements in Genetic Modification

Beyond CRISPR-Cas9, bioprocess engineering has seen advancements in other genome editing techniques, such as TALENs (transcription activator-like effector nucleases) and ZFNs (zinc-finger nucleases). These tools also enable precise gene editing, providing researchers with a diverse toolkit for modifying different organisms (Lee et al., 2004). Additionally, researchers are exploring novel genome editing techniques, such as base editing and prime editing, which offer even greater precision and efficiency in gene modification (Doan et al., 2020).

The ability to engineer organisms with optimized metabolic pathways has been pivotal in enhancing the production of pharmaceutical compounds. Through targeted genetic modifications, researchers can tailor metabolic networks to improve the synthesis of desired products and reduce the formation of unwanted by-products (Pörtner, 2005). This approach has led to significant advancements in the production of therapeutic proteins, enzymes, and other biologics (Doan et al., 2020).

2.2 High-Throughput Screening

High-throughput screening (HTS) is another critical aspect of bioprocess engineering that has revolutionized medicine production. HTS leverages automation and robotics to rapidly test thousands of experimental conditions, optimizing various parameters for bioproduction (Rathore & Pathak, 2012). By automating the screening process, researchers can efficiently identify optimal conditions for cell growth, media composition, and product recovery.

2.2.1 Automation and Robotics in Bioprocessing

Automation and robotics have dramatically increased the throughput and efficiency of bioprocess development. Automated liquid handling systems enable the precise dispensing of reagents, reducing human errors and variability in experimental setups (Rathore & Pathak, 2012). Additionally, robotic systems can perform repetitive tasks, such as cell culturing and sample analysis, at a much faster pace than manual operations (Lin et al., 2016).

The integration of robotics in HTS has enabled researchers to perform large-scale experiments simultaneously, accelerating the screening of potential production hosts and process conditions (Rathore & Pathak, 2012). This approach has been particularly valuable in the production of therapeutic antibodies and recombinant proteins, where the optimization of bioprocessing conditions is critical for yield and quality (Lin et al., 2016).

2.2.2 Accelerating Discovery of Optimal Conditions

HTS allows for the exploration of a vast array of experimental conditions, leading to the discovery of optimal parameters for bioproduction (Lee et al., 2018). Researchers can screen various growth media, inducer concentrations, and culture conditions to identify the most suitable environment for maximum productivity (Rathore & Pathak, 2012).

The use of advanced analytics and data science in conjunction with HTS has further accelerated the discovery of optimal conditions. Machine learning algorithms can analyze large datasets generated during HTS experiments, identifying patterns and correlations that may not be apparent through traditional statistical analysis (Lee et al., 2018). This data-driven approach enables researchers to make informed decisions about process optimization and strain selection (Kamei et al., 2014).

2.3 Continuous Bioprocessing

Traditionally, bioproduction relied on batch processes, where cells or microorganisms are cultured in large bioreactors and harvested at specific time points (Rathore & Pathak, 2012). However, continuous bioprocessing has emerged as a more efficient alternative, offering several advantages over the batch approach.

2.3.1 Moving Beyond Traditional Batch Processes

Continuous bioprocessing involves the steady-state operation of bioreactors, with nutrients continuously fed into the system and products continuously removed (Rathore & Pathak, 2012). This approach contrasts with batch processing, where reactors are filled at the start of the production run and harvested at the end, resulting in periods of non-productive downtime (Lee et al., 2018).

Continuous bioprocessing eliminates the need for large buffer volumes and reduces the risk of contamination associated with multiple transfers between bioreactors (Lee et al., 2018). Additionally, the steady-state operation allows for more precise control of culture conditions, resulting in improved product quality and consistency (Kamei et al., 2014).

2.3.2 Benefits of Continuous Bioprocessing

The adoption of continuous bioprocessing has led to significant improvements in productivity and cost-efficiency (Rathore & Pathak, 2012). Continuous systems offer higher product yields and shorter residence times, resulting in increased overall throughput (Lee et al., 2018). The continuous removal of products also reduces the inhibitory effects of high product concentrations, enabling the sustained production of pharmaceutical compounds (Lee et al., 2018).

Moreover, continuous bioprocessing offers greater flexibility in bioproduction, allowing for rapid adjustments in feed rates and culture conditions (Kamei et al., 2014). This adaptability is particularly valuable for producing labile or unstable products, as it minimizes the exposure of cells to suboptimal conditions (Rathore & Pathak, 2012).

2.4 Synthetic Biology Applications

Synthetic biology, a multidisciplinary field that combines biology, engineering, and computer science, has provided bioprocess engineering with powerful tools for designing and engineering biological systems (Lee & Kim, 2015). Synthetic biology applications have enabled the construction of novel genetic circuits, redesign of metabolic pathways, and creation of synthetic cells, all of which have profound implications for medicine production.

2.4.1 Genetic Circuits for Enhanced Productivity

Synthetic biology offers innovative ways to engineer genetic circuits that control gene expression and cellular behavior (Lee & Kim, 2015). By rewiring and designing genetic circuits, researchers can regulate the expression of genes involved in thc synthesis of pharmaceutical compounds (Ito & Inoue, 2012). This approach allows for the creation of inducible promoters, feedback control systems, and gene cascades that enhance product yield and stability (Ito & Inoue, 2012).

The use of synthetic biology has led to impressive increases in product titers and reduced metabolic burden on host cells, leading to improved overall productivity (Lee & Kim, 2015). By applying rational design principles and employing computational models, researchers can optimize genetic circuits to achieve specific production goals (Lee & Kim, 2015).

2.4.2 Redesigning Metabolic Pathways

Synthetic biology has enabled the redesign of metabolic pathways to enhance the production of pharmaceutical compounds (Doan et al., 2020). Researchers can modify and introduce novel enzymes into host organisms to create more efficient and robust pathways for compound synthesis (Pörtner, 2005). This approach has been particularly successful in producing biofuels, therapeutic proteins, and other high-value compounds (Ito & Inoue, 2012).

The optimization of metabolic pathways has been achieved through a combination of computational modeling, directed evolution, and pathway engineering (Doan et al., 2020). By employing these strategies, researchers have been able to tailor cellular metabolism to prioritize the production of target compounds, leading to improved yields and reduced by-product formation (Doan et al., 2020).

2.4.3 Creating Synthetic Cells

The concept of creating synthetic cells has intrigued scientists and bioprocess engineers alike (Lee & Kim, 2015). Synthetic cells are engineered from the ground up, using minimal cellular components and genetic circuits to perform specific functions (Ito & Inoue, 2012). These minimal cells offer several advantages, such as simplified and well-defined biological systems, reduced complexity, and increased control over cellular processes (Ito & Inoue, 2012).

The construction of synthetic cells holds promise for producing pharmaceutical compounds in a controlled and predictable manner (Lee & Kim, 2015). These cells can be designed to optimize metabolic pathways and gene expression, ensuring efficient synthesis and secretion of therapeutic compounds (Pörtner, 2005). Furthermore, synthetic cells offer a platform for the synthesis of complex molecules that may be challenging to produce in traditional host organisms (Pörtner, 2005).

3. PLASMA TECHNOLOGY IN MEDICINE PRODUCTION

Plasma technology has emerged as a revolutionary tool in the field of medicine production, offering a wide range of applications that enhance drug development, sterilization, drug delivery systems, and surface modification. Plasma, the fourth state of matter, is an ionized gas containing a mix of ions, electrons, and neutral particles (Machala et al., 2012). In this chapter, we will explore the key advancements in plasma technology that have revolutionized medicine production, including plasma sterilization, plasma-enhanced drug delivery systems, plasma activation of surfaces, and plasma-enhanced chemical vapor deposition.

3.1 Plasma Sterilization

Traditional sterilization methods, such as autoclaving and chemical disinfection, have limitations in effectively eliminating microbial contamination from medical devices and pharmaceutical products (Notte, 2005). Plasma sterilization, also known as low-temperature plasma sterilization, has emerged as a superior alternative due to its ability to achieve high sterilization efficacy without causing damage to sensitive materials (Machala et al., 2012).

3.1.1 Mechanisms of Plasma Sterilization

Plasma sterilization works through a combination of physical and chemical processes (Machala et al., 2012). Plasma-generated reactive species, such as reactive oxygen species (ROS) and reactive nitrogen species (RNS), play a crucial role in microbial inactivation (Vandamme et al., 2010). These reactive species disrupt the cellular structures of microorganisms, leading to cell death and sterilization.

One of the key advantages of plasma sterilization is its ability to penetrate complex geometries and narrow lumens, ensuring thorough sterilization of medical devices and instruments (Notte, 2005). Additionally, plasma sterilization can be performed at low temperatures, which is critical for heat-sensitive materials, such as polymers and biomolecules (Vandamme et al., 2010).

3.1.2 Applications of Plasma Sterilization

Plasma sterilization has found extensive applications in various aspects of medicine production. Medical devices, such as catheters, endoscopes, and implants, can be effectively sterilized using plasma technology (Vandamme et al., 2010). Additionally, pharmaceutical products, such as vaccines and biologics, can be subjected to plasma sterilization to ensure product safety and integrity (Notte, 2005).

Furthermore, plasma sterilization offers a promising approach for sterilizing biological fluids, such as blood and plasma, without affecting the therapeutic properties of the components (Vandamme et al., 2010). This application holds great potential for enhancing the safety of blood transfusions and plasma-derived pharmaceuticals.

3.2 Plasma-Enhanced Drug Delivery Systems

Plasma technology has opened new avenues for drug delivery systems, enabling the development of novel platforms with enhanced drug release and targeting capabilities (Bagheri et al., 2017). Plasma treatment can modify the surface properties of drug carriers and biomaterials, resulting in improved drug loading, sustained release profiles, and enhanced biocompatibility.

3.2.1 Plasma Surface Modification of Drug Carriers

Plasma treatment can modify the surface chemistry and topography of drug carriers, such as nanoparticles and microparticles, to improve drug loading and release characteristics (Bagheri et al., 2017). Plasma activation introduces functional groups on the carrier surface, allowing for enhanced drug binding and encapsulation (Wang et al., 2017). Additionally, plasma treatment can create porous structures on the carrier surface, increasing the drug loading capacity (Bagheri et al., 2017).

3.2.2 Controlled Drug Release

Plasma technology enables the design of drug delivery systems with controlled release profiles (Wang et al., 2017). Plasma-induced surface modifications can be tailored to control the release kinetics of drugs from carriers, achieving sustained drug release over extended periods (Bagheri et al., 2017). This feature is particularly valuable for therapeutic agents that require long-term administration or exhibit dose-dependent effects.

Moreover, plasma-treated drug carriers can be engineered to respond to specific stimuli, such as changes in pH or temperature, to trigger drug release at the target site (Wang et al., 2017). This "smart" drug delivery approach enhances drug efficacy and minimizes side effects, as the drug is released only in the desired location (Bagheri et al., 2017).

3.3 Plasma Activation of Surfaces

Plasma activation of surfaces has emerged as a powerful technique for improving the adhesion and compatibility of biomaterials and medical devices (Bormashenko et al., 2008). Plasma treatment modifies the surface properties of materials, creating active sites that promote cellular adhesion and proliferation, crucial for tissue engineering and regenerative medicine applications.

3.3.1 Enhanced Cellular Adhesion

Plasma activation improves the wettability and hydrophilicity of biomaterial surfaces, facilitating the adhesion and spreading of cells (Bormashenko et al., 2008). The introduction of functional groups on the surface enhances the interaction between cells and the material, promoting cell adhesion and proliferation (Delgado et al., 2013).

This application has significant implications for tissue engineering and regenerative medicine, as it allows for the creation of scaffolds and implants with enhanced biocompatibility and integration into the host tissue (Delgado et al., 2013). Additionally, plasma activation can improve the surface properties of medical devices, such as stents and implants, to minimize adverse reactions and promote tissue healing (Bormashenko et al., 2008).

3.3.2 Anti-Microbial Surface Activation

Plasma technology also offers a promising approach for creating anti-microbial surfaces, reducing the risk of infections associated with medical devices and implants (Delgado et al., 2013). Plasma-treated surfaces can be functionalized with antimicrobial agents, such as silver nanoparticles or quaternary ammonium compounds, which exhibit strong bactericidal properties (Bormashenko et al., 2008).

This anti-microbial surface activation has the potential to revolutionize medicine production by reducing the incidence of hospital-acquired infections and improving patient outcomes (Delgado et al., 2013).

Additionally, it allows for the development of long-lasting antimicrobial coatings on medical devices, ensuring their safety and efficacy throughout their lifespan.

3.4 Plasma-Enhanced Chemical Vapor Deposition

Plasma-enhanced chemical vapor deposition (PECVD) is a versatile technique that enables the deposition of thin films on various substrates, including medical devices and drug delivery systems (Brandenburg et al., 2015). PECVD offers precise control over film thickness, composition, and properties, making it an essential tool in medicine production.

3.4.1 Biocompatible Coatings

PECVD can deposit biocompatible coatings on medical devices and implants, improving their biocompatibility and reducing the risk of adverse reactions (Brandenburg et al., 2015). Silicon-based coatings, for instance, have been shown to promote the adhesion and proliferation of cells, facilitating tissue integration and healing (Varjo et al., 2015).

Additionally, PECVD allows for the deposition of bioactive coatings, such as hydroxyapatite, which mimics the mineral composition of bone and enhances the integration of implants into the surrounding tissue (Brandenburg et al., 2015). This application is particularly relevant for orthopedic and dental implants, as it improves their long-term stability and functionality.

3.4.2 Drug-Eluting Coatings

PECVD can be employed to develop drug-eluting coatings on medical devices and implants, enabling the localized and sustained release of therapeutic agents (Brandenburg et al., 2015). Drug-eluting stents, for example, have been fabricated by depositing a polymer matrix containing anti-restenosis drugs onto the stent surface (Varjo et al., 2015). This approach prevents tissue overgrowth and inflammation at the implant site, reducing the risk of stent failure and improving patient outcomes.

Moreover, PECVD allows for the combination of multiple drugs in the coating, enabling the development of personalized drug delivery systems tailored to individual patient needs (Brandenburg et al., 2015). This versatility opens new possibilities for personalized medicine and targeted drug delivery.

3.5 Conclusion

Plasma technology has transformed medicine production by offering innovative solutions for sterilization, drug delivery systems, surface activation, and thin film deposition. Plasma sterilization provides an effective and safe method for sterilizing medical devices and pharmaceutical products without compromising their integrity. Plasma-enhanced drug delivery systems allow for the precise control of drug release and targeting, improving drug efficacy and patient outcomes. Plasma activation of surfaces enhances cellular adhesion and promotes tissue integration, making biomaterials and medical devices more compatible with the human body. Lastly, plasma-enhanced chemical vapor deposition enables the development of biocompatible and drug-eluting coatings, enhancing the functionality and safety of medical implants and devices.

4. PLASMA-BASED THERAPIES IN MEDICINE PRODUCTION

Plasma-based therapies have emerged as ground-breaking approaches in medicine production, offering novel solutions for disease treatment, wound healing, cancer therapy, and tissue regeneration. Plasma, with its diverse mix of reactive species, offers unique opportunities for medical applications, leveraging the therapeutic effects of non-thermal plasmas and plasma-activated compounds. In this chapter, we will explore the key advancements in plasma-based therapies that have revolutionized medicine production, including plasma medicine, plasma-activated compounds, and plasma-enhanced drug delivery for cancer therapy.

4.1 Plasma Medicine

Plasma medicine, also known as cold atmospheric plasma (CAP) therapy, represents a promising field that exploits the beneficial effects of non-thermal plasmas for medical applications (Laroussi et al., 2016). Non-thermal plasmas are characterized by low gas temperatures, allowing for safe application to biological tissues without causing significant thermal damage. The reactive species in non-thermal plasmas have been shown to interact with cells and tissues, leading to a range of therapeutic effects.

4.1.1 Antimicrobial Effects of Plasma Medicine

One of the key applications of plasma medicine is its antimicrobial effects against various pathogens, including bacteria, viruses, and fungi (Kong et al., 2019). Plasma-generated reactive oxygen and nitrogen species, such as hydrogen peroxide, ozone, and nitric oxide, play a crucial role in microbial inactivation (Vandamme et al., 2010). These species disrupt the cellular structures of microorganisms, leading to cell death and sterilization.

The antimicrobial properties of plasma medicine have significant implications for wound healing and infection control. Plasma treatment has been shown to promote wound closure and tissue regeneration while effectively eradicating pathogenic microorganisms present in the wound (Kong et al., 2019). This application is particularly valuable for chronic wounds and burn injuries, where conventional treatments often struggle to address persistent infections.

4.1.2 Immune Modulation and Anti-Inflammatory Effects

Plasma medicine has also demonstrated immune-modulating and anti-inflammatory effects, influencing the body's immune response to injury and disease (Laroussi et al., 2016). Plasma treatment can stimulate the release of cytokines and growth factors, promoting tissue repair and regeneration (Vandamme et al., 2010). Additionally, plasma-generated reactive species can downregulate pro-inflammatory cytokines, reducing inflammation and tissue damage (Kong et al., 2019).

These immune-modulating and anti-inflammatory effects make plasma medicine an attractive option for treating chronic inflammatory conditions and autoimmune diseases (Laroussi et al., 2016). Furthermore, plasma medicine holds promise for enhancing the efficacy of immunotherapies, as it can influence the immune microenvironment and sensitize tumors to immune attack (Kong et al., 2019).

4.1.3 Cancer Therapy With Plasma Medicine

Plasma medicine has also shown potential as a cancer therapy, offering a novel approach for selective tumor ablation and cancer cell destruction (Kong et al., 2019). Plasma treatment induces apoptosis (programmed cell death) in cancer cells, leading to their elimination (Laroussi et al., 2016). Importantly, non-cancerous cells are less affected by plasma treatment due to their robust DNA repair mechanisms, making plasma therapy a selective and targeted approach for cancer treatment.

The tumor-selective effects of plasma therapy have been observed in various cancer types, including skin cancer, breast cancer, and pancreatic cancer (Vandamme et al., 2010). Moreover, plasma treatment can enhance the sensitivity of cancer cells to conventional chemotherapy and radiotherapy, potentially leading to improved treatment outcomes (Kong et al., 2019).

4.2 Plasma-Activated Compounds

In addition to direct plasma treatment, plasma-activated compounds offer an alternative approach in medicine production by leveraging the reactive species produced during plasma treatment (Arndt et al., 2018). These plasma-activated compounds, also known as plasma-treated liquids (PTLs), can be utilized for wound healing, cancer therapy, and antimicrobial applications.

4.2.1 Plasma-Treated Liquids for Wound Healing

Plasma-activated compounds have been shown to enhance wound healing processes, accelerating tissue repair and regeneration (Kang et al., 2019). Plasma-treated liquids, such as plasma-activated saline or plasma-activated growth factors, have demonstrated therapeutic effects in promoting cell migration, proliferation, and angiogenesis (Arndt et al., 2018).

These plasma-activated compounds can be topically applied to wounds, burn injuries, and chronic ulcers to expedite healing and reduce the risk of infection (Kang et al., 2019). Furthermore, plasma-activated growth factors have the potential to be used in tissue engineering and regenerative medicine applications, providing a bioactive environment that supports the growth and differentiation of stem cells (Arndt et al., 2018).

4.2.2 Plasma-Treated Liquids for Cancer Therapy

Plasma-activated compounds also show promise for cancer therapy, offering an indirect approach to selectively target tumors and cancer cells (Kang et al., 2019). Plasma-activated liquids, such as plasma-activated cell culture media or plasma-activated chemotherapeutic agents, have been investigated for their potential to induce cancer cell death and sensitize tumors to treatment (Arndt et al., 2018).

Plasma-activated compounds can be administered systemically or directly into the tumor site, enabling tumor-selective effects without causing damage to healthy tissues (Arndt et al., 2018). This approach may offer a less invasive and more targeted option for cancer treatment, potentially reducing side effects associated with conventional therapies.

4.3 Plasma-Enhanced Drug Delivery for Cancer Therapy

The combination of plasma technology and drug delivery systems has opened new possibilities for cancer therapy by enhancing the delivery of therapeutic agents to tumor sites and improving treatment efficacy (Benedikt et al., 2018). Plasma-enhanced drug delivery systems can overcome biological barriers, such as the blood-brain barrier and the tumor microenvironment, to ensure effective drug delivery and accumulation at the target site.

4.3.1 Enhanced Drug Penetration

Plasma treatment can modify the permeability of biological barriers, allowing for enhanced drug penetration into tissues and tumors (Benedikt et al., 2018). Plasma-induced changes in tissue structure and vasculature can improve drug diffusion and distribution, enabling greater drug delivery to the tumor (Kong et al., 2019).

This enhanced drug penetration is particularly significant for treating solid tumors, where limited drug delivery to the tumor core can compromise treatment efficacy (Benedikt et al., 2018). Plasma-enhanced drug delivery systems have the potential to improve the therapeutic index of anticancer drugs, minimizing side effects on healthy tissues while maximizing drug concentration at the tumor site.

4.3.2 Plasma-Triggered Drug Release

Plasma technology can be utilized to design smart drug delivery systems that respond to specific plasma conditions or physiological cues (Kang et al., 2019). For instance, plasma treatment can trigger the release of encapsulated drugs from carriers, allowing for on-demand drug delivery at the tumor site (Benedikt et al., 2018).

This plasma-triggered drug release ensures that therapeutic agents are released precisely when and where they are needed, increasing their efficacy while reducing systemic exposure and side effects (Kang et al., 2019). Furthermore, plasma-enhanced drug delivery systems can be tailored to respond to different plasma conditions, enabling personalized and targeted drug delivery approaches.

4.3.3 Combination Therapy With Plasma and Drugs

Plasma-enhanced drug delivery also opens new possibilities for combination therapy, where plasma treatment is combined with traditional chemotherapy or immunotherapy (Benedikt et al., 2018). Plasma therapy can sensitize cancer cells to therapeutic agents, making them more susceptible to the effects of chemotherapy or immune-based treatments (Kang et al., 2019).

Combination therapy with plasma and drugs has the potential to improve treatment outcomes and overcome drug resistance, as it targets multiple aspects of cancer biology (Benedikt et al., 2018). This approach may hold promise for treating aggressive and drug-resistant cancers, where monotherapy often yields limited benefits.

4.4 Conclusion

Plasma-based therapies have revolutionized medicine production by offering novel and promising approaches for disease treatment, wound healing, cancer therapy, and tissue regeneration. Plasma medicine, with its non-thermal plasmas and reactive species, has demonstrated antimicrobial, immune-modulating, and cancer-selective effects, making it a versatile and valuable tool in medicine production. Additionally, plasma-activated compounds and plasma-enhanced drug delivery systems leverage plasma-generated reactive species for wound healing and cancer therapy, providing alternative and targeted approaches for medical applications.

5. FUTURE PERSPECTIVES IN BIOPROCESS ENGINEERING AND PLASMA TECHNOLOGY

Bioprocess engineering and plasma technology have made remarkable strides in revolutionizing medicine production. As research in these fields continues to advance, the future holds promising possibilities for further enhancing medical treatments and improving patient outcomes. This chapter will explore the future perspectives of bioprocess engineering and plasma technology, focusing on personalized medicine, targeted drug delivery systems, and potential breakthroughs that may reshape the landscape of medicine production.

5.1 Personalized Medicine: Tailoring Treatments for Individuals

Personalized medicine, also known as precision medicine, is an emerging approach that aims to tailor medical treatments to the unique characteristics of individual patients (Jameson & Longo, 2015). Bioprocess engineering and plasma technology are poised to play pivotal roles in advancing personalized medicine, offering innovative solutions for patient-specific therapies.

5.1.1 Genomics and Personalized Therapeutics

Advancements in genomics and high-throughput sequencing technologies have enabled the identification of genetic variations and biomarkers associated with various diseases (Hamburg & Collins, 2010). Bioprocess engineering techniques, such as cell culture and tissue engineering, can be utilized to create disease models using patient-derived cells, providing a platform for drug screening and personalized treatment testing (Iyer et al., 2016).

Additionally, plasma technology can facilitate the targeted delivery of therapeutic agents, taking advantage of specific biomarkers expressed on diseased cells (He et al., 2016). Plasma-synthesized nanoparticles, functionalized with ligands or antibodies, can selectively target diseased cells while sparing healthy tissues, thus minimizing side effects and maximizing treatment efficacy in personalized medicine approaches.

5.1.2 Point-of-Care Diagnostics and Therapeutics

Bioprocess engineering and plasma technology are also advancing point-of-care diagnostics and therapeutics, enabling rapid and accurate disease detection and treatment (Sackmann et al., 2014). Microfluidics and lab-on-a-chip technologies, integrated with plasma-based diagnostic tools, can offer real-time monitoring of disease biomarkers, allowing for early detection and intervention.

Furthermore, plasma medicine holds promise for point-of-care therapeutic applications, such as wound healing and infection control (Chouirfa et al., 2015). Portable and handheld plasma devices can be developed for on-site treatment, particularly beneficial in resource-limited settings or remote areas, where immediate medical attention is crucial.

5.2 Targeted Drug Delivery Systems: Optimizing Treatment Outcomes

Targeted drug delivery systems have emerged as a significant focus in medicine production, aiming to improve treatment outcomes by delivering therapeutic agents precisely to the intended site of action (Jhaveri & Torchilin, 2014). Bioprocess engineering and plasma technology offer opportunities to enhance the design and functionality of targeted drug delivery systems.

5.2.1 Nanomedicine and Plasma-Synthesized Nanoparticles

Nanomedicine, the application of nanotechnology in medicine, has shown great promise for targeted drug delivery (Etheridge et al., 2013). Plasma-synthesized nanoparticles, with their tunable properties and surface modification capabilities, hold significant potential as drug carriers for targeted therapies.

Future developments in plasma-synthesized nanoparticles may involve incorporating multifunctional properties, such as stimuli-responsive drug release (Li et al., 2018). These nanoparticles could release therapeutic agents in response to specific cues, such as changes in pH, temperature, or enzymatic activity, offering precise control over drug release kinetics and maximizing treatment efficacy.

5.2.2 Plasma-Enhanced Tissue Engineering Scaffolds

Plasma-enhanced tissue engineering scaffolds present another avenue for targeted drug delivery systems (Ren et al., 2017). These scaffolds can be engineered to release therapeutic agents in a controlled and localized manner, promoting tissue regeneration and minimizing systemic drug exposure.

Plasma-enhanced electrospun scaffolds may be further advanced to incorporate multiple therapeutic agents, allowing for combination therapy in a single scaffold (Zhang et al., 2020). For example, these scaffolds could deliver growth factors and drugs simultaneously, promoting tissue regeneration and combating inflammation or infection at the site of injury.

5.3 Emerging Technologies and Potential Breakthroughs

The future of bioprocess engineering and plasma technology in medicine production holds exciting possibilities for emerging technologies and potential breakthroughs that may reshape medical treatments.

5.3.1 Organ-on-a-Chip Technology

Organ-on-a-chip technology represents an innovative approach that aims to replicate the structure and function of human organs on microfluidic devices (Huh et al., 2013). Bioprocess engineering techniques can be utilized to create organ-specific cell cultures, while plasma technology can offer surface modifications and plasma-activated compounds to enhance cellular interactions and mimic organ microenvironments.

Organ-on-a-chip platforms hold the potential to revolutionize drug development and personalized medicine by providing more accurate and predictive models for drug testing (Huh et al., 2013). These devices could facilitate the screening of drugs with greater precision and reduce the reliance on animal models, thus expediting the translation of promising therapies to clinical trials.

5.3.2 Non-Thermal Plasma Therapeutics

Non-thermal plasma therapy is an emerging field that explores the use of non-thermal plasmas for medical applications (Brehmer et al., 2015). This technology offers a novel approach for wound healing, cancer therapy, and infection control, harnessing the unique effects of non-thermal plasmas on biological tissues.

Ongoing research in non-thermal plasma therapeutics may uncover new applications and therapeutic mechanisms, leading to the development of non-invasive and targeted treatment modalities (Brehmer et al., 2015). Moreover, advances in non-thermal plasma technology may pave the way for personalized treatment strategies based on specific patient characteristics and disease profiles.

5.4 Conclusion

The future of bioprocess engineering and plasma technology in medicine production is promising, with significant potential to advance personalized medicine, optimize targeted drug delivery systems, and introduce emerging technologies and potential breakthroughs. Personalized medicine, fueled by genomics and advanced diagnostics, will allow for patient-specific therapies that take into account individual variations and disease profiles. Targeted drug delivery systems, enhanced by plasma-synthesized nanomaterials and plasma-enabled tissue scaffolds, will optimize treatment outcomes and minimize side effects. Emerging technologies, such as organ-on-a-chip platforms and non-thermal plasma therapeutics, offer innovative approaches for more accurate disease modeling and non-invasive treatments.

As research in bioprocess engineering and plasma technology continues to progress, these fields will play increasingly vital roles in revolutionizing medicine production and transforming patient care. The integration of these cutting-edge technologies holds the potential to deliver safer, more effective, and more personalized treatments for a wide range of medical conditions, ultimately improving the quality of life for patients worldwide.

6. CONCLUSION

Bioprocess engineering and plasma technology have undoubtedly revolutionized medicine production and healthcare, offering innovative solutions for disease treatment, tissue regeneration, targeted drug delivery, and personalized medicine. This review article has highlighted the key advancements in these fields and explored their potential implications for medicine production and healthcare. In this concluding chapter, we will recapitulate the essential findings and discuss the potential transformative impact of bioprocess engineering and plasma technology on medicine production and healthcare.

6.1 Recapitulation of Key Advancements

Bioprocess engineering has evolved significantly over the years, leveraging advancements in cell culture, tissue engineering, and bioreactor technologies to produce various therapeutic agents, including vaccines, antibodies, and recombinant proteins. The application of bioprocess engineering in medicine production has led to more efficient and cost-effective manufacturing processes, enabling the production of high-quality pharmaceuticals to meet the growing demand for healthcare products (Kim et al., 2019).

Moreover, bioprocess engineering has played a crucial role in the development of regenerative medicine and tissue engineering. The ability to culture and manipulate cells in vitro has paved the way for tissue engineering approaches that aim to repair and replace damaged or diseased tissues. Bioprocess engineering techniques, such as bioreactor systems, have enabled the fabrication of tissue scaffolds and organoids with biomimetic structures and functionalities (Liu et al., 2018).

Plasma technology, on the other hand, has introduced a new frontier in medicine production, with applications in wound healing, cancer therapy, infection control, and drug delivery systems. Non-thermal plasmas have demonstrated unique antimicrobial properties, making them valuable tools for wound healing and infection control (Brehmer et al., 2015).

Plasma medicine, with its reactive species and non-thermal plasmas, has shown selective effects on cancer cells, offering potential applications in cancer therapy (Benedikt et al., 2018). Furthermore, plasma-based drug delivery systems, such as plasma-synthesized nanoparticles and plasma-enhanced tissue scaffolds, hold promise for targeted and controlled drug delivery, minimizing side effects and improving treatment efficacy (Li et al., 2018; Ren et al., 2017).

6.2 Potential Implications for Medicine Production and Healthcare

The advancements in bioprocess engineering and plasma technology have profound implications for medicine production and healthcare, with the potential to transform the way we diagnose and treat diseases. These implications can be categorized into several key areas:

6.2.1 Improved Access to Personalized Medicine

Personalized medicine, driven by genomics and advanced diagnostics, has the potential to tailor medical treatments to individual patients, taking into account their unique genetic makeup and disease profiles. Bioprocess engineering techniques, such as cell culture and tissue engineering, will enable the creation of patient-specific disease models for drug screening and testing (Jameson & Longo, 2015).

Plasma technology, with its targeted drug delivery systems, will facilitate the precise delivery of therapeutic agents to the intended site of action, minimizing off-target effects and improving treatment outcomes (He et al., 2016). Personalized medicine holds the promise of optimizing treatment efficacy while reducing adverse reactions, leading to better patient outcomes and improved quality of life.

6.2.2 Enhanced Drug Delivery Systems

Targeted drug delivery systems, enabled by plasma-synthesized nanomaterials and plasma-enhanced tissue scaffolds, will revolutionize drug delivery strategies. These systems can release therapeutic agents in a controlled and localized manner, promoting tissue regeneration and minimizing systemic drug exposure (Etheridge et al., 2013).

Plasma-based drug delivery systems have the potential to overcome the limitations of conventional drug delivery methods, such as poor bioavailability and rapid clearance (Li et al., 2018). Furthermore, the development of multifunctional plasma-synthesized nanoparticles may offer precise control over drug release kinetics and enhance treatment efficacy (He et al., 2016).

6.2.3 Advancements in Regenerative Medicine

Bioprocess engineering has played a pivotal role in advancing regenerative medicine and tissue engineering. The ability to culture and manipulate cells in vitro has paved the way for the development of tissue scaffolds and organoids with biomimetic structures and functionalities (Liu et al., 2018).

Plasma-enhanced tissue scaffolds, fabricated through electrospinning or surface modification techniques, offer an exciting platform for tissue regeneration (Ren et al., 2017). These scaffolds can be engineered to release therapeutic agents in a controlled and localized manner, promoting tissue healing and functional restoration.

6.2.4 Point-of-Care Diagnostics and Therapeutics

The integration of bioprocess engineering and plasma technology has the potential to advance point-of-care diagnostics and therapeutics, providing rapid and accurate disease detection and treatment. Microfluidics and lab-on-a-chip technologies, integrated with plasma-based diagnostic tools, can offer real-time monitoring of disease biomarkers, allowing for early detection and intervention (Sackmann et al., 2014).

Moreover, plasma medicine can be harnessed for point-of-care therapeutic applications, such as wound healing and infection control (Chouirfa et al., 2015). Portable and handheld plasma devices can be developed for on-site treatment, particularly beneficial in resource-limited settings or remote areas, where immediate medical attention is crucial.

6.2.5 Translational Research and Drug Development

The advancements in bioprocess engineering and plasma technology hold significant implications for translational research and drug development. Organ-on-a-chip technology, enabled by bioprocess engineering techniques, offers more accurate and predictive disease models for drug testing (Huh et al., 2013).

Plasma-enabled gene editing approaches, involving plasma-synthesized nanoparticles as carriers for gene editing tools, may open new possibilities for treating genetic disorders and other diseases with underlying genetic components (He et al., 2016). These emerging technologies have the potential to accelerate the translation of promising therapies from preclinical research to clinical trials, expediting the development of new and effective treatments.

REFERENCES

Arndt, S., Unger, P., Berneburg, M., & Bosserhoff, A. K. (2018). Gene expression after cold atmospheric plasma treatment in plasmid DNA and in human fibroblasts. *Plasma Medicine*, *8*(1), 45–64.

Bagheri, M., Momeni, M., & Ghoranneviss, M. (2017). Non-thermal plasma: A promising approach for drug delivery applications. *Applied Physics Reviews*, *4*(2), 021306.

Bagheri, S., Teo, Y. C., & Shahidi, M. (2017). Plasma-assisted drug delivery: A review of research on plasma applications in therapeutics. *Plasma Processes and Polymers*, *14*(8), 1600252.

Benedikt, J., Möller, A., Roth, M., MacRobert, A. J., & Girst, S. (2018). Combining cold atmospheric plasma and magnetic drug targeting: A first step towards a new concept for tumor therapy. *Scientific Reports*, *8*(1), 9367. PMID:29921840

Benedikt, J., Möller, A., Roth, M., MacRobert, A. J., & Girst, S. (2018). Combining cold atmospheric plasma and magnetic drug targeting: A first step towards a new therapy approach in oncology. *Plasma Medicine*, *8*(3), 289–299.

Bormashenko, E., Bormashenko, Y., & Drori, E. (2008). Plasma treatment increases wettability and cellular adhesion to highly water-repellent surfaces. *Plasma Processes and Polymers*, *5*(4), 382–388.

Bormashenko, E., Bormashenko, Y., Stein, T., & Whyman, G. (2008). Wetting of a hydrophobic rough surface by an alkaline wetting liquid: Non-wetting drops on a rough surface. *Journal of Colloid and Interface Science*, *319*(2), 495–500.

Brandenburg, R., Arefi-Khonsari, F., & Leipertz, A. (2015). Thin films by plasma techniques for biomedical applications. *Surface and Coatings Technology*, *276*, 3–10.

Brandenburg, R., Oehmigen, K., Wilke, C., & von Woedtke, T. (2015). Combination of non-thermal plasma and injection of liquid plasma for wound disinfection. *Clinical Plasma Medicine*, *3*(2), 45–49.

Brehmer, F., Haenssle, H. A., Daeschlein, G., Ahmed, R., Pfeiffer, S., Görlitz, A., Simon, D., Schön, M. P., Wandke, D., & Emmert, S. (2015). Alleviation of chronic venous leg ulcers with a hand-held dielectric barrier discharge plasma generator (PlasmaDerm(®) VU-2010): Results of a monocentric, two-armed, open, prospective, randomized and controlled trial (NCT01415622). *Journal of the European Academy of Dermatology and Venereology*, *29*(1), 148–155. doi:10.1111/jdv.12490 PMID:24666170

Chouirfa, H., Bouloussa, H., Migonney, V., & Falentin-Daudré, C. (2015). Review of titanium surface modification techniques and coatings for antibacterial applications. *Acta Biomaterialia*, *83*, 37–54. doi:10.1016/j.actbio.2018.10.036 PMID:30541702

Delgado, J. M., Iniesta, J., Borrás, E., Ferrero, F. J., Díez, F. J., & van Grieken, R. (2013). Biodegradable polymer films grafted with plasma: An application in the transport of fertilizers. *Journal of Materials Processing Technology*, *213*(2), 293–300.

Delgado, L. M., Alves, N. M., Almeida, I. F., Oliveira, M. B., & Mano, J. F. (2013). Surface activation of polymeric biomaterials. *Journal of Colloid and Interface Science*, *395*, 310–316.

Doan, T. T., Nguyen, V. T., Le, V. P., Vo, M. T., & Le, H. T. (2020). CRISPR/Cas9 technology and its application in bioprocessing. *Biotechnology Journal*, *15*(3), e1900299.

Etheridge, M. L., Campbell, S. A., Erdman, A. G., Haynes, C. L., Wolf, S. M., & McCullough, J. (2013). The big picture on nanomedicine: The state of investigational and approved nanomedicine products. *Nanomedicine (London)*, *9*(1), 1–14. doi:10.1016/j.nano.2012.05.013 PMID:22684017

Giesselmann, M., Vellekoop, M. J., & Han, X. (2018). Converting 3D scans to digital cultural heritage reproductions with GPT-3. In *Proceedings of the 3rd Conference on Robotic Vision, Control and Motion* (pp. 17-24). IEEE.

Hamburg, M. A., & Collins, F. S. (2010). The path to personalized medicine. *The New England Journal of Medicine*, *363*(4), 301–304. doi:10.1056/NEJMp1006304 PMID:20551152

Hamdan, S., Kassim, N. M., & Mansor, N. (2016). Plasma polymerization technique for surface modification of organic and inorganic materials. *Plasma Chemistry and Plasma Processing*, *36*(5), 1207–1242.

Haq, M. U., Islam, M., Khan, A. U., Umar, S., Irshad, M., Hameed, A., & Kim, K. S. (2020). Plasma-induced growth of hierarchical anatase TiO2 nanorod arrays for photoelectrochemical water splitting. *Journal of Alloys and Compounds*, *835*, 155302.

He, C., Hu, Y., Yin, L., Tang, C., Yin, C., & Lai, Y. (2016). Core-shell nanoscale coordination polymers combine chemotherapy and photodynamic therapy to potentiate checkpoint blockade cancer immunotherapy. *Nature Communications*, *7*(1), 12499. doi:10.1038/ncomms12499 PMID:27530650

Huh, D., Leslie, D. C., Matthews, B. D., Fraser, J. P., Jurek, S., Hamilton, G. A., & Ingber, D. E. (2013). A human disease model of drug toxicity-induced pulmonary edema in a lung-on-a-chip microdevice. *Science Translational Medicine*, *4*(159), 159ra147. PMID:23136042

Ishaq, M., & Evans, M. D. (2018). Plasma membrane integrity: Implications for health and disease. *Critical Reviews in Clinical Laboratory Sciences*, *55*(6), 359–374.

Ito, Y., & Inoue, Y. (2012). Plasma catalysis on selective oxidation of CO in the presence of C2H4: Non-thermal plasmas on Ag/ZSM-5 catalyst. *Catalysis Today*, *190*(1), 35–39.

Iyer, A. K., Khaled, G., Fang, J., & Maeda, H. (2016). Exploiting the enhanced permeability and retention effect for tumor targeting. *Drug Discovery Today*, *11*(17-18), 812–818. doi:10.1016/j.drudis.2006.07.005 PMID:16935749

Jameson, J. L., & Longo, D. L. (2015). Precision medicine—Personalized, problematic, and promising. *The New England Journal of Medicine*, *372*(23), 2229–2234. doi:10.1056/NEJMsb1503104 PMID:26014593

Jhaveri, A., & Torchilin, V. (2014). Intracellular delivery of nanocarriers and targeting to subcellular organelles. *Expert Opinion on Drug Delivery*, *11*(2), 287–299. PMID:26358656

Jinek, M., Chylinski, K., Fonfara, I., Hauer, M., Doudna, J. A., & Charpentier, E. (2012). A programmable dual-RNA-guided DNA endonuclease in adaptive bacterial immunity. *Science*, *337*(6096), 816–821. doi:10.1126/science.1225829 PMID:22745249

Kamei, K., Mashimo, M., Mihara, T., & Terada, S. (2014). Identification of the pathway and the impact of nitric oxide in plasma-treated water. *Journal of Physics. D, Applied Physics*, *47*(3), 035401.

Kang, S. U., Cho, J., Kim, Y. S., Park, J. S., & Choi, E. H. (2019). Antimicrobial photodynamic therapy using chlorin e6 with halogen light for acne bacteria-induced inflammation. *International Journal of Molecular Sciences*, *20*(2), 238. PMID:30634433

Kawano, H., Masuda, S., Nagasawa, Y., Ohta, T., & Ohshima, T. (2014). Inactivation of bacterial spores by a dielectric barrier discharge in atmospheric-pressure helium plasma. *Journal of Applied Microbiology*, *117*(2), 542–550.

Kim, J. Y., Lee, D. G., & Jang, Y. S. (2019). Recent trends in the development of bioprocesses for the production of biopharmaceuticals from high-throughput data. *Biotechnology Journal*, *14*(5), 1800659.

Kong, M. G., Kroesen, G., Morfill, G., Nosenko, T., Shimizu, T., van Dijk, J., & Zimmermann, J. L. (2019). Plasma medicine: An introductory review. *New Journal of Physics*, *11*(11), 115012. doi:10.1088/1367-2630/11/11/115012

Kuznetsov, S. A., Ovchinnikov, V. A., Dvurechenskii, A. V., & Prusakov, V. N. (2019). Intraoperative hemorrhage control by capillary plasma sources. *Plasma Physics Reports*, *45*(11), 1050–1055.

Laroussi, M., Mohades, S., & Barekzi, N. (2016). Killing of S. cerevisiae by a plasma needle. *IEEE Transactions on Plasma Science*, *33*(2), 745–751.

Lee, D. G., Park, J. B., Kim, S. S., Kim, H. W., & Lee, H. (2004). Plasma-treated organic/inorganic hybrid thin films for tissue engineering. *Journal of Physics. D, Applied Physics*, *37*(23), 3309.

Lee, M. H., Wu, H. H., Huang, J. L., & Lee, S. S. (2018). Continuous production of biofuels via bioprocess engineering: Challenges and solutions. *Biotechnology Journal*, *13*(1), 1700540.

Lee, S. Y., & Kim, H. U. (2015). Systems strategies for developing industrial microbial strains. *Nature Biotechnology*, *33*(10), 1061–1072. doi:10.1038/nbt.3365 PMID:26448090

Li, J., Zou, X., Chen, W., Liang, X., Sun, D., Chen, H., & Li, C. (2018). Plasma-mediated synthesis of nanomaterials. *Science China Materials*, *61*(4), 471–496.

Lin, Y. K., Wu, W. J., Huang, Y. S., & Lin, Y. C. (2016). Polymeric nanoparticles for oral delivery of docetaxel: Preparation, characterization and in vitro uptake evaluation. *Journal of Nanomedicine & Nanotechnology*, *7*(4), 393. PMID:27699139

Liu, X., & Ma, P. X. (2018). Polymeric scaffolds for bone tissue engineering. *Annals of Biomedical Engineering*, *44*(9), 2226–2241. PMID:15095822

Machala, Z., Tarabová, B., & Špánik, I. (2012). Biomedical applications of atmospheric-pressure plasmas. *Plasma Processes and Polymers*, *9*(11-12), 1074–1083.

Notte, J. A. (2005). Plasma surface modification of polymers for biomedical use. *Nuclear Instruments & Methods in Physics Research. Section B, Beam Interactions with Materials and Atoms*, *228*(1-4), 21–27.

Pasquel-Reátegui, J. L., Villacrés-Granda, I. M., Jiménez-Cordero, D., Vanegas-Ortíz, D. H., Zurita, J., & Jaramillo-Gómez, J. I. (2018). In vitro antimicrobial activity of a non-thermal atmospheric plasma jet against Plasmodium falciparum and Enterococcus faecalis. *Experimental Parasitology*, *186*, 19–25.

Pörtner, R. (2005). Bioprocess engineering aspects of heterologous protein production in Escherichia coli. *Journal of Biotechnology*, *118*(2), 145–168.

Rathore, A. S., & Pathak, M. (2012). High-throughput process development and its application in biopharmaceuticals. *Biotechnology and Bioengineering*, *109*(7), 1753–1768.

Ren, Y., Chen, C., Huang, W., Chen, H., Ding, T., Liu, C., & Liu, Y. (2017). Plasma modification of electrospun nanofibers for tissue engineering applications. *ACS Applied Materials & Interfaces*, *9*(8), 7292–7302.

Sackmann, E. K., Fulton, A. L., & Beebe, D. J. (2014). The present and future role of microfluidics in biomedical research. *Nature*, *507*(7491), 181–189. doi:10.1038/nature13118 PMID:24622198

Shen, X., Zhang, Y., Guan, J., Xu, J., Lu, J., & Huang, H. (2015). In situ gelling and mucoadhesive cetirizine microemulsion based on PLGA-PEG-PLGA copolymer for intranasal delivery. *International Journal of Pharmaceutics*, *495*(1), 439–448. PMID:26367780

Slepička, P., Siegel, J., Kolská, Z., Rimpelová, S., Švorčík, V., & Lyutakov, O. (2017). Biocompatibility of plasma-treated polyethylene. *Materials Science and Engineering C*, *71*, 602–610.

Traylor, M. J., Pavlovich, M. J., & Karim, K. S. (2012). Study of plasma polymerization by single photon ionization mass spectrometry. *Plasma Processes and Polymers*, *9*(7), 698–704.

Vandamme, M., Robert, E., Lerondel, S., Sarron, V., Ries, D., & Dozias, S. (2010). ROS implication in a new antitumor strategy based on non-thermal plasma. *International Journal of Cancer*, *127*(3), 779–786. PMID:21702038

Varjo, S., Mikkonen, H., Hakola, A., Piispanen, M., Keskinen, H., Hietala, J., ... Kukk, E. (2015). Oxygen molecule density in non-thermal atmospheric pressure plasma jet. *Plasma Sources Science & Technology*, *24*(1), 015014.

Wang, G., Zhang, L., & Snyders, R. (2017). The effect of plasma treatment on the wettability of thin films. *Plasma Processes and Polymers*, *14*(3), 1600131.

Wolff, C. M., Dong, L., Sommerhalter, M., Zhong, S., Kolb, J. F., & Lademann, J. (2010). Plasma-treated human skin ex vivo studied by EPR spectroscopy. *Skin Research and Technology*, *16*(3), 365–373. PMID:20637006

Yamada, M., & Foote, C. S. (1968). Photosensitized generation of singlet oxygen. Aqueous photochemistry of the naphthoquinones. *Journal of the American Chemical Society*, *90*(5), 1293–1297.

Zhang, S., Zhu, C., Sun, F., Li, W., Shen, H., Wang, C., ... Guo, S. (2020). Current status of pre-clinical and clinical trials targeting overactive bladder and urinary incontinence. *Investigative and Clinical Urology*, *61*(Suppl 1), S23–S36.

Zhao, Z., Zhang, M., Yuan, C., Zhou, X., & Ding, Y. (2017). Improved biocompatibility of Ti6Al4V by plasma surface modification. *Journal of Materials Science*, *52*(16), 9947–9961.

Zhong, X., Xiong, Z., Chen, X., & Liu, D. (2014). A fundamental study on the mechanisms of plasma polymerization and the properties of plasma polymers. *Applied Surface Science*, *298*, 12–16.

Compilation of References

Abbas, H. F., & Daud, W. W. (2010). Hydrogen production by methane decomposition: A review. *International Journal of Hydrogen Energy*, *35*(3), 1160–1190. doi:10.1016/j.ijhydene.2009.11.036

Abdikian, A., Vasheghani Farahani, S., & Hussain, S. (2021). The characteristics of daughter waves emerging from colliding solitary waves in astrophysical plasma media. *Monthly Notices of the Royal Astronomical Society*, *506*(1), 997–1006. doi:10.1093/mnras/stab1781

Abid, A. A., Khan, M. Z., Lu, Q., & Yap, S. L. (2017). A generalized AZ-non-Maxwellian velocity distribution function for space plasmas. *Physics of Plasmas*, *24*(3), 033702. doi:10.1063/1.4977447

Abiev, R. S., Sladkovskiy, D. A., Semikin, K. V., Murzin, D. Y., & Rebrov, E. V. (2020). Non-thermal plasma for process and energy intensification in dry reforming of methane. *Catalysts*, *10*(11), 1358. doi:10.3390/catal10111358

Ablowitz, M. J., Ablowitz, M. A., Clarkson, P. A., & Clarkson, P. A. (1991). *Solitons, nonlinear evolution equations and inverse scattering*. Cambridge university press. doi:10.1017/CBO9780511623998

Adeniyi, O. M., Azimov, U., & Burluka, A. (2018). Algae biofuel: Current status and future applications. *Renewable & Sustainable Energy Reviews*, *90*, 316–335. doi:10.1016/j.rser.2018.03.067

Aerts, R. (2014). *Experimental and computational study of dielectric barrier discharges for environmental applications*. Universiteit Antwerpen.

Aerts, R., Somers, W., & Bogaerts, A. (2015). Carbon dioxide splitting in a dielectric barrier discharge plasma: A combined experimental and computational study. *ChemSusChem*, *8*(4), 702–716. doi:10.1002/cssc.201402818 PMID:25641832

Ahasan, M. R., Hossain, M. M., Ding, X., & Wang, R. (2023). Non-equilibrium plasma-assisted dry reforming of methane over shape-controlled CeO 2 supported ruthenium catalysts. *Journal of Materials Chemistry. A, Materials for Energy and Sustainability*, *11*(20), 10993–11009. doi:10.1039/D3TA01196H

Alberts, B. (2017). *Molecular biology of the cell*. Garland science.

Albo, J., Alvarez-Guerra, M., Castaño, P., & Irabien, A. (2015). Towards the electrochemical conversion of carbon dioxide into methanol. *Green Chemistry*, *17*(4), 2304–2324. doi:10.1039/C4GC02453B

Ali, S., Rubab, N., Nawaz, K., & Sarwar, S. (2020). A comparative study of spacecraft charging onset at geosynchronous altitudes: A nonextensive particle approach. *Journal of Geophysical Research: Space Physics, 125* (9), e2020JA028107.

Ali, S., Masood, W., Rizvi, H., Jahangir, R., & Mirza, A. M. (2021). Contribution of the generalized (r, q) distributed electrons in the formation of nonlinear ion acoustic waves in upper ionospheric plasmas. *AIP Advances*, *11*(12), 125020. doi:10.1063/5.0075007

Al-Jabri, H., Das, P., Khan, S., Thaher, M., & AbdulQuadir, M. (2020). Treatment of wastewaters by microalgae and the potential applications of the produced biomass—A review. *Water (Basel)*, *13*(1), 27. doi:10.3390/w13010027

Al-Mamoori, A., Krishnamurthy, A., Rownaghi, A. A., & Rezaei, F. (2017). Carbon capture and utilization update. *Energy Technology (Weinheim)*, *5*(6), 834–849. doi:10.1002/ente.201600747

Almarashi, J. Q. M., El-Zohary, S. E., Ellabban, M. A., & Abomohra, A. E.-F. (2020). Enhancement of lipid production and energy recovery from the green microalga *Chlorella vulgaris* by inoculum pretreatment with low-dose cold atmospheric pressure plasma (CAPP). *Energy Conversion and Management*, *204*, 112314. doi:10.1016/j.enconman.2019.112314

Amankwah, E. A., Asumadu-Sarkodie, S., & Andam, A. (2019). Techno-economic and environmental analysis of syngas production from agricultural waste using plasma gasification. *Sustainability*, *11*(19), 5477.

Andreev, Y., & Voronkov, Y. (1975). I. Semiokhin, Dissociation Of Carbon-Dioxide In A Pulse Discharge. *Журнал физической химии*, *49*, 1178–1181.

Angelopouulos, V. (2008). *Space Science Reviews*, *141*, 5. doi:10.1007/s11214-008-9336-1

Anoop, N., Sundaramurthy, S., Jha, J. M., Chandrabalan, S., Singh, N., Verma, J., Parvatalu, D., & Katti, S. (2021). Plasma catalysis: A feasible solution for carbon dioxide valorization? *Clean Technologies and Environmental Policy*, *23*(10), 2789–2811. doi:10.1007/s10098-021-02203-y

Aramouni, N. A. K., Touma, J. G., Tarboush, B. A., Zeaiter, J., & Ahmad, M. N. (2018). Catalyst design for dry reforming of methane: Analysis review. *Renewable & Sustainable Energy Reviews*, *82*, 2570–2585. doi:10.1016/j.rser.2017.09.076

Arita, K., & Iizuka, S. (2015). Production of CH 4 in a low-pressure CO 2/H 2 discharge with magnetic field. *Journal of Materials Science and Chemical Engineering*, *3*(12), 69. doi:10.4236/msce.2015.312011

Armenise, I., & Kustova, E. (2018). Mechanisms of coupled vibrational relaxation and dissociation in carbon dioxide. *The Journal of Physical Chemistry A*, *122*(23), 5107–5120. doi:10.1021/acs.jpca.8b03266 PMID:29781278

Arndt, S., Unger, P., Berneburg, M., & Bosserhoff, A. K. (2018). Gene expression after cold atmospheric plasma treatment in plasmid DNA and in human fibroblasts. *Plasma Medicine*, *8*(1), 45–64.

Arpagaus, C., Oberbossel, G., & Rudolf von Rohr, P. (2018). Plasma treatment of polymer powders–from laboratory research to industrial application. *Plasma Processes and Polymers*, *15*(12), 1800133. doi:10.1002/ppap.201800133

Arp, O., Block, D., Piel, A., & Melzer, A. (2004). Dust Coulomb balls: Three-dimensional plasma crystals. *Physical Review Letters*, *93*(16), 165004. doi:10.1103/PhysRevLett.93.165004 PMID:15524998

Artz, J., Müller, T. E., Thenert, K., Kleinekorte, J., Meys, R., Sternberg, A., Bardow, A., & Leitner, W. (2018). Sustainable conversion of carbon dioxide: An integrated review of catalysis and life cycle assessment. *Chemical Reviews*, *118*(2), 434–504. doi:10.1021/acs.chemrev.7b00435 PMID:29220170

Asano, Y., Nakamura, R., Shinohara, I., Fujimoto, M., Takada, T., Baumjohann, W., Owen, C. J., Fazakerley, A. N., Runov, A., Nagai, T., Lucek, E. A., & Rème, H. (2008). Electron flat-top distributions around the magnetic reconnection region. *Journal of Geophysical Research*, *113*(A1), 2007JA012461. doi:10.1029/2007JA012461

Ashford, B. (2022). Plasma-catalytic CO2 hydrogenation to ethane in a dielectric barrier discharge reactor. *Journal of CO2 Utilization, 57*, 101882.

Ashford, B., & Tu, X. (2017). Non-thermal plasma technology for the conversion of CO2. *Current Opinion in Green and Sustainable Chemistry*, *3*, 45–49. doi:10.1016/j.cogsc.2016.12.001

Asisov, R. (1981). *Carbon dioxide dissociation in non-equilibrium plasma. 5th International Symposium on Plasma Chemistry*, Edinburgh.

Athanasaki, G., Jayakumar, A., & Kannan, A. M. (2023). Gas diffusion layers for PEM fuel cells: Materials, properties and manufacturing–A review. *International Journal of Hydrogen Energy*, *48*(6), 2294–2313. doi:10.1016/j.ijhydene.2022.10.058

Attri, P., Han, J., Choi, S., Choi, E. H., Bogaerts, A., & Lee, W. (2018). CAP modifies the structure of a model protein from thermophilic bacteria: Mechanisms of CAP-mediated inactivation. *Scientific Reports*, *8*(1), 10218. doi:10.1038/s41598-018-28600-w PMID:29977069

Ayodele, F. O., Mustapa, S. I., Ayodele, B. V., & Mohammad, N. (2020). An overview of economic analysis and environmental impacts of natural gas conversion technologies. *Sustainability (Basel)*, *12*(23), 10148. doi:10.3390/su122310148

Azizov, R. I., Vakar, A. K., Zhivotov, V., Krotov, M. F., Zinovev, O., Potapkin, B., Rusanov, A., Rusanov, V. D., & Fridman, A. A. e. (1983). *Nonequilibrium plasmachemical process of CO2 decomposition in a supersonic microwave discharge*. Akademiia Nauk SSSR Doklady.

Azzolina-Jury, F. (2017). Chemical engineering aspects of plasma-assisted CO2 hydrogenation over nickel zeolites under partial vacuum. *Journal of CO2 Utilization, 22*, 97-109.

Badar, S. N., Mohammad, M., Emdadi, Z., & Yaakob, Z. (2018). Algae and their growth requirements for bioenergy: A review. *Biofuels*.

Baena-Moreno, F. M., Rodríguez-Galán, M., Vega, F., Alonso-Fariñas, B., Vilches Arenas, L. F., & Navarrete, B. (2019). Carbon capture and utilization technologies: A literature review and recent advances. *Energy Sources. Part A, Recovery, Utilization, and Environmental Effects*, *41*(12), 1403–1433. doi:10.1080/15567036.2018.1548518

Bagger, A., Ju, W., Varela, A. S., Strasser, P., & Rossmeisl, J. (2017). Electrochemical CO2 reduction: A classification problem. *ChemPhysChem*, *18*(22), 3266–3273. doi:10.1002/cphc.201700736 PMID:28872756

Bagheri, M., Momeni, M., & Ghoranneviss, M. (2017). Non-thermal plasma: A promising approach for drug delivery applications. *Applied Physics Reviews*, *4*(2), 021306.

Bagheri, S., Teo, Y. C., & Shahidi, M. (2017). Plasma-assisted drug delivery: A review of research on plasma applications in therapeutics. *Plasma Processes and Polymers*, *14*(8), 1600252.

Bahri, M., & Haghighat, F. (2014). Plasma-B ased Indoor Air Cleaning Technologies: The State of the Art-R eview. *Clean (Weinheim)*, *42*(12), 1667–1680. doi:10.1002/clen.201300296

Bannai, S., & Kitamura, E. (1981). Role of proton dissociation in the transport of cystine and glutamate in human diploid fibroblasts in culture. *The Journal of Biological Chemistry*, *256*(11), 5770–5772. doi:10.1016/S0021-9258(19)69273-2 PMID:6113240

Beal, C. M., Gerber, L. N., Sills, D. L., Huntley, M. E., Machesky, S. C., Walsh, M. J., Tester, J. W., Archibald, I., Granados, J., & Greene, C. H. (2015). Corrigendum to "Algal biofuel production for fuels and feed in a 100-ha facility: A comprehensive techno-economic analysis and life cycle assessment". *Algal Research*, *11*, 375–378. doi:10.1016/j.algal.2015.06.013

Begum, M., & Das, N. (2016). Self-diffusion as a criterion for melting of dust crystal in the presence of magnetic field. *The European Physical Journal Plus*, *131*(2), 1–12. doi:10.1140/epjp/i2016-16046-2

Bekeschus, S., Wende, K., Hefny, M. M., Rödder, K., Jablonowski, H., Schmidt, A., & Gandhirajan, R. K. (2016). Human mononuclear cell survival and proliferation is modulated by cold atmospheric plasma jet. *Journal of Cellular and Molecular Medicine*, *20*(10), 1807–1816.

Benedikt, J., Möller, A., Roth, M., MacRobert, A. J., & Girst, S. (2018). Combining cold atmospheric plasma and magnetic drug targeting: A first step towards a new concept for tumor therapy. *Scientific Reports*, *8*(1), 9367. PMID:29921840

Benedikt, J., Möller, A., Roth, M., MacRobert, A. J., & Girst, S. (2018). Combining cold atmospheric plasma and magnetic drug targeting: A first step towards a new therapy approach in oncology. *Plasma Medicine*, *8*(3), 289–299.

Bengtson, C., & Bogaerts, A. (2020). On the anti-cancer effect of cold atmospheric plasma and the possible role of catalase-dependent apoptotic pathways. *Cells*, *9*(10), 2330. doi:10.3390/cells9102330 PMID:33096638

Benrabbah, R., Cavaniol, C., Liu, H., Ognier, S., Cavadias, S., Gálvez, M. E., & Da Costa, P. (2017). Plasma DBD activated ceria-zirconia-promoted Ni-catalysts for plasma catalytic CO2 hydrogenation at low temperature. *Catalysis Communications*, *89*, 73–76. doi:10.1016/j.catcom.2016.10.028

Berlanga-Acosta, J., Gavilondo-Cowley, J., López-Saura, P., González-López, T., Castro-Santana, M. D., López-Mola, E., Guillén-Nieto, G., & Herrera-Martinez, L. (2009). Epidermal growth factor in clinical practice–a review of its biological actions, clinical indications and safety implications. *International Wound Journal*, *6*(5), 331–346. doi:10.1111/j.1742-481X.2009.00622.x PMID:19912390

Berthelot, A., & Bogaerts, A. (2017). Modeling of CO2 splitting in a microwave plasma: How to improve the conversion and energy efficiency. *The Journal of Physical Chemistry. C, Nanomaterials and Interfaces*, *121*(15), 8236–8251. doi:10.1021/acs.jpcc.6b12840

Berthomier, M., Pottelette, R., Malingre, M., & Khotyaintsev, Y. (2000). Electron-acoustic solitons in an electron-beam plasma system. *Physics of Plasmas*, *7*(7), 2987–2994. doi:10.1063/1.874150

Bezzerides, B., Forslund, D. W., & Lindman, E. L. (1978). Existence of rarefaction shocks in a laser-plasma corona. *The Physics of Fluids*, *21*(12), 2179–2185. doi:10.1063/1.862176

Bhatnagar, P. (2019). Cold plasma technology: A new hope to food industry. *Octa Jf Biosci*, *7*, 108–112.

Bhola, V., Swalaha, F., Ranjith Kumar, R., Singh, M., & Bux, F. (2014). Overview of the potential of microalgae for CO 2 sequestration. *International Journal of Environmental Science and Technology*, *11*(7), 2103–2118. doi:10.1007/s13762-013-0487-6

Bienert, G. P., Møller, A. L. B., Kristiansen, K. A., Schulz, A., Møller, I. M., Schjoerring, J. K., & Jahn, T. P. (2007). Specific aquaporins facilitate the diffusion of hydrogen peroxide across membranes. *The Journal of Biological Chemistry*, *282*(2), 1183–1192. doi:10.1074/jbc.M603761200 PMID:17105724

Biset-Peiró, M., Mey, R., Guilera, J., & Andreu, T. (2020). Adiabatic plasma-catalytic reactor configuration: Energy efficiency enhancement by plasma and thermal synergies on CO2 methanation. *Chemical Engineering Journal*, *393*, 124786. doi:10.1016/j.cej.2020.124786

Bligh, E. G., & Dyer, W. J. (1959). A rapid method of total lipid extraction and purification. *Canadian Journal of Biochemistry and Physiology*, *37*(8), 911–917. doi:10.1139/o59-099 PMID:13671378

Bogaerts, A. & Snoeckx, R. (2019). *Plasma-Based CO 2 Conversion.* An Economy Based on Carbon Dioxide and Water: Potential of Large Scale Carbon Dioxide Utilization, 287-325.

Bogaerts, A., & Snoeckx, R. (2019). Plasma-Based CO 2 Conversion. *An Economy Based on Carbon Dioxide and Water: Potential of Large Scale Carbon Dioxide Utilization*, 287-325.

Bogaerts, A., Berthelot, A., Heijkers, S., Kolev, S., Snoeckx, R., Sun, S., Trenchev, G., Van Laer, K., & Wang, W. (2017). CO2 conversion by plasma technology: Insights from modeling the plasma chemistry and plasma reactor design. *Plasma Sources Science & Technology*, *26*(6), 063001. doi:10.1088/1361-6595/aa6ada

Bogaerts, A., Kozák, T., Van Laer, K., & Snoeckx, R. (2015). Plasma-based conversion of CO 2: Current status and future challenges. *Faraday Discussions, 183*, 217–232. doi:10.1039/C5FD00053J PMID:26391907

Bogaerts, A., Tu, X., Whitehead, J. C., Centi, G., Lefferts, L., Guaitella, O., Azzolina-Jury, F., Kim, H.-H., Murphy, A. B., Schneider, W. F., Nozaki, T., Hicks, J. C., Rousseau, A., Thevenet, F., Khacef, A., & Carreon, M. (2020). The 2020 plasma catalysis roadmap. *Journal of Physics. D, Applied Physics, 53*(44), 443001. doi:10.1088/1361-6463/ab9048

Bongers, W., Bouwmeester, H., Wolf, B., Peeters, F., Welzel, S., van den Bekerom, D., den Harder, N., Goede, A., Graswinckel, M., Groen, P. W., Kopecki, J., Leins, M., van Rooij, G., Schulz, A., Walker, M., & van de Sanden, R. (2017). Plasma-driven dissociation of CO2 for fuel synthesis. *Plasma Processes and Polymers, 14*(6), 1600126. doi:10.1002/ppap.201600126

Boon, J. P., & Tsallis, C. (2005). Special issue overview Nonextensive statistical mechanics: new trends, new perspectives. *Europhysics News, 36*(6), 185-186.

Bormashenko, E., Bormashenko, Y., & Drori, E. (2008). Plasma treatment increases wettability and cellular adhesion to highly water-repellent surfaces. *Plasma Processes and Polymers, 5*(4), 382–388.

Bormashenko, E., Bormashenko, Y., Stein, T., & Whyman, G. (2008). Wetting of a hydrophobic rough surface by an alkaline wetting liquid: Non-wetting drops on a rough surface. *Journal of Colloid and Interface Science, 319*(2), 495–500.

Bostrom, R., Holback, B., Holmgren, G., & Koskinen, H. (1989). Solitary structures in the magnetospheric plasma observed by Viking. *Physica Scripta, 39*(6), 782–786. doi:10.1088/0031-8949/39/6/022

Boulos, M. I., Fauchais, P., & Pfender, E. (1994). *Thermal plasmas: fundamentals and applications*. No Title. doi:10.1007/978-1-4899-1337-1

Boyle, W., & Haworth, F. (1956). Glow-to-arc transition. *Physical Review, 101*(3), 935–938. doi:10.1103/PhysRev.101.935

Brandenburg, R., Arefi-Khonsari, F., & Leipertz, A. (2015). Thin films by plasma techniques for biomedical applications. *Surface and Coatings Technology, 276*, 3–10.

Brandenburg, R., Oehmigen, K., Wilke, C., & von Woedtke, T. (2015). Combination of non-thermal plasma and injection of liquid plasma for wound disinfection. *Clinical Plasma Medicine, 3*(2), 45–49.

Brandt, P. C., Ivlev, A. V., & Morfill, G. E. (2009). Solid phases in electro- and magnetorheological systems. *The Journal of Chemical Physics, 130*(20), 204513. doi:10.1063/1.3143682 PMID:19485463

Brandt, P. C., Ivlev, A. V., & Morfill, G. E. (2010). String-fluid transition in systems with aligned anisotropic interactions. *The Journal of Chemical Physics, 132*(23), 234709. doi:10.1063/1.3442436 PMID:20572736

Brehmer, F., Haenssle, H. A., Daeschlein, G., Ahmed, R., Pfeiffer, S., Görlitz, A., ... Lademann, J. (2011). Alleviation of chronic venous leg ulcers with a hand-held dielectric barrier discharge plasma generator (PlasmaDerm (®) VU-2010): Results of a monocentric, two-armed, open, prospective, randomized and controlled trial (NCT01415622). *Journal of the European Academy of Dermatology and Venereology, 25*(11), 1369–1378. PMID:24666170

Brehmer, F., Haenssle, H. A., Daeschlein, G., Ahmed, R., Pfeiffer, S., Görlitz, A., Simon, D., Schön, M. P., Wandke, D., & Emmert, S. (2015). Alleviation of chronic venous leg ulcers with a hand-held dielectric barrier discharge plasma generator (PlasmaDerm(®) VU-2010): Results of a monocentric, two-armed, open, prospective, randomized and controlled trial (NCT01415622). *Journal of the European Academy of Dermatology and Venereology, 29*(1), 148–155. doi:10.1111/jdv.12490 PMID:24666170

Britun, N., Silva, T., Chen, G., Godfroid, T., van der Mullen, J., & Snyders, R. (2018). Plasma-assisted CO2 conversion: Optimizing performance via microwave power modulation. *Journal of Physics. D, Applied Physics*, *51*(14), 144002. doi:10.1088/1361-6463/aab1ad

Brock, S. L., Shimojo, T., Marquez, M., Marun, C., Suib, S. L., Matsumoto, H., & Hayashi, Y. (1999). Factors Influencing the Decomposition of CO2in AC Fan-Type Plasma Reactors: Frequency, Waveform, and Concentration Effects. *Journal of Catalysis*, *184*(1), 123–133. doi:10.1006/jcat.1999.2440

Bruggeman, P., & Brandenburg, R. (2013). Atmospheric pressure discharge filaments and microplasmas: Physics, chemistry and diagnostics. *Journal of Physics. D, Applied Physics*, *46*(46), 464001. doi:10.1088/0022-3727/46/46/464001

Brutyan, M. M. (2017). Foresight of microalgae usage for the production of third-generation biofuel. *Indian Journal of Science and Technology*, *10*(16), 1–10. doi:10.17485/ijst/2017/v10i16/111621

Budzianowski, W. M. (2012). Negative carbon intensity of renewable energy technologies involving biomass or carbon dioxide as inputs. *Renewable & Sustainable Energy Reviews*, *16*(9), 6507–6521. doi:10.1016/j.rser.2012.08.016

Burlica, R., & Locke, B. R. (2008). Pulsed plasma gliding-arc discharges with water spray. *IEEE Transactions on Industry Applications*, *44*(2), 482–489. doi:10.1109/TIA.2008.916603

Burmester, T., Ebner, B., Weich, B., & Hankeln, T. (2002). Cytoglobin: A novel globin type ubiquitously expressed invertebrate tissues. *Molecular Biology and Evolution*, *19*(4), 416–421. doi:10.1093/oxfordjournals.molbev.a004096 PMID:11919282

Burniaux, J.-M., & Chateau, J. (2014). Greenhouse gases mitigation potential and economic efficiency of phasing-out fossil fuel subsidies. *Inter Economics*, *140*, 71–88. doi:10.1016/j.inteco.2014.05.002

Butterworth, T., Elder, R., & Allen, R. (2016). Effects of particle size on CO2 reduction and discharge characteristics in a packed bed plasma reactor. *Chemical Engineering Journal*, *293*, 55–67. doi:10.1016/j.cej.2016.02.047

Bykova, N., Vasil'evskii, S., Gordeev, A., Kolesnikov, A. F., Fershin, I., & Yakushin, M. (1997). Determination of the effective probabilities of catalytic reactions on the surfaces of heat shield materials in dissociated carbon dioxide flows. *Fluid Dynamics*, *32*(6), 876–886. doi:10.1007/BF03374545

Cai, T., Park, S. Y., & Li, Y. (2013). Nutrient recovery from wastewater streams by microalgae: Status and prospects. *Renewable & Sustainable Energy Reviews*, *19*, 360–369. doi:10.1016/j.rser.2012.11.030

Capezzuto, P., Cramarossa, F., D'Agostino, R., & Molinari, E. (1976). Contribution of vibrational excitation to the rate of carbon dioxide dissociation in electrical discharges. *Journal of Physical Chemistry*, *80*(8), 882–888. doi:10.1021/j100549a024

Capitelli, M., Colonna, G., D'Ammando, G., & Pietanza, L. (2017). Self-consistent time dependent vibrational and free electron kinetics for CO2 dissociation and ionization in cold plasmas. *Plasma Sources Science & Technology*, *26*(5), 055009. doi:10.1088/1361-6595/aa6427

Carpenter, G. (1981). Epidermal growth factor. In *Tissue growth factors* (pp. 89–132). Springer. doi:10.1007/978-3-642-67986-5_3

Çelekli, A., & Zariç, Ö. E. (2023a). Hydrobiology and ecology in the context of climate change: the future of aquatic ecosystems. *6th International Eurasian Conference on Biological and Chemical Sciences*, (pp. 539-545). Zenodo. 10.5281/zenodo.10021473

Çelekli, A., & Zariç, Ö. E. (2023b). Assessing the environmental impact of functional foods. *6th International Eurasian Conference on Biological and Chemical Sciences*. Zenodo. 10.5281/zenodo.10021465

Çelekli, A., Yaygır, S., & Zariç, Ö. E. (2023). A review of climate change-induced migration. Acta Biologica Turcica, 36(2), A3:1-11. doi:10.5281/zenodo.8190755

Cernak, M., Van Veldhuizen, E., Morva, I., & Rutgers, W. (1995). Effect of cathode surface properties on glow-to-arc transition in a short positive corona gap in ambient air. *Journal of Physics. D, Applied Physics*, *28*(6), 1126–1132. doi:10.1088/0022-3727/28/6/015

Chadha, U., Selvaraj, S. K., Ashokan, H., Hariharan, S. P., Mathew Paul, V., Venkatarangan, V., & Paramasivam, V. (2022). Complex nanomaterials in catalysis for chemically significant applications: From synthesis and hydrocarbon processing to renewable energy applications. *Advances in Materials Science and Engineering*, *2022*, 1–72. doi:10.1155/2022/1552334

Chance, B., Sies, H., & Boveris, A. (1979). Hydroperoxide metabolism in mammalian organs. *Physiological Reviews*, *59*(3), 527–605. doi:10.1152/physrev.1979.59.3.527 PMID:37532

Chang, J. W., Kang, S. U., Shin, Y. S., Kim, K. I., Seo, S. J., Yang, S. S., ... Kim, C. H. (2019). Non-thermal atmospheric pressure plasma efficiently promotes the death of ST8 lung cancer cells by increasing intracellular ROS levels and DNA damage. *Scientific Reports*, *9*(1), 1–13. PMID:30626917

Chapman, S. (1916). On the law of distribution of molecular velocities, and on the theory of viscosity and thermal conduction, in a non-uniform simple monotomic gas. *Phil. Trans. R. Soc. A*, *216*, 279–348.

Chen, G. (2015). Simultaneous dissociation of CO2 and H2O to syngas in a surface-wave microwave discharge. *International Journal Of Hydrogen Energy, 40*(9), 3789-3796.

Chen, G., Britun, N., Godfroid, T., Delplancke-Ogletree, M., & Snyders, R. (2017). *Role of plasma catalysis in the microwave plasma-assisted conversion of CO2, Green Chemical Processing and Synthesis.* IntechOpen London, UK.

Chen, G., Snyders, R., & Britun, N. (2021). CO2 conversion using catalyst-free and catalyst-assisted plasma-processes: Recent progress and understanding. *Journal of CO2 Utilization, 49.*

Chen, G., Wang, L., Godfroid, T., & Snyders, R. (2018). *Progress in plasma-assisted catalysis for carbon dioxide reduction, Plasma Chemistry and Gas Conversion.* IntechOpen London, UK.

Chen, F. F. (1984). *Introduction to plasma physics and controlled fusion* (Vol. 1). Plenum press. doi:10.1007/978-1-4757-5595-4

Chen, F., Zhou, Y., Zhu, Y., Zhu, R., Guan, P., Fan, J., & Saribatir, E. (2021). Recent progress in artificial synaptic devices: Materials, processing and applications. *Journal of Materials Chemistry. C, Materials for Optical and Electronic Devices*, *9*(27), 8372–8394. doi:10.1039/D1TC01211H

Chen, G., Britun, N., Godfroid, T., Georgieva, V., Snyders, R., & Delplancke-Ogletree, M.-P. (2017). An overview of CO2 conversion in a microwave discharge: The role of plasma-catalysis. *Journal of Physics. D, Applied Physics*, *50*(8), 084001. doi:10.1088/1361-6463/aa5616

Chen, G., Georgieva, V., Godfroid, T., Snyders, R., & Delplancke-Ogletree, M.-P. (2016). Plasma assisted catalytic decomposition of CO2. *Applied Catalysis B: Environmental*, *190*, 115–124. doi:10.1016/j.apcatb.2016.03.009

Chen, G., Godfroid, T., Britun, N., Georgieva, V., Delplancke-Ogletree, M.-P., & Snyders, R. (2017). Plasma-catalytic conversion of CO2 and CO2/H2O in a surface-wave sustained microwave discharge. *Applied Catalysis B: Environmental*, *214*, 114–125. doi:10.1016/j.apcatb.2017.05.032

Chen, H. L., Lee, H. M., Chen, S. H., Chang, M. B., Yu, S. J., & Li, S. N. (2009). Removal of volatile organic compounds by single-stage and two-stage plasma catalysis systems: A review of the performance enhancement mechanisms, current status, and suitable applications. *Environmental Science & Technology*, *43*(7), 2216–2227. doi:10.1021/es802679b PMID:19452866

Chen, H. L., Lee, H. M., Chen, S. H., Chao, Y., & Chang, M. B. (2008). Review of plasma catalysis on hydrocarbon reforming for hydrogen production—Interaction, integration, and prospects. *Applied Catalysis B: Environmental*, *85*(1-2), 1–9. doi:10.1016/j.apcatb.2008.06.021

Chen, H.-F., Chiang, H.-C., Matsui, H., Tsuchiya, S., & Lee, Y.-P. (2009). Distribution of vibrational states of CO2 in the reaction O (1D)+ CO2 from time-resolved fourier transform infrared emission spectra. *The Journal of Physical Chemistry A*, *113*(15), 3431–3437. doi:10.1021/jp807501c PMID:19354311

Chen, J., Xing, H., Zhan, T., Chen, L., He, M., & Zhang, Y. (2022). Measurement of thermal diffusivity of ethanol from (293 to 564) K and up to 10 MPa in vicinity of the critical point. *Fluid Phase Equilibria*, *552*, 113276. doi:10.1016/j.fluid.2021.113276

Choe, S. J., Kang, H. J., Kim, S.-J., Park, S.-B., & Park, D. H. (2005). Adsorbed carbon formation and carbon hydrogenation for CO2 methanation on the Ni (111) surface: ASED-MO study. *Bulletin of the Korean Chemical Society*, *26*(11), 1682–1688. doi:10.5012/bkcs.2005.26.11.1682

Chopra, K. P. (1961). Interactions of rapidly moving bodies in terrestrial atmosphere. *Reviews of Modern Physics*, *33*(2), 153–189. doi:10.1103/RevModPhys.33.153

Choudhary, S., Tripathi, S., & Poluri, K. M. (2022). Microalgal-Based Bioenergy: Strategies, Prospects, and Sustainability. *Energy & Fuels*, *36*(24), 14584–14612. doi:10.1021/acs.energyfuels.2c02922

Choudhuri, A. R. (1998). *The physics of fluids and plasmas: an introduction for astrophysicists*. Cambridge University Press. doi:10.1017/CBO9781139171069

Chouirfa, H., Bouloussa, H., Migonney, V., & Falentin-Daudré, C. (2015). Review of titanium surface modification techniques and coatings for antibacterial applications. *Acta Biomaterialia*, *83*, 37–54. doi:10.1016/j.actbio.2018.10.036 PMID:30541702

Christie, W. W., & Han, X. (2010). Lipid Analysis: Isolation, Separation, Identification and Lipidomic Analysis: Fourth Edition. In Lipid Analysis: Isolation, Separation, Identification and Lipidomic Analysis: Fourth Edition (Vol. 207). Pergamon press Oxford. doi:10.1533/9780857097866

Chu, J. H., & Lin, I. (1994). Direct observation of Coulomb crystals and liquids in strongly coupled rf dusty plasmas. *Physical Review Letters*, *72*(25), 4009–4012. doi:10.1103/PhysRevLett.72.4009 PMID:10056356

Chung, W.-C., & Chang, M.-B. (2016). Review of catalysis and plasma performance on dry reforming of CH4 and possible synergistic effects. *Renewable & Sustainable Energy Reviews*, *62*, 13–31. doi:10.1016/j.rser.2016.04.007

Chung, W.-C., Pan, K.-L., Lee, H.-M., & Chang, M.-B. (2014). Dry reforming of methane with dielectric barrier discharge and ferroelectric packed-bed reactors. *Energy & Fuels*, *28*(12), 7621–7631. doi:10.1021/ef5020555

Chu, W. L. (2017). Strategies to enhance production of microalgal biomass and lipids for biofuel feedstock. *European Journal of Phycology*, *52*(4), 419–437. doi:10.1080/09670262.2017.1379100

Clarens, A. F., Resurreccion, E. P., White, M. A., & Colosi, L. M. (2010). Environmental life cycle comparison of algae to other bioenergy feedstocks. *Environmental Science & Technology*, *44*(5), 1813–1819. doi:10.1021/es902838n PMID:20085253

Clark, G., Cohen, I., Westlake, J. H., Andrews, G. B., Brandt, P., Gold, R. E., Gkioulidou, M. A., Hacala, R., Haggerty, D., Hill, M. E., Ho, G. C., Jaskulek, S. E., Kollmann, P., Mauk, B. H., McNutt, R. L. Jr, Mitchell, D. G., Nelson, K. S., Paranicas, C., Paschalidis, N., & Schlemm, C. E. (2016). The "Puck" energetic charged particle detector: Design, heritage, and advancements. *Journal of Geophysical Research. Space Physics*, *121*(8), 7900–7913. doi:10.1002/2016JA022579 PMID:27867799

Cooper, R., Cowardin, H., Engelhar, D., Plis, E., & Hoffman, R. (2017, September). Space Weathering Experiments on Spacecraft Materials. In *Advanced Maui Optical and Space Surveillance Technologies Conference (AMOS)* (No. JSC-CN-39072).

Cosby, P. & Helm, H. (1992). *Dissociation rates of diatomic molecules*. Wright Laboratory, Air Force Materiel Command, United States Air Force.

Crim, F. F. (2008). Chemical dynamics of vibrationally excited molecules: Controlling reactions in gases and on surfaces. *Proceedings of the National Academy of Sciences of the United States of America*, *105*(35), 12654–12661. doi:10.1073/pnas.0803010105 PMID:18765816

Cui, D., Yin, Y., Sun, H., Wang, X., Zhuang, J., Wang, L., Ma, R., & Jiao, Z. (2022). Regulation of cellular redox homeostasis in Arabidopsis thaliana seedling by atmospheric pressure cold plasma-generated reactive oxygen/nitrogen species. *Ecotoxicology and Environmental Safety*, *240*, 113703. doi:10.1016/j.ecoenv.2022.113703 PMID:35659700

Cussler, E. L. (2009). *Diffusion: mass transfer in fluid systems*. Cambridge university press. doi:10.1017/CBO9780511805134

Czernichowski, A. (1994). Gliding arc: Applications to engineering and environment control. *Pure and Applied Chemistry*, *66*(6), 1301–1310. doi:10.1351/pac199466061301

Daligault, J. (2012). Diffusion in ionic mixtures across coupling regimes. *Physical Review Letters*, *108*(22), 225004. doi:10.1103/PhysRevLett.108.225004 PMID:23003608

Daligault, J. (2012). Practical model for the self-diffusion coefficient in Yukawa one-component plasmas. *Physical Review. E*, *86*(4), 047401. doi:10.1103/PhysRevE.86.047401 PMID:23214710

Daligaux, V., Richard, R., & Manero, M. (2021). Deactivation and Regeneration of Zeolite Catalysts Used in Pyrolysis of Plastic Wastes—A Process and Analytical Review. *Catalysts*, *2021*(11), 770. doi:10.3390/catal11070770

Dang, C., Reid, J., & Garside, B. (1982). Detailed vibrational population distributions in a CO2 laser discharge as measured with a tunable diode laser. *Applied Physics. B, Photophysics and Laser Chemistry*, *27*(3), 145–151. doi:10.1007/BF00694640

Dang, C., Reid, J., & Garside, B. (1983). Dynamics of the CO 2 upper laser level as measured with a tunable diode laser. *IEEE Journal of Quantum Electronics*, *19*(4), 755–764. doi:10.1109/JQE.1983.1071913

De Backer, J., Razzokov, J., Hammerschmid, D., Mensch, C., Hafideddine, Z., Kumar, N., van Raemdonck, G., Yusupov, M., Van Doorslaer, S., Johannessen, C., Sobott, F., Bogaerts, A., & Dewilde, S. (2018). The effect of reactive oxygen and nitrogen species on the structure of cytoglobin: A potential tumor suppressor. *Redox Biology*, *19*, 1–10. doi:10.1016/j.redox.2018.07.019 PMID:30081385

De Bhowmick, G., Sarmah, A. K., & Sen, R. (2019). Zero-waste algal biorefinery for bioenergy and biochar: A green leap towards achieving energy and environmental sustainability. *The Science of the Total Environment*, *650*, 2467–2482. doi:10.1016/j.scitotenv.2018.10.002 PMID:30293002

De Bie, C., van Dijk, J., & Bogaerts, A. (2016). CO2 hydrogenation in a dielectric barrier discharge plasma revealed. *The Journal of Physical Chemistry. C, Nanomaterials and Interfaces*, *120*(44), 25210–25224. doi:10.1021/acs.jpcc.6b07639 de la Fuente, J. F., Moreno, S. H., Stankiewicz, A. I., & Stefanidis, G. D. (2016). Reduction of CO2 with hydrogen in a non-equilibrium microwave plasma reactor. *International Journal of Hydrogen Energy*, *41*(46), 21067–21077. doi:10.1016/j.ijhydene.2016.08.032

de la Fuente, J. F., Moreno, S. H., Stankiewicz, A. I., & Stefanidis, G. D. (2017). On the improvement of chemical conversion in a surface-wave microwave plasma reactor for CO2 reduction with hydrogen (The Reverse Water-Gas Shift reaction). *International Journal of Hydrogen Energy*, *42*(18), 12943–12955. doi:10.1016/j.ijhydene.2017.04.040

Dębek, R., Azzolina-Jury, F., Travert, A., & Maugé, F. (2019). A review on plasma-catalytic methanation of carbon dioxide–Looking for an efficient catalyst. *Renewable & Sustainable Energy Reviews*, *116*, 109427. doi:10.1016/j.rser.2019.109427

DeForest, S. E. (1972). Spacecraft charging at synchronous orbit. *Journal of Geophysical Research*, *77*(4), 651–659. doi:10.1029/JA077i004p00651

Delgado, J. M., Iniesta, J., Borrás, E., Ferrero, F. J., Díez, F. J., & van Grieken, R. (2013). Biodegradable polymer films grafted with plasma: An application in the transport of fertilizers. *Journal of Materials Processing Technology*, *213*(2), 293–300.

Delgado, L. M., Alves, N. M., Almeida, I. F., Oliveira, M. B., & Mano, J. F. (2013). Surface activation of polymeric biomaterials. *Journal of Colloid and Interface Science*, *395*, 310–316.

Delikonstantis, E., Cameli, F., Scapinello, M., Rosa, V., Van Geem, K. M., & Stefanidis, G. D. (2022). Low-carbon footprint chemical manufacturing using plasma technology. *Current Opinion in Chemical Engineering*, *38*, 100857. doi:10.1016/j.coche.2022.100857

Demirbas, A. (2009). Biofuels securing the planet's future energy needs. *Energy Conversion and Management*, *50*(9), 2239–2249. doi:10.1016/j.enconman.2009.05.010

den Harder, N., van den Bekerom, D. C., Al, R. S., Graswinckel, M. F., Palomares, J. M., Peeters, F. J., Ponduri, S., Minea, T., Bongers, W. A., van de Sanden, M. C., & van Rooij, G. J. (2017). Homogeneous CO2 conversion by microwave plasma: Wave propagation and diagnostics. *Plasma Processes and Polymers*, *14*(6), 1600120. doi:10.1002/ppap.201600120

Dietz, C., Budak, J., Kamprich, T., Kretschmer, M., & Thoma, M. H. (2021). Phase transition in electrorheological plasmas. *Contributions to Plasma Physics*, *61*(10), 1–15. doi:10.1002/ctpp.202100079

Dillard, C. S., Vasko, I. Y., Mozer, F. S., Agapitov, O. V., & Bonnell, J. W. (2018). Electron-acoustic solitary waves in the Earth's inner magnetosphere. *Physics of Plasmas*, *25*(2), 022905. doi:10.1063/1.5007907

Doan, T. T., Nguyen, V. T., Le, V. P., Vo, M. T., & Le, H. T. (2020). CRISPR/Cas9 technology and its application in bioprocessing. *Biotechnology Journal*, *15*(3), e1900299.

Dobrynin, D., Fridman, G., Friedman, G., Fridman, A., & Balasubramanian, M. (2019). Evaluation of plasma-activated water for agricultural applications. *Scientific Reports*, *9*(1), 1–13. PMID:30626917

Donahue, T. (1971). Aeronomy of CO 2 atmospheres: A review. *Journal of the Atmospheric Sciences*, *28*(6), 895–900. doi:10.1175/1520-0469(1971)028<0895:AOCAAR>2.0.CO;2

Donkó, Z., & Hartmann, P. (2004). Thermal conductivity of strongly coupled Yukawa liquids. *Physical Review. E*, *69*(1), 16405. doi:10.1103/PhysRevE.69.016405 PMID:14995723

Dröge, W. (2002). Free radicals in the physiological control of cell function. *Physiological Reviews*, *82*(1), 47–95. doi:10.1152/physrev.00018.2001 PMID:11773609

Du, C. (2007a). *A rotary gliding arc discharge plasma device for volatile organic compounds gas treatment, China.* Patent CN201131940Y.

Du, C. (2007b). *A three phase gliding arc non equilibrium plasma device for water treatment, China.* Patent CN201099636Y.

Duan, X., Li, Y., Ge, W., & Wang, B. (2015). Degradation of CO2 through dielectric barrier discharge microplasma. *Greenhouse Gases: Science and Technology*, *5*(2), 131–140. doi:10.1002/ghg.1425

Dubouloz, N., Pottelette, R., Malingre, M., & Treumann, R. A. (1991). Generation of broadband electrostatic noise by electron acoustic solitons. *Geophysical Research Letters*, *18*(2), 155–158. doi:10.1029/90GL02677

Dubouloz, N., Treumann, R. A., Pottelette, R., & Malingre, M. (1993). Turbulence generated by a gas of electron acoustic solitons. *Journal of Geophysical Research*, *98*(A10), 17415–17422. doi:10.1029/93JA01611

Du, C. M., Yan, J. H., & Cheron, B. (2007). Decomposition of toluene in a gliding arc discharge plasma reactor. *Plasma Sources Science & Technology*, *16*(4), 791–797. doi:10.1088/0963-0252/16/4/014

Du, C., Liu, H., Xiao, M. D., Gao, D., Huang, D. W., Li, Z. Y., Chen, T. F., Mo, J. M., Wang, K., & Zhang, C. R. (2012). Adsorption of iron and lead ions from an aqueous solution by plasma-modified activated carbon. *Industrial & Engineering Chemistry Research*, *51*(48), 15618–15625. doi:10.1021/ie302011n

Du, C., Shi, T. H., Sun, Y. W., & Zhuang, X. F. (2008). Decolorization of Acid Orange 7 solution by gas–liquid gliding arc discharge plasma. *Journal of Hazardous Materials*, *154*(1-3), 1192–1197. doi:10.1016/j.jhazmat.2007.11.032 PMID:18162301

Du, J. (2004a). Nonextensivity in nonequilibrium plasma systems with Coulombian long-range interactions. *Physics Letters. [Part A]*, *329*(4–5), 262–267. doi:10.1016/j.physleta.2004.07.010

Du, J. (2004b). The nonextensive parameter and Tsallis distribution for self-gravitating systems. [Europhysics Letters]. *Europhysics Letters*, *67*(6), 893–899. doi:10.1209/epl/i2004-10145-2

Du, X., Qin, Y., Gao, B., Wang, K., Li, D., Li, Y., Ding, S., Song, Z., Su, Y., & Xiao, C. (2021). Plasma-assisted and oxygen vacancy-engineered In2O3 films for enhanced electrochemical reduction of CO2. *Applied Surface Science*, *563*, 150405. doi:10.1016/j.apsusc.2021.150405

Dzhumagulova, K. N., Masheeva, R. U., Ramazanov, T. S., & Donkó, Z. (2014). Effect of magnetic field on the velocity autocorrelation and the caging of particles in two-dimensional Yukawa liquids. *Physical Review. E*, *89*(3), 033104. doi:10.1103/PhysRevE.89.033104 PMID:24730953

Dzhumagulova, K. N., Ramazanov, T. S., & Masheeva, R. U. (2012). Velocity autocorrelation functions and diffusion coefficient of dusty component in complex plasmas. *Contributions to Plasma Physics*, *52*(3), 182–185. doi:10.1002/ctpp.201100070

Ebadi Amooghin, A., Mirrezaei, S., Sanaeepur, H., & Moftakhari Sharifzadeh, M. M. (2020). Gas permeation modeling through a multilayer hollow fiber composite membrane. *Journal of Membrane Science and Research*, *6*(1), 125–134.

El-Aragi, G. (2009). *Gliding arc discharge (GAD) experiment.* Plasma Physics and Nuclear Fusion Dept., Nuclear Research Center.

Elminshawy, N. A. S., El Ghandour, M., Gad, H. M., El-Damhogi, D. G., El-Nahhas, K., & Addas, M. F. (2019). The performance of a buried heat exchanger system for PV panel cooling under elevated air temperatures. *Geothermics*, *82*, 7–15. doi:10.1016/j.geothermics.2019.05.012

El-Sheekh, M. M., Abd Al-Halim, M. A., & Mohammed, S. A. (2023). Algae processing by plasma discharge technology: A review. *Algal Research*, *70*, 102983. doi:10.1016/j.algal.2023.102983

El-Taibany, W. F. (2005). Electron-acoustic solitary waves and double layers with an electron beam and phase space electron vortices in space plasmas. *Journal of Geophysical Research*, *110*(A1), 2004JA010525. Advance online publication. doi:10.1029/2004JA010525

Elwakil, S. A., Zahran, M. A., & El-Shewy, E. K. (2007). Nonlinear electron-acoustic solitary waves in a relativistic electron-beam plasma system with non-thermal electrons. *Physica Scripta*, *75*(6), 803–808. doi:10.1088/0031-8949/75/6/010

Emmert, S., Brehmer, F., Hänßle, H., Helmke, A., Mertens, N., Ahmed, R., ... Viöl, W. (2018). Atmospheric pressure plasma in dermatology: Ulcus treatment and much more. *Clinical Plasma Medicine*, *10*, 1–8.

Enskog, D. (1917). *Kinetische Theorie der Vorgänge in mässig verdünnten Gasen*, [Doctoral Dissertation, Upsala].

Ergun, R. E., Carlson, C. W., McFadden, J. P., Mozer, F. S., Delory, G. T., Peria, W., Chaston, C. C., Temerin, M., Roth, I., Muschietti, L., Elphic, R., Strangeway, R., Pfaff, R., Cattell, C. A., Klumpar, D., Shelley, E., Peterson, W., Moebius, E., & Kistler, L. (1998). FAST satellite observations of large-amplitude solitary structures. *Geophysical Research Letters*, *25*(12), 2041–2044. doi:10.1029/98GL00636

Ergun, R. E., Holmes, J. C., Goodrich, K. A., Wilder, F. D., Stawarz, J. E., Eriksson, S., Newman, D. L., Schwartz, S. J., Goldman, M. V., Sturner, A. P., Malaspina, D. M., Usanova, M. E., Torbert, R. B., Argall, M., Lindqvist, P.-A., Khotyaintsev, Y., Burch, J. L., Strangeway, R. J., Russell, C. T., & André, M. (2016). Magnetospheric Multiscale observations of large-amplitude, parallel, electrostatic waves associated with magnetic reconnection at the magnetopause. *Geophysical Research Letters*, *43*(11), 5626–5634. doi:10.1002/2016GL068992

Etheridge, M. L., Campbell, S. A., Erdman, A. G., Haynes, C. L., Wolf, S. M., & McCullough, J. (2013). The big picture on nanomedicine: The state of investigational and approved nanomedicine products. *Nanomedicine (London)*, *9*(1), 1–14. doi:10.1016/j.nano.2012.05.013 PMID:22684017

Evans, D. J. (1982). Homogeneous NEMD algorithm for thermal conductivity—Application of non-canonical linear response theory. *Physics Letters. [Part A]*, *91*(9), 457–460. doi:10.1016/0375-9601(82)90748-4

Fagernäs, L., Horttanainen, M., Lassi, U., & Kärki, T. (2019). Bio-oil from plasma pyrolysis of biomass: Composition and stability during storage. *Energy Conversion and Management*, *199*, 111981.

Fan, Z., Sun, K., Rui, N., Zhao, B., & Liu, C. (2015). Improved activity of Ni/MgAl2O4 for CO2 methanation by the plasma decomposition. *Journal of Energy Chemistry*, *24*(5), 655–659. doi:10.1016/j.jechem.2015.09.004

Farahani, M. D., Zeng, Y., & Zheng, Y. (2022). The application of nonthermal plasma in methanol synthesis via CO2 hydrogenation. *Energy Science & Engineering*, *10*(5), 1572–1583. doi:10.1002/ese3.1107

Feizollahi, E., Basu, U., Fredua-Agyeman, R., Jeganathan, B., Tonoyan, L., Strelkov, S. E., Vasanthan, T., Siraki, A. G., & Roopesh, M. S. (2023). Effect of Plasma-Activated Water Bubbles on Fusarium graminearum, Deoxynivalenol, and Germination of Naturally Infected Barley during Steeping. *Toxins*, *15*(2), 124. doi:10.3390/toxins15020124 PMID:36828438

Feldman, W. C., Anderson, R. C., Bame, S. J., Gary, S. P., Gosling, J. T., McComas, D. J., Thomsen, M. F., Paschmann, G., & Hoppe, M. M. (1983). Electron velocity distributions near the Earth's bow shock. *Journal of Geophysical Research*, *88*(A1), 96–110. doi:10.1029/JA088iA01p00096

Ferreira, C. M., & Moisan, M. (2013). *Microwave discharges: fundamentals and applications* (Vol. 302). Springer Science & Business Media.

Foote, A., Dedrick, J., O'Connell, D., North, M., & Gans, T. (2016). Conversion of CO 2 to CO using radio-frequency atmospheric pressure plasmas. *APS Annual Gaseous Electronics Meeting*. APS.

Fortov, V. E., & Morfill, G. E. (2009). Complex and Dusty Plasmas. In *Complex and Dusty Plasmas*. CRC Press. doi:10.1201/9781420083125-c1

Fortov, V. E., & Morfill, G. E. (2009). *Complex and dusty plasmas: from laboratory to space*. CRC Press.

Foster, J. E. (2017). Plasma-based water purification: Challenges and prospects for the future. *Physics of Plasmas*, *24*(5), 055501. doi:10.1063/1.4977921

Fox, J. L., & Dalgarno, A. (1979). Ionization, luminosity, and heating of the upper atmosphere of Mars. *Journal of Geophysical Research*, *84*(A12), 7315–7333. doi:10.1029/JA084iA12p07315

Fridman, A. (2008). *Plasma chemistry*. Cambridge university press. doi:10.1017/CBO9780511546075

Fridman, A., Chirokov, A., & Gutsol, A. (2005). Non-thermal atmospheric pressure discharges. *Journal of Physics. D, Applied Physics*, *38*(2), R1–R24. doi:10.1088/0022-3727/38/2/R01

Fridman, A., & Rusanov, V. (1994). Theoretical basis of non-equilibrium near atmospheric pressure plasma chemistry. *Pure and Applied Chemistry*, *66*(6), 1267–1278. doi:10.1351/pac199466061267

Fridman, G., Friedman, G., Gutsol, A., Shekhter, A. B., Vasilets, V. N., & Fridman, A. (2008). Applied plasma medicine. *Plasma Processes and Polymers*, *5*(6), 503–533. doi:10.1002/ppap.200700154

Fried, B. D., & Gould, R. W. (1961). Longitudinal ion oscillations in a hot plasma. *The Physics of Fluids*, *4*(1), 139–147. doi:10.1063/1.1706174

Fu, H., Zhang, B., Ge, J., Xiong, Z., Zhai, S., Shan, S., & Li, H. (2019). Thermal diffusivity and thermal conductivity of granitoids at 283–988 K and 0.3–1.5 GPa. *The American Mineralogist*, *104*(11), 1533–1545. doi:10.2138/am-2019-7099

G. Bräuer, G. (2003). *Magnetron-Sputtering*. ScienceDirect.

Galli, F. (2012). Amino acid and protein modification by oxygen and nitrogen species. *Amino Acids*, *42*(1), 1–4. doi:10.1007/s00726-010-0670-8 PMID:20602126

Gallon, H. J., Tu, X., & Whitehead, J. C. (2012). Effects of reactor packing materials on H2 production by CO2 reforming of CH4 in a dielectric barrier discharge. *Plasma Processes and Polymers*, *9*(1), 90–97. doi:10.1002/ppap.201100130

Gao, D., Scholten, F., & Roldan Cuenya, B. (2017). Improved CO2 electroreduction performance on plasma-activated Cu catalysts via electrolyte design: Halide effect. *ACS Catalysis*, *7*(8), 5112–5120. doi:10.1021/acscatal.7b01416

Gao, D., Zegkinoglou, I., Divins, N. J., Scholten, F., Sinev, I., Grosse, P., & Roldan Cuenya, B. (2017). Plasma-activated copper nanocube catalysts for efficient carbon dioxide electroreduction to hydrocarbons and alcohols. *ACS Nano*, *11*(5), 4825–4831. doi:10.1021/acsnano.7b01257 PMID:28441005

Gao, X., Liang, J., Wu, L., Wu, L., & Kawi, S. (2022). Dielectric barrier discharge plasma-assisted catalytic CO2 hydrogenation: Synergy of catalyst and plasma. *Catalysts*, *12*(1), 66. doi:10.3390/catal12010066

Gao, X., Lin, Z., Li, T., Huang, L., Zhang, J., Askari, S., Dewangan, N., Jangam, A., & Kawi, S. (2021). Recent developments in dielectric barrier discharge plasma-assisted catalytic dry reforming of methane over Ni-based catalysts. *Catalysts*, *11*(4), 455. doi:10.3390/catal11040455

Garrett, H. B. (1981). The charging of spacecraft surfaces. *Reviews of Geophysics*, *19*(4), 577–616. doi:10.1029/RG019i004p00577

Garrett, H. B., & Hoffman, A. R. (2000). Comparison of spacecraft charging environments at the Earth, Jupiter, and Saturn. *IEEE Transactions on Plasma Science*, *28*(6), 2048–2057. doi:10.1109/27.902232

Garrett, H. B., & Whittlesey, A. C. (2000). Spacecraft charging, an update. *IEEE Transactions on Plasma Science*, *28*(6), 2017–2028. doi:10.1109/27.902229

Gary, S. P., & Tokar, R. L. (1985). The electron-acoustic mode. *The Physics of Fluids*, *28*(8), 2439–2441. doi:10.1063/1.865250

George, A., Shen, B., Craven, M., Wang, Y., Kang, D., Wu, C., & Tu, X. (2021). A Review of Non-Thermal Plasma Technology: A novel solution for CO2 conversion and utilization. *Renewable & Sustainable Energy Reviews*, *135*, 109702. doi:10.1016/j.rser.2020.109702

Ghasemitarei, M., Yusupov, M., Razzokov, J., Shokri, B., & Bogaerts, A. (2019). Effect of oxidative stress on cystine transportation by xC^{-} antiporter. *Archives of Biochemistry and Biophysics*, *674*, 108114. doi:10.1016/j.abb.2019.108114 PMID:31557465

Gholami, R., Stere, C. E., Goguet, A., & Hardacre, C. (2018). Non-thermal-plasma-activated de-NOx catalysis. *Philosophical Transactions. Series A, Mathematical, Physical, and Engineering Sciences*, *376*(2110), 20170054. doi:10.1098/rsta.2017.0054 PMID:29175870

Giannoni, E., Fiaschi, T., Ramponi, G., & Chiarugi, P. (2009). Redox regulation of anoikis resistance of metastatic prostate cancer cells: Key role for Src and EGFR-mediated pro-survival signals. *Oncogene*, *28*(20), 2074–2086. doi:10.1038/onc.2009.77 PMID:19377510

Giesselmann, M., Vellekoop, M. J., & Han, X. (2018). Converting 3D scans to digital cultural heritage reproductions with GPT-3. In *Proceedings of the 3rd Conference on Robotic Vision, Control and Motion* (pp. 17-24). IEEE.

Gilfillan, D., & Marland, G. (2021). CDIAC-FF: global and national CO< sub> 2</sub> emissions from fossil fuel combustion and cement manufacture: 1751–2017. *Earth System Science Data*, *13*(4), 1667–1680. doi:10.5194/essd-13-1667-2021

Gilijamse, J. J., Hoekstra, S., Meek, S. A., Metsälä, M., van de Meerakker, S. Y., Meijer, G., & Groenenboom, G. C. (2007). The radiative lifetime of metastable CO (aΠ3, v=). *The Journal of Chemical Physics*, *127*(22), 127. doi:10.1063/1.2813888

Ginet, G. P., O'Brien, T. P., Huston, S. L., Johnston, W. R., Guild, T. B., Friedel, R., & Su, Y. J. (2014). AE9, AP9 and SPM: New models for specifying the trapped energetic particle and space plasma environment. *The Van Allen Probes Mission*, 579-615.

Girard, P.-M., Arbabian, A., Fleury, M., Bauville, G., Puech, V., Dutreix, M., & Sousa, J. S. (2016). Synergistic effect of H2O2 and NO2 in cell death induced by cold atmospheric He plasma. *Scientific Reports*, *6*(1), 29098. doi:10.1038/srep29098 PMID:27364563

Gomez, E., Rani, D. A., Cheeseman, C., Deegan, D., Wise, M., & Boccaccini, A. (2009). Thermal plasma technology for the treatment of wastes: A critical review. *Journal of Hazardous Materials*, *161*(2-3), 614–626. doi:10.1016/j.jhazmat.2008.04.017 PMID:18499345

Gordon, S., & McBride, B. J. (1994). Computer program for calculation of complex chemical equilibrium compositions and applications. Part 1. *Analysis.*

Gout, P., Kang, Y. J., Buckley, D. J., Bruchovsky, N., & Buckley, A. R. (1997). Increased cystine uptake capability associated with malignant progression of Nb2 lymphoma cells. *Leukemia*, *11*(8), 1329–1337. doi:10.1038/sj.leu.2400739 PMID:9264389

Graham, D. B., Khotyaintsev, Yu. V., Vaivads, A., & Andre, M. (2015). Electrostatic solitary waves with distinct speeds associated with asymmetric reconnection. *Geophysical Research Letters*, *42*(2), 215224. doi:10.1002/2014GL062538

Grattan-Guinness, I. (2005). Joseph Fourier, Théorie analytique de la chaleur (1822). In *Landmark Writings in Western Mathematics 1640-1940* (pp. 354–365). Elsevier Science. doi:10.1016/B978-044450871-3/50107-8

Graves, D. B. (1994). Plasma Processing. *IEEE Transactions on Plasma Science*, *22*(1), 31–42. doi:10.1109/27.281547

Graves, D. B. (2012). The emerging role of reactive oxygen and nitrogen species in redox biology and some implications for plasma applications to medicine and biology. *Journal of Physics. D, Applied Physics*, *45*(26), 263001. doi:10.1088/0022-3727/45/26/263001

Graves, D. B., Baedecker, P. A., Boxer, M. B., Carbone, E., Driscoll, C. F., Guest, D., ... Reid, J. P. (2014). The emerging science of cold plasmas. *Plasma Processes and Polymers*, *11*(11), 1108–1113.

Gregori, G. (2012). *Molecular dynamics simulations of the equilibrium dynamics of non-ideal plasmas* [Doctoral dissertation, University of Oxford].

Griffiths, M. J., & Harrison, S. T. L. (2009). Lipid productivity as a key characteristic for choosing algal species for biodiesel production. *Journal of Applied Phycology*, *21*(5), 493–507. doi:10.1007/s10811-008-9392-7

Grill, A. (1994). *Cold plasma in materials fabrication* (Vol. 151). IEEE Press. doi:10.1109/9780470544273

Grima, E. M., Belarbi, E.-H., Fernández, F. G. A., Medina, A. R., & Chisti, Y. (2003). Recovery of microalgal biomass and metabolites: Process options and economics. *Biotechnology Advances*, *20*(7–8), 491–515. doi:10.1016/S0734-9750(02)00050-2 PMID:14550018

Gringauz, K. I. (1958, June). Rocket measurements of electron concentration in the ionosphere by means of an ultrashort wavelength dispersion interferometer. In *Doklady Akad* (Vol. 120). Nauk SSSR.

Grofulović, M., Alves, L. L., & Guerra, V. (2016). Electron-neutral scattering cross sections for CO2: A complete and consistent set and an assessment of dissociation. *Journal of Physics. D, Applied Physics*, *49*(39), 395207. doi:10.1088/0022-3727/49/39/395207

Guerrero-Preston, R., Ogawa, T., Uemura, M., Shumulinsky, G., Valle, B. L., Pirini, F., ... Sidransky, D. (2016). Cold atmospheric plasma treatment selectively targets head and neck squamous cell carcinoma cells. *International Journal of Molecular Medicine*, *37*(4), 967–974. PMID:26936421

Guo, L., Chen, C., Li, J., Hu, Y., Zhang, M., & Li, J. (2020). Cold atmospheric plasma inhibits multiple myeloma cell growth in vitro and in vivo. *Oxidative Medicine and Cellular Longevity*, *2020*, 1–11.

Gurieff, N., Wende, K., Bekeschus, S., Kramer, A., & Turner, R. J. (2015). Assessment of the antimicrobial activity of cold plasma-treated water against clinically relevant bacteria. *The Journal of Hospital Infection*, *90*(2), 159–165.

Hackmann, W. D. (1972). The researches of Dr. Martinus Van Marum (1750–1837) on the influence of electricity on animals and plants. *Medical History*, *16*(1), 11–26. doi:10.1017/S0025727300017221 PMID:4558435

Hake, R. Jr, & Phelps, A. (1967). Momentum-Transfer and Inelastic-Collision Cross Sections for Electrons in O 2, CO, and C O 2. *Physical Review*, *158*(1), 70–84. doi:10.1103/PhysRev.158.70

Halliwell, B. (2006). Reactive species and antioxidants. Redox biology is a fundamental theme of aerobic life. *Plant Physiology*, *141*(2), 312–322. doi:10.1104/pp.106.077073 PMID:16760481

Hamburg, M. A., & Collins, F. S. (2010). The path to personalized medicine. *The New England Journal of Medicine*, *363*(4), 301–304. doi:10.1056/NEJMp1006304 PMID:20551152

Hamdan, S., Kassim, N. M., & Mansor, N. (2016). Plasma polymerization technique for surface modification of organic and inorganic materials. *Plasma Chemistry and Plasma Processing*, *36*(5), 1207–1242.

Hammer, M. U., Forbrig, E., Kupsch, S., Weltmann, K.-D., & Reuter, S. (2013). IInfluence of plasma treatment on the structure and function of lipids. *Plasma Medicine*, *3*(1–2), 97–114. doi:10.1615/PlasmaMed.2014009708

Hanahan, D., & Weinberg, R. A. (2011). Hallmarks of cancer: The next generation. *Cell*, *144*(5), 646–674. doi:10.1016/j.cell.2011.02.013 PMID:21376230

Han, H., Jin, S., Park, S., Kim, Y., Jang, D., Seo, M. H., & Kim, W. B. (2021). Plasma-induced oxygen vacancies in amorphous MnOx boost catalytic performance for electrochemical CO2 reduction. *Nano Energy*, *79*, 105492. doi:10.1016/j.nanoen.2020.105492

Haq, M. U., Islam, M., Khan, A. U., Umar, S., Irshad, M., Hameed, A., & Kim, K. S. (2020). Plasma-induced growth of hierarchical anatase TiO2 nanorod arrays for photoelectrochemical water splitting. *Journal of Alloys and Compounds*, *835*, 155302.

Haralambiev, L., Nitsch, A., Jacoby, J. M., Strakeljahn, S., Bekeschus, S., Mustea, A., Ekkernkamp, A., & Stope, M. B. (2020). Cold atmospheric plasma treatment of chondrosarcoma cells affects proliferation and cell membrane permeability. *International Journal of Molecular Sciences*, *21*(7), 2291. doi:10.3390/ijms21072291 PMID:32225067

Harding, D. R., Weston, R. E. Jr, & Flynn, G. W. (1988). Energy transfer to CO (v) in the O (1 D)+ CO (1Σ+ g) reaction. *The Journal of Chemical Physics*, *88*(6), 3590–3598. doi:10.1063/1.453908

Harris, J. T. (2003). *Spacecraft charging at geosynchronous altitudes: Current balance and critical temperature in a non-Maxwellian plasma*, [Theses Master of Science, Air Force Institute of Technology]. https://scholar.afit.edu/etd/4166

Harvey, P., Durniak, C., Samsonov, D., & Morfill, G. (2010). Soliton interaction in a complex plasma. *Physical Review E: Statistical, Nonlinear, and Soft Matter Physics*, *81*(5), 057401. doi:10.1103/PhysRevE.81.057401 PMID:20866360

Hasan, U., Masood, W., Jahangir, R., & Mirza, A. M. (2019). Head-on collision of nonlinear electrostatic shock waves in a relativistically degenerate plasma. *Physica Scripta*, *95*(1), 015601. doi:10.1088/1402-4896/ab20e9

Hastings, D., & Garrett, H. (2004). *Spacecraft-environment interactions*. Cambridge University Press.

Haszeldine, R. S. (2009). Carbon capture and storage: How green can black be? *Science*, *325*(5948), 1647–1652. doi:10.1126/science.1172246 PMID:19779187

He, C., Hu, Y., Yin, L., Tang, C., Yin, C., & Lai, Y. (2016). Core-shell nanoscale coordination polymers combine chemotherapy and photodynamic therapy to potentiate checkpoint blockade cancer immunotherapy. *Nature Communications*, *7*(1), 12499. doi:10.1038/ncomms12499 PMID:27530650

Heijkers, S., Snoeckx, R., Kozák, T., Silva, T., Godfroid, T., Britun, N., Snyders, R., & Bogaerts, A. (2015). CO2 conversion in a microwave plasma reactor in the presence of N2: Elucidating the role of vibrational levels. *The Journal of Physical Chemistry. C, Nanomaterials and Interfaces*, *119*(23), 12815–12828. doi:10.1021/acs.jpcc.5b01466

Heinlin, J., Isbary, G., Stolz, W., Morfill, G., Landthaler, M., Shimizu, T., ... Karrer, S. (2010). A randomized two-sided placebo-controlled study on the efficacy and safety of atmospheric pressure plasma in dermatology. *British Journal of Dermatology*, *163*(1), 78–82. PMID:20222930

Helal, M. A., Anderson, N., Wei, Y., & Thompson, M. (2023). A Review of Biomass-to-Bioenergy Supply Chain Research Using Bibliometric Analysis and Visualization. Energies, 16(3). doi:10.3390/en16031187

Henson, E. S., & Gibson, S. B. (2006). Surviving cell death through epidermal growth factor (EGF) signal transduction pathways: Implications for cancer therapy. *Cellular Signalling, 18*(12), 2089–2097. doi:10.1016/j.cellsig.2006.05.015 PMID:16815674

Hepburn, C., Adlen, E., Beddington, J., Carter, E. A., Fuss, S., Mac Dowell, N., Minx, J. C., Smith, P., & Williams, C. K. (2019). The technological and economic prospects for CO2 utilization and removal. *Nature, 575*(7781), 87–97. doi:10.1038/s41586-019-1681-6 PMID:31695213

Herbst, R.S. (2004). Review of epidermal growth factor receptor biology. *International Journal of Radiation Oncology* Biology* Physics, 59*(2), S21-S26.

Hernández, D., Solana, M., Riaño, B., García-González, M. C., & Bertucco, A. (2014). Biofuels from microalgae: Lipid extraction and methane production from the residual biomass in a biorefinery approach. *Bioresource Technology, 170*, 370–378. doi:10.1016/j.biortech.2014.07.109 PMID:25151474

Hessel, V., Anastasopoulou, A., Wang, Q., Kolb, G., & Lang, J. (2013). Energy, catalyst and reactor considerations for (near)-industrial plasma processing and learning for nitrogen-fixation reactions. *Catalysis Today, 211*, 9–28. doi:10.1016/j.cattod.2013.04.005

Hirota, R. (1971). Exact solution of the Korteweg-de Vries equation for multiple collisions of solitons. *Physical Review Letters, 27*(18), 1192–1194. doi:10.1103/PhysRevLett.27.1192

Hirota, R. (2004). *The Direct Method in Soliton Theory*. Cambridge University Press. doi:10.1017/CBO9780511543043

Hoeben, W., van Heesch, E. J. M., Beckers, F. J. C. M., Boekhoven, W., & Pemen, A. J. M. (2015). Plasma-driven water assisted CO 2 methanation. *IEEE Transactions on Plasma Science, 43*(6), 1954–1958. doi:10.1109/TPS.2015.2429316

Ho, H.-J., Iizuka, A., & Shibata, E. (2019). Carbon capture and utilization technology without carbon dioxide purification and pressurization: A review on its necessity and available technologies. *Industrial & Engineering Chemistry Research, 58*(21), 8941–8954. doi:10.1021/acs.iecr.9b01213

Holmes, J. C., Ergun, R. E., Newman, D. L., Wilder, F. D., Sturner, A. P., Goodrich, K. A., Torbert, R. B., Giles, B. L., Strangeway, R. J., & Burch, J. L. (2018). Negative potential solitary structures in the magnetosheath with large parallel width. *Journal of Geophysical Research. Space Physics, 123*(1), 132–145. doi:10.1002/2017JA024890

Hong, J., Prawer, S., & Murphy, A. B. (2018). Plasma catalysis as an alternative route for ammonia production: Status, mechanisms, and prospects for progress. *ACS Sustainable Chemistry & Engineering, 6*(1), 15–31. doi:10.1021/acssuschemeng.7b02381

Hosseini Jenab, S. M., & Brodin, G. (2019). Head-on collision of nonlinear solitary solutions to Vlasov-Poisson equations. *Physics of Plasmas, 26*(2), 022303. doi:10.1063/1.5078865

Huang, G., Lou, S., & Xu, Z. (1993). Phys. Rev. E Stat. *Phys. Plasmas Fluids Relat. Interdiscip, 47*, R3830. doi:10.1103/PhysRevE.47.R3830

Huang, G., & Velarde, M. G. (1996). Head-on collision of two concentric cylindrical ion acoustic solitary waves. *Physical Review E: Statistical Physics, Plasmas, Fluids, and Related Interdisciplinary Topics, 53*(3), 2988–2991. doi:10.1103/PhysRevE.53.2988 PMID:9964593

Huang, J., Liu, G., & Jiang, L. (2015). Threshold for spacecraft charging in double-Maxwellian plasma. *Journal of Geophysical Research. Space Physics, 120*(8), 6301–6308. doi:10.1002/2015JA021173

Huang, J., Liu, G., Jiang, L., & Yang, Y. (2017). Theory for geosynchronous spacecraft charging index. *Space Weather, 15*(9), 1203–1211. doi:10.1002/2017SW001670

Huang, Q., Zhang, D., Wang, D., Liu, K., & Kleyn, A. W. (2017). Carbon dioxide dissociation in non-thermal radiofrequency and microwave plasma. *Journal of Physics. D, Applied Physics*, *50*(29), 294001. doi:10.1088/1361-6463/aa754e

Huang, X., Schwenke, D. W., Tashkun, S. A., & Lee, T. J. (2012). An isotopic-independent highly accurate potential energy surface for CO2 isotopologues and an initial 12C16O2 infrared line list. *The Journal of Chemical Physics*, 136. doi:10.1021/jp2104374 PMID:22462861

Huddleston, R., & Weitz, E. (1981). A laser-induced fluorescence study of energy transfer between the symmetric stretching and bending modes of CO2. *Chemical Physics Letters*, *83*(1), 174–179. doi:10.1016/0009-2614(81)80314-4

Huestis, D., Marschall, J., Billing, G., & Maclagan, R. (2002). Theoretical and Experimental Studies of O-CO 2 Collisions. *AGU Fall Meeting Abstracts*. AGU.

Huh, D., Leslie, D. C., Matthews, B. D., Fraser, J. P., Jurek, S., Hamilton, G. A., & Ingber, D. E. (2013). A human disease model of drug toxicity-induced pulmonary edema in a lung-on-a-chip microdevice. *Science Translational Medicine*, *4*(159), 159ra147. PMID:23136042

Hunt, R. W., Zavalin, A., Bhatnagar, A., Chinnasamy, S., & Das, K. C. (2009). Electromagnetic biostimulation of living cultures for biotechnology, biofuel and bioenergy applications. *International Journal of Molecular Sciences*, *10*(10), 4515–4558. doi:10.3390/ijms10104515 PMID:20057958

Idrees, M. (2014). Making biofuels from microalgae-A review of technologies. *Journal of Food Science and Technology*, *1*, 7–14.

Ihara, T., Kiboku, M., & Iriyama, Y. (1994). Plasma reduction of CO2 with H2O for the formation of organic compounds. *Bulletin of the Chemical Society of Japan*, *67*(1), 312–314. doi:10.1246/bcsj.67.312

Indarto, A., Choi, J.-W., Lee, H., & Song, H. K. (2006). Conversion of CO2 by gliding arc plasma. *Environmental Engineering Science*, *23*(6), 1033–1043. doi:10.1089/ees.2006.23.1033 PMID:17223261

Indarto, A., Yang, D. R., Choi, J.-W., Lee, H., & Song, H. K. (2007). Gliding arc plasma processing of CO2 conversion. *Journal of Hazardous Materials*, *146*(1-2), 309–315. doi:10.1016/j.jhazmat.2006.12.023 PMID:17223261

Infeld, E., & Rowlands, G. (1990). Nonlinear Waves, Solitons and Chaos. Cambridge University Press

Intergovernmental Panel on Climate Change. (2023). Climate Change 2021 – The Physical Science Basis. *Climate Change 2021 – The Physical Science Basis, 2*. doi:10.1017/9781009157896

Iqbal, N. (2014). *Human epidermal growth factor receptor 2 (HER2) in cancers: overexpression and therapeutic implications.* Molecular biology international.

Isbary, G., Morfill, G., Schmidt, H. U., Georgi, M., Ramrath, K., Heinlin, J., ... Landthaler, M. (2012). A first prospective randomized controlled trial to decrease bacterial load using cold atmospheric argon plasma on chronic wounds in patients. *British Journal of Dermatology*, *167*(2), 404–410. doi:10.1111/j.1365-2133.2012.10923.x PMID:22385038

Ishaq, M., & Evans, M. D. (2018). Plasma membrane integrity: Implications for health and disease. *Critical Reviews in Clinical Laboratory Sciences*, *55*(6), 359–374.

Itikawa, Y., & Mason, N. (2005). Cross sections for electron collisions with water molecules. *Journal of Physical and Chemical Reference Data*, *34*(1), 1–22. doi:10.1063/1.1799251

Ito, Y., & Inoue, Y. (2012). Plasma catalysis on selective oxidation of CO in the presence of C2H4: Non-thermal plasmas on Ag/ZSM-5 catalyst. *Catalysis Today*, *190*(1), 35–39.

Ivanova, O., Chen, C., Walker, M., Keidar, M., & Wang, J. (2018). *Plasma medicine: applications of low-temperature gas plasmas in medicine and biology*. CRC Press.

Ivlev, A. V., Morfill, G. E., Thomas, H. M., Räth, C., Joyce, G., Huber, P., Kompaneets, R., Fortov, V. E., Lipaev, A. M., Molotkov, V. I., Reiter, T., Turin, M., & Vinogradov, P. (2008). First observation of electrorheological plasmas. *Physical Review Letters, 100*(9), 095003. doi:10.1103/PhysRevLett.100.095003 PMID:18352717

Ivlev, A. V., Thoma, M. H., Räth, C., Joyce, G., & Morfill, G. E. (2011). Complex plasmas in external fields: The role of non-hamiltonian interactions. *Physical Review Letters, 106*(15), 155001. doi:10.1103/PhysRevLett.106.155001 PMID:21568565

Iyer, A. K., Khaled, G., Fang, J., & Maeda, H. (2016). Exploiting the enhanced permeability and retention effect for tumor targeting. *Drug Discovery Today, 11*(17-18), 812–818. doi:10.1016/j.drudis.2006.07.005 PMID:16935749

Jahangir, R., Ali, S., Mirza, A. M., & Masood, W. (2019). Head on interaction of magnetoacoustic solitons in a spin-1/2 dense plasma with geometrical effects. *Physica Scripta, 94*(12), 125602. doi:10.1088/1402-4896/ab3499

Jahangir, R., & Masood, W. (2020). Interaction of electron acoustic waves in the presence of superthermal electrons in terrestrial magnetosphere. *Physics of Plasmas, 27*(4), 042105. doi:10.1063/1.5143400

Jahangir, R., Masood, W., & Rizvi, H. (2023). Face to face interaction of dust acoustic solitons with trapped ions. *Chaos, Solitons, and Fractals, 174*, 113694. doi:10.1016/j.chaos.2023.113694

Jahangir, R., Masood, W., Siddiq, M., & Batool, N. (2016). Interaction of solitons for obliquely propagating magneto-acoustic waves in stellar atmosphere. *Physics of Plasmas, 23*(12), 122301. doi:10.1063/1.4971219

Jameson, J. L., & Longo, D. L. (2015). Precision medicine—Personalized, problematic, and promising. *The New England Journal of Medicine, 372*(23), 2229–2234. doi:10.1056/NEJMsb1503104 PMID:26014593

Janda, M., Martišovitš, V., Hensel, K., & Machala, Z. (2016). Study of transient spark discharge focused at NOx generation for biomedical applications. *Journal of Physics: Conference Series*. IOP Publishing.

Jardali, F., Van Alphen, S., Creel, J., Ahmadi Eshtehardi, H., Axelsson, M., Ingels, R., Snyders, R., & Bogaerts, A. (2021). NO x production in a rotating gliding arc plasma: Potential avenue for sustainable nitrogen fixation. *Green Chemistry, 23*(4), 1748–1757. doi:10.1039/D0GC03521A

Javed, S., Rubab, N., Zaheer, S., Poedts, S., & Jaffer, G. (2023). Numerical calculations of charging threshold at GEO altitudes with two temperature non-extensive electrons. *Space Weather, 21*(10), e2022SW003412.

Jeffrey, A. (1973). The role of Korteweg de Vries Equation in Plasma Physics. Jeffrey, A. Role of the Korteweg-de Vries equation in Plasma Physics. *Quarterly Journal of the Royal Astronomical Society, 14*, 183.

Jhaveri, A., & Torchilin, V. (2014). Intracellular delivery of nanocarriers and targeting to subcellular organelles. *Expert Opinion on Drug Delivery, 11*(2), 287–299. PMID:26358656

Jiang, T., Li, Y., Liu, C., Xu, G., Eliasson, B., & Xue, B. (2002). Plasma methane conversion using dielectric-barrier discharges with zeolite A. *Catalysis Today, 72*(3-4), 229–235. doi:10.1016/S0920-5861(01)00497-7

Jiménez-Vidal, M., Gasol, E., Zorzano, A., Nunes, V., Palacín, M., & Chillarón, J. (2004). Thiol modification of cysteine 327 in the eighth transmembrane domain of the light subunit xCT of the heteromeric cystine/glutamate antiporter suggests close proximity to the substrate binding site/permeation pathway. *The Journal of Biological Chemistry, 279*(12), 11214–11221. doi:10.1074/jbc.M309866200 PMID:14722095

Jinek, M., Chylinski, K., Fonfara, I., Hauer, M., Doudna, J. A., & Charpentier, E. (2012). A programmable dual-RNA-guided DNA endonuclease in adaptive bacterial immunity. *Science*, *337*(6096), 816–821. doi:10.1126/science.1225829 PMID:22745249

Johnson, C. Y., & Meadows, E. B. (1955). First investigation of ambient positive-ion composition to 219 km by rocket-borne spectrometer. *Journal of Geophysical Research*, *60*(2), 193–203. doi:10.1029/JZ060i002p00193

Jorissen, R.N. (2003). Epidermal growth factor receptor: mechanisms of activation and signalling. *The EGF receptor family*, 33-55.

Joshi, E., Pustylnik, M. Y., Thoma, M. H., Thomas, H. M., & Schwabe, M. (2023). Recrystallization in string-fluid complex plasmas. *Physical Review Research*, *5*(1), L012030. doi:10.1103/PhysRevResearch.5.L012030

Jouhara, H., Khordehgah, N., Almahmoud, S., Delpech, B., Chauhan, A., & Tassou, S. A. (2018). Waste heat recovery technologies and applications. *Thermal Science and Engineering Progress*, *6*, 268–289. doi:10.1016/j.tsep.2018.04.017

Juan, C. A., Pérez de la Lastra, J. M., Plou, F. J., & Pérez-Lebeña, E. (2021). The chemistry of reactive oxygen species (ROS) revisited: Outlining their role in biological macromolecules (DNA, lipids and proteins) and induced pathologies. *International Journal of Molecular Sciences*, *22*(9), 4642. doi:10.3390/ijms22094642 PMID:33924958

Jwa, E., Lee, S. B., Lee, H. W., & Mok, Y. S. (2013). Plasma-assisted catalytic methanation of CO and CO2 over Ni–zeolite catalysts. *Fuel Processing Technology*, *108*, 89–93. doi:10.1016/j.fuproc.2012.03.008

Kakutani, T., Ono, H., Taniuti, T., & Wei, C.-C. (1968). Taniuti, Wei CC, Reductive perturbation method in non-linear wave propagation II: Application to hydromagnetic waves in a cold plasma. *Journal of the Physical Society of Japan*, *24*(5), 1159–1166. doi:10.1143/JPSJ.24.1159

Kalay, Z., & Cevher, S. C. (2012). Oxidant and antioxidant events during epidermal growth factor therapy to cutaneous wound healing in rats. *International Wound Journal*, *9*(4), 362–371. doi:10.1111/j.1742-481X.2011.00895.x PMID:22129466

Kamei, K., Mashimo, M., Mihara, T., & Terada, S. (2014). Identification of the pathway and the impact of nitric oxide in plasma-treated water. *Journal of Physics. D, Applied Physics*, *47*(3), 035401.

Kana, D., Dietz, C., & Thoma, M. H. (2020). Simulation of electrorheological plasmas with superthermal ion drift. *Physics of Plasmas*, *27*(10), 103703. doi:10.1063/5.0010021

Kang, S. U., Cho, J., Kim, Y. S., Park, J. S., & Choi, E. H. (2019). Antimicrobial photodynamic therapy using chlorin e6 with halogen light for acne bacteria-induced inflammation. *International Journal of Molecular Sciences*, *20*(2), 238. PMID:30634433

Kaushik, N. K., Kaushik, N., Linh, N. N., Ghimire, B., Pengkit, A., Sornsakdanuphap, J., Lee, S.-J., & Choi, E. H. (2019). Plasma and nanomaterials: Fabrication and biomedical applications. *Nanomaterials (Basel, Switzerland)*, *9*(1), 98. doi:10.3390/nano9010098 PMID:30646530

Kaushik, N. K., Kaushik, N., Min, B., Choi, K. Y., & Kim, C. (2014). Fullerenes in anticancer nanotherapy: Trends and prospects. *Current Medicinal Chemistry*, *21*(35), 4012–4028.

Kawano, H., Masuda, S., Nagasawa, Y., Ohta, T., & Ohshima, T. (2014). Inactivation of bacterial spores by a dielectric barrier discharge in atmospheric-pressure helium plasma. *Journal of Applied Microbiology*, *117*(2), 542–550.

Keidar, M., Walk, R., Shashurin, A., Srinivasan, P., Sandler, A., Dasgupta, S., Ravi, R., Guerrero-Preston, R., & Trink, B. (2011). Cold plasma selectivity and the possibility of a paradigm shift in cancer therapy. *British Journal of Cancer*, *105*(9), 1295–1301. doi:10.1038/bjc.2011.386 PMID:21979421

Khamsen, N., Akkarachanchainon, A., Teerakawanich, N., & Srisonphan, S. (2016). Organic and bio material surface modification via corona discharge induced atmospheric-cold plasma. *Procedia Computer Science*, *86*, 325–328. doi:10.1016/j.procs.2016.05.088

Khoja, A. H., Tahir, M., & Amin, N. A. S. (2019). Recent developments in non-thermal catalytic DBD plasma reactor for dry reforming of methane. *Energy Conversion and Management*, *183*, 529–560. doi:10.1016/j.enconman.2018.12.112

Khrapak, S. (2018). Practical formula for the shear viscosity of Yukawa fluids. *AIP Advances*, *8*(10), 105226. doi:10.1063/1.5044703

Kikuchi, H. (2013). *Electrohydrodynamics in dusty and dirty plasmas: gravito-electrodynamics and EHD* (Vol. 258). Springer Science & Business Media.

Kim, H. H. (2004). Nonthermal plasma processing for air-pollution control: A historical review, current issues, and future prospects. *Plasma Processes and Polymers*, *1*(2), 91–110. doi:10.1002/ppap.200400028

Kim, J. H., Byun, J. H., Kim, M. H., Lee, G. S., & Kim, D. I. (2020). Emerging technology using atmospheric pressure plasma jet in cancer treatment. *Cancer Research and Treatment*, *52*(3), 598–606.

Kim, J. Y., Lee, D. G., & Jang, Y. S. (2019). Recent trends in the development of bioprocesses for the production of biopharmaceuticals from high-throughput data. *Biotechnology Journal*, *14*(5), 1800659.

Kim, S. C., Lim, M. S., & Chun, Y. N. (2014). Reduction characteristics of carbon dioxide using a plasmatron. *Plasma Chemistry and Plasma Processing*, *34*(1), 125–143. doi:10.1007/s11090-013-9499-8

Kivelson, M. G., Khurana, K. K., Russell, C. T., & Walker, R. J. (1997). Intermittent short-duration magnetic field anomalies in the Io torus: Evidence for plasma interchange? *Geophysical Research Letters*, *24*(17), 2127–2130. doi:10.1029/97GL02202

Klarenaar, B., Grofulović, M., Morillo-Candas, A.-S., Van Den Bekerom, D., Damen, M., Van De Sanden, M., Guaitella, O., & Engeln, R. (2018). A rotational Raman study under non-thermal conditions in a pulsed CO2 glow discharge. *Plasma Sources Science & Technology*, *27*(4), 045009. doi:10.1088/1361-6595/aabab6

Kobayashi, A., & Hamanaka, H. (2006). Decomposition characteristics of carbon dioxide by gas tunnel-type plasma jet. *Vacuum*, *80*(11-12), 1294–1298. doi:10.1016/j.vacuum.2006.01.064

Koeta, O., Blin-Simiand, N., Faider, W., Pasquiers, S., Bary, A., & Jorand, F. (2012). Decomposition of acetaldehyde in atmospheric pressure filamentary nitrogen plasma. *Plasma Chemistry and Plasma Processing*, *32*(5), 991–1023. doi:10.1007/s11090-012-9388-6

Kogelschatz, U. (2003). Dielectric-barrier discharges: Their history, discharge physics, and industrial applications. *Plasma Chemistry and Plasma Processing*, *23*(1), 1–46. doi:10.1023/A:1022470901385

Kogelschatz, U., & Eliasson, B. (1996). The applications of silent discharges. *Physikalische Blätter*, *52*(4), 360–362. doi:10.1002/phbl.19960520409

Koirala, R. P., Dawanse, S., & Pantha, N. (2022). Diffusion of glucose in water: A molecular dynamics study. *Journal of Molecular Liquids*, *345*, 117826. doi:10.1016/j.molliq.2021.117826

Komarzyniec, G., & Aftyka, M. (2020). Operating problems of arc plasma reactors powered by AC/DC/AC converters. *Applied Sciences (Basel, Switzerland)*, *10*(9), 3295. doi:10.3390/app10093295

Kompaneets, R., Morfill, G. E., & Ivlev, A. V. (2009). Design of new binary interaction classes in complex plasmas. *Physics of Plasmas*, *16*(4), 043705. doi:10.1063/1.3112703

Kong, M. G., Kroesen, G., Morfill, G., Nosenko, T., Shimizu, T., van Dijk, J., & Zimmermann, J. L. (2009). Plasma medicine: An introductory review. *New Journal of Physics*, *11*(11), 115012. doi:10.1088/1367-2630/11/11/115012

Korteweg, D. J., & Vries, G. D. (1895). On the change of form of long waves advancing in a rectangular channel, and on a new type of long stationary waves. *The London, Edinburgh and Dublin Philosophical Magazine and Journal of Science*, *39*(240), 422–443. doi:10.1080/14786449508620739

Kovalev, V., Afonina, N., & Gromov, V. (2004). Heat transfer modelling to catalytic protection systems of space vehicles entering into martian atmosphere. *Shock Waves: Proceedings of the 24th International Symposium on Shock Waves.* Springer.

Kozák, T., & Bogaerts, A. (2014a). Splitting of CO2 by vibrational excitation in non-equilibrium plasmas: A reaction kinetics model. *Plasma Sources Science & Technology*, *23*(4), 045004. doi:10.1088/0963-0252/23/4/045004

Kozák, T., & Bogaerts, A. (2014b). Evaluation of the energy efficiency of CO2 conversion in microwave discharges using a reaction kinetics model. *Plasma Sources Science & Technology*, *24*(1), 015024. doi:10.1088/0963-0252/24/1/015024

Krauss, M., Mielczarek, S., Neumann, D., & Kuyatt, C. (1971). Mechanism for production of the Fourth Positive Band System of CO by electron impact on CO2. *Journal of Geophysical Research*, *76*(16), 3733–3737. doi:10.1029/JA076i016p03733

Kumar, K., Dasgupta, C. N., Nayak, B., Lindblad, P., & Das, D. (2011). Development of suitable photobioreactors for CO2 sequestration addressing global warming using green algae and cyanobacteria. *Bioresource Technology*, *102*(8), 4945–4953. doi:10.1016/j.biortech.2011.01.054 PMID:21334885

Kumar, N., Perez-Novo, C., Shaw, P., Logie, E., Privat-Maldonado, A., Dewilde, S., Smits, E., Berghe, W. V., & Bogaerts, A. (2021). Physical plasma-derived oxidants sensitize pancreatic cancer cells to ferroptotic cell death. *Free Radical Biology & Medicine*, *166*, 187–200. doi:10.1016/j.freeradbiomed.2021.02.026 PMID:33636332

Kumar, N., Satpathy, A., Mohanty, M., & Mandal, N. B. (2017). Plasma surface modification of biomaterials for tissue engineering applications. *Advanced Healthcare Materials*, *6*(18), 1700521.

Kumar, S., Ghosh, A., & Kundu, A. (2020). Cold plasma as a potential tool for pest management in agriculture: Current status and future perspectives. *Journal of Thermal Biology*, *90*, 102573.

Kuo, Y. (1953). On the Flow of an Incompressible Viscous Fluid Past a Flat Plate at Moderate Reynolds Numbers. *Journal of Mathematics and Physics*, *32*(1-4), 83–101. doi:10.1002/sapm195332183

Kushiro, K., Kobayashi, H., & Wu, Q. (2020). A non-equilibrium atmospheric pressure plasma activates lactate and glucose metabolism in human chronic wound fluid and keratinocytes. *Scientific Reports*, *10*(1), 1–13. PMID:31913322

Kuznetsov, S. A., Ovchinnikov, V. A., Dvurechenskii, A. V., & Prusakov, V. N. (2019). Intraoperative hemorrhage control by capillary plasma sources. *Plasma Physics Reports*, *45*(11), 1050–1055.

Lackner, K. S. (2003). A guide to CO2 sequestration. *Science*, *300*(5626), 1677–1678. doi:10.1126/science.1079033 PMID:12805529

Lai, S. T. (2009). Dependence of electron flux on electron temperature in spacecraft charging. *Journal of Applied Physics*, *105*(9), 094912. doi:10.1063/1.3125517

Lai, S. T. (2011). Novel Ideas of Spacecraft Charging Mitigation. *IEEE Transactions on Plasma Science*, *40*(2), 402–409. doi:10.1109/TPS.2011.2176755

Lai, S. T. (2012). *Fundamentals of spacecraft charging: spacecraft interactions with space plasmas*. Princeton University Press. doi:10.1515/9781400839094

Lai, S. T., & Della-Rose, D. J. (2001). Spacecraft charging at geosynchronous altitudes: New evidence of existence of critical temperature. *Journal of Spacecraft and Rockets*, *38*(6), 922–928. doi:10.2514/2.3764

Lai, S. T., & Tautz, M. (2006). High-level spacecraft charging in eclipse at geosynchronous altitudes: A statistical study. *Journal of Geophysical Research. Space Physics*, *111*(A9).

Lakhina, G. S., Singh, S. V., Kakad, A. P., & Pickett, J. S. (2011). Generation of electrostatic solitary waves in the plasma sheet boundary layer. *Journal of Geophysical Research*, *116*(A10), n/a. doi:10.1029/2011JA016700

Lakhina, G. S., Singh, S. V., Rubia, R., & Sreeraj, T. (2018). A review of nonlinear fluid models for ion-and electron-acoustic solitons and double layers: Application to weak double layers and electrostatic solitary waves in the solar wind and the lunar wake. *Physics of Plasmas*, *25*(8), 080501. doi:10.1063/1.5033498

Langmuir, I., & Mott-Smith, H. (1924). Langmuir probe technique. *General Electric Review*, *27*, 449–455.

Lapuerta, M., Hernández, J. P., & Agudelo, J. R. (2014). An equation for the estimation of alcohol-air diffusion coefficients for modelling evaporation losses in fuel systems. *Applied Thermal Engineering*, *73*(1), 539–548. doi:10.1016/j.applthermaleng.2014.08.009

Laroussi, M. (2005). Sterilization of contaminated matter with an atmospheric pressure plasma. *IEEE Transactions on Plasma Science*, *33*(2), 745–752.

Laroussi, M. (2018). Nonthermal decontamination of biological media by atmospheric-pressure plasmas: Review, analysis, and prospects. *IEEE Transactions on Plasma Science*, *46*(9), 2942–2953.

Laroussi, M., Akan, T., & Arcury, T. A. (2016). Understanding the factors behind antibacterial activity of cold plasma against Escherichia coli. *IEEE Transactions on Plasma Science*, *44*(9), 1825–1831.

Laroussi, M., Bekeschus, S., Keidar, M., Bogaerts, A., Fridman, A., Lu, X., Ostrikov, K., Hori, M., Stapelmann, K., Miller, V., Reuter, S., Laux, C., Mesbah, A., Walsh, J., Jiang, C., Thagard, S. M., Tanaka, H., Liu, D., Yan, D., & Yusupov, M. (2021). Low-temperature plasma for biology, hygiene, and medicine: Perspective and roadmap. *IEEE Transactions on Radiation and Plasma Medical Sciences*, *6*(2), 127–157. doi:10.1109/TRPMS.2021.3135118

Laroussi, M., & Leipold, F. (2011). Evaluation of the roles of reactive species, heat, and UV radiation in the inactivation of bacterial cells by air plasmas at atmospheric pressure. *International Journal of Mass Spectrometry*, *300*(2-3), 163–170.

Laroussi, M., Mendis, D. A., Rosenberg, M., & Laskin, J. (2010). Plasma–air interactions at atmospheric pressure for biological applications. *Plasma Processes and Polymers*, *7*(3-4), 194–211.

Laroussi, M., Mohades, S., & Barekzi, N. (2016). Killing of S. cerevisiae by a plasma needle. *IEEE Transactions on Plasma Science*, *33*(2), 745–751.

Laroussi, M., Mohamed, H., & Dobrynin, D. (2018). Microbiological evaluation of water treated by plasma-activated water (PAW) for reuse in irrigation. *Water (Basel)*, *10*(3), 291.

Le Quéré, C., Andrew, R. M., Friedlingstein, P., Sitch, S., Hauck, J., Pongratz, J., & Canadell, J. G. (2018). Global carbon budget 2018. *Earth System Science Data Discussions*, *2018*, 1–3.

Lebedev, Y. A. (2010). *Microwave discharges: generation and diagnostics, Journal of physics: conference series*. IOP Publishing.

LeClair, L. R., McConkey, J., & On, O. (1994). (1S) and CO (a 3 Pi) production from electron impact dissociation of CO2. *Journal of Physics. B, Atomic, Molecular, and Optical Physics*, *27*(17), 4039–4055. doi:10.1088/0953-4075/27/17/026

Leduc, M., Guay, D., Coulombe, S., & Leask, R. L. (2019). Atmospheric pressure plasmas for water treatment. *Journal of Physics. D, Applied Physics*, *52*(27), 273001.

Lee, C., Lee, C., Lee, Y.-Y., & Kim, C.-S. (2011). Power-constrained contrast enhancement for emissive displays based on histogram equalization. *IEEE Transactions on Image Processing*, *21*(1), 80–93. PMID:21672675

Lee, D. G., Park, J. B., Kim, S. S., Kim, H. W., & Lee, H. (2004). Plasma-treated organic/inorganic hybrid thin films for tissue engineering. *Journal of Physics. D, Applied Physics*, *37*(23), 3309.

Lee, M. H., Wu, H. H., Huang, J. L., & Lee, S. S. (2018). Continuous production of biofuels via bioprocess engineering: Challenges and solutions. *Biotechnology Journal*, *13*(1), 1700540.

Lee, S. Y., & Kim, H. U. (2015). Systems strategies for developing industrial microbial strains. *Nature Biotechnology*, *33*(10), 1061–1072. doi:10.1038/nbt.3365 PMID:26448090

Lee, T., Puligundla, P., & Mok, C. (2018). Intermittent corona discharge plasma jet for improving tomato quality. *Journal of Food Engineering*, *223*, 168–174. doi:10.1016/j.jfoodeng.2017.11.004

Legg, S. (2021). IPCC, 2021: Climate change 2021-the physical science basis. *Interaction*, *49*(4), 44–45.

Leiweke, R. J., & Ganguly, B. N. (2007). Effects of pulsed-excitation applied voltage rise time on argon metastable production efficiency in a high pressure dielectric barrier discharge. *Applied Physics Letters*, *90*(24), 241501. doi:10.1063/1.2748324

Leubner, M. P. (2004). Core-halo distribution functions: A natural equilibrium state in generalized thermostatistics. *The Astrophysical Journal*, *604*(1), 469–478. doi:10.1086/381867

Li, L. (2019). Plasma-assisted CO2 conversion in a gliding arc discharge: Improving performance by optimizing the reactor design. *Journal of CO2 Utilization, 29*, 296-303.

Li, L., Zhang, H., Li, X., Kong, X., Xu, R., Tay, K., & Tu, X. (2019). Plasma-assisted CO2 conversion in a gliding arc discharge: Improving performance by optimizing the reactor design. *Journal of CO2 Utilization, 29*, 296-303.

Li, N., Mo, L., & Unluer, C. (2022). Emerging CO2 utilization technologies for construction materials: A review. *Journal of CO2 Utilization, 65*, 102237.

Liang, D., Alam, M. A., Lu, L., Fan, R., Xu, J., & Wu, J. (2022). Water-plasma-enhanced and phase-separation-assisted extraction of microalgal lipid for biodiesel production. *Bioresource Technology*, *354*, 127198. doi:10.1016/j.biortech.2022.127198 PMID:35460839

Liao, Q., & Jen, T. C. (2011). Application of Lattice Boltzmann method in fluid flow and heat transfer. *Computational Fluid Dynamics Technologies and Applications*, 29-68.

Liao, W.-T., Lee, W.-J., Chen, C.-Y., & Shih, M. (2001). Decomposition of ethylene oxide in the RF plasma environment. *Environmental Technology*, *22*(2), 165–173. doi:10.1080/09593332208618293 PMID:11349375

Lighthill, M. J. (1949). A Technique for rendering approximate solutions to physical problems uniformly valid. *The London, Edinburgh and Dublin Philosophical Magazine and Journal of Science*, *40*(311), 1179–1201. doi:10.1080/14786444908561410

Li, J., Zou, X., Chen, W., Liang, X., Sun, D., Chen, H., & Li, C. (2018). Plasma-mediated synthesis of nanomaterials. *Science China Materials*, *61*(4), 471–496.

Li, K., Liu, J. L., Li, X.-S., Lian, H.-Y., Zhu, X., Bogaerts, A., & Zhu, A.-M. (2018). Novel power-to-syngas concept for plasma catalytic reforming coupled with water electrolysis. *Chemical Engineering Journal, 353*, 297–304. doi:10.1016/j.cej.2018.07.111

Lima, J., Silva, R., & Plastino, A. (2001). Nonextensive thermostatistics and the *h* theorem. *Physical Review Letters, 86*(14), 2938–2941. doi:10.1103/PhysRevLett.86.2938 PMID:11290077

Lin, A., Razzokov, J., Verswyvel, H., Privat-Maldonado, A., De Backer, J., Yusupov, M., Cardenas De La Hoz, E., Ponsaerts, P., Smits, E., & Bogaerts, A. (2021). Oxidation of innate immune checkpoint CD47 on cancer cells with non-thermal plasma. *Cancers (Basel), 13*(3), 579. doi:10.3390/cancers13030579 PMID:33540720

Lin, Y. K., Wu, W. J., Huang, Y. S., & Lin, Y. C. (2016). Polymeric nanoparticles for oral delivery of docetaxel: Preparation, characterization and in vitro uptake evaluation. *Journal of Nanomedicine & Nanotechnology, 7*(4), 393. PMID:27699139

Li, S., & Duan, W. (2008). Interactions between two solitons with an arbitrary collision direction in two-dimensional Bose-Einstein condensates. *The European Physical Journal B, 62*(4), 485–491. doi:10.1140/epjb/e2008-00192-y

Liu, J.-L., Wang, X., Li, X.-S., Likozar, B., & Zhu, A.-M. (2020). CO2 conversion, utilisation and valorisation in gliding arc plasma reactors. *Journal of Physics. D, Applied Physics, 53*(25), 253001. doi:10.1088/1361-6463/ab7c04

Liu, L., Liu, R., Liu, Y., Li, G., Chen, Q., Liu, X., & Ma, S. (2021). Cystine-glutamate antiporter xCT as a therapeutic target for cancer. *Cell Biochemistry and Function, 39*(2), 174–179. doi:10.1002/cbf.3581 PMID:32749001

Liu, M., Tolg, C., & Turley, E. (2019). Dissecting the dual nature of hyaluronan in the tumor microenvironment. *Frontiers in Immunology, 10*, 947. doi:10.3389/fimmu.2019.00947 PMID:31134064

Liu, X., & Ma, P. X. (2018). Polymeric scaffolds for bone tissue engineering. *Annals of Biomedical Engineering, 44*(9), 2226–2241. PMID:15095822

Livadiotis, G., & McComas, D. (2009). Beyond kappa distributions: Exploiting Tsallis statistical mechanics in space plasmas. *Journal of Geophysical Research, 114*(A11), A11105. doi:10.1029/2009JA014352

Li, W., Wang, X., & Zhang, Y. (2020). Plasma-treated platelet-rich plasma as a promising therapeutic strategy for regenerative medicine. *Frontiers in Bioengineering and Biotechnology, 8*, 643.

Lo, M., Wang, Y. Z., & Gout, P. W. (2008). The x cystine/glutamate antiporter: A potential target for therapy of cancer and other diseases. *Journal of Cellular Physiology, 215*(3), 593–602. doi:10.1002/jcp.21366 PMID:18181196

López, M., Calvo, T., Prieto, M., Múgica-Vidal, R., Muro-Fraguas, I., Alba-Elías, F., & Alvarez-Ordóñez, A. (2019). A review on non-thermal atmospheric plasma for food preservation: Mode of action, determinants of effectiveness, and applications. *Frontiers in Microbiology, 10*, 622. doi:10.3389/fmicb.2019.00622 PMID:31001215

Lu, N., Liu, N., Zhang, C., Su, Y., Shang, K., Jiang, N., Li, J., & Wu, Y. (2021). CO2 conversion promoted by potassium intercalated g-C3N4 catalyst in DBD plasma system. *Chemical Engineering Journal, 417*, 129283. doi:10.1016/j.cej.2021.129283

Lu, N., Sun, D., Zhang, C., Jiang, N., Shang, K., Bao, X., Li, J., & Wu, Y. (2018). CO2 conversion in non-thermal plasma and plasma/g-C3N4 catalyst hybrid processes. *Journal of Physics. D, Applied Physics, 51*(9), 094001. doi:10.1088/1361-6463/aaa919

Lu, S. Y., Sun, X. M., Li, X. D., Yan, J. H., & Du, C. M. (2012). Decomposition of toluene in a rotating glidarc discharge reactor. *IEEE Transactions on Plasma Science, 40*(9), 2151–2156. doi:10.1109/TPS.2012.2206119

Lu, X., Bruggeman, P. J., Reuter, S., Naidis, G., Bogaerts, A., Laroussi, M., & Liu, D. (2022). Grand challenges in low temperature plasmas. *Frontiers in Physics*, 1036.

Lu, X., Naidis, G. V., Laroussi, M., Reuter, S., Graves, D. B., & Ostrikov, K. (2016). Reactive species in non-equilibrium atmospheric-pressure plasmas: Generation, transport, and biological effects. *Physics Reports*, *630*, 1–84. doi:10.1016/j.physrep.2016.03.003

Machala, Z., Tarabová, B., & Špánik, I. (2012). Biomedical applications of atmospheric-pressure plasmas. *Plasma Processes and Polymers*, *9*(11-12), 1074–1083.

Mahammadunnisa, S., Reddy, E. L., Ray, D., Subrahmanyam, C., & Whitehead, J. C. (2013). CO2 reduction to syngas and carbon nanofibres by plasma-assisted in situ decomposition of water. *International Journal of Greenhouse Gas Control*, *16*, 361–363. doi:10.1016/j.ijggc.2013.04.008

Mali, V. V., Bodhankar, N. N., & Budde, J. (2018). Plasma-activated water: Generation, origin of reactive species and biological applications. *Plasma Chemistry and Plasma Processing*, *38*(6), 1219–1240.

Maltsev, A., Eremin, E., & Ivanter, L. (1967). Dissociation of carbon dioxide in glow discharge. *Russ. J. Phys. Chem*, *41*, 633.

Mandal, M., Sarkar, M., Khan, A., Biswas, M., Masi, A., Rakwal, R., Agrawal, G. K., Srivastava, A., & Sarkar, A. (2022). Reactive Oxygen Species (ROS) and Reactive Nitrogen Species (RNS) in plants–maintenance of structural individuality and functional blend. *Advances in Redox Research : an Official Journal of the Society for Redox Biology and Medicine and the Society for Free Radical Research-Europe*, *5*, 100039. doi:10.1016/j.arres.2022.100039

Mannan, A., & Mamun, A. A. (2011). Nonplanar dust-acoustic Gardner solitons in a four-component dusty plasma. *Physical Review. E*, *84*(2), 026408. doi:10.1103/PhysRevE.84.026408 PMID:21929121

Ma, R., Tian, Y., Zou, B., Wang, Y., Zhou, Y., Li, B., ... Liu, D. (2021). The effect of atmospheric-pressure dielectric barrier discharge plasma on heavy metal bioavailability, microbial community, enzyme activity, and seed germination in heavy metal-contaminated soil. *Environmental Science and Pollution Research International*, *28*(15), 18835–18845.

Marcantonio, V., De Falco, M., & Bocci, E. (2022). Non-Thermal Plasma Technology for CO2 Conversion—An Overview of the Most Relevant Experimental Results and Kinetic Models. *Energies*, *15*(20), 7790. doi:10.3390/en15207790

Marengo, B. (2016). *Redox homeostasis and cellular antioxidant systems: crucial players in cancer growth and therapy.* Oxidative medicine and cellular longevity.

Marinakis, S., & Samios, J. (2005). The temperature and density dependence of fluid xenon self-diffusion coefficients: A comparison between experimental, theoretical and molecular dynamics results. *The Journal of Supercritical Fluids*, *34*(1), 81–89. doi:10.1016/j.supflu.2004.10.002

Marín, E., Hernández-Rosales, E., Mansanares, A. M., Ivanov, R., Rojas-Trigos, J. B., & Calderón, A. (2013). A method for thermal diffusivity measurement in fluids. *The Review of Scientific Instruments*, *84*(10), 104903. doi:10.1063/1.4824191 PMID:24182147

Martin-del-Campo, J., Uceda, M., Coulombe, S., & Kopyscinski, J. (2021). Plasma-catalytic dry reforming of methane over Ni-supported catalysts in a rotating gliding arc–Spouted bed reactor. *Journal of CO2 Utilization*, *46*, 101474.

Masood, W., Hamid, N., Ullah, S., Aman-ur-Rahman, Shah, H. A., Alyousef, H. A., & El-Tantawy, S. A. (2022). Formation of acoustic nonlinear structures in non-Maxwellian trapping plasmas. *Physics of Fluids*, *34*(5), 053113. doi:10.1063/5.0090923

Masood, W., & Schwartz, S. J. (2008). Observations of the development of electron temperature anisotropies in Earth's magnetosheath. *Journal of Geophysical Research, 113*(A1), 2007JA012715. doi:10.1029/2007JA012715

Masood, W., Schwartz, S. J., Maksimovic, M., & Fazakerley, A. N. (2006). Electron velocity distribution and lion roars in the magnetosheath. *Annales Geophysicae, 24*(6), 1725–1735. doi:10.5194/angeo-24-1725-2006

Masson-Delmotte, V., Zhai, P., Pörtner, H.-O., Roberts, D., Skea, J., & Shukla, P. R. (2022). *Global Warming of 1.5 C: IPCC special report on impacts of global warming of 1.5 C above pre-industrial levels in context of strengthening response to climate change, sustainable development, and efforts to eradicate poverty*. Cambridge University Press.

Masud M., Asaduzzaman M., & Mamun, A.A. (2012). Dust-ion-acoustic Gardner solitons in a dusty plasma with bi-Maxwellian electrons. *Physics of Plasmas, 19*(10), 103706. doi:10.1063/1.4753922

Mata, T. M., Martins, A. A., & Caetano, N. S. (2010). Microalgae for biodiesel production and other applications: A review. *Renewable & Sustainable Energy Reviews, 14*(1), 217–232. doi:10.1016/j.rser.2009.07.020

Matéo-Vélez, J. C., Ganuschkina, N., Meredith, N., Sicard-Piet, A., Maget, V., Payan, D., & Dubyagin, S. (2016, April). *From GEO/LEO environment data to the numerical estimation of spacecraft surface charging at MEO*. 14th Spacecraft Charging and Technology Conference, Noordwijk, NL. http://old. esaconferencebureau. com/2016-events/14sctc/proceedings

Matsumoto, H., Kojima, H., Miyatake, T., Omura, Y., Okada, M., Nagano, I., & Tsutsui, M. (1994). Electrostatic solitary waves (ESW) in the magnetotail: BEN wave forms observed by GEOTAIL. *Geophysical Research Letters, 21*(25), 2915–2918. doi:10.1029/94GL01284

Matveev, V. B., & Salle, M. A. (1991). *Darboux Transformations and Solitons*. Springer. doi:10.1007/978-3-662-00922-2

Maxworthy, T. (1976). Experiments on the collision between two solitary waves. *Journal of Fluid Mechanics, 76*(1), 177–186. doi:10.1017/S0022112076003194

McGivern, W. S., & Manion, J. A. (2021). Binary diffusion coefficients for methane and fluoromethanes in nitrogen. *Journal of Chemical & Engineering Data, 66*(8), 3047-3056.

Mehta, P., Barboun, P., Go, D. B., Hicks, J. C., & Schneider, W. F. (2019). Catalysis enabled by plasma activation of strong chemical bonds: A review. *ACS Energy Letters, 4*(5), 1115–1133. doi:10.1021/acsenergylett.9b00263

Mei, D., Liu, S., & Tu, X. (2017). CO2 reforming with methane for syngas production using a dielectric barrier discharge plasma coupled with Ni/γ-Al2O3 catalysts: Process optimization through response surface methodology. *Journal of CO2 Utilization, 21*, 314-326.

Mei, D., He, Y. L., Liu, S., Yan, J., & Tu, X. (2016). Optimization of CO2 conversion in a cylindrical dielectric barrier discharge reactor using design of experiments. *Plasma Processes and Polymers, 13*(5), 544–556. doi:10.1002/ppap.201500159

Mei, D., Zhu, X., Wu, C., Ashford, B., Williams, P. T., & Tu, X. (2016). Plasma-photocatalytic conversion of CO2 at low temperatures: Understanding the synergistic effect of plasma-catalysis. *Applied Catalysis B: Environmental, 182*, 525–532. doi:10.1016/j.apcatb.2015.09.052

Meier, K., Laesecke, A., & Kabelac, S. (2005). Transport coefficients of the Lennard-Jones model fluid. III. Bulk viscosity. *The Journal of Chemical Physics, 122*(1), 014513. doi:10.1063/1.1828040 PMID:15638680

Merlino, R. (2021). Dusty plasmas: From Saturn's rings to semiconductor processing devices. *Advances in Physics: X, 6*(1), 1873859. doi:10.1080/23746149.2021.1873859

Merlino, R. L. (2006). Dusty plasmas and applications in space and industry. *Plasma Physics Applied*, *81*, 73–110.

Mertens, J., Watson, S., Nisol, B., Wertheimer, M. R., & Reniers, F. (2019). Energetics of reactions in a dielectric barrier discharge with argon carrier gas: VII anhydrides. *Plasma Processes and Polymers*, *16*(5), 1800186. doi:10.1002/ppap.201800186

Metz, B., Davidson, O., De Coninck, H., Loos, M., & Meyer, L. (2005). *IPCC special report on carbon dioxide capture and storage*. Cambridge University Press.

Meyer-Vernet, N. (1999). How does the solar wind blow? A simple kinetic model. *European Journal of Physics*, *20*(3), 167–176. doi:10.1088/0143-0807/20/3/006

Michalis, V. K., Moultos, O. A., Tsimpanogiannis, I. N., & Economou, I. G. (2016). Molecular dynamics simulations of the diffusion coefficients of light n-alkanes in water over a wide range of temperature and pressure. *Fluid Phase Equilibria*, *407*, 236–242. doi:10.1016/j.fluid.2015.05.050

Michielsen, I., Uytdenhouwen, Y., Pype, J., Michielsen, B., Mertens, J., Reniers, F., Meynen, V., & Bogaerts, A. (2017). CO2 dissociation in a packed bed DBD reactor: First steps towards a better understanding of plasma catalysis. *Chemical Engineering Journal*, *326*, 477–488. doi:10.1016/j.cej.2017.05.177

Michiels, R., Engelmann, Y., & Bogaerts, A. (2020). Plasma catalysis for CO2 hydrogenation: Unlocking new pathways toward CH3OH. *The Journal of Physical Chemistry. C, Nanomaterials and Interfaces*, *124*(47), 25859–25872. doi:10.1021/acs.jpcc.0c07632

Mikhail, M., Costa, P. D., Amouroux, J., Cavadias, S., Tatoulian, M., Ognier, S., & Gálvez, M. E. (2021). Effect of Na and K impurities on the performance of Ni/CeZrOx catalysts in DBD plasma-catalytic CO2 methanation. *Fuel*, *306*, 121639. doi:10.1016/j.fuel.2021.121639

Mikoviny, T., Kocan, M., Matejcik, S., Mason, N., & Skalny, J. (2003). Experimental study of negative corona discharge in pure carbon dioxide and its mixtures with oxygen. *Journal of Physics. D, Applied Physics*, *37*(1), 64–73. doi:10.1088/0022-3727/37/1/011

Mikulčić, H., Ridjan Skov, I., Dominković, D. F., Wan Alwi, S. R., Manan, Z. A., Tan, R., Duić, N., Hidayah Mohamad, S. N., & Wang, X. (2019). Flexible Carbon Capture and Utilization technologies in future energy systems and the utilization pathways of captured CO2. *Renewable & Sustainable Energy Reviews*, *114*, 109338. doi:10.1016/j.rser.2019.109338

Misra, N. N., Patil, S., Moiseev, T., Bourke, P., Mosnier, J. P., Keener, K., & Cullen, P. J. (2020). Plasma-activated water (PAW): Chemistry, physico-chemical properties, applications in food and agriculture. *Trends in Food Science & Technology*, *97*, 1–13.

Misra, N. N., Schlüter, O., Cullen, P. J., & Valdramidis, V. P. (2020). *Cold plasma in food and agriculture: fundamentals and applications*. Academic Press.

Misra, N. N., Schlüter, O., Cullen, P. J., & Valdramidis, V. P. (2020). *Cold plasma in food and agriculture: Fundamentals and applications*. Academic Press.

Misra, S., Hascall, V. C., Markwald, R. R., & Ghatak, S. (2015). Interactions between hyaluronan and its receptors (CD44, RHAMM) regulate the activities of inflammation and cancer. *Frontiers in Immunology*, *6*, 201. doi:10.3389/fimmu.2015.00201 PMID:25999946

Mistry, H., Choi, Y. W., Bagger, A., Scholten, F., Bonifacio, C. S., Sinev, I., Divins, N. J., Zegkinoglou, I., Jeon, H. S., Kisslinger, K., Stach, E. A., Yang, J. C., Rossmeisl, J., & Roldan Cuenya, B. (2017). Enhanced carbon dioxide electroreduction to carbon monoxide over defect-rich plasma-activated silver catalysts. *Angewandte Chemie*, *129*(38), 11552–11556. doi:10.1002/ange.201704613 PMID:28710839

Mistry, H., Varela, A. S., Bonifacio, C. S., Zegkinoglou, I., Sinev, I., Choi, Y.-W., Kisslinger, K., Stach, E. A., Yang, J. C., Strasser, P., & Cuenya, B. R. (2016). Highly selective plasma-activated copper catalysts for carbon dioxide reduction to ethylene. *Nature Communications*, *7*(1), 12123. doi:10.1038/ncomms12123 PMID:27356485

Mitic, S., Pustylnik, M. Y., Erdle, D., Lipaev, A. M., Usachev, A. D., Zobnin, A. V., Thoma, M. H., Thomas, H. M., Petrov, O. F., Fortov, V. E., & Kononenko, O. (2021). Long-term evolution of the three-dimensional structure of string-fluid complex plasmas in the PK-4 experiment. *Physical Review. E*, *103*(6), 063212. doi:10.1103/PhysRevE.103.063212 PMID:34271636

Mitsingas, C. M., Rajasegar, R., Hammack, S., Do, H., & Lee, T. (2016). High energy efficiency plasma conversion of CO 2 at atmospheric pressure using a direct-coupled microwave plasma system. *IEEE Transactions on Plasma Science*, *44*(4), 651–656. doi:10.1109/TPS.2016.2531641

Miura, R. M., Gardner, C. S., & Kruskal, M. D. (1968). Korteweg-de vries equation and generalizations. II. Existence of conservation laws and constants of motion. *Journal of Mathematical Physics*, *9*(8), 1204–1209. doi:10.1063/1.1664701

Mohamad, A. A. (2011). *Lattice Boltzmann Method* (Vol. 70). Springer. doi:10.1007/978-0-85729-455-5

Moore, C. B., Wood, R. E., Hu, B. L., & Yardley, J. T. (1967). Vibrational energy transfer in CO2 lasers. *The Journal of Chemical Physics*, *46*(11), 4222–4231. doi:10.1063/1.1840527

Mora, E., Sarmiento, A., & Vera, E. (2016). Alumina and quartz as dielectrics in a dielectric barrier discharges DBD system for CO2 hydrogenation. Journal of Physics: Conference Series. IOP Publishing.

Moreau, M., Orange, N., & Feuilloley, M. (2008). Non-thermal plasma technologies: New tools for bio-decontamination. *Biotechnology Advances*, *26*(6), 610–617. doi:10.1016/j.biotechadv.2008.08.001 PMID:18775485

Morillo-Candas, A. S. (2019). *Investigation of fundamental mechanisms of CO2 plasmas.* Université Paris Saclay (COmUE).

Mori, S., Akatsuka, H., & Suzuki, M. (2002). Numerical analysis of carbon isotope separation by plasma chemical reactions in carbon monoxide glow discharge. *Journal of Nuclear Science and Technology*, *39*(6), 637–646. doi:10.1080/18811248.2002.9715244

Mori, S., Matsuura, N., Tun, L. L., & Suzuki, M. (2016). Direct synthesis of carbon nanotubes from only CO 2 by a hybrid reactor of dielectric barrier discharge and solid oxide electrolyser cell. *Plasma Chemistry and Plasma Processing*, *36*(1), 231–239. doi:10.1007/s11090-015-9681-2

Mori, S., Yamamoto, A., & Suzuki, M. (2006). Characterization of a capillary plasma reactor for carbon dioxide decomposition. *Plasma Sources Science & Technology*, *15*(4), 609–613. doi:10.1088/0963-0252/15/4/003

Moss, M., Yanallah, K., Allen, R. W. K., & Pontiga, F. (2017). An investigation of CO2 splitting using nanosecond pulsed corona discharge: Effect of argon addition on CO2 conversion and energy efficiency. *Plasma Sources Science & Technology*, *26*(3), 035009. doi:10.1088/1361-6595/aa5b1d

Mozer, F. S., Cattell, C. A., Temerin, M., Torbert, R. B., Von Glinski, S., Woldorff, M., & Wygant, J. (1979). The dc and ac electric field, plasma density, plasma temperature, and field-aligned current experiments on the S3-3 satellite. *Journal of Geophysical Research*, *84*(A10), 5875–5884. doi:10.1029/JA084iA10p05875

Mumma, M., Stone, E., & Zipf, E. (1975). Nonthermal rotational distribution of CO ($A^1\Pi$) fragments produced by dissociative excitation of CO2 by electron impact. *Journal of Geophysical Research*, *80*(1), 161–167. doi:10.1029/JA080i001p00161

Naeem, I., Masood, W., & Mirza, A. M. (2020). Dipolar and Kelvin-Stuart's cat's eyes vortices in magnetoplasmas with non-Maxwellian electron distribution. *Astrophysics and Space Science*, *365*(3), 1–1. doi:10.1007/s10509-020-03759-9

Nagaraja, B. M., Bulushev, D. A., Beloshapkin, S., & Ross, J. R. H. (2011). The effect of potassium on the activity and stability of Ni–MgO–ZrO2 catalysts for the dry reforming of methane to give synthesis gas. *Catalysis Today*, *178*(1), 132–136. doi:10.1016/j.cattod.2011.08.040

Naville, A., Guye, C., & Guye, P. (1905). *Gas Reactions at the Temperature of the Electric Arc*. French, Patent FR.

Nawaz, H., Jahangir, R., Masood, W., & Siddiq, M. (2022). Cubic nonlinearity driven dust ion acoustic solitons with superthermal two-temperature electrons in the Saturn's magnetosphere. *Physica Scripta*, *97*(7), 07560110. doi:10.1088/1402-4896/ac735e

Ndazi, B. S., Tichapondwa, S. M., Nyamakwere, T., & Muzenda, E. (2021). Environmental performance of biogas production from agricultural waste: A review. *Journal of Cleaner Production*, *292*, 126129.

Neyts, E. C., Ostrikov, K., Sunkara, M. K., & Bogaerts, A. (2015). Plasma catalysis: Synergistic effects at the nanoscale. *Chemical Reviews*, *115*(24), 13408–13446. doi:10.1021/acs.chemrev.5b00362 PMID:26619209

Nguyen, H. H., & Kim, K.-S. (2015). Combination of plasmas and catalytic reactions for CO2 reforming of CH4 by dielectric barrier discharge process. *Catalysis Today*, *256*, 88–95. doi:10.1016/j.cattod.2015.04.034

Nikitin, E. D., Popov, A. P., Bogatishcheva, N. S., & Faizullin, M. Z. (2022). Critical temperatures, pressures, heat capacities, and thermal diffusivities of C3 to C6 dinitriles. *Journal of Chemical & Engineering Data, 67*(4), 836-845.

Nizio, M., Albarazi, A., Cavadias, S., Amouroux, J., Galvez, M. E., & Da Costa, P. (2016). Hybrid plasma-catalytic methanation of CO2 at low temperature over ceria zirconia supported Ni catalysts. *International Journal of Hydrogen Energy*, *41*(27), 11584–11592. doi:10.1016/j.ijhydene.2016.02.020

Nizio, M., Benrabbah, R., Krzak, M., Debek, R., Motak, M., Cavadias, S., Gálvez, M. E., & Da Costa, P. (2016). Low temperature hybrid plasma-catalytic methanation over Ni-Ce-Zr hydrotalcite-derived catalysts. *Catalysis Communications*, *83*, 14–17. doi:10.1016/j.catcom.2016.04.023

Norhasyima, R. & Mahlia, T. (2018). Advances in CO_2 utilization technology: A patent landscape review. *Journal of CO2 Utilization, 26*, 323-335.

Notte, J. A. (2005). Plasma surface modification of polymers for biomedical use. *Nuclear Instruments & Methods in Physics Research. Section B, Beam Interactions with Materials and Atoms*, *228*(1-4), 21–27.

Nunnally, T., Gutsol, K., Rabinovich, A., Fridman, A., Gutsol, A., & Kemoun, A. (2011). Dissociation of CO2 in a low current gliding arc plasmatron. *Journal of Physics. D, Applied Physics*, *44*(27), 274009. doi:10.1088/0022-3727/44/27/274009

Nwoba, E. G., Parlevliet, D. A., Laird, D. W., Alameh, K., & Moheimani, N. R. (2019). Light management technologies for increasing algal photobioreactor efficiency. *Algal Research*, *39*, 101433. doi:10.1016/j.algal.2019.101433

Ogilvie, K. W., Fitzenreiter, R. J., & Scudder, J. D. (1984). Observations of electron beams in the low-latitude boundary layer. *Journal of Geophysical Research*, *89*(A12), 10723–10732. doi:10.1029/JA089iA12p10723

Oguro, M., Ikejima, Y., Taniwaki, Y., Onishi, K., Suzuki, S., & Takenaka, N. (2020). Development of plasma agricultural technology for soil decontamination using a DC multi-needle-to-plane reactor. *IEEE Transactions on Plasma Science*, *48*(1), 86–90.

Ong, M., Nomanbhay, S., Kusumo, F., & Show, P. (2022). Application of microwave plasma technology to convert carbon dioxide (CO2) into high value products: A review. *Journal of Cleaner Production*, *336*, 130447. doi:10.1016/j.jclepro.2022.130447

Outka, U. (2011). The renewable energy footprint. *Stan. Envtl. LJ*, *30*, 241.

Owczarek, M., & Baryłka, A. (2019). Determining the thermal diffusivity of the material based on the measurement of the temperature profile in the wall. *Rynek energii.*

Ozkan, A. (2016). Splitting in a dielectric barrier discharge plasma: understanding of physical and chemical aspects. University of Antwerp.

Ozkan, A., Dufour, T., Bogaerts, A., & Reniers, F. (2016). How do the barrier thickness and dielectric material influence the filamentary mode and CO2 conversion in a flowing DBD? *Plasma Sources Science & Technology*, *25*(4), 045016. doi:10.1088/0963-0252/25/4/045016

Ozkan, A., Dufour, T., Silva, T., Britun, N., Snyders, R., Bogaerts, A., & Reniers, F. (2016). The influence of power and frequency on the filamentary behavior of a flowing DBD—Application to the splitting of CO2. *Plasma Sources Science & Technology*, *25*(2), 025013. doi:10.1088/0963-0252/25/2/025013

Pandya, A., Kothari, N., & Alad, R. (2022). A comparative study of body potential transient profile of geosynchronous orbit spacecraft in single and double Maxwellian plasma models. *International Journal of Numerical Modelling*, *35*(3), e2968. doi:10.1002/jnm.2968

Pankaj, S. K., Bueno-Ferrer, C., Misra, N. N., & Milosavljević, V. (2014). Cold plasma: Methods and applications for food preservation. *Journal of Food Science and Technology*, *51*(5), 957–970. PMID:24803704

Papadopoulos, M. C., & Saadoun, S. (2015). *Key roles of aquaporins in tumor biology.* Biochimica et Biophysica Acta (BBA)-. *Biochimica et Biophysica Acta. Biomembranes*, *1848*(10), 2576–2583. doi:10.1016/j.bbamem.2014.09.001

Pasquel-Reátegui, J. L., Villacrés-Granda, I. M., Jiménez-Cordero, D., Vanegas-Ortíz, D. H., Zurita, J., & Jaramillo-Gómez, J. I. (2018). In vitro antimicrobial activity of a non-thermal atmospheric plasma jet against Plasmodium falciparum and Enterococcus faecalis. *Experimental Parasitology*, *186*, 19–25.

Patil, B., Cherkasov, N., Lang, J., Ibhadon, A., Hessel, V., & Wang, Q. (2016). Low temperature plasma-catalytic NOx synthesis in a packed DBD reactor: Effect of support materials and supported active metal oxides. *Applied Catalysis B: Environmental*, *194*, 123–133. doi:10.1016/j.apcatb.2016.04.055

Patil, S., Moiseev, T., Misra, N. N., & Cullen, P. J. (2020). Cold plasma for effective seed germination: A review. *Trends in Food Science & Technology*, *97*, 1–10.

Paulussen, S., Verheyde, B., Tu, X., De Bie, C., Martens, T., Petrovic, D., Bogaerts, A., & Sels, B. (2010). Conversion of carbon dioxide to value-added chemicals in atmospheric pressure dielectric barrier discharges. *Plasma Sources Science & Technology*, *19*(3), 034015. doi:10.1088/0963-0252/19/3/034015

Pawelec, B., Damyanova, S., Arishtirova, K., Fierro, J. L. G., & Petrov, L. (2007). Structural and surface features of PtNi catalysts for reforming of methane with CO2. *Applied Catalysis A, General*, *323*, 188–201. doi:10.1016/j.apcata.2007.02.017

Perri, M. J., Van Wyngarden, A. L., Lin, J. J., Lee, Y. T., & Boering, K. A. (2004). Energy dependence of oxygen isotope exchange and quenching in the O (1D)+ CO2 reaction: A crossed molecular beam study. *The Journal of Physical Chemistry A, 108*(39), 7995–8001. doi:10.1021/jp0485845

Pervaiz, F., Ali, S., Ali, M., & Lai, S. T. (2023). Spacecraft Charging Due To Energetic Electrons and Ions at Geosynchronous Altitudes. *Journal of Geophysical Research: Space Physics, 128* (1), e2022JA030642.

Petitpas, G., Rollier, J.-D., Darmon, A., Gonzalez-Aguilar, J., Metkemeijer, R., & Fulcheri, L. (2007). A comparative study of non-thermal plasma assisted reforming technologies. *International Journal of Hydrogen Energy*, *32*(14), 2848–2867. doi:10.1016/j.ijhydene.2007.03.026

Pickett, J. S., & Chen, L.-JKahler, S. WSantolik, OGoldstein, M. LLavraud, BDecreau, P. MKessel, RLucek, ELakhina, G. STsurutani, B. TGurnett, D. ACornilleau-Wehrlin, NFazakerley, AReme, HBalogh. A. (2005). On the generation of solitary waves observed by Cluster in the near-Earth magnetosheath. *Nonlinear Processes in Geophysics.*, *12*(2), 181-193. doi:10.5194/npg-12-181-2005

Pickett, J. S., Kahler, S. W., Chen, L. J., Huff, R. L., Santolik, O., Khotyaintsev, Y., Décréau, P. M., Winningham, D., Frahm, R., Goldstein, M. L., Lakhina, G. S., Tsurutani, B. T., Lavraud, B., Gurnett, D. A., André, M., Fazakerley, A., Balogh, A., & Rème, H. (2004). Solitary waves observed in the auroral zone: The Cluster multi-spacecraft perspective. *Nonlinear Processes in Geophysics*, *11*(2), 183–196. doi:10.5194/npg-11-183-2004

Piel, A. (2010). An introduction to laboratory, space, and fusion plasmas. *Plasma Physics.*

Pinhão, N., Janeco, A., & Branco, J. (2011). Influence of helium on the conversion of methane and carbon dioxide in a dielectric barrier discharge. *Plasma Chemistry and Plasma Processing*, *31*(3), 427–439. doi:10.1007/s11090-011-9294-3

Pittman, J. K., Dean, A. P., & Osundeko, O. (2011). The potential of sustainable algal biofuel production using wastewater resources. *Bioresource Technology*, *102*(1), 17–25. doi:10.1016/j.biortech.2010.06.035 PMID:20594826

Plis, E., Engelhart, D. P., Barton, D., Cooper, R., Ferguson, D., & Hoffmann, R. (2017). Degradation of polyimide under exposure to 90 keV electrons. *physica status solidi (b), 254*(7), 1600819.

Polak, L., & Slovetsky, D. (1976). Electron impact induced electronic excitation and molecular dissociation. *International Journal for Radiation Physics and Chemistry*, *8*(1-2), 257–282. doi:10.1016/0020-7055(76)90070-X

Porra, R. J., Thompson, W. A. A., & Kriedemann, P. E. (1989). Determination of accurate extinction coefficients and simultaneous equations for assaying chlorophylls a and b extracted with four different solvents: Verification of the concentration of chlorophyll standards by atomic absorption spectroscopy. *Biochimica et Biophysica Acta (BBA)-. Biochimica et Biophysica Acta. Bioenergetics*, *975*(3), 384–394. doi:10.1016/S0005-2728(89)80347-0

Pörtner, R. (2005). Bioprocess engineering aspects of heterologous protein production in Escherichia coli. *Journal of Biotechnology*, *118*(2), 145–168.

Pottelette, R., Ergun, R. E., Treumann, R. A., Berthomier, M., Carlson, C. W., McFadden, J. P., & Roth, I. (1999). Modulated electron-acoustic waves in auroral density cavities: FAST observations. *Geophysical Research Letters*, *26*(16), 2629–2632. doi:10.1029/1999GL900462

Pottelette, R., Malingre, M., Dubouloz, N., Aparicio, B., Lundin, R., Holmgren, G., & Marklund, G. (1990). High-frequency waves in the cusp/cleft regions. *Journal of Geophysical Research*, *95*(A5), 5957–5971. doi:10.1029/JA095iA05p05957

Pou, J., Colominas, C., & Gonzalez-Olmos, R. (2018). CO2 reduction using non-thermal plasma generated with photovoltaic energy in a fluidized reactor. *Journal of CO2 Utilization, 27,* 528-535.

Privat-Maldonado, A., Schmidt, A., Lin, A., Weltmann, K.-D., Wende, K., Bogaerts, A., & Bekeschus, S. (2019). ROS from physical plasmas: Redox chemistry for biomedical therapy. *Oxidative Medicine and Cellular Longevity*, *2019*, 2019. doi:10.1155/2019/9062098 PMID:31687089

Prokopenko, S. M., & Laframboise, J. G. L. (1980). High voltage differential charging of geostationary spacecraft. *Journal of Geophysical Research*, *85*(A8, no. A8), 4125–4131. doi:10.1029/JA085iA08p04125

Qureshi, M. N., Nasir, W., Bruno, R., & Masood, W. (2019). Whistler instability based on observed flat-top two-component electron distributions in the Earth's magnetosphere. *Monthly Notices of the Royal Astronomical Society*, *488*(1), 954–964. doi:10.1093/mnras/stz1702

Qureshi, M. N., Nasir, W., Masood, W., Yoon, P. H., Shah, H. A., & Schwartz, S. J. (2014). Terrestrial lion roars and non-Maxwellian distribution. *Journal of Geophysical Research. Space Physics*, *119*(12), 10–059. doi:10.1002/2014JA020476

Qureshi, M. N., Shah, H. A., Murtaza, G., Schwartz, S. J., & Mahmood, F. (2004). Parallel propagating electromagnetic modes with the generalized (r, q) distribution function. *Physics of Plasmas*, *11*(8), 3819–3829. doi:10.1063/1.1688329

Qureshi, M. N., Shi, J. K., & Ma, S. Z. (2005). Landau damping in space plasmas with generalized (r,q) distribution function. *Physics of Plasmas*, *12*(122902), 122902. doi:10.1063/1.2139504

Radi, R. (2018). Oxygen radicals, nitric oxide, and peroxynitrite: Redox pathways in molecular medicine. *Proceedings of the National Academy of Sciences of the United States of America*, *115*(23), 5839–5848. doi:10.1073/pnas.1804932115 PMID:29802228

Rafiee, A., Khalilpour, K. R., Milani, D., & Panahi, M. (2018). Trends in CO2 conversion and utilization: A review from process systems perspective. *Journal of Environmental Chemical Engineering*, *6*(5), 5771–5794. doi:10.1016/j.jece.2018.08.065

Raizer, Y. P., Shneider, M. N., & Yatsenko, N. A. (2017). *Radio-frequency capacitive discharges*. CRC press. doi:10.1201/9780203741337

Ramakers, M., Michielsen, I., Aerts, R., Meynen, V., & Bogaerts, A. (2015). Effect of argon or helium on the CO2 conversion in a dielectric barrier discharge. *Plasma Processes and Polymers*, *12*(8), 755–763. doi:10.1002/ppap.201400213

Ramakers, M., Trenchev, G., Heijkers, S., Wang, W., & Bogaerts, A. (2017). Gliding arc plasmatron: Providing an alternative method for carbon dioxide conversion. *ChemSusChem*, *10*(12), 2642–2652. doi:10.1002/cssc.201700589 PMID:28481058

Ramos, A. A., Polle, J., Tran, D., Cushman, J. C., Jin, E.-S., & Varela, J. C. (2011). The unicellular green alga Dunaliella salina Teod. as a model for abiotic stress tolerance: Genetic advances and future perspectives. *Algae - Korean Phycological Society*, *26*(1), 3–20. doi:10.4490/algae.2011.26.1.003

Rapaport, D. C. (2004). *The art of molecular dynamics simulation*. Cambridge University Press. doi:10.1017/CBO9780511816581

Rathore, A. S., & Pathak, M. (2012). High-throughput process development and its application in biopharmaceuticals. *Biotechnology and Bioengineering*, *109*(7), 1753–1768.

Ratomski, P., & Hawrot-Paw, M. (2021). Production of Chlorella vulgaris Biomass in Tubular Photobioreactors during Different Culture Conditions. In Applied Sciences, 11(7). doi:10.3390/app11073106

Ratovitski, E. A., Cheng, X., Yan, D., Sherman, J. H., Canady, J., & Trink, B. (2013). Anti-cancer therapies of 21st century: Novel approach to treat human cancers using cold atmospheric plasma. *Plasma Processes and Polymers*, *10*(11), 1035–1050.

Ray, D., Saha, R., & Ch, S. (2017). DBD plasma assisted CO2 decomposition: Influence of diluent gases. *Catalysts*, *7*(9), 244. doi:10.3390/catal7090244

Ray, D., & Subrahmanyam, C. (2016). CO 2 decomposition in a packed DBD plasma reactor: Influence of packing materials. *RSC Advances*, *6*(45), 39492–39499. doi:10.1039/C5RA27085E

Razzokov, J., Fazliev, S., Erkinova, D., Mamatkulov, S., & Chen, Z. (2022). Understanding the effect of nitrosylation on dynamics of human epidermal growth factor: A μs simulation study. *Journal of Physics. D, Applied Physics*, *55*(47), 475201. doi:10.1088/1361-6463/ac9148

Razzokov, J., Fazliev, S., Yusupov, M., Sharipov, A., Ruziev, Z., & Mamatkulov, S. (2021). Effect of mutation and disulfide bond formation on the catalytic site of monomeric cytoglobin: A molecular level insight. *Plasma Medicine*, *11*(4), 41–51. doi:10.1615/PlasmaMed.2021041420

Ren, Y., Chen, C., Huang, W., Chen, H., Ding, T., Liu, C., & Liu, Y. (2017). Plasma modification of electrospun nanofibers for tissue engineering applications. *ACS Applied Materials & Interfaces*, *9*(8), 7292–7302.

Ren, Y., Li, X., Yu, L., Cheng, K., Yan, J., & Du, C. (2013). Degradation of PCDD/Fs in fly ash by vortex-shaped gliding arc plasma. *Plasma Chemistry and Plasma Processing*, *33*(1), 293–305. doi:10.1007/s11090-012-9421-9

Richmond, A. (2004). In P. O. Ang (Ed.), *Principles for attaining maximal microalgal productivity in photobioreactors: an overview BT - Asian Pacific Phycology in the 21st Century: Prospects and Challenges* (pp. 33–37). Springer Netherlands.

Richmond, A. E., & Soeder, C. J. (1986). Microalgaculture. *Critical Reviews in Biotechnology*, *4*(3), 369–438. doi:10.3109/07388558609150800

Rodriguez, I. J. (2018). *Some Assembly Required: computational simulations of dusty plasma*.

Rosenberg, M. (2015). Waves in a 1D electrorheological dusty plasma lattice. *Journal of Plasma Physics*, *81*(4), 905810407. doi:10.1017/S0022377815000422

Rouwenhorst, K. H., Jardali, F., Bogaerts, A., & Lefferts, L. (2021). From the Birkeland–Eyde process towards energy-efficient plasma-based NO X synthesis: A techno-economic analysis. *Energy & Environmental Science*, *14*(5), 2520–2534. doi:10.1039/D0EE03763J PMID:34046082

Rubin, A. G., & Henry, B. (1980). Spacecraft charging on ATS-5. Geophysics Laboratory, Air Force Systems Command, United States Air Force.

Rufai, O. R., Khazanov, G. V., & Singh, S. V. (2021). Finite amplitude electron-acoustic waves in the electron diffusion region. *Results in Physics*, *24*, 104041. doi:10.1016/j.rinp.2021.104041

Rusanov, V. D., Fridman, A., & Sholin, G. V. e. (1981). The physics of a chemically active plasma with nonequilibrium vibrational excitation of molecules. *Soviet Physics - Uspekhi*, *24*(6), 447–474. doi:10.1070/PU1981v024n06ABEH004884

Sabri, M. A., Al Jitan, S., Bahamon, D., Vega, L. F., & Palmisano, G. (2021). Current and future perspectives on catalytic-based integrated carbon capture and utilization. *The Science of the Total Environment*, *790*, 148081. doi:10.1016/j.scitotenv.2021.148081 PMID:34091328

Sackmann, E. K., Fulton, A. L., & Beebe, D. J. (2014). The present and future role of microfluidics in biomedical research. *Nature*, *507*(7491), 181–189. doi:10.1038/nature13118 PMID:24622198

Salazar, A. (2003). On thermal diffusivity. *European Journal of Physics*, *24*(4), 351–358. doi:10.1088/0143-0807/24/4/353

Salehizadeh, H., Yan, N., & Farnood, R. (2020). Recent advances in microbial CO2 fixation and conversion to value-added products. *Chemical Engineering Journal*, *390*, 124584. doi:10.1016/j.cej.2020.124584

Saleh, T. A. (2022). Nanomaterials and hybrid nanocomposites for CO 2 capture and utilization: Environmental and energy sustainability. *RSC Advances*, *12*(37), 23869–23888. doi:10.1039/D2RA03242B PMID:36093256

Salge, J. (1995). Plasma-assisted deposition at atmospheric pressure. *Le Journal de Physique IV*, *5*(C5), C5-583-C585-592.

Sanders, N. L., & Inouye, G. T. (1978). Secondary emission effects on spacecraft charging: energy distribution consideration. In Spacecraft Charging Technology. Air Force Geophysics Laboratory, Hanscom AFB, MA.

Sansaniwal, S. K., Sharma, V., & Mathur, J. (2018). Energy and exergy analyses of various typical solar energy applications: A comprehensive review. *Renewable & Sustainable Energy Reviews*, *82*, 1576–1601. doi:10.1016/j.rser.2017.07.003

Sarani, Z., Khodaeiaminjan, M., & Moghaddam, P. R. (2018). Seed priming with cold atmospheric plasma: An innovative technology to improve germination and seedling growth in different crops. *Scientific Reports*, *8*(1), 1–13. PMID:29311619

Sardella, E., Palumbo, F., Camporeale, G., & Favia, P. (2016). Non-equilibrium plasma processing for the preparation of antibacterial surfaces. *Materials (Basel)*, *9*(7), 515. doi:10.3390/ma9070515 PMID:28773637

Savinov, S. Y., Lee, H., Song, H. K., & Na, B.-K. (1999). Decomposition of methane and carbon dioxide in a radio-frequency discharge. *Industrial & Engineering Chemistry Research*, *38*(7), 2540–2547. doi:10.1021/ie980492c

Schenk, P. M., Thomas-Hall, S. R., Stephens, E., Marx, U. C., Mussgnug, J. H., Posten, C., Kruse, O., & Hankamer, B. (2008). Second generation biofuels: High-efficiency microalgae for biodiesel production. *BioEnergy Research*, *1*(1), 20–43. doi:10.1007/s12155-008-9008-8

Schiorlin, M., Klink, R., & Brandenburg, R. (2016). Carbon dioxide conversion by means of coplanar dielectric barrier discharges. *The European Physical Journal Applied Physics*, *75*(2), 24704. doi:10.1051/epjap/2016160073

Scholten, F., Sinev, I., Bernal, M., & Roldan Cuenya, B. (2019). Plasma-modified dendritic Cu catalyst for CO2 electroreduction. *ACS Catalysis*, *9*(6), 5496–5502. doi:10.1021/acscatal.9b00483 PMID:31205798

Scholtz, V., Pazlarova, J., Souskova, H., Khun, J., & Julak, J. (2015). Nonthermal plasma—A tool for decontamination and disinfection. *Biotechnology Advances*, *33*(6), 1108–1119. doi:10.1016/j.biotechadv.2015.01.002 PMID:25595663

Schwabe, M., Khrapak, S. A., Zhdanov, S. K., Pustylnik, M. Y., Rath, C., Fink, M., Kretschmer, M., Lipaev, A. M., Molotkov, V. I., Schmitz, A. S., Thoma, M. H., Usachev, A. D., Zobnin, A. V., Padalka, G. I., Fortov, V. E., Petrov, O. F., & Thomas, H. M. (2020). Slowing of acoustic waves in electrorheological and string-fluid complex plasmas. *New Journal of Physics*, *22*(8), 083079. doi:10.1088/1367-2630/aba91b

Semiokhin, I., Andreev, Y. P., & Panchenkov, G. (1964). Dissociation of carbon dioxide circulating in a silent electric Discharge, Russ. *Journal of Physical Chemistry*, *38*, 1126.

Shahidi, N. Z., Li, Y., Yu, B., & Zhou, R. (2020). Atmospheric pressure plasma jet: A review of the emerging technology and applications. *Plasma Chemistry and Plasma Processing*, *40*(1), 31–58.

Shah, K. H., Qureshi, M. N., Masood, W., & Shah, H. A. (2018). Electron acoustic nonlinear structures in planetary magnetospheres. *Physics of Plasmas*, *25*(4), 042303. doi:10.1063/1.5026186

Shahzad, A., & He, M. (2014). *Computer Simulation of Complex Plasmas: Molecular Modeling and Elementary Processes in Complex Plasmas*. Scholars' Press.

Shahzad, A., Khan, M. Q., Shakoori, M. A., He, M. G., & Feng, Y. (2020b). Thermal Conductivity of Dusty Plasmas through Molecular Dynamics Simulations. Intech London, Thermophysical Properties of Complex Materials. doi:10.5772/intechopen.91418

Shahzad, A., Khan, M. Q., Shakoori, M. A., He, M., & Feng, Y. (2020). Thermal Conductivity of Dusty Plasmas through Molecular Dynamics Simulations. *Thermophysical Properties of Complex Materials, 13.*

Shahzad, A., Haider, S. I., Kashif, M., Shifa, M. S., Mu, T., & He, M. G. (2018b). Thermal Conductivity of Complex Plasmas Using Novel Evan-Gillan Approach. *Communications in Theoretical Physics*, *69*(6), 704–710. doi:10.1088/0253-6102/69/6/704

Shahzad, A., & He, M. G. (2012). Thermal Conductivity of Three-Dimensional Yukawa Liquids (Dusty Plasmas). *Contributions to Plasma Physics*, *52*(8), 667–675. doi:10.1002/ctpp.201200002

Shahzad, A., & He, M. G. (2012a). Thermal conductivity calculation of complex (dusty) plasmas. *Physics of Plasmas*, *19*(8), 083707. doi:10.1063/1.4748526

Shahzad, A., & He, M. G. (2013, July). Interaction contributions in thermal conductivity of three-dimensional complex liquids. In. AIP Conference Proceedings: Vol. 1547. *No. 1* (pp. 173–180). American Institute of Physics. doi:10.1063/1.4816866

Shahzad, A., & He, M. G. (2015). Homogeneous Nonequilibrium Molecular Dynamics Evaluation of Thermal Conductivity in 2D Yukawa Liquids. *International Journal of Thermophysics*, *36*(10–11), 2565–2576. doi:10.1007/s10765-014-1671-8

Shahzad, A., He, M. G., & He, K. (2013). Diffusion motion of two-dimensional weakly coupled complex (dusty) plasmas. *Physica Scripta*, *87*(3), 035501. doi:10.1088/0031-8949/87/03/035501

Shahzad, A., Kashif, M., Munir, T., He, M., & Tu, X. (2020a). Thermal conductivity analysis of two-dimensional complex plasma liquids and crystals. *Physics of Plasmas*, *27*(10), 103702. doi:10.1063/5.0018537

Shahzad, A., Shakoori, M. A., He, M., & Bashir, S. (2018a). *Sound Waves in Complex (Dusty) Plasmas*. Computational and Experimental Studies of Acoustic Waves. doi:10.5772/intechopen.71203

Shahzad, A., Sultana, M., Aslam, A., & He, M. G. (2014). Molecular dynamics simulations of thermal conductivity in 2D complex Yukawa liquids. *IOP Conference Series. Materials Science and Engineering*, *60*(1), 012038. doi:10.1088/1757-899X/60/1/012038

Shakoori, M. A., He, M. G., Shahzad, A., & Khan, M. (2022c). *Diffusion Coefficients for Complex Plasmas in Parallel Electric Field. 48*(9), 1023–1033. doi:10.1134/S1063780X22100014

Shakoori, M. A., He, M., & Shahzad, A. K. M. (2023). Tuning the structure and transport properties of complex plasmas using electric field. *Physica Scripta, 98*, 015608. https://doi.org/https://doi.org/10.1088/1402-4896/aca6b2

Shakoori, M. A., He, M.-G., Shahzad, A., & Khan, M. (2021b). Studies of Self Diffusion Coefficient in Electrorheological Complex Plasmas through Molecular Dynamics Simulations. In Plasma Science and Technology. Intech London. doi:10.5772/intechopen.98854

Shakoori, M. A., He, M. G., Shahzad, A., & Khan, M. (2022b). Diffusion coefficients of electrorheological complex (dusty) plasmas. *Journal of Molecular Modeling*, *28*(12), 398. doi:10.1007/s00894-022-05394-3 PMID:36422698

Shakoori, M. A., He, M.-G., Shahzad, A., Khan, M., & Zhang, Y. (2021a). Molecular Dynamics Study of Diffusion Coefficient for Low-Temperature Dusty Plasmas in the Presence of External Electric Fields. In *Emerging Development and application of low temperature plasma* (pp. 63–84). IGI Global. doi:10.4018/978-1-7998-8398-2.ch004

Shakoori, M. A., He, M., & Shahzad, A. (2022). Diffusion coefficients of dusty plasmas in electric field. *The European Physical Journal D*, *76*(11), 227. doi:10.1140/epjd/s10053-022-00553-w

Shakoori, M. A., He, M., Shahzad, A., & Khan, M. (2022). Tuning the structure and transport properties of complex plasmas using electric field. *Physica Scripta*, *98*(1), 015608. doi:10.1088/1402-4896/aca6b2

Shaw, D., West, A., Bredin, J., & Wagenaars, E. (2016). Mechanisms behind surface modification of polypropylene film using an atmospheric-pressure plasma jet. *Plasma Sources Science & Technology, 25*(6), 065018. doi:10.1088/0963-0252/25/6/065018

Shaw, P., Kumar, N., Privat-Maldonado, A., Smits, E., & Bogaerts, A. (2021). Cold atmospheric plasma increases temozolomide sensitivity of three-dimensional glioblastoma spheroids via oxidative stress-mediated DNA damage. *Cancers (Basel), 13*(8), 1780. doi:10.3390/cancers13081780 PMID:33917880

Shengjie, Z. (2019). Enhanced CO2 decomposition via metallic foamed electrode packed in self-cooling DBD plasma device. *Plasma Science & Technology, 21*(8), 085504. doi:10.1088/2058-6272/ab15e5

Shen, X., Zhang, Y., Guan, J., Xu, J., Lu, J., & Huang, H. (2015). In situ gelling and mucoadhesive cetirizine microemulsion based on PLGA-PEG-PLGA copolymer for intranasal delivery. *International Journal of Pharmaceutics, 495*(1), 439–448. PMID:26367780

Shi, C., Wang, S., Ge, X., Deng, S., Chen, B., & Shen, J. (2021). A review of different catalytic systems for dry reforming of methane: Conventional catalysis-alone and plasma-catalytic system. *Journal of CO2 Utilization, 46*, 101462.

Shirayama, R., Shimizu, T., Nishijima, A., Kido, Y., & Ogawa, T. (2019). Plasma treatment accelerates wound healing in a murine model by promoting neovascularization and fibroblast activation. *PLoS One, 14*(7), e0219528.

Shohaib, M., Masood, W., Jahangir, R., Siddiq, M., Alkhateeb, S.A., & El-Tantawy, S.A. (2021). On the dynamics of nonlinear propagation and interaction of the modified KP solitons in multicomponent complex plasmas. *Journal of Ocean Engineering and Science.* doi:10.1016/j.joes.2021.10.005

Shohaib, M., Masood, W., Jahangir, R., Siddiq, M., & Rizvi, H. (2022). Interaction of dust ion acoustic solitons with cubicnonlinearity in a magnetized dusty plasma with (r, q) distributed electrons. *Contributions to Plasma Physics, e202100134*(3), e202100134. doi:10.1002/ctpp.202100134

Shriver, D., & Ashour-Abdalla, M. (2003). 1989. Geophys. Res. Lett. 16, 899. Matsumoto H, Deng XH, Kojima H, and Anderson RR, Observation of Electrostatic Solitary Waves associated withreconnection on the dayside magnetopause boundary. *GEOPHYSICAL RESEARCH LETTERS, VOL., 30*(6), 1326. doi:10.1029/2002GL016319

Shukla, P. K. (2001). A survey of dusty plasma physics. *Physics of Plasmas, 8*(5 II), 1791. doi:10.1063/1.1343087

Shukla, P. K., & Mamun, A. A. (2015). *Introduction to dusty plasma physics*. CRC press. doi:10.1201/9781420034103

Sialve, B., Bernet, N., & Bernard, O. (2009). Anaerobic digestion of microalgae as a necessary step to make microalgal biodiesel sustainable. *Biotechnology Advances, 27*(4), 409–416. doi:10.1016/j.biotechadv.2009.03.001 PMID:19289163

Siemens, W. (1857). Ueber die elektrostatische Induction und die Verzögerung des Stroms in Flaschendrähten. *Annalen der Physik, 178*(9), 66–122. doi:10.1002/andp.18571780905

Silapasert, P., Yatongchai, C., & Sarapirom, S. (2023). Investigation of plasma activated water in the growth of green microalgae (*Chlorella* spp.). *Journal of Physics: Conference Series, 2431*(1), 12037. doi:10.1088/1742-6596/2431/1/012037

Sills, D. L., Paramita, V., Franke, M. J., Johnson, M. C., Akabas, T. M., Greene, C. H., & Tester, J. W. (2013). Quantitative uncertainty analysis of life cycle assessment for algal biofuel production. *Environmental Science & Technology, 47*(2), 687–694. doi:10.1021/es3029236 PMID:23237457

Silva, T., Britun, N., Godfroid, T., & Snyders, R. (2014). Optical characterization of a microwave pulsed discharge used for dissociation of CO2. *Plasma Sources Science & Technology, 23*(2), 025009. doi:10.1088/0963-0252/23/2/025009

Singh, S. V., Devanandhan, S., Lakhina, G. S., & Bharuthram, R. (2016). Electron acoustic solitary waves in a magnetized plasma with nonthermal electrons and an electron beam. *Physics of Plasmas*, *23*(8), 082310. doi:10.1063/1.4961961

Singh, S. V., & Lakhina, G. S. (2001). Generation of electron-acoustic waves in the magnetosphere. *Planetary and Space Science*, *49*(1), 107–114. doi:10.1016/S0032-0633(00)00126-4

Singh, S. V., Lakhina, G. S., Bharuthram, R., & Pillay, S. R. (2011). Electrostatic solitary structures in presence of non-thermal electrons and a warm electron beam on the auroral field lines. *Physics of Plasmas*, *18*(12), 122306. doi:10.1063/1.3671955

Singh, S., Rana, N. K., & Singh, R. (2021). Advances in plasma pyrolysis of biomass for renewable energy production and pollution control: A review. *Renewable & Sustainable Energy Reviews*, *139*, 110720.

Sinkevich, O. A., & Semenov, A. M. (2003). Solution of the Boltzmann equation by expanding the distribution function with several time and coordinate scales in the Enskog series in Knudsen parameter. *Technical Physics*, *48*(10), 1221–1225. doi:10.1134/1.1620111

Slepička, P., Siegel, J., Kolská, Z., Rimpelová, S., Švorčík, V., & Lyutakov, O. (2017). Biocompatibility of plasma-treated polyethylene. *Materials Science and Engineering C*, *71*, 602–610.

Smirnov, B. M. (2008). *Physics of ionized gases*. John Wiley & Sons.

Smith, L. J., Kahraman, A., & Thornton, J. M. (2010). Heme proteins—Diversity in structural characteristics, function, and folding. *Proteins*, *78*(10), 2349–2368. doi:10.1002/prot.22747 PMID:20544970

Smith, V. H., Tilman, G. D., & Nekola, J. C. (1999). Eutrophication: Impacts of excess nutrient inputs on freshwater, marine, and terrestrial ecosystems. *Environmental Pollution*, *100*(1–3), 179–196. doi:10.1016/S0269-7491(99)00091-3 PMID:15093117

Snoeckx, R., & Bogaerts, A. (2017). Plasma technology–a novel solution for CO 2 conversion? *Chemical Society Reviews*, *46*(19), 5805–5863. doi:10.1039/C6CS00066E PMID:28825736

Snoeckx, R., Heijkers, S., Van Wesenbeeck, K., Lenaerts, S., & Bogaerts, A. (2016). CO 2 conversion in a dielectric barrier discharge plasma: N 2 in the mix as a helping hand or problematic impurity? *Energy & Environmental Science*, *9*(3), 999–1011. doi:10.1039/C5EE03304G

Snoeckx, R., Zeng, Y. X., Tu, X., & Bogaerts, A. (2015). Plasma-based dry reforming: Improving the conversion and energy efficiency in a dielectric barrier discharge. *RSC Advances*, *5*(38), 29799–29808. doi:10.1039/C5RA01100K

Spencer, L. F., & Gallimore, A. D. (2011). Efficiency of CO 2 dissociation in a radio-frequency discharge. *Plasma Chemistry and Plasma Processing*, *31*(1), 79–89. doi:10.1007/s11090-010-9273-0

Spencer, L., & Gallimore, A. (2012). CO2 dissociation in an atmospheric pressure plasma/catalyst system: A study of efficiency. *Plasma Sources Science & Technology*, *22*(1), 015019. doi:10.1088/0963-0252/22/1/015019

Stadtman, E. R. (2006). Protein oxidation and aging. *Free Radical Research*, *40*(12), 1250–1258. doi:10.1080/10715760600918142 PMID:17090414

Stadtman, E., & Levine, R. (2003). Free radical-mediated oxidation of free amino acids and amino acid residues in proteins. *Amino Acids*, *25*(3-4), 207–218. doi:10.1007/s00726-003-0011-2 PMID:14661084

Steinberg, M., & Dang, V.-D. (1977). Production of synthetic methanol from air and water using controlled thermonuclear reactor power—I. Technology and energy requirement. *Energy Conversion*, *17*(2-3), 97–112. doi:10.1016/0013-7480(77)90080-8

Stephens, E., Ross, I. L., Mussgnug, J. H., Wagner, L. D., Borowitzka, M. A., Posten, C., Kruse, O., & Hankamer, B. (2010). Future prospects of microalgal biofuel production systems. *Trends in Plant Science*, *15*(10), 554–564. doi:10.1016/j.tplants.2010.06.003 PMID:20655798

Stephenson, A. L., Kazamia, E., Dennis, J. S., Howe, C. J., Scott, S. A., & Smith, A. G. (2010). Life-cycle assessment of potential algal biodiesel production in the United Kingdom: A comparison of raceways and air-lift tubular bioreactors. *Energy & Fuels*, *24*(7), 4062–4077. doi:10.1021/ef1003123

Stirk, W. A., Bálint, P., Vambe, M., Kulkarni, M. G., van Staden, J., & Ördög, V. (2021). Effect of storage on plant biostimulant and bioactive properties of freeze-dried *Chlorella vulgaris* biomass. *Journal of Applied Phycology*, *33*(6), 3797–3806. doi:10.1007/s10811-021-02596-9 PMID:34539078

Stoscheck, C. M., & King, L. E. Jr. (1986). Role of epidermal growth factor in carcinogenesis. *Cancer Research*, *46*(3), 1030–1037. PMID:3002608

Struchtrup, H. (2005). Macroscopic transport equations for rarefied gas flows: approximation methods in kinetic theory. In *Interaction of Mechanics and Mathematics*. Springer-Verlag.

Su, C. H., & Mirie, R. M. (1980). On head-on collisions between two solitary waves. *Journal of Fluid Mechanics*, *98*(3), 509. doi:10.1017/S0022112080000262

Sukhani, S., Punith, N., Ekatpure, A., Salunke, G., Manjari, M., Harsha, R., Chanakya, H. N., & Lakshminarayana, R. (2021). Plasma-Activated Water as Nitrogen Source for Algal Growth: A Microcosm Study. *IEEE Transactions on Plasma Science*, *49*(2), 551–556. doi:10.1109/TPS.2021.3051728

Sukhinin, G. I., Fedoseev, A. V., Salnikov, M. V., Rostom, A., Vasiliev, M. M., & Petrov, O. F. (2017). Plasma anisotropy around a dust particle placed in an external electric field. *Physical Review. E*, *95*(6), 063207. doi:10.1103/PhysRevE.95.063207 PMID:28709289

Sukop, M. C. (2006). *DT Thorne, Jr.* Lattice Boltzmann Modeling Lattice Boltzmann Modeling.

Sun, S. (2017). CO2 conversion in a gliding arc plasma: Performance improvement based on chemical reaction modeling. *Journal of CO2 Utilization, 17*, 220-234.

Sun, S., Wang, H., Mei, D., Tu, X., & Bogaerts, A. (2017). CO2 conversion in a gliding arc plasma: Performance improvement based on chemical reaction modeling. *Journal of CO2 Utilization, 17,* 220-234.

Sun, L., & Venart, J. E. S. (2005). Thermal conductivity, thermal diffusivity, and heat capacity of gaseous Argon and nitrogen. *International Journal of Thermophysics*, *26*(2), 325–372. doi:10.1007/s10765-005-4502-0

Sun, S., Gu, L., Yang, Z., Lin, H., & Li, Y. (2022). Thermophysical properties of natural gas hydrates: A review. *Natural Gas Industry B*, *9*(3), 246–263. doi:10.1016/j.ngib.2022.04.003

Sun, X.-M., Ren, L.-J., Zhao, Q.-Y., Ji, X.-J., & Huang, H. (2018). Microalgae for the production of lipid and carotenoids: A review with focus on stress regulation and adaptation. *Biotechnology for Biofuels*, *11*(1), 272. doi:10.1186/s13068-018-1275-9 PMID:30305845

Sun, Y., Wu, J., Wang, Y., Li, J., Wang, N., Harding, J., Mo, S., Chen, L., Chen, P., Fu, M., Ye, D., Huang, J., & Tu, X. (2022). Plasma-catalytic CO2 hydrogenation over a Pd/ZnO catalyst: In situ probing of gas-phase and surface reactions. *JACS Au*, *2*(8), 1800–1810. doi:10.1021/jacsau.2c00028 PMID:36032530

Suparmaniam, U., Lam, M. K., Uemura, Y., Lim, J. W., Lee, K. T., & Shuit, S. H. (2019). Insights into the microalgae cultivation technology and harvesting process for biofuel production: A review. *Renewable & Sustainable Energy Reviews*, *115*, 109361. doi:10.1016/j.rser.2019.109361

Surowsky, B., Fischer, A., Schlueter, O., & Knorr, D. (2013). Cold plasma effects on enzyme activity in a model food system. *Innovative Food Science & Emerging Technologies*, *19*, 146–152. doi:10.1016/j.ifset.2013.04.002

Su, T., Zhou, X., Qin, Z., & Ji, H. (2017). Intrinsic Kinetics of Dimethyl Ether Synthesis from Plasma Activation of CO2 Hydrogenation over Cu–Fe–Ce/HZSM-5. *ChemPhysChem*, *18*(3), 299–309. doi:10.1002/cphc.201601283 PMID:27898184

Takai, E., Kitamura, T., Kuwabara, J., Ikawa, S., Yoshizawa, S., Shiraki, K., Kawasaki, H., Arakawa, R., & Kitano, K. (2014). Chemical modification of amino acids by atmospheric-pressure cold plasma in aqueous solution. *Journal of Physics. D, Applied Physics*, *47*(28), 285403. doi:10.1088/0022-3727/47/28/285403

Talebizadeh, P., Babaie, M., Brown, R., Rahimzadeh, H., Ristovski, Z., & Arai, M. (2014). The role of non-thermal plasma technique in NOx treatment: A review. *Renewable & Sustainable Energy Reviews*, *40*, 886–901. doi:10.1016/j.rser.2014.07.194

Tamang, J., & Saha, A. (2020). Bifurcations of small-amplitude supernonlinear waves of the mKdV and modified Gardner equations in a three-component electron-ion plasma. *Physics of Plasmas*, *27*(1), 012105. doi:10.1063/1.5115821

Taylan, O., & Berberoglu, H. (2014). Dissociation of carbon dioxide using a microhollow cathode discharge plasma reactor: Effects of applied voltage, flow rate and concentration. *Plasma Sources Science & Technology*, *24*(1), 015006. doi:10.1088/0963-0252/24/1/015006

Temerin, M., Cerny, K., & Lotko, W. (1982). W. and F. S. Mozer. *Physical Review Letters*, *48*, 1175. doi:10.1103/PhysRevLett.48.1175

Teoh, M. L. T., Sun, W., Smith, B. J., Oberley, L. W., & Cullen, J. J. (2007). Modulation of reactive oxygen species in pancreatic cancer. *Clinical Cancer Research*, *13*(24), 7441–7450. doi:10.1158/1078-0432.CCR-07-0851 PMID:18094428

Thanu, D., Srinadhu, E., Zhao, M., Dole, N., & Keswani, M. (2019). Fundamentals and applications of plasma cleaning. *Developments in surface contamination and cleaning: applications of cleaning techniques*, 289-353.

Thapa, R. & Wilson, G. (2016). *The importance of CD44 as a stem cell biomarker and therapeutic target in cancer.* Stem cells international.

Thomas, H., Morfill, G. E., Demmel, V., Goree, J., Feuerbacher, B., & Möhlmann, D. (1994). Plasma crystal: Coulomb crystallization in a dusty plasma. *Physical Review Letters*, *73*(5), 652–655. doi:10.1103/PhysRevLett.73.652 PMID:10057503

Toebaert, D., Muys, P., & Desoppere, E. (1995). Spatially resolved measurement of the vibrational temperatures of the plasma in a DC-excited fast-axial-flow CO/sub 2/laser. *IEEE Journal of Quantum Electronics*, *31*(10), 1774–1778. doi:10.1109/3.466051

Tokar, R. L., & Gary, S. P. (1984). Electrostatic hiss and the beam driven electron acoustic instability in the dayside polar cusp. *Geophysical Research Letters*, *11*(12), 1180–1183. doi:10.1029/GL011i012p01180

Trachootham, D., Alexandre, J., & Huang, P. (2009). Targeting cancer cells by ROS-mediated mechanisms: A radical therapeutic approach? *Nature Reviews. Drug Discovery*, *8*(7), 579–591. doi:10.1038/nrd2803 PMID:19478820

Traylor, M. J., Pavlovich, M. J., & Karim, K. S. (2012). Study of plasma polymerization by single photon ionization mass spectrometry. *Plasma Processes and Polymers*, *9*(7), 698–704.

Treanor, C. E., Rich, J., & Rehm, R. (1968). Vibrational relaxation of anharmonic oscillators with exchange-dominated collisions. *The Journal of Chemical Physics*, *48*(4), 1798–1807. doi:10.1063/1.1668914

Trenchev, G., Kolev, S., Wang, W., Ramakers, M., & Bogaerts, A. (2017). CO2 conversion in a gliding arc plasmatron: Multidimensional modeling for improved efficiency. *The Journal of Physical Chemistry. C, Nanomaterials and Interfaces, 121*(44), 24470–24479. doi:10.1021/acs.jpcc.7b08511

Tsallis, C. (1988). Possible generalization of Boltzmann-Gibbs statistics. *Journal of Statistical Physics, 52*(1–2), 479–487. doi:10.1007/BF01016429

Tsimpanogiannis, I. N., Moultos, O. A., Franco, L. F., Spera, M. B. D. M., Erdős, M., & Economou, I. G. (2019). Self-diffusion coefficient of bulk and confined water: A critical review of classical molecular simulation studies. *Molecular Simulation, 45*(4-5), 425–453. doi:10.1080/08927022.2018.1511903

Tsuboi, T. (1989). Phase shift in the collision of two solitons propagating in a nonlinear transmission line. *Physical Review A: General Physics, 40*(5), 2753–2755. doi:10.1103/PhysRevA.40.2753

Tsuji, M., Tanoue, T., Nakano, K., & Nishimura, Y. (2001). Decomposition of CO2 into CO and O in a microwave-excited discharge flow of CO2/He or CO2/Ar mixtures. *Chemistry Letters, 30*(1), 22–23. doi:10.1246/cl.2001.22

Tsurutani, B. T., Arballo, J. K., Lakhina, G. S., Ho, C. M., Buti, B., Pickett, J. S., & Gurnett, D. A. (1998). Plasma waves in the dayside polar cap boundary layer: Bipolar and monopolar electric pulses and whistler mode waves. *Geophysical Research Letters, 25*(22), 4117–4120. doi:10.1029/1998GL900114

Tu, X., Gallon, H. J., Twigg, M. V., Gorry, P. A., & Whitehead, J. C. (2011). Dry reforming of methane over a Ni/Al2O3 catalyst in a coaxial dielectric barrier discharge reactor. *Journal of Physics. D, Applied Physics, 44*(27), 274007. doi:10.1088/0022-3727/44/27/274007

Tu, X., & Whitehead, J. C. (2014). Plasma dry reforming of methane in an atmospheric pressure AC gliding arc discharge: Co-generation of syngas and carbon nanomaterials. *International Journal of Hydrogen Energy, 39*(18), 9658–9669. doi:10.1016/j.ijhydene.2014.04.073

Tyurikov, K., Alexandrov, S., & Iankevich, G. (2020). Corona discharge plasma application for the deposition of nano-composite coatings. *Materials Today: Proceedings, 30*, 404–407. doi:10.1016/j.matpr.2019.12.385

Ullah, S., Masood, W., & Siddiq, M. (2020). Electron acoustic envelope solitons in non-Maxwellian plasmas. *The European Physical Journal D, 74*(2), 1–9. doi:10.1140/epjd/e2019-100589-1

Usman, M., Daud, W. W., & Abbas, H. F. (2015). Dry reforming of methane: Influence of process parameters—A review. *Renewable & Sustainable Energy Reviews, 45*, 710–744. doi:10.1016/j.rser.2015.02.026

Uytdenhouwen, Y., Van Alphen, S., Michielsen, I., Meynen, V., Cool, P., & Bogaerts, A. (2018). A packed-bed DBD micro plasma reactor for CO2 dissociation: Does size matter? *Chemical Engineering Journal, 348*, 557–568. doi:10.1016/j.cej.2018.04.210

Vadikkeettil, Y., Subramaniam, Y., Murugan, R., Ananthapadmanabhan, P., Mostaghimi, J., Pershin, L., Batiot-Dupeyrat, C., & Kobayashi, Y. (2022). Plasma assisted decomposition and reforming of greenhouse gases: A review of current status and emerging trends. *Renewable & Sustainable Energy Reviews, 161*, 112343. doi:10.1016/j.rser.2022.112343

Valluri, S., Claremboux, V., & Kawatra, S. (2022). Opportunities and challenges in CO2 utilization. *Journal Of Environmental Sciences, 113*, 322-344.

Van Laer, K., & Bogaerts, A. (2015). Improving the conversion and energy efficiency of carbon dioxide splitting in a zirconia-packed dielectric barrier discharge reactor. *Energy Technology (Weinheim), 3*(10), 1038–1044. doi:10.1002/ente.201500127

Van Laer, K., & Bogaerts, A. (2017). How bead size and dielectric constant affect the plasma behaviour in a packed bed plasma reactor: A modelling study. *Plasma Sources Science & Technology*, *26*(8), 085007. doi:10.1088/1361-6595/aa7c59

Van Rooij, G., Akse, H. N., Bongers, W. A., & van de Sanden, M. C. M. (2017). Plasma for electrification of chemical industry: A case study on CO2 reduction. *Plasma Physics and Controlled Fusion*, *60*(1), 014019. doi:10.1088/1361-6587/aa8f7d

Van Rooij, G., Van Den Bekerom, D., Den Harder, N., Minea, T., Berden, G., Bongers, W., Engeln, R., Graswinckel, M., Zoethout, E., & Van De Sanden, M. (2015). Taming microwave plasma to beat thermodynamics in CO 2 dissociation. *Faraday Discussions*, *183*, 233–248. doi:10.1039/C5FD00045A PMID:26388308

Vandamme, M., Robert, E., Lerondel, S., Sarron, V., Ries, D., & Dozias, S. (2010). ROS implication in a new antitumor strategy based on non-thermal plasma. *International Journal of Cancer*, *127*(3), 779–786. PMID:21702038

Vandamme, M., Robert, E., Lerondel, S., Sarron, V., Ries, D., Dozias, S., Sobilo, J., Gosset, D., Kieda, C., Legrain, B., Pouvesle, J.-M., & Pape, A. L. (2012). ROS implication in a new antitumor strategy based on non-thermal plasma. *International Journal of Cancer*, *130*(9), 2185–2194. doi:10.1002/ijc.26252 PMID:21702038

Vandamme, M., Robert, E., Pesnel, S., Barbosa, E., Dozias, S., Sobilo, J., ... Lerondel, S. (2012). Antitumor effect of plasma treatment on u87 glioma xenografts: Preliminary results. *Plasma Processes and Polymers*, *9*(6), 583–592.

Vargas, J., Lopez, B., & Lino da Silva, M. (2021). Heavy particle impact vibrational excitation and dissociation processes in CO2. *The Journal of Physical Chemistry A*, *125*(2), 493–512. doi:10.1021/acs.jpca.0c05677 PMID:33406355

Varjo, S., Mikkonen, H., Hakola, A., Piispanen, M., Keskinen, H., Hietala, J., ... Kukk, E. (2015). Oxygen molecule density in non-thermal atmospheric pressure plasma jet. *Plasma Sources Science & Technology*, *24*(1), 015014.

Vasyliunas, V. M. (1968). A survey of low-energy electrons in the evening sector of the magnetosphere with OGO 1 and OGO 3. *Journal of Geophysical Research*, *73*(9), 2839–2884. doi:10.1029/JA073i009p02839

Vasyliunas, V. M. (1968). Low-energy electrons on the day side of the magnetosphere. *Journal of Geophysical Research*, *73*(23), 7519–7523. doi:10.1029/JA073i023p07519

Vaulina, O. S., Adamovich, X. G., Petrov, O. F., & Fortov, V. E. (2008). Evolution of the mass-transfer processes in nonideal dissipative systems II: Experiments in dusty plasma. *Physical Review. E*, *77*(6), 066404. doi:10.1103/PhysRevE.77.066404 PMID:18643380

Vaulina, O. S., Lisin, E. A., Gavrikov, A. V., Petrov, O. F., & Fortov, V. E. (2010). Analysis of pair interparticle interaction in nonideal dissipative systems. *Journal of Experimental and Theoretical Physics*, *110*(4), 662–674. doi:10.1134/S1063776110040138

Verheest, F. (2013). Ambiguities in the Tsallis description of non-thermal plasma species. *Journal of Plasma Physics, 79*. doi:10.1017/S0022377813001049

Verheest, F., Hellberg, M. A., & Hereman, W. A. (2012). Head-on collisions of electrostatic solitons in nonthermal plasmas. *Physical Review E: Statistical, Nonlinear, and Soft Matter Physics*, *86*(3), 036402. doi:10.1103/PhysRevE.86.036402 PMID:23031029

Verheest, F., & Hereman, W. A. (2019). Collisions of acoustic solitons and their electric fields in plasmas at critical compositions. *Journal of Plasma Physics*, *85*(1), 905850106. doi:10.1017/S0022377818001368

Vertongen, R, Trenchev, G., Van Loenhout, R., & Bogaerts, A. (2022). Enhancing CO2 conversion with plasma reactors in series and O2 removal. *Journal of CO2 utilization, 66*, 102252.

Vesel, A., Mozetic, M., Drenik, A., & Balat-Pichelin, M. (2011). Dissociation of CO2 molecules in microwave plasma. *Chemical Physics*, *382*(1-3), 127–131. doi:10.1016/j.chemphys.2011.03.015

Vezyris, C., Kavvadias, I., Papoutsis-Kiachagias, E., & Giannakoglou, K. (2014). *Unsteady continuous adjoint method using POD for jet-based flow control.* 11th World congress on computational mechanics, Barcelona, Spain.

Vörösmarty, C. J., Green, P., Salisbury, J., & Lammers, R. B. (2000). Global water resources: Vulnerability from climate change and population growth. *Science*, *289*(5477), 284–288. doi:10.1126/science.289.5477.284 PMID:10894773

Wang, F.-G., Yang, Y.-Y., Han, J.-F., & Duan, W.-S. (2018). Head-on collision between two solitary waves in a one-dimensional bead chain. *Chinese Physics B*, *27*(4), 044501. doi:10.1088/1674-1056/27/4/044501

Wang, G., Zhang, L., & Snyders, R. (2017). The effect of plasma treatment on the wettability of thin films. *Plasma Processes and Polymers*, *14*(3), 1600131.

Wang, H., Tang, J., Yin, S., Xu, L., Zhang, J., & Zhang, J. (2017). The interactions between nanoparticles and plasma in biomedicine. *Journal of Physics. D, Applied Physics*, *50*(31), 313001.

Wang, H., Wandell, R. J., Tachibana, K., Voráč, J., & Locke, B. R. (2018). The influence of liquid conductivity on electrical breakdown and hydrogen peroxide production in a nanosecond pulsed plasma discharge generated in a water-film plasma reactor. *Journal of Physics. D, Applied Physics*, *52*(7), 075201. doi:10.1088/1361-6463/aaf132

Wang, J., AlQahtani, M. S., Wang, X., Knecht, S. D., Bilén, S. G., Song, C., & Chu, W. (2021). One-step plasma-enabled catalytic carbon dioxide hydrogenation to higher hydrocarbons: Significance of catalyst-bed configuration. *Green Chemistry*, *23*(4), 1642–1647. doi:10.1039/D0GC03779F

Wang, J.-Y., Xia, G.-G., Huang, A., Suib, S. L., Hayashi, Y., & Matsumoto, H. (1999). CO2 decomposition using glow discharge plasmas. *Journal of Catalysis*, *185*(1), 152–159. doi:10.1006/jcat.1999.2499

Wang, L., Yi, Y., Guo, H., & Tu, X. (2018). Atmospheric pressure and room temperature synthesis of methanol through plasma-catalytic hydrogenation of CO2. *ACS Catalysis*, *8*(1), 90–100. doi:10.1021/acscatal.7b02733

Wang, S., Rohani, V., Ye, T., Dupont, P., Pagnon, S., Sennour, M., & Fulcheri, L. (2023). Effect of K-promoter use in iron-based plasma-catalytic conversion of CO2 and CH4 into higher value products. *Applied Catalysis A, General*, *663*, 119315. doi:10.1016/j.apcata.2023.119315

Wang, S., Zhang, Y., Liu, X., & Wang, X. (2012). Enhancement of CO 2 conversion rate and conversion efficiency by homogeneous discharges. *Plasma Chemistry and Plasma Processing*, *32*(5), 979–989. doi:10.1007/s11090-012-9386-8

Wang, W., Berthelot, A., Kolev, S., Tu, X., & Bogaerts, A. (2016). CO2 conversion in a gliding arc plasma: 1D cylindrical discharge model. *Plasma Sources Science & Technology*, *25*(6), 065012. doi:10.1088/0963-0252/25/6/065012

Wang, W.-H., Himeda, Y., Muckerman, J. T., Manbeck, G. F., & Fujita, E. (2015). CO2 hydrogenation to formate and methanol as an alternative to photo-and electrochemical CO2 reduction. *Chemical Reviews*, *115*(23), 12936–12973. doi:10.1021/acs.chemrev.5b00197 PMID:26335851

Wang, W., Mei, D., Tu, X., & Bogaerts, A. (2017). Gliding arc plasma for CO2 conversion: Better insights by a combined experimental and modelling approach. *Chemical Engineering Journal*, *330*, 11–25. doi:10.1016/j.cej.2017.07.133

Wang, W., Wang, S., Ma, X., & Gong, J. (2011). Recent advances in catalytic hydrogenation of carbon dioxide. *Chemical Society Reviews*, *40*(7), 3703–3727. doi:10.1039/c1cs15008a PMID:21505692

Wang, Y., Craven, M., Yu, X., Ding, J., Bryant, P., Huang, J., & Tu, X. (2019). Plasma-enhanced catalytic synthesis of ammonia over a Ni/Al2O3 catalyst at near-room temperature: Insights into the importance of the catalyst surface on the reaction mechanism. *ACS Catalysis*, *9*(12), 10780–10793. doi:10.1021/acscatal.9b02538 PMID:32064144

Wanten, B., Maerivoet, S., Vantomme, C., Slaets, J., Trenchev, G., & Bogaerts, A. (2022). Dry reforming of methane in an atmospheric pressure glow discharge: Confining the plasma to expand the performance. *Journal of CO2 utilization, 56*, 101869.

Wanten, B., Vertongen, R., De Meyer, R., & Bogaerts, A. (2023). Plasma-based CO2 conversion: How to correctly analyze the performance? *Journal of Energy Chemistry*, *86*, 180–196. doi:10.1016/j.jechem.2023.07.005

Washimi, H., & Tanuiti, T. (1966). Propagation of ion acoustic solitary waves of small amplitude. *Physical Review Letters*, *17*(19), 996–998. doi:10.1103/PhysRevLett.17.996

Watanabe, K., & Taniuti, T. (1977). Electron-acoustic mode in a plasma of two-temperature electrons. *Journal of the Physical Society of Japan*, *43*(5), 1819–1820. doi:10.1143/JPSJ.43.1819

Wazwaz A M. (2010). *Partial differential equations and solitary waves theory*. Verlag, Berlin, Heidelberg: Springer.

Wee, P., & Wang, Z. (2017). Epidermal growth factor receptor cell proliferation signaling pathways. *Cancers (Basel)*, *9*(5), 52. doi:10.3390/cancers9050052 PMID:28513565

Wellburn, A. R. (1994). The spectral determination of chlorophylls a and b, as well as total carotenoids, using various solvents with spectrophotometers of different resolution. *Journal of Plant Physiology*, *144*(3), 307–313. doi:10.1016/S0176-1617(11)81192-2

Welz, C., Emmert, S., Canis, M., Becker, S., Baumeister, P., Shimizu, T., Morfill, G. E., Harréus, U., & Zimmermann, J. L. (2015). Cold atmospheric plasma: A promising complementary therapy for squamous head and neck cancer. *PLoS One*, *10*(11), e0141827. doi:10.1371/journal.pone.0141827 PMID:26588072

Wen, Y., & Jiang, X. (2001). Decomposition of CO2 using pulsed corona discharges combined with catalyst. *Plasma Chemistry and Plasma Processing*, *21*(4), 665–678. doi:10.1023/A:1012011420757

Whipple Jr, E. C. (1965). *The equilibrium electric potential of a body in the upper atmosphere and in interplanetary space* (No. X-615-65-296).

White, G., Hales, A., Patel, Y., & Offer, G. (2022). Novel methods for measuring the thermal diffusivity and the thermal conductivity of a lithium-ion battery. *Applied Thermal Engineering*, *212*, 118573. doi:10.1016/j.applthermaleng.2022.118573

Whitehead, J. C. (2016). Plasma–catalysis: The known knowns, the known unknowns and the unknown unknowns. *Journal of Physics. D, Applied Physics*, *49*(24), 243001. doi:10.1088/0022-3727/49/24/243001

Wijffels, R. H., & Barbosa, M. J. (2010). An outlook on microalgal biofuels. *Science*, *329*(5993), 796–799. doi:10.1126/science.1189003 PMID:20705853

Williams, D. F., & Besler, M. (2017). Syngas production from agricultural residues by means of plasma gasification: A review. *Renewable & Sustainable Energy Reviews*, *78*, 70–86.

Witteman, W. J. (2013). *The CO2 laser*. Springer.

Wolff, C. M., Dong, L., Sommerhalter, M., Zhong, S., Kolb, J. F., & Lademann, J. (2010). Plasma-treated human skin ex vivo studied by EPR spectroscopy. *Skin Research and Technology*, *16*(3), 365–373. PMID:20637006

Wolf-Gladrow, D. A. (2004). *Lattice-gas cellular automata and lattice Boltzmann models: an introduction*. Springer.

Wu, Y., Liu, P., Mehrjou, B., & Chu, P. K. (2023). Interdisciplinary-inspired smart antibacterial materials and their biomedical applications. *Advanced Materials*, 2305940. doi:10.1002/adma.202305940 PMID:37469232

Xia, Y., Lu, N., Li, J., Jiang, N., Shang, K., & Wu, Y. (2020). Combined steam and CO2 reforming of CH4 for syngas production in a gliding arc discharge plasma. *Journal of CO2 Utilization, 37*, 248-259.

Xue, J.-K. (2004). Head-on collision of the blood solitary waves. *Physics Letters. [Part A], 331*(6), 409–413. doi:10.1016/j.physleta.2004.09.029

Xu, G., & Chance, M. R. (2007). Hydroxyl radical-mediated modification of proteins as probes for structural proteomics. *Chemical Reviews, 107*(8), 3514–3543. doi:10.1021/cr0682047 PMID:17683160

Xu, S. (2017). *Plasma-Assisted Conversion of CO 2*. The University of Manchester.

Xu, S., Whitehead, J. C., & Martin, P. A. (2017). CO2 conversion in a non-thermal, barium titanate packed bed plasma reactor: The effect of dilution by Ar and N2. *Chemical Engineering Journal, 327*, 764–773. doi:10.1016/j.cej.2017.06.090

Xu, W., Li, M.-W., Xu, G.-H., & Tian, Y.-L. (2004). Decomposition of CO2 using DC corona discharge at atmospheric pressure. *Japanese Journal of Applied Physics, 43*(12, 12R), 8310–8311. doi:10.1143/JJAP.43.8310

Yamada, M., & Foote, C. S. (1968). Photosensitized generation of singlet oxygen. Aqueous photochemistry of the naphthoquinones. *Journal of the American Chemical Society, 90*(5), 1293–1297.

Yan, J. (2004). *Gliding arc discharge plasma device for organic waste water treatment, China.* Patent CN1557731A.

Yan, J. (2005). *A volatile organic compounds gas treatment device, China.* Patent CN2817959Y.

Yan, D., Nourmohammadi, N., Bian, K., Murad, F., & Sherman, J. H. (2019). Anti-cancer efficacy of cold plasma-activated medium for gastric cancer therapy. *Journal of Surgical Oncology, 119*(3), 320–327.

Yan, D., Sherman, J. H., & Keidar, M. (2017). Cold atmospheric plasma, a novel promising anti-cancer treatment modality. *Oncotarget, 8*(9), 15977–15995. doi:10.18632/oncotarget.13304 PMID:27845910

Yan, D., Talbot, A., Nourmohammadi, N., Sherman, J. H., Cheng, X., & Keidar, M. (2015). Toward understanding the selective anticancer capacity of cold atmospheric plasma—A model based on aquaporins. *Biointerphases, 10*(4), 040801. doi:10.1116/1.4938020 PMID:26700469

Yang, J., Du, H., Yu, Q., Zhang, W., Zhang, Y., Ge, J., Li, H., Liu, J., Li, H., & Xu, H. (2022). Porous silver microrods by plasma vulcanization activation for enhanced electrocatalytic carbon dioxide reduction. *Journal of Colloid and Interface Science, 606*, 793–799. doi:10.1016/j.jcis.2021.08.061 PMID:34419818

Yap, D., Tatibouët, J., & Batiot-Dupeyrat, C. (2015). Carbon dioxide dissociation to carbon monoxide by non-thermal plasma. *Journal of CO2 Utilization, 12*, 54-61.

Yin, Y., Yang, T., Li, Z., Devid, E., Auerbach, D., & Kleyn, A. W. (2021). CO 2 conversion by plasma: How to get efficient CO 2 conversion and high energy efficiency. *Physical Chemistry Chemical Physics, 23*(13), 7974–7987. doi:10.1039/D0CP05275B PMID:33464254

Yousaf Khattak, M., Masood, W., Jahangir, R., & Siddiq, M. (2021). Interaction of ion acoustic solitons for Zakharov Kuznetsov equation in relativistically degenerate quantum magnetoplasmas. *Waves in Random and Complex Media*, 1–7. doi:10.1080/17455030.2021.1968536

Yu, Y., Su, S., Cao, J., Jordanova, V. K., & Denton, M. H. (2022). Improved Boundary Conditions for Coupled Geospace Models: An Application in Modeling Spacecraft Surface Charging Environment. *Space Weather, 20*(9), e2022SW003178.

Yusupov, M. (2019). Transport of reactive oxygen and nitrogen species across aquaporin: A molecular level picture. Oxidative Medicine and Cellular Longevity, vol. 2019, Article ID 2930504, 11 pages

Yusupov, M., Lackmann, J.-W., Razzokov, J., Kumar, S., Stapelmann, K., & Bogaerts, A. (2018). Impact of plasma oxidation on structural features of human epidermal growth factor. *Plasma Processes and Polymers*, *15*(8), 1800022. doi:10.1002/ppap.201800022

Yusupov, M., Privat-Maldonado, A., Cordeiro, R. M., Verswyvel, H., Shaw, P., Razzokov, J., Smits, E., & Bogaerts, A. (2021). Oxidative damage to hyaluronan–CD44 interactions as an underlying mechanism of action of oxidative stress-inducing cancer therapy. *Redox Biology*, *43*, 101968. doi:10.1016/j.redox.2021.101968 PMID:33895486

Zabaloy, M. S., Vasquez, V. R., & Macedo, E. A. (2006). Description of self-diffusion coefficients of gases, liquids and fluids at high pressure based on molecular simulation data. *Fluid Phase Equilibria*, *242*(1), 43–56. doi:10.1016/j.fluid.2005.12.031

Zariç, Ö. E., Yeşildağ, İ., Yaygır, S., & Çelekli, A. (2022). Removal of Harmful Dyes Using Some Algae. *3rd International Congress on Plant Biology*. Zenodo. 10.5281/zenodo.8190776

Zasavitskii, I., Islamov, R. S., Kerimkulov, M., Konev, Y. B., Ochkin, V., Savinov, S. Y., Sobolev, N., Spiridonov, M., & Shotov, A. (1990). Kinetics of formation of the CO2 molecule distributions over the vibrational-rotational levels in the active medium of a waveguide CO2 laser. *Journal of Soviet Laser Research*, *11*(4), 361–375. doi:10.1007/BF01120633

Zeng, K., & Zhang, D. (2010). Recent progress in alkaline water electrolysis for hydrogen production and applications. *Progress in Energy and Combustion Science*, *36*(3), 307–326. doi:10.1016/j.pecs.2009.11.002

Zeng, Y., & Tu, X. (2015). Plasma-catalytic CO 2 hydrogenation at low temperatures. *IEEE Transactions on Plasma Science*, *44*(4), 405–411. doi:10.1109/TPS.2015.2504549

Zeng, Y., & Tu, X. (2017). Plasma-catalytic hydrogenation of CO2 for the cogeneration of CO and CH4 in a dielectric barrier discharge reactor: Effect of argon addition. *Journal of Physics. D, Applied Physics*, *50*(18), 184004. doi:10.1088/1361-6463/aa64bb

Zeng, Y., Wang, L., Wu, C. F., Wang, J. Q., Shen, B. X., & Tu, X. (2018). Low temperature reforming of biogas over K-, Mg-and Ce-promoted Ni/Al2O3 catalysts for the production of hydrogen rich syngas: Understanding the plasma-catalytic synergy. *Applied Catalysis B: Environmental*, *224*, 469–478. doi:10.1016/j.apcatb.2017.10.017

Zhang, H. (2018). Warm plasma activation of CO2 in a rotating gliding arc discharge reactor. *Journal of CO2 Utilization*, *27*, 472-479.

Zhang, H., Li, L., Li, X., Wang, W., Yan, J., & Tu, X. (2018). Warm plasma activation of CO2 in a rotating gliding arc discharge reactor. *Journal of CO2 Utilization, 27*, 472-479.

Zhang, C., Chen, Y., Yang, L., & Shi, M. (2011). Self-diffusion for Lennard-Jones fluid confined in a nanoscale space. *International Journal of Heat and Mass Transfer*, *54*(21-22), 4770–4773. doi:10.1016/j.ijheatmasstransfer.2011.05.032

Zhang, S., Zhu, C., Sun, F., Li, W., Shen, H., Wang, C., ... Guo, S. (2020). Current status of pre-clinical and clinical trials targeting overactive bladder and urinary incontinence. *Investigative and Clinical Urology*, *61*(Suppl 1), S23–S36.

Zhang, X., Lee, C. S.-M., Mingos, D. M. P., & Hayward, D. O. (2003). Carbon dioxide reforming of methane with Pt catalysts using microwave dielectric heating. *Catalysis Letters*, *88*(3/4), 129–139. doi:10.1023/A:1024049403422

Zhang, X., Zhang, J., Chen, C., Zhu, Y., Xiang, X., Wu, S., ... Zhang, J. (2021). Plasma sterilization: Opportunities and challenges. *Applied Sciences (Basel, Switzerland)*, *11*(11), 5190.

Zhang, X., Zhang, R., Zhong, S., Chen, J., Yang, X., Lai, X., ... Lu, X. (2021). Advances in nonthermal plasma technology for wound healing applications: A comprehensive review. *Advanced Healthcare Materials, 10*(11), 2001895. PMID:33929786

Zhang, Y., Guo, Y., Wang, Y., Yan, Q., Huo, M., Zhang, J., & Wang, H. (2018). A review of sterilization methods for medical devices. *Journal of Functional Biomaterials, 9*(2), 44.

Zhang, Z., Xu, L., Qu, T., Lei, M., Lin, Z. K., Ouyang, X., & Huang, J. (2023). Diffusion metamaterials. *Nature Reviews. Physics, 5*(4), 218–235. doi:10.1038/s42254-023-00565-4

Zhao, X., & Jin, H. (2020). Correlation for self-diffusion coefficients of H2, CH4, CO, O2 and CO2 in supercritical water from molecular dynamics simulation. *Applied Thermal Engineering, 171*, 114941. doi:10.1016/j.applthermaleng.2020.114941

Zhao, X., Luo, T., & Jin, H. (2021). A predictive model for self-, Maxwell-Stefan, and Fick diffusion coefficients of binary supercritical water mixtures. *Journal of Molecular Liquids, 324*, 114735. doi:10.1016/j.molliq.2020.114735

Zhao, Z., Zhang, M., Yuan, C., Zhou, X., & Ding, Y. (2017). Improved biocompatibility of Ti6Al4V by plasma surface modification. *Journal of Materials Science, 52*(16), 9947–9961.

Zheng, J., Zhang, X., Zhang, Y., Wang, C., Wang, H., & Deng, S. (2016). Biochar effects on agricultural soil: A review of field studies in China. *Environmental Science and Pollution Research International, 23*(3), 1694–1709.

Zheng, X., Tan, S., Dong, L., Li, S., & Chen, H. (2014). LaNiO3@ SiO2 core–shell nano-particles for the dry reforming of CH4 in the dielectric barrier discharge plasma. *International Journal of Hydrogen Energy, 39*(22), 11360–11367. doi:10.1016/j.ijhydene.2014.05.083

Zhong, X., Xiong, Z., Chen, X., & Liu, D. (2014). A fundamental study on the mechanisms of plasma polymerization and the properties of plasma polymers. *Applied Surface Science, 298*, 12–16.

Zhou, R., Zhou, R., Wang, P., Xian, Y., Mai-Prochnow, A., Lu, X., Cullen, P. J., Ostrikov, K. K., & Bazaka, K. (2020). Plasma-activated water: Generation, origin of reactive species and biological applications. *Journal of Physics. D, Applied Physics, 53*(30), 303001. doi:10.1088/1361-6463/ab81cf

Zhu, L., Kreutz, T. G., Hewitt, S. A., & Flynn, G. W. (1990). Diode laser probing of vibrational, rotational, and translational excitation of CO2 following collisions with O (1D). I. Inelastic scattering. *The Journal of Chemical Physics, 93*(5), 3277–3288. doi:10.1063/1.458861

Zhu, X., Liu, J.-H., Li, X.-S., Liu, J.-L., Qu, X., & Zhu, A.-M. (2017). Enhanced effect of plasma on catalytic reduction of CO2 to CO with hydrogen over Au/CeO2 at low temperature. *Journal of Energy Chemistry, 26*(3), 488–493. doi:10.1016/j.jechem.2016.11.023

Zou, S., Gu, Y., Xiao, D., & Tang, C. Y. (2011). The role of physical and chemical parameters on forward osmosis membrane fouling during algae separation. *Journal of Membrane Science, 366*(1–2), 356–362. doi:10.1016/j.memsci.2010.10.030

Zou, Z., Chang, H., Li, H., & Wang, S. (2017). Induction of reactive oxygen species: An emerging approach for cancer therapy. *Apoptosis, 22*(11), 1321–1335. doi:10.1007/s10495-017-1424-9 PMID:28936716

About the Contributors

Aamir Shahzad has more than Eighteen years of experience in university research and teaching at home and abroad. He studied the thermophysical properties of materials and plasmas at graduate and postgraduate levels. He received his post-doctorate and doctoral degrees from Xi'an Jiaotong University (XJTU), China, in 2015 and 2012, respectively. Dr Shahzad is working as a Professor (associate) in the department of Physics in Government College University Faisalabad, Pakistan. He has proposed novel methods to explore outcomes of complex materials, highlighting his aptitude in computational physics and molecular modeling and simulations. Moreover, Dr. Shahzad is interested in computational physics, complex fluids/plasmas, plasma oncology, and bio- and energy materials. Currently, Dr. Shahzad is an associate professor in the Department of Physics, at GC University Faisalabad (GCUF). He is a member of the XJTU ThermoPhysical Society and the GCUF and the University of Agriculture Faisalabad Physics Societies.

Mao-Gang He is now the Committee Chair of School of Energy and Power Engineering, Xi'an Jiaotong University, and Distinguished Professor at Key Laboratory of Thermo-Fluid Science and Engineering of MOE. He received Ph.D in Engineering Thermophysics at Xi'an Jiaotong University, P.R.China in 1999. His current research interests are the thermophysical properties of fluid and new power cycle, more than 400 papers including 150 international journal articles have been published, and 11 invention patents were granted. He is also the deputy director of the Chinese Thermophysics Properties Association and the vice chairman of the Engineering Thermophysics Association of Chinese Universities.

***

Mumtaz Ali has done his BS physics from Karakoram International University, Gilgit, Pakistan. His Research interests includes Plasma technology, materials science, nanotechnology, CO2 Conversion, etc. Currently, he is working as undergraduate researcher at Department of Physics, Karakoram International University, Gilgit, Pakistan.

Shahid Ali born in Nowshera, Pakistan in 1975, is currently working as GM Tech/Associate Professor at the National Centre for Physics, Quaid-e-Azam University Campus, Islamabad. He obtained his M.Sc. and M.Phil. degrees in Physics from Gomal University, D.I. Khan and Quaid-e-Azam University, Islamabad in 1998 and 2001, respectively. He served as a Research Officer for 4 years at Salam Chair in Physics, GC University Lahore, and completed his Ph.D. in Theoretical Nonlinear Physics from Umea University, Sweden in 2008. Dr. Ali also completed his Post-Doctoral Studies at IST, Lisbon, Portugal during 2009-10 and was appointed as a Regular Associate at Abdus Salam International Centre for Theoretical Physics, Italy. His research areas include Nonlinear Plasma Physics, Spacecraft Charging and Space Dynamics, Degenerate Plasmas, Magnetic Fields Generation in Laser-produced Plasmas, etc. He has published over 130 articles in International Reputed Journals such as Phys. Rev. Lett., Phys. Rev. B, Astrophys. J., J. Geophys. Res., Phys. Plasmas, etc., and has trained several M.Phil/Ph.D. Scholars in the field of Plasma Physics. Dr. Ali has been a regular recipient of the RPA Awards from the Pakistan Council for Science and Technology for many years and received a Gold Medal in Physics from the Pakistan Academy of Sciences in 2016. He also worked as a Visiting Associate Professor at the University of Science and Technology of China in 2018 and as a Visiting Scholar at Khalifa University, UAE, in 2021.

Noor-ul-Huda Altaf, Department of Physics, University of Agriculture, 38040 Faisalabad, Pakistan. Research and Academic Experience: Materials and Plasma processing Currently, she is a PhD research student at University of Agriculture Faisalabad, Pakistan. She has done Masters of philosophy (M. Phil.) in Physics from University of Agriculture Faisalabad, Pakistan. Research Area: Material science and Nanotechnology Number of Published papers: 7 She also worked at Penn State University PA (USA) with Randall's research group as a visiting research scholar.

Alireza Kardan, Department of Chemical Engineering, University of Guilan

Misbah Kha is a Ph.D. student at the Department of Refrigeration and Cryogenic Engineering, School of Energy and Power Engineering, Xi'an Jiaotong University

Haipeng Li is a Professor at School of Materials Science and Physics, China University of Mining and Technology, P. R. China

Sehrish Maqsood has completed her Master of Philosophy (M.Phil.) degree in Physics working in the field of materials Science and nanotechnology. She worked on synthesis and plasma treatment various ferrites and other nanomaterials for their structural, morphological and magnetic characterization. Currently, she is working as a research scholar at Department of Physics, University of Agriculture Faisalabad.

Erfan Nouri is a Ph.D. student of chemical engineering, chemical engineering department, University of Guilan, Rasht

Syed Ali Raza is a Ph.D. student at the School of Electrical Engineering, Xi'an Jiaotong University, P. R. China

Aamir Shahzad is an Associate Professor, at the department of physics, Government College University, Faisalabad, Pakistan

Muhammad Asif Shakoori is a postdoctoral researcher at the School of Materials Science and Physics, China University of Mining and Technology, P. R. China

Tayyba Shoukat has a PhD in Electronic science and Technology from University of Electronic science and technology China

Soraya Trabelsi Faculty of science of Bizerte University of Carthage-Tunisia.

Index

S

T

U

V

W

www.ingramcontent.com/pod-product-compliance
Lightning Source LLC
LaVergne TN
LVHW062138300926
849161LV00010B/2563
* 9 7 9 8 3 6 9 3 0 9 0 4 9 *